Numerical Modeling in Open Channel Hydraulics

Water Science and Technology Library

VOLUME 83

For further volumes: http:// www.springer.com/series/6689

Numerical Modeling in Open Channel Hydraulics

Romuald Szymkiewicz

Faculty of Civil and Environmental Engineering,
Gdańsk University of Technology, Poland

Springer

Romuald Szymkiewicz
Faculty of Civil and Environmental
Engineering
Gdańsk University of Technology
ul. Narutowicza 11/12
80-233 Gdańsk
Poland
rszym@pg.gda.pl

ISBN 978-90-481-3673-5 e-ISBN 978-90-481-3674-2
DOI 10.1007/978-90-481-3674-2
Springer Dordrecht Heidelberg London New York

Library of Congress Control Number: 2009942268

Printed on acid-free paper

Springer is part of Springer Science+Business Media (www.springer.com)

Preface

Open channel hydraulics has always been a very interesting domain of scientific and engineering activity because of the great importance of water for human living. The free surface flow, which takes place in the oceans, seas and rivers, can be still regarded as one of the most complex physical processes in the environment. The first source of difficulties is the proper recognition of physical flow processes and their mathematical description. The second one is related to the solution of the derived equations. The equations arising in hydrodynamics are rather complicated and, except some much idealized cases, their solution requires application of the numerical methods. For this reason the great progress in open channel flow modeling that took place during last 40 years paralleled the progress in computer technique, informatics and numerical methods. It is well known that even typical hydraulic engineering problems need applications of computer codes. Thus, we witness a rapid development of ready-made packages, which are widely disseminated and offered for engineers. However, it seems necessary for their users to be familiar with some fundamentals of numerical methods and computational techniques applied for solving the problems of interest. This is helpful for many reasons. The ready-made packages can be effectively and safely applied on condition that the users know their possibilities and limitations. For instance, such knowledge is indispensable to distinguish in the obtained solutions the effects coming from the considered physical processes and those caused by numerical artifacts. This is particularly important in the case of hyperbolic equations, like the Saint-Venant equations or the advection equation.

In principle, numerical open channel hydraulics can be regarded as a sub-domain of Computational Fluid Dynamics (CFD) and the general methods and experiences of CFD are applicable in open channel flow modeling. Moreover, the open channel flow can be often treated as one-dimensional, which makes it relatively easy to solve compared to multidimensional flows considered in geophysics and industrial engineering. On the other hand, due to a range of specific issues, numerical open channel hydraulics developed into a branch of its own as early as in the years 1960s and 1970s. There exist a number of very good books on the subject, written at that time and later. A non-exhaustive list includes "Unsteady flow in open channel" edited by K. Mahmood and V. Yevjevich and containing the papers written by recognized experts, "Practical aspects of computational river hydraulics" by J.A. Cunge, F.M.

Holly and A. Verwey, "Computational hydraulics-Elements of the theory of free surface flow" by M.B. Abbott, "Dynamic hydrology" by P.S. Eagleson, "Open channel hydraulics" by V.T. Chow, "Open channel flow" by F.M. Henderson, "Kinematic wave modeling in water resources: surface water hydrology" by V.P. Singh, "The hydraulics of open channel flow: An introduction" by H. Chanson. These books cover most of the theoretical and practical issues related to open channel flow modeling and can be recommended for any engineer working in this field.

As far as the computational techniques are considered, one can recommend the following books: "Incompressible flow and the finite-element method" by P.M. Gresho and R.L. Sani, "Computational fluid dynamics" by M.B. Abbott and D.R. Basco, "Numerical head transfer and fluid flow" by S.V. Patankar, "Computational physics" by D. Potter, "Computational techniques for fluid dynamics" by C.A.J. Fletcher, "Finite volume methods for hyperbolic problems" by R.J. LeVeque, "The finite element method in engineering science" by O. C. Zienkiewicz. These books covering large area of the fluid dynamics and other engineering sciences are useful for open channel flow modeling as well.

In view of the continuous advance in numerical techniques, the present book is an attempt to complement the existing works with a more detailed and up-to-date discussion of selected numerical aspects of open channel hydraulics. It is largely based on author's own research and focuses on one-dimensional models of steady and unsteady flow and transport in open channels and their networks.

The book is organized in nine chapters. Chapter 1 presents the background information on the open channel hydraulics and the derivation of the governing equations for both steady and unsteady flow, as well as for the transport of the constituents dissolved in the flowing water including the transport of thermal energy.

The next two chapters cover the basic numerical methods applicable for solving nonlinear equations and systems of linear and nonlinear equations (Chapter 2) and ordinary differential equations and their systems (Chapter 3). Implementation of the presented methods for solution the steady gradually varied flow in a single channel and in channel network is given in Chapter 4. These methods are also the basic building blocks for more complex numerical algorithms described in the following chapters.

Chapter 5 is an introduction to the partial differential equations of hyperbolic and parabolic types, frequently occurring in open channel hydraulic. It covers the classification of equations, formulation of solution problem, and introduction to the finite difference and element methods. This chapter ends by discussion of convergence, consistency and stability of the numerical methods.

In Chapters 6 and 7 the advection and advection-diffusion equations are considered. Basing on the solution using the finite difference box scheme, the main problems of numerical integration of the hyperbolic equations are discussed. The modified equation approach for the accuracy analysis of the numerical solution of hyperbolic equations is described. Besides the standard methods of solution, the splitting technique for the advection-diffusion transport equation is presented.

Chapter 8 is entirely devoted to the unsteady flow. Solution of the system of Saint Venant equations using both finite difference and element methods is described.

Solution of unsteady flow with moveable bed and the problem of propagation of steep waves are briefly described.

Chapter 9 covers the simplified models and theirs application for flood routing. Particular attention is focused on the close relation between the spatially lumped models and the discrete forms of distributed models. The conservative properties of the non-linear and linear simplified models are discussed.

The book includes numerous computational examples and step-by-step descriptions of numerical algorithms. It is the hope of the author that the reader will find it useful and easy to follow.

I am grateful to Springer and to the members of the Editorial Advisory Board of series Water Science and Technology Library for the possibility of publishing my work. My thanks are to Prof. Witold Strupczewski, for initiating the idea of this book, to the Editor-in-Chief Prof. Vijay. P. Singh for his valuable suggestions on its contents, and to Ms. Petra van Steenbergen and Ms. Cynthia de Jonge for their kind assistance in submitting the manuscript. I would also like to acknowledge the support received from Prof. Ireneusz Kreja, Dean of the Faculty of Civil and Environmental Engineering of the Gdańsk University of Technology. I owe a lot to the persons who assisted me in the work on the manuscript: Dr. Dariusz Gąsiorowski, who prepared many numerical examples, Ms. Katarzyna Olszonowicz, who prepared all the figures and my son Dr. Adam Szymkiewicz, whose remarks and suggestions helped to improve the text. Finally, I highly appreciate the effort of all members of staff involved at all stages of the editorial process of my book.

Gdansk, Poland Romuald Szymkiewicz

Contents

Chapter 1
Open Channel Flow Equations

1.1 Basic Definitions

Open channel is a conduit in which a part of the cross-section of the flowing stream, taken perpendicularly to the flow velocity vector, is exposed atmosphere. Open channels can be divided into natural and artificial ones. Natural channels, like rivers or creeks, result from the geophysical processes acting at the Earth surface, without essential participation of human activity. Conversely, artificial channels are built by men and comprise, among others, navigable, energetic and irrigation channels.

Another possible division of open channels is between prismatic and non-prismatic ones. A prismatic channel has constant cross-sectional shape and constant longitudinal bottom slope. Of course, such properties can be possessed only by artificial channels. All natural channels are non-prismatic. A schematic representation of a non-prismatic channel is shown in Fig. 1.1.

In natural channels the sections can be either compacted (Fig. 1.1a) or composite (Fig. 1.1b). In composite cross-section distinct parts exist (1, 2 and 3 in Fig. 1.1b), characterized by different bed elevations and possibly different bed roughness. This case is representative, for instance, of a river flowing over flooded terrace and calculation of the flow parameters requires appropriate treatment. Figure 1.1a also shows some basic variables used in open channel hydraulics. They are listed below.

Water stage, denoted by h, is the elevation of the free surface of water at a specific cross-section of the channel, measured with respect to the assumed datum. The elevation of the bottom measured with respect to the same datum is denoted by Z.

The depth of flow H is the vertical distance between the water surface and the channel bottom. In natural channels the depth and bed elevation have local meaning only, since they vary along the axis x, which is parallel to the flow direction and y axis, which is horizontal and normal to x. Conversely, in prismatic channels both Z and H are uniquely defined for each cross-section and depend on x only. Thus, the following relation holds:

$$h(x) = Z(x) + H(x). \tag{1.1}$$

R. Szymkiewicz, *Numerical Modeling in Open Channel Hydraulics*, Water Science and Technology Library 83, DOI 10.1007/978-90-481-3674-2_1,

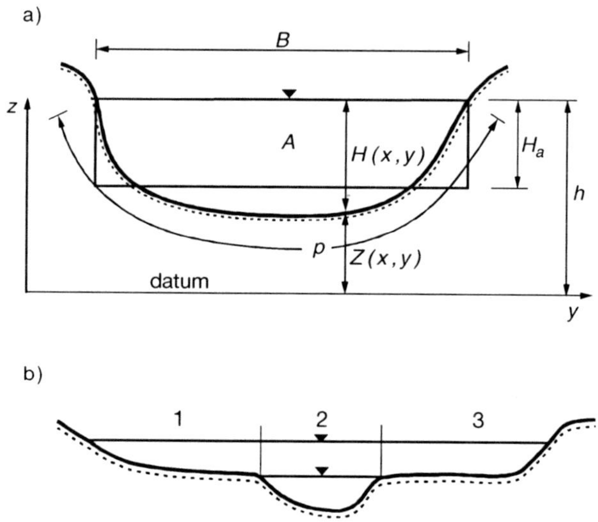

Fig. 1.1 Possible shapes of cross-sections of the non-prismatic channel (**a**) compacted, (**b**) composite

The *longitudinal bed slope* is defined as:

$$s = -\frac{\partial Z}{\partial x}, \tag{1.2}$$

Its value is usually small, which allows for considerable simplifications of the flow equations. In Fig. 1.2 a longitudinal profile of channel is shown.

The slope is uniquely defined for prismatic channels only. For natural channels the bed slope has rather local meaning, however its average value along channel axis taken over longer distances of practical interest is rough but important information.

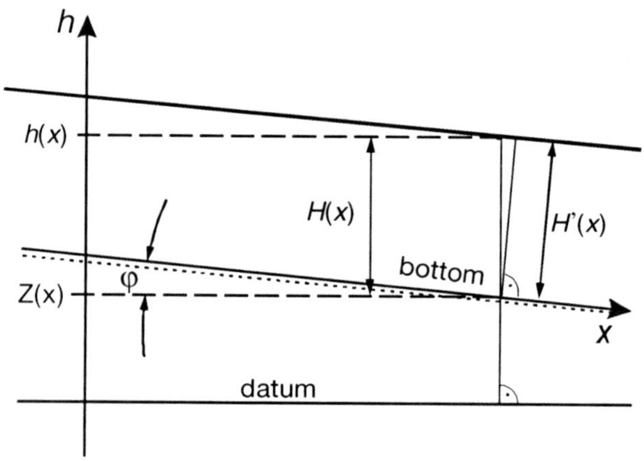

Fig. 1.2 Definition of the cross-section parameters for prismatic channel

The value of $H(x)$ should be distinguished from so-called depth of flow of section $H'(x)$, which is measured perpendicularly to the bottom (Fig. 1.2). Since both depths are related to each other by means of the bed slope $H'(x) = H(x)\cos(\varphi)$, then for small values of the longitudinal bed slope, when $\cos(\varphi) \approx 1$, one can assume that $\mathrm{H}(\mathrm{x}) \approx H'(x)$. Only in particular situations, for steep channels, this difference is appreciable (Akan 2006, French 1985). Consequently, usually it is reasonable to assume that the flow depth is measured vertically from bottom to the water surface, whereas the distance along channel axis is considered as horizontal one. The problem of co-ordinates system is summarized by Liggett (1975) as follows: "For open channel flow the system of co-ordinates is not entirely orthogonal in that x lies in the bed of the channel and z is vertical. This arrangement assumes that the cosine of the channel slope is approximately unity."

The cross-section can be characterized by the following geometric parameters:

- *Top width B* is the width of the channel at the level of water surface (Fig. 1.1).
- *Flow area A* is the wetted cross-sectional area measured perpendicularly to the vector of channel flow velocity.
- *Wetted perimeter P* is the length of the interface between the water and the channel bed.
- *Hydraulic radius R* is the ratio of the wetted flow area and the wetted perimeter:

$$R = \frac{A}{P}. \tag{1.3}$$

  This parameter is meaningful only for compact cross-sections as shown in Fig. 1.1a.
- *Hydraulic depth H_a* is the ratio of the wetted flow area A and the top width B:

$$H_a = \frac{A}{B}. \tag{1.4}$$

  H_a and B represent the dimensions of a rectangle of area A equivalent to the natural cross-section (Fig. 1.1a).

The cross-sections of prismatic channels are typically triangular, rectangular or trapezoidal so the parameters listed above can be expressed as analytical functions of the depth H. Consider for example a trapezoidal channel (Fig. 1.3) characterized by the channel width at the level of bottom b, and the side slope m, which is the cotangent of the angle ψ between the side and horizontal plane.

For such type of channel the cross-section parameters A, P and B are given by the following formulas:

$$A = b \cdot H + m \cdot H^2 \tag{1.5}$$

$$P = b + 2H\sqrt{1 + m^2} \tag{1.6}$$

$$B = b + 2m \cdot H \tag{1.7}$$

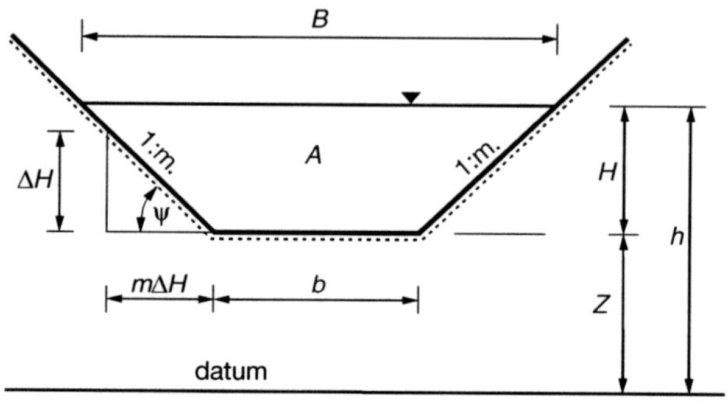

Fig. 1.3 Trapezoidal channel cross-section

For $m = 0$ the trapezoidal cross-section becomes rectangular, whereas for $b = 0$ it becomes triangular. For natural channels, the geometric parameters are usually expressed as function of the water stage, presented in tabularized form.

The basic quantities characterizing liquid flow are: pressure, density, velocity and acceleration. The flow is described by the scalar or vector fields of these variable parameters. In general all those variables depend on three spatial coordinates and time. However, in open channel flow the component of the velocity vector parallel to the channel axis is typically dominating and in addition it has relatively uniform distribution over the cross-section area. Thus, it is assumed that the flow parameters are functions of two variables only: x and t, which represent the spatial coordinate related to the channel axis and time.

The flowing stream is conveniently characterized by the following averaged quantities:

- *Discharge*, which represents the mass or volume of water flowing through considered cross-section per unit time. The volume discharge is defined as:

$$Q = \iint_A u \cdot dA \tag{1.8}$$

where:

u – normal velocity to the cross-section,
A – wetted cross-sectional area,
Q – flow discharge.

- *Average flow velocity* in the cross-section, which ensures the same discharge as actual velocity distribution over cross-section. It is defined as follows:

$$U = \frac{Q}{A} = \frac{1}{A} \iint_A u \cdot dA \tag{1.9}$$

where: U – average velocity in a cross-section.

- *Volume flux per unit time and unit width* of the vertical cross-section measured between the water surface and the bottom, given by the formula:

$$q = \int_{Z}^{h} u \cdot dz = U \cdot H \tag{1.10}$$

where q is flow discharge related to the width unit of a channel. For channel having a rectangular cross-section q is equal:

$$q = \frac{Q}{B} \tag{1.11}$$

The total energy of water particle traveling along a streamline (a line constructed in the velocity field, so that at its every point the velocity vector is tangential to this line) is given by the Bernoulli equation:

$$E_{sl} = z_{sl} + \frac{p_{sl}}{\gamma} + \frac{u_{sl}^2}{2g} \tag{1.12}$$

where:

E_{sl} – total energy of traveling particle along a streamline,
z_{sl} – elevation of the streamline above the assumed datum,
p_{sl} – local pressure,
γ – specific weight,
u_{sl} – local velocity,
g – local acceleration due to gravity.

In Eq. (1.12) the term (p_{sl}/γ) is the pressure head, whereas the term $(u_{sl}^2/2g)$ represents velocity head. The term $(p_{sl}/\gamma + z_{sl})$ is called the piezometric head (Chanson 2004).

When the flow in open channel with cross-sectional average velocity is considered, the total energy of flow referred to the assumed datum is:

$$E = Z + H + \frac{\alpha \cdot U^2}{2g} = h + \frac{\alpha \cdot U^2}{2g}, \tag{1.13}$$

whereas the specific energy of the open channel flow relative to the bottom of a channel is defined by formula (Fig. 1.4):

$$E_S = H + \frac{\alpha \cdot U^2}{2g} \tag{1.14}$$

where:

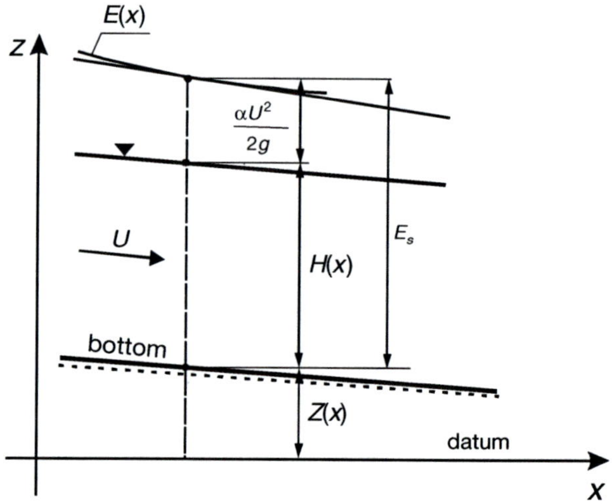

Fig. 1.4 Representation of the energy equation's terms

E – total energy related to the assumed datum,
E_s – specific energy related to the bottom,
Z – elevation of the bed above the assumed datum,
H – depth,
h – elevation of the water surface above the assumed datum,
U – average flow velocity,
α – kinetic energy correction factor (Coriolis coefficient).

The Coriolis coefficient α is introduced to correct the kinetic energy calculated using the average flow velocity U instead of the actual velocities, variable over wetted cross-sectional area A. Comparison of both energies yields:

$$\alpha = \frac{1}{U^3 \cdot A} \iint_A u^3 \cdot dA \tag{1.15}$$

Coefficient α is never less than 1. Its value increases with the increasing differentiation of flow velocity distribution over cross-section.

Analysis of Eq. (1.13) for horizontal channel with $\alpha = 1$ shows that the specific energy is a function of the flow depth H only, i.e. $E_s = E_s\,(H)$. From its examination results that E_s tends to infinity for H tending either to 0 or to infinity. This means that in the interval $\langle 0, \infty)$ the function $E_s(H)$ has an extreme minimum point. At this point the following condition resulting from Eq. (1.14) is valid:

$$\frac{\alpha \cdot Q^2}{g} = \frac{A^3}{B} \tag{1.16}$$

Then except this point, there are two alternate depths H for which $E_s(H)$ takes the same values. The depth corresponding to the extreme minimum is *critical depth.* This depth, designed as H_c, ensures that the flowing stream with flow rate Q has least specific energy or for given specific energy the flow rate is greatest. The critical depth divides the interval $\langle 0, \infty)$ into two parts. The channel flow with $H < H_c$ is so called supercritical or torrential flow, whereas the flow with $H > H_c$ is called subcritical or fluvial one. To distinguish these cases the following dimensionless Froude number is introduced:

$$F_r = \frac{U}{\sqrt{g \cdot H}} \tag{1.17}$$

This number, representing the ratio of inertial and gravity forces, is expressed by the average flow velocity U and the celerity of gravity wave in shallow water $(g\ H)^{1/2}$. Using the Froude number one can distinguish:

- critical flow when $F_r = 1$,
- supercritical flow when $F_r > 1$,
- subcritical flow when $F_r < 1$.

The subcritical flow is typical form of channel flow. It is characterized by relatively high depths and small velocities. This form of flow is required as the most suitable one in natural channel. The supercritical flow is characterized by relatively small depths and great velocities. In natural conditions it can occur in upper parts of rivers. This kind of flow is typically observed on the weirs of dams.

Since Eq. (1.13) represents the total mechanical energy related to the assumed datum, then its derivative with regard to x represents local energy grade line slope S. Therefore one can write:

$$\frac{dE}{dx} = -S \tag{1.18}$$

where S is energy grade line slope or friction slope. The negative sign at right side of Eq. (1.18) takes into account the fact that energy decreases with increasing of x (Fig. 1.4).

The water flow in open channel can be classified with regard to various criteria. Taking into account the time variability one can distinguish:

- steady flow, when the flow parameters do not vary in time,
- unsteady flow when the flow parameters vary in time.

With the time variation of flow is connected the notion of time-invariant process or system. A channel reach can be considered as a physical system, which is capable to transform the flood wave occurring at the upstream end to the one observed at the downstream end. If the same waves entering the channel at given time intervals produce the same response, the considered channel reach is a time-invariant system.

It means that its main properties affecting the flow process, such as the shape of cross-sections and hydraulic roughness do not vary in time. In the opposite case, when the channel properties vary in time, the system is called time-variable.

Steady flows can be further divided with respect to the spatial variability of the parameters. From this point of view one can distinguish uniform and non-uniform flows. A uniform flow is characterized by spatially constant flow parameters such as velocities and depths, so that the water surface and the energy line are parallel to the bottom. Thus uniform flow can occur in prismatic channels only, in which the shape of cross-sections does not vary and the bed slope is constant. In natural channels such kind of flow does not exist. For steady uniform flow the average velocity is given by the well known empirical formulas:

- The Chézy equation:

$$U = C_C (R \cdot s)^{1/2} \tag{1.19}$$

where:

C_C – the Chézy coefficient,
R – hydraulic radius,
s – longitudinal channel bed slope, equal to the energy line slope S.

- The Manning equation

$$U = \frac{1}{n_M} R^{2/3} \cdot s^{1/2} \tag{1.20}$$

This equation can be considered as the Chézy equation with C_C defined as follows:

$$C_C = \frac{1}{n_M} R^{1/6} \tag{1.21}$$

where n_M is the Manning roughness coefficient. In SI units this coefficient is expressed in $s/m^{1/3}$.

The depth H_n which satisfies Eqs. (1.19) or (1.20) is called *normal depth*. In the case of uniform flow, when the flow velocities and wetted cross-sectional areas do not vary, the energy grade line, the hydraulic grade line and bottom line are parallel one another.

In particular cases it can occur that the normal depth is equal to the critical depth. The channel slope, for which the uniform flow is critical, is called the *critical slope* (Chanson 2004, French 1985). Usually it is denoted s_c. Relation between the normal and critical depths can be used to characterize the channel slope. One can distinguish (Chanson 2004):

- mild slope when $H_n > H_c$,
- critical slope when $H_n = H_c$,
- steep slope when $H_n < H_c$.

The non-uniform flow does not satisfy the conditions of uniform flow, so the depths and velocities vary along channel axis. Consequently the bottom, water surface and energy line are not parallel to each other. Depending on the degree of the spatial variability one can distinguish gradually and rapidly varied flows. To explain the essential difference between both types of flow, let us recall the Bernoulli equation (Eq. 1.12). In open channel gradually varied flow the water particles move along nearly straight streamlines. Since the streamlines have insignificant curvature, the cross-sectional surface, orthogonal to the streamline, are nearly flat. It means that they are almost normal to the velocity vectors and the vertical components of velocity do not exist. These assumptions imply that the sum $(z_{sl} + p/\gamma)$ representing the elevation of the hydraulic grade line above the assumed datum, can be considered as constant. Consequently the pressure distribution in the stream flowing in a channel with small bed slope, in which the streamlines are nearly parallel to the bottom, is governed by the hydrostatic law. Conversely, when rapid flow is considered, one needs to take into account the curvature of the streamlines and cross-sections. The law of hydrostatic distribution of pressure along the depth is no longer valid.

In non-uniform flow the friction slope S is usually also expressed with the Manning formula. Although this formula has been derived for steady uniform flow, it is applicable for non-uniform flow as well. It is assumed that the error generated due to this approximation is small comparing with the one resulting from the estimation of the coefficient n_M. The reformed formula (1.20) is used:

$$S = \frac{n_M^2 \cdot U^2}{R^{4/3}} = \frac{n_M^2 \cdot Q^2}{R^{4/3} \cdot A^2} \tag{1.22}$$

For steady uniform flow the friction slope is constant, whereas for non-uniform flow it varies along channel axis.

To solve practical open channel flow problems the energy equations written for two neighboring stations is usually applied. For a channel reach presented in Fig. 1.5 it takes the following form:

$$h_i + \frac{\alpha \cdot U_i^2}{2g} = h_{i+1} + \frac{\alpha \cdot U_{i+1}^2}{2g} + \Delta x_i \cdot \overline{S} \tag{1.23}$$

where:

h_i, h_{i+1} – elevations of the water surface above the assumed datum in cross-section i and i+1 respectively,
U_1, U_{i+1} – average flow velocities in cross-section i and i+1 respectively,
$\overline{S}$ – average friction slope between both cross-sections,
Δx_i – distance between considered stations.

The average value of the friction slope is determined using its local values in both cross-sections calculated using Eq. (1.22). Equation (1.23) allows us to compute the water surface profile in open channel for steady gradually varied flow.

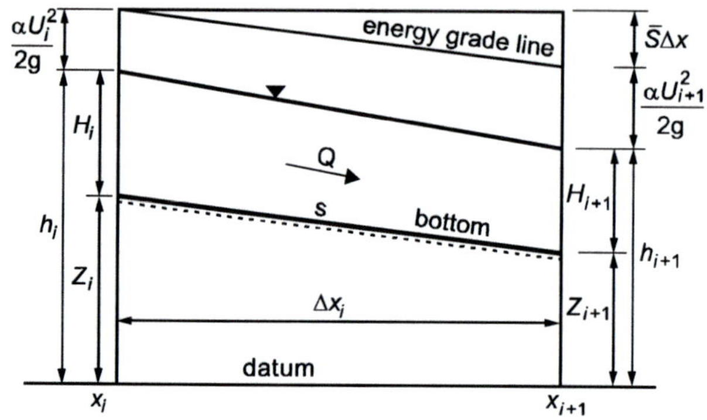

Fig. 1.5 Sketch of a channel reach

This section covers only basic information on the open channel flow, which is necessary for further presentation of numerical applications. For more comprehensive discussion of the theoretical background for the open channel hydraulics the reader is directed to the works of Akan (2006), Chanson (2004), Chow (1959), French (1985), Henderson (1966), Jain (2000), Singh (1996) and others.

1.2 General Equations for Incompressible Liquid Flow

Domination of one component of the velocity vector is a typical attribute of the open channel flow. This feature allows us to consider the flow process as spatially one-dimensional phenomenon. Similarly to the other types of surface water flows, the governing equations are derived from two principles of conservation:

- momentum conservation law,
- mass conservation law.

The derivation can be performed in various ways. The approaches found in the literature differ with respect to the formulation of the principle of conservation, and to the stage at which the one-dimensionality assumption is introduced. For example, the open channel flow equations are very often derived by balancing the fluxes and forces acting on the considered control volume, with an a priori assumption of uniform velocity flow distribution over a channel cross-section. Next, when the governing equations are derived, some additional factors are introduced to correct the balanced quantity. In such a way the correction parameters α or β appear in the 1D dynamic equation.

A more consistent approach is to derive the unsteady open channel flow equations from general equations of hydrodynamics. It allows us to show clearly that this kind of flow is governed by equations being a particular case of general equations for 3D unsteady flow. In addition, the effects of introduced simplifications and assumptions can be easily followed. This point of view coincides with the suggestion given by Abbott and Basco (1989). These authors cite Bird, Steward and Lightfoot (1960), who stated: "It is not... necessary to formulate a momentum balance (or mass balance for continuity) whenever one begins to work on a new flow problem. In fact, it is seldom desirable to do so. It is quicker, easier and safer to start with the equations of conservation of mass and momentum in general form and to simplify these equations to fit the problem at hand". Moreover Abbott and Basco (1989) state: "The advantage of this procedure is that when we are finished discarding terms and simplifying the equations (based on intuition, experiments, field data, experience, etc.) we will have available a complete list of assumptions." This approach will be followed here.

It is well known that in general case the flow of viscous and incompressible fluid is described by the Navier–Stokes equations and the continuity equation. These equations are derived from the momentum and mass conservation principles, respectively. Although the Navier–Stokes equations describe all forms of flow, in practice they are useful for laminar flow only, while every large scale geophysical surface flow is turbulent (Eagleson 1970). Therefore the open channel flow must be considered as turbulent. In such kind of flow random fluctuations of velocities and pressure are present. Since it is impossible to describe these fluctuations, the instantaneous velocity is expressed in terms of a time-averaged velocity and its random part, according to the Reynolds hypothesis. The same is applied for pressure. Inserting these relations into the Navier–Stokes equations one obtains the Reynolds equations. As a result of this operation an additional tensor of turbulent stresses appears. Applying the Boussinesq concept of the eddy viscosity one obtains the following system of equations (French 1985):

$$\frac{\partial u}{\partial t} + u\frac{\partial u}{\partial x} + v\frac{\partial u}{\partial y} + w\frac{\partial u}{\partial z} = F_x - \frac{1}{\rho}\frac{\partial p}{\partial x} + \frac{1}{\rho}(\mu + \mu^T)\Delta u, \tag{1.24}$$

$$\frac{\partial v}{\partial t} + u\frac{\partial v}{\partial x} + v\frac{\partial v}{\partial y} + w\frac{\partial v}{\partial z} = F_y - \frac{1}{\rho}\frac{\partial p}{\partial y} + \frac{1}{\rho}(\mu + \mu^T)\Delta v, \tag{1.25}$$

$$\frac{\partial w}{\partial t} + u\frac{\partial w}{\partial x} + v\frac{\partial w}{\partial y} + w\frac{\partial w}{\partial z} = F_z - \frac{1}{\rho}\frac{\partial p}{\partial z} + \frac{1}{\rho}(\mu + \mu^T)\Delta w, \tag{1.26}$$

$$\frac{\partial u}{\partial x} + \frac{\partial v}{\partial y} + \frac{\partial w}{\partial z} = 0, \tag{1.27}$$

where:

t – time,
x, y, z – spatial coordinates,
u, v, w – components of the velocity vector in x, y and z direction respectively,
ρ – water density,
μ – coefficient of dynamic viscosity,
μ^T – coefficient of turbulent viscosity (eddy viscosity),
p – pressure,

F_x, F_y, F_z – components of gravitational forces in x, y and z direction respectively,
Δ – Laplace operator.

The system of Eqs. (1.24), (1.25), (1.26) and (1.27) is formally very similar the system of Navier–Stokes equations. However, it is interpreted in a different way. Eqs. (1.24), (1.25), (1.26) and (1.27) describe the relations between time averaged values of the dependent variable, while the Navier–Stokes equations relate their instantaneous values. Conversely to the coefficient μ depending on the physical properties of the water, the coefficient μ^T is a function of the flow turbulence only. As the turbulent diffusion is much more important than the molecular one because $\mu^T \gg \mu$ (Martin and McCutcheon 1999), is reasonable to neglect the dynamic viscosity.

The system of Eqs. (1.24), (1.25), (1.26) and (1.27) representing the momentum and mass conservation equations written in general form will be used to derive the equations for open channel flow.

1.3 Derivation of 1D Dynamic Equation

Equations (1.23), (1.24), (1.25) and (1.26) can be simplified when the specific features of the open channel flow problem are taken into account. Usually open channel flow can be considered as a propagation of long waves in shallow water. These waves have relatively small amplitudes compared to their lengths so the accelerations and velocities in vertical direction are negligibly small in relation to the accelerations and velocities in horizontal directions. Let us assume the co-ordinate system as shown in Fig. 1.6, in which after Liggett (1975), x corresponds to the direction of the primary flow, y is the horizontal direction normal to primary flow, whereas z is the vertical direction.

The gravity force is the only force acting along z axis:

$$F_z = -g, \tag{1.28}$$

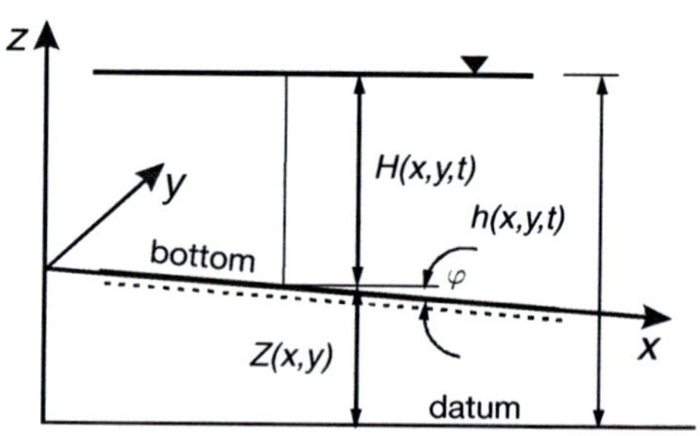

Fig. 1.6 Assumed system of co-ordinates

Consequently Eq. (1.26) takes the following form:

$$-g-\frac{1}{\rho}\frac{\partial p}{\partial z}=0. \tag{1.29}$$

Integration of this equation with regard to z from the bottom to the water surface (Fig. 1.6)

$$\int_{Z}^{h}\frac{\partial p}{\partial z}dz=-\int_{Z}^{h}\rho\cdot g\cdot dz \tag{1.30}$$

yields:

$$p\,(h)-p\,(Z)=-\rho\cdot g\,(h-Z)\,. \tag{1.31}$$

Since at the water surface the pressure is equal to the atmospheric one $p(h)=P_a$ then the following formula representing pressure variation along z axis accordingly to the hydrostatic pressure law, is obtained:

$$p\,(z)=P_a+\rho\cdot g\,(h-z)\ \text{for}\ Z\leq z\leq h \tag{1.32}$$

where:

P_a – atmospheric pressure,
ρ – water density,
g – acceleration due to gravity,
h – elevation of the water surface above the datum,
Z – elevation of the bottom above the datum,
z – vertical co-ordinate.

Let us assume that the velocity vector is parallel to x axis – it has one component only $u=u(z,x,t)$. Consequently Eq. (1.25) for the component v disappears, whereas Eq. (1.24) becomes:

$$\frac{\partial u}{\partial t}+u\frac{\partial u}{\partial x}=F_x-\frac{1}{\rho}\frac{\partial p}{\partial x}+\frac{\mu^T}{\rho}\left(\frac{\partial^2 u}{\partial x^2}+\frac{\partial^2 u}{\partial z^2}\right). \tag{1.33}$$

In addition, let us assume that the channel bed is inclined towards x axis and its slope is given by the angle φ (Fig. 1.6). The only force acting along x axis is the component of the gravitational force parallel to the bottom:

$$F_x=-g\cdot\sin\left(\varphi\right). \tag{1.34}$$

For a small bottom slope which is usually assumed for open channels (see Section 1.1), one has $\sin(\varphi)\approx \mathrm{tg}(\varphi)$. Then the component F_x is expressed as:

$$F_x = -g\frac{\partial Z}{\partial x}, \tag{1.35}$$

where $\partial Z/\partial x$ accordingly to Eq. (1.2) is the longitudinal channel bed slope. With Eqs. (1.32) and (1.35), Eq. (1.33) is rewritten as follows:

$$\frac{\partial u}{\partial t} + u\frac{\partial u}{\partial x} + g\frac{\partial Z}{\partial x} + \frac{1}{\rho}\frac{\partial}{\partial x}(P_a + \rho \cdot g \cdot H) - \frac{\mu^T}{\rho}\left(\frac{\partial^2 u}{\partial x^2} + \frac{\partial^2 u}{\partial z^2}\right) = 0. \tag{1.36}$$

In turbulent open channel flow the vertical velocity distribution is rather uniform, so the actual velocity $u(x,z,t)$ can be replaced by the depth-averaged one $U(x,t)$. Let us introduce:

- the average horizontal velocity $U(x,t)$:

$$U(x,t) = \frac{1}{H}\int_Z^h u(x,z,t)\,dz, \tag{1.37}$$

  where $H = h-Z$ designates the depth;
- the deviation $u''(z)$ of the actual velocity from the average one (Fig. 1.7):

$$u''(z) = u(z) - U. \tag{1.38}$$

From Eq. (1.37) results the following condition:

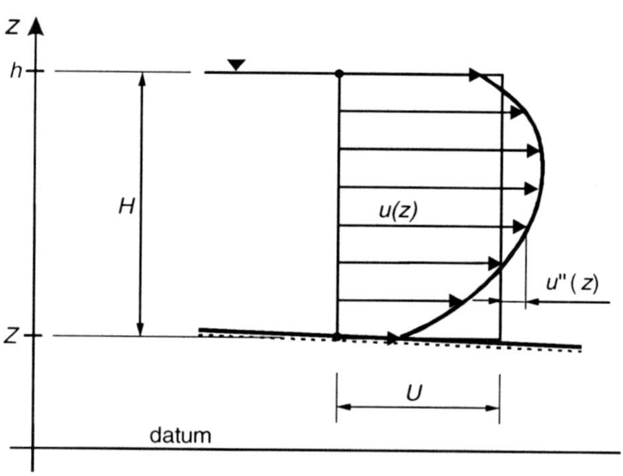

Fig. 1.7 Actual and averaged distribution of horizontal velocity u along the vertical axis

$$\int_{Z}^{h} u''(z)\,dz = 0. \tag{1.39}$$

Averaging the horizontal velocity over depth allows us to eliminate the variability in z direction and consequently to reduce the number of dimensions. To do it, Eq. (1.36) must be integrated over depth:

$$\int_{Z}^{h} \left(\frac{\partial u}{\partial t} + u\frac{\partial u}{\partial x} + g\frac{\partial Z}{\partial x} + \frac{1}{\rho}\frac{\partial}{\partial x}(P_a + \rho \cdot g \cdot H) - \frac{\mu^T}{\rho}\left(\frac{\partial^2 u}{\partial x^2} + \frac{\partial^2 u}{\partial z^2}\right)\right) dz = 0. \tag{1.40}$$

The integration limits, i.e. the water stage h and the bottom elevation Z, in a general case are functions of the horizontal spatial coordinates. Therefore while integrating Eq. (1.40) we will face a problem of differentiation inside the integral with variable limits. According to the Leibniz rule (Korn and Korn 1968, McQuarrie 2003) in such a case the differentiation of any function ϕ is carried out as follows:

$$\int_{Z(x)}^{h(x)} \frac{\partial \phi}{\partial x} dz = \frac{\partial}{\partial x}\int_{Z(x)}^{h(x)} \phi \cdot dz - \phi(h)\frac{\partial h}{\partial x} + \phi(Z)\frac{\partial Z}{\partial x}. \tag{1.41}$$

Let us integrate subsequent terms of Eq. (1.40). The first term is integrated as follows:

$$I_1 = \int_{Z}^{h} \frac{\partial u}{\partial t} dz = \frac{\partial}{\partial t}\int_{Z}^{h} u \cdot dz - u(h)\frac{\partial h}{\partial t} + u(Z)\frac{\partial Z}{\partial t}. \tag{1.42}$$

If we:

- make use of the definition (1.37),
- assume that the channel bed does not change its position with time,
- assume that the velocity at the water surface is equal to the average one then Eq. (1.42) will take the following final form:

$$I_1 = \frac{\partial}{\partial t}(U \cdot H) - U\frac{\partial H}{\partial t} = H\frac{\partial U}{\partial t}. \tag{1.43}$$

The integral of the second term with Eq. (1.38) is rearrangered to the form:

$$I_2 = \int_Z^h u \frac{\partial u}{\partial x} dz = \frac{1}{2} \int_Z^h \frac{\partial (U + u'')^2}{\partial x} dz$$
$$= \frac{1}{2} \left[\frac{\partial}{\partial x} \int_Z^h (U + u'')^2 dz - (U + u'')^2_{z=h} \frac{\partial h}{\partial x} + (U + u'')^2_{z=Z} \frac{\partial Z}{\partial x} \right]. \tag{1.44}$$

Developing the square of sum and assuming that the flow velocities at the water surface and at the channel bottom are equal to the average velocity U, one obtains:

$$I_2 = \frac{1}{2} \left[\frac{\partial}{\partial x} \left(\int_Z^h U^2 \cdot dz + \int_Z^h 2U \cdot u'' \cdot dz + \int_Z^h (u'')^2 dz \right) - U^2 \frac{\partial (h - Z)}{\partial x} \right]. \tag{1.45}$$

Integrals in the brackets are calculated term by term as follows:

$$\int_Z^h U^2 \cdot dz = U^2 \cdot H, \tag{1.46}$$

$$\int_Z^h 2U \cdot u'' \cdot dz = 2U \int_Z^h u'' \cdot dz = 0. \tag{1.47}$$

The last integral at the right hand side of Eq. (1.45) $\int_Z^h (u'')^2 dz$ cannot be calculated since the actual distribution of the velocity over the depth is unknown. However, a rough estimation of this integral in relation to the integral (1.46) can be attempted. To this end let us assume a hypothetical vertical distribution of $u(z)$ as presented in Fig. 1.8.

Over the entire depth the deviation of actual velocity is taken as $u''(z) = \pm 0.2U$. Then we have:

$$\int_Z^h (u'')^2 dz = \int_Z^h (0.2U)^2 dz = 0.04\, U^2 \int_Z^h dz = 0.04\, U \cdot H. \tag{1.48}$$

The result of integration shows that even for the assumed significant non-uniformity of the vertical distribution of velocity, the value of integral (1.48) constitutes only 4% of the value of integral (1.46). To take into account these integral we can consider it as a fraction of integral (1.46) and to add its value to (1.46). This allows us to write:

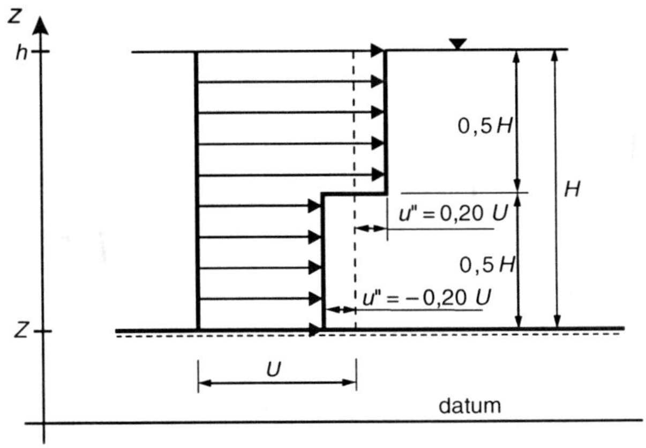

Fig. 1.8 Assumed hypothetical vertical distribution of flow velocity

$$\int_Z^h U^2 \cdot dz + \int_Z^h (u'')^2 dz = U^2 \cdot H + \int_Z^h (u'')^2 dz = \left(1 + \frac{\int_Z^h (u'')^2 dz}{U^2 \cdot H}\right) U^2 \cdot H = \beta \cdot U^2 \cdot H. \tag{1.49}$$

In this equation we recognize

$$\beta = 1 + \frac{\int_Z^h (u'')^2 dz}{U^2 \cdot H}. \tag{1.50}$$

as a correction parameter. Coming back to Eq. (1.45) we have:

$$I_2 = \frac{1}{2}\left[\frac{\partial}{\partial x}\left(\beta \cdot U^2 \cdot H\right) - U^2 \frac{\partial H}{\partial x}\right]. \tag{1.51}$$

Developing of the derivative in Eq. (1.51) yields:

$$I_2 = \frac{1}{2}\left[H \frac{\partial\left(\beta \cdot U^2\right)}{\partial x} + \beta \cdot U^2 \frac{\partial H}{\partial x} - U^2 \frac{\partial H}{\partial x}\right]. \tag{1.52}$$

Neglecting the difference of the last two terms as small value, which additionally can be corrected by chosen value of β, the integral of the second term of Eq. (1.40) takes its following final form:

$$I_2 = \beta \cdot U \frac{\partial U}{\partial x} H. \tag{1.53}$$

Introduction of the correction factor β requires a comment. Assumption of the uniform flow velocity distribution instead of the actual one always requires a correction of the conservative quantity, which depends on the velocity. In the case of the discrete energy equation (1.23) applied for channel flow, comparison of the actual kinetic energy with that calculated with the average velocity shows that it is necessary to introduce the correction factor α, given by Eq. (1.15). This coefficient was introduced into the dynamic equation of the Saint Venant system by Chow (1959). However, Abbott (1979) in this same equation introduces the factor β, which corrects the momentum. It is worth emphasizing that usually at first the unsteady flow equations are derived assuming a priori one dimensionality and afterwards the effect of averaging is corrected. In the derivation from general equations as we have just applied, the correction factor appears naturally as a result of the averaging process. Such a way of derivation ensures a simple interpretation of the corrective parameter β and explains the real reasons of its appearance. It corrects the balance of momentum affected by the average flow velocity introduced instead of the actual velocity distribution over the cross-sectional area. This parameter allows us to include the estimated value of integral (1.48), impossible to be calculated directly. For $u(z)=$ const., i.e. when $u''(z) = 0$ for $Z \leq z \leq h$ one obtains $\beta = 1$.

The terms of Eq. (1.40) representing gravitation and pressure are integrated as follows:

$$I_3 = \int_Z^h g\frac{\partial Z}{\partial x}dz = g\frac{\partial Z}{\partial x}H, \tag{1.54}$$

$$I_4 = \frac{1}{\rho}\int_Z^h \frac{\partial}{\partial x}(P_a + \rho \cdot g \cdot H)\,dz = \frac{1}{\rho}\frac{\partial P_a}{\partial x}H + g\frac{\partial H}{\partial x}H. \tag{1.55}$$

The term describing the turbulent diffusion of momentum in x direction takes the form:

$$\begin{aligned} I_5 &= \int_Z^h \frac{\mu^T}{\rho}\frac{\partial^2 u}{\partial x^2}dz = \frac{\mu^T}{\rho}\int_Z^h \frac{\partial^2}{\partial x^2}(U + u'')\,dz = \\ &= \frac{\mu^T}{\rho}\frac{\partial^2 U}{\partial x^2}H + \frac{\mu^T}{\rho}\int_Z^h \frac{\partial^2 u''}{\partial x^2}dz = \frac{\mu^D}{\rho}\frac{\partial^2 U}{\partial x^2}H, \end{aligned} \tag{1.56}$$

where μ^D is the coefficient of dispersive transport of momentum in x direction. It represents the combined effect of the turbulent viscosity with coefficient μ^T and vertical velocity averaging. This is an analogous situation to the one discussed previously, where integrating the term of velocity advection led to the correction factor β (see Eq. 1.50).

The term describing the momentum transfer in vertical is integrated as follows:

$$I_6 = \int_Z^h \frac{\mu^T}{\rho} \frac{\partial^2 u}{\partial z^2} dz = \frac{1}{\rho}\left(\mu^T \left.\frac{\partial u}{\partial z}\right|_h - \mu^T \left.\frac{\partial u}{\partial z}\right|_Z\right). \tag{1.57}$$

The terms in parentheses are the tangential stresses at the water surface and at the bottom, respectively. The former ones are caused by the action of the wind blowing above the water surface in direction parallel to x axis. It generates the stresses which can be estimated using the following formula (Hervouet 2007, Singh 1996):

$$\tau(h) = a\,|W|\,W \tag{1.58}$$

where:

- a – empirical coefficient depending on the density and viscosity of air and on the roughness of the water surface,
- W – component of the wind velocity vector acting along the channel axis.

In many practical cases the wind-generated stresses are negligible. They can be important for example in lower parts of river, where the bed slope is typically small.

The second term of Eq. (1.57) describing the stresses at the bottom is expressed by the well known formula (Chanson 2004):

$$\tau(Z) = \frac{\rho \cdot g \cdot n^2}{H^{1/3}}\,|U|\,U. \tag{1.59}$$

Inserting Eqs. (1.58) and (1.59). In Eq. (1.57) yields:

$$I_6 = \frac{a}{\rho}\,|W|\,W - \frac{g \cdot n^2}{H^{1/3}}\,|U|\,U. \tag{1.60}$$

Finally, the integration of Eq. (1.40) allows us to present it in the form:

$$\begin{aligned} H\frac{\partial U}{\partial t} &+ \beta \cdot U\frac{\partial U}{\partial x}H + g\frac{\partial Z}{\partial x}H + \frac{1}{\rho}\frac{\partial P_a}{\partial x}H + g\frac{\partial H}{\partial x}H + \\ &- \frac{\mu^D}{\rho}\frac{\partial^2 U}{\partial x^2}H - \frac{a}{\rho}\,|W|\,W + \frac{g n^2}{H^{1/3}}\,|U|\,U = 0, \end{aligned} \tag{1.61}$$

After dividing both sides by the depth H, which is always non-zero, one obtains:

$$\begin{aligned} \frac{\partial U}{\partial t} &+ \beta \cdot U\frac{\partial U}{\partial x} + g\frac{\partial}{\partial x}(H+Z) + \\ &+ \frac{g \cdot n_M^2}{H^{4/3}}\,|U|\,U + \frac{1}{\rho}\frac{\partial P_a}{\partial x} - \frac{a}{\rho \cdot H}\,|W|\,W - \frac{\mu^D}{\rho}\frac{\partial^2 U}{\partial x^2} = 0. \end{aligned} \tag{1.62}$$

Equation (1.62) is the dynamic equation for 1D unsteady open channel gradually varied flow. This equation, derived from Eqs. (1.23), (1.24), (1.25) and (1.26), contains all the significant components which determine the flow process. The only process omitted in the derivation is the lateral water inflow distributed along the channel.

1.4 Derivation of 1D Continuity Equation

Similarly to the dynamic equation, the continuity equation for one-dimensional flow can be derived from the general three-dimensional from. Let us assume a uniform motion in y direction. Then Eq. (1.27) takes a simpler form, which must be integrated over the depth from the bottom $Z(x)$ to the surface $h(x)$:

$$\int_{Z}^{h}\left(\frac{\partial u}{\partial x}+\frac{\partial w}{\partial z}\right)dz=0. \tag{1.63}$$

The terms in parentheses are integrated separately giving:

$$\begin{aligned} I_1 &= \int_{Z}^{h}\frac{\partial u}{\partial x}dz=\frac{\partial}{\partial x}\int_{Z}^{h}u\cdot dz-u(h)\frac{\partial h}{\partial x}+u(Z)\frac{\partial Z}{\partial x}= \\ &= \frac{\partial}{\partial x}(U\cdot H)-u(h)\frac{\partial h}{\partial x}+u(Z)\frac{\partial Z}{\partial x}, \end{aligned} \tag{1.64}$$

$$I_2=\int_{Z}^{h}\frac{\partial w}{\partial x}dz=w(h)-w(Z), \tag{1.65}$$

where $w(h)$ and $w(Z)$ are the vertical velocities at the water surface and at the bottom, respectively. These velocities can be found from the kinematic condition at each of the limits, which says that a particle laying at the limit moves together with the limit. Then:

$$w(h)=\frac{Dh}{Dt}, \tag{1.66}$$

$$w(Z)=\frac{DZ}{Dt}. \tag{1.67}$$

The total derivatives Dh/Dt and DZ/Dt represent the velocities of traveling water surface and bottom in the vertical direction, respectively:

$$\frac{Dh}{Dt}=\frac{\partial h}{\partial t}+u(h)\frac{\partial h}{\partial x}, \tag{1.68}$$

$$\frac{DZ}{Dt}=\frac{\partial Z}{\partial t}+u(Z)\frac{\partial Z}{\partial x} \tag{1.69}$$

In many geophysical flows the motion of the water surface in the vertical direction is caused by external sources like rain, evaporation or lateral inflow. This fact must be taken into account in Eq. (1.68), which should be modified as follows:

$$\frac{Dh}{Dt} = \frac{\partial h}{\partial t} + u(h)\frac{\partial h}{\partial x} - \delta \tag{1.70}$$

where δ is the source term representing additional inflow.

Substitution of Eqs. (1.69) and (1.70) in Eq. (1.65) yields:

$$I_2 = \frac{\partial h}{\partial t} + u(h)\,\frac{\partial h}{\partial x} - \delta - \frac{\partial Z}{\partial t} - u(Z)\frac{\partial Z}{\partial x}. \tag{1.71}$$

The sum $I_1 + I_2$, is equivalent to the integral (1.63) and can be written as:

$$\begin{aligned}&\frac{\partial}{\partial x}(UH) - u(h)\,\frac{\partial h}{\partial x} + u(Z)\frac{\partial Z}{\partial x} + \\ &\quad + \frac{\partial h}{\partial t} + u(h)\frac{\partial h}{\partial x} - \delta - \frac{\partial Z}{\partial t} - u(Z)\frac{\partial Z}{\partial x} = 0.\end{aligned} \tag{1.72}$$

Since accordingly to Eq. (1.1) we have $h = H{+}Z$ then Eq. (1.72) becomes:

$$\frac{\partial H}{\partial t} + \frac{\partial}{\partial x}(UH) = \delta. \tag{1.73}$$

In such a way the 1D continuity equation (1.73) for open channel flow was derived from 2D equation via its integration along the vertical direction.

1.5 System of Equations for Unsteady Gradually Varied Flow in Open Channel

Integration of the system of equations (1.24), (1.25), (1.26) and (1.27) over the depth with the assumption of uniform flow in y direction allowed us to derive 1D dynamic and continuity equations:

$$\begin{aligned}&\frac{\partial U}{\partial t} + \beta \cdot U\frac{\partial U}{\partial x} + g\frac{\partial}{\partial x}(H + Z) + \frac{g \cdot n^2}{H^{4/3}}\,|U|\,U + \\ &\quad + \frac{1}{\rho}\frac{\partial P_a}{\partial x} - \frac{a}{\rho \cdot H}\,|W|\,W - \frac{\mu^D}{\rho}\frac{\partial^2 U}{\partial x^2} = 0,\end{aligned} \tag{1.74}$$

$$\frac{\partial H}{\partial t} + \frac{\partial}{\partial x}(U \cdot H) = \delta. \tag{1.75}$$

The above system of partially differential equations describes unsteady flow in an open channel. Let us recall the major assumptions introduced during the derivation:

- Motion of water is gradually varied,
- Channel bed slope is small,
- Flow is considered as one dimensional process,
- Velocity distribution over depth is uniform,
- Pressure is governed by the hydrostatic law,
- Friction slope is estimated as for the steady flow,
- Lateral inflow does not influence the momentum balance.

Since Eqs. (1.74) and (1.75) are written per unit channel width, this system is valid for a wide rectangular channel. The parameter β in dynamic equation (1.74) occurred as a result of the integration over depth. It corrects the error in momentum balance introduced by velocity averaging. It appeared naturally from the Leibniz rule applied to calculate the integrals with variable limits.

Usually, in Eq. (1.74) the last three terms are neglected as less important compared to the others. Dimensionless analysis shows that longitudinal dispersion of momentum is insignificant (French 1985). Similarly, in most cases the temporal and spatial variability of the atmospheric pressure and the wind velocity do not influence remarkably the river flow. Therefore the terms representing them can be omitted. If in addition we assume $\beta = 1$ and $\delta = 0$, introduce the bottom slope s defined by Eq. (1.7) and the friction slope S, which for flow per unit width is:

$$S = \frac{n_M^2}{H^{4/3}} \, |U| \, U. \tag{1.76}$$

then Eqs. (1.74) and (1.75) become:

$$\frac{\partial U}{\partial t} + U \frac{\partial U}{\partial x} + g \frac{\partial H}{\partial x} = g(s - S), \tag{1.77}$$

$$\frac{\partial H}{\partial t} + \frac{\partial}{\partial x}(U \cdot H) = 0, \tag{1.78}$$

In practical applications it is more convenient to use the flow rate and water stage instead of depth and velocity as the dependent variables. The equations with new variables can be derived as previously from Eqs. (1.24), (1.25), (1.26) and (1.27), with the same assumptions. However, they should be integrated not only with regard to depth but with regard to the channel width as well. Taking into account that the velocity is averaged over the entire wetted cross-sectional area A, the following system of equations is obtained:

$$\frac{1}{g} \frac{\partial U}{\partial t} + \frac{\partial}{\partial x} \left(\frac{U^2}{2\,g} \right) + \frac{\partial h}{\partial x} + S = 0, \tag{1.79}$$

$$\frac{\partial A}{\partial t} + \frac{\partial Q}{\partial x} = q, \tag{1.80}$$

where:

U – cross-sectional average flow velocity,
h – water level above assumed datum defined by Eq. (1.1),
A – cross-sectional area,
Q – flow rate
q – lateral inflow per unit length of a channel.

Omitting the lateral inflow q we obtain the unsteady open channel flow equations in the form proposed in 1871 by Barré de Saint-Venant to describe the process of propagation of the flood waves (Chanson 2004). After him, these equations are known in literature as the system of Saint Venant equations.

The dynamic equation (1.79) is similar to Eq. (1.77). However, the variables involved in Eq. (1.79) are related to the whole cross-section, and not to the unit width. For this reason, their interpretation is different. Now U is the velocity averaged over the cross-section:

$$U(x,t) = \frac{Q}{A} = \frac{1}{A} \iint_A u(x,y,z,t)\, dA, \tag{1.81}$$

whereas S is the friction slope, which for a channel cross-section is given by formula:

$$S = \frac{n^2}{R^{4/3}} |U|\, U, \tag{1.82}$$

where R is hydraulic radius.

Similarly, the correction parameter β is related to the entire cross-section. Now, its general definition resulting from integration of the convective term $u\partial u/\partial x$ over the whole cross-section is following:

$$\beta = 1 + \frac{\iint_A (u'')^2 dA}{U^2 \cdot A}, \tag{1.83}$$

where $u''=u - U$ is the difference between the actual velocity u at any point of the cross-section and the averaged flow velocity U.

The correction factor β, usually applied in dynamic equation of the Saint Venant system, is given as (Abbott 1979, Liggett 1975):

$$\beta = \frac{\iint_A u^2 \cdot dA}{U^2 \cdot A} \tag{1.84}$$

This factor called the Boussinesq coefficient (Chanson 2004) represents the ratio of the actual momentum and the momentum calculated with the averaged velocity U. If instead of u'' we substitute in Eq. (1.83) the expression $u''=u - U$ then

$$\beta = 1 + \frac{\iint\limits_A (u-U)^2 dA}{U^2 \cdot A} = \frac{\iint\limits_A u^2 dA}{U^2 \cdot A} + \frac{2U^2 \cdot A - 2U \iint\limits_A u \cdot dA}{U^2 \cdot A}. \tag{1.85}$$

Since with Eq. (1.81) the second term at the right hand side vanishes, then Eq. (1.85) becomes Eq. (1.84). As one can see, integration of the momentum equation over depth using the Leibnitz rule gives another possible viewpoint on introduction of the momentum correction while averaging of the flow velocity.

Eliminating both depth and average flow velocity from Eqs. (1.79) and (1.80) can be carried out using Eqs. (1.1) and (1.10). In addition, a constant position of the channel bed is assumed so $\partial Z/\partial t = 0$. Since the wetted cross-section area A is a function of the water stage h, then one can write:

$$\frac{\partial A}{\partial t} = \frac{\partial A}{\partial h}\frac{\partial h}{\partial t} = B\frac{\partial h}{\partial t} \tag{1.86}$$

These assumptions allow us to reform both equations of unsteady flow to obtain:

$$\frac{\partial Q}{\partial t} + \frac{\partial}{\partial x}\left(\frac{\beta \cdot Q^2}{A}\right) + g \cdot A\frac{\partial h}{\partial x} + \frac{g \cdot n^2}{R^{4/3}}\frac{|Q|\,Q}{A} = 0, \tag{1.87}$$

$$\frac{\partial h}{\partial t} + \frac{1}{B}\frac{\partial Q}{\partial x} = \frac{q}{B}. \tag{1.88}$$

where B is channel width at the water surface level. Equations (1.87) and (1.88) constitute the most popular mathematical model of the unsteady open channel flow. It should be added that the system of Saint Venant equations is expressed in many forms depending on the assumed dependent variables, geometry of a channel and other factors taken into account. For a systematic review of these forms see Singh (1996).

1.6 Steady Gradually Varied Flow in Open Channel

In many practical applications the steady and gradually varied flow in open channels must be considered. Typical problem connected with this kind of flow is the determination of the flow profile behind a dam.

The governing equation describing steady gradually varied flow can be obtained in different ways. The manner of derivation depends on our preferences. We can write balance equation directly for the steady gradually varied flow with introduced a priori simplifying assumptions or to derive them from the Saint Venant equations, describing a more general case of unsteady gradually varied flow. Both approaches will be shown in the following sections.

1.6.1 Derivation of Governing Equation from the Energy Equation

Let us consider two neighboring cross-sections enclosing a channel reach of length dx as shown in Fig. 1.9. At the upstream end the depth is equal to H and the average flow velocity is equal to U, whereas at the downstream end we have the depth $H + dH$ and the velocity $U + dU$, respectively. The Bernoulli equation takes the following form:

$$Z + H + \frac{\alpha \cdot U^2}{2g} = Z - s \cdot dx + H + dH + \frac{\alpha\,(U + dU)^2}{2g} + S \cdot dx \qquad (1.89)$$

where:

E – total mechanical energy above the assumed datum,
Z – bed elevation with regard to the datum,
H – depth,
U – average flow velocity,
s – bottom slope,
S – friction slope,
g – gravitational acceleration,
α – energy correction factor.

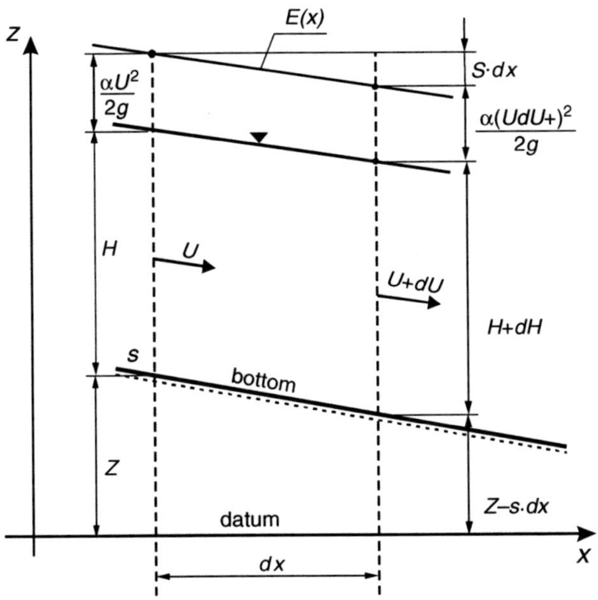

Fig. 1.9 Schematic representations of a channel reach

The square velocity at the downstream end can be expressed using an approximate formula, since the term $(dU)^2$ can be omitted as significantly smaller than other terms in the expression $(U+dU)^2$:

$$(U + dU)^2 = U^2 + 2U \cdot dU + (dU)^2 \approx U^2 + 2U \cdot dU$$

Using this relation Eq. (1.89) can be rearranged to the following form:

$$s = \frac{dH}{dx} + \frac{\alpha}{g}\frac{U \cdot dU}{dx} + S \tag{1.90}$$

The second term at the right-hand side of Eq. (1.90) can be expressed as follows:

$$\frac{\alpha}{g}\frac{U \cdot dU}{dx} = \frac{\alpha}{2g}\frac{dU^2}{dx} \tag{1.91}$$

Substitution of Eq. (1.91) in Eq. (1.90) and using the definitions (1.1) and (1.2) yields:

$$\frac{d}{dx}\left(h + \frac{\alpha \cdot Q^2}{2g \cdot A^2}\right) = -S \tag{1.92}$$

This is differential representation of the energy conservation principle. Equation (1.92) has general character, since it holds for both prismatic and non-prismatic channels.

Equation (1.92) can be reduced to the standard form of the steady gradually flow equation for a prismatic channel, in which the depth is dependent variable. To this end let us introduce relation (1.1), which allows us to rewrite Eq. (1.92) as follows:

$$\frac{d}{dx}\left(Z + H + \frac{\alpha \cdot Q^2}{2g \cdot A^2}\right) = -S \tag{1.93}$$

Let us differentiate the left-hand side of Eq. (1.93). Assuming that Q = const. one obtains:

$$\frac{d}{dx}\left(Z + H + \frac{\alpha \cdot Q^2}{2g \cdot A^2}\right) = \frac{dZ}{dx} + \frac{dH}{dx} - \frac{\alpha \cdot Q^2 \cdot B}{g \cdot A^3}\frac{dH}{dx} \tag{1.94}$$

where B is channel width at the water surface. Equations (1.94) and (1.22) are substituted in Eq. (1.93). Simple rearrangement yields:

$$\left(1 - \frac{\alpha \cdot Q^2 \cdot B}{g \cdot A^3}\right)\frac{dH}{dx} = -\frac{dZ}{dx} - \frac{n_M^2 \cdot |Q|\, Q}{R^{4/3} \cdot A^2} \tag{1.95}$$

One can assume that in a single channel the direction of the steady flow coincides with the positive direction of x axis. Then the modulus in last term of Eq. (1.95) can be omitted. The next assumption concerns the flow velocity distribution over the cross-section of the channel. It is considered as quasi-uniform, so that one can assume $\alpha = 1$. Moreover, for a rectangular cross-section the second term in brackets represents the square of the Froude number since:

$$\frac{Q^2 \cdot B}{g \cdot A^3} = \frac{1}{g}\frac{Q^2}{A^2}\frac{1}{H} = \frac{U^2}{g \cdot H} = F_r^2 \tag{1.96}$$

Substitution of Eqs. (1.2), (1.22) and (1.96) in Eq. (1.95) yields:

$$\frac{dH}{dx} = \frac{s - S}{1 - F_r^2}. \tag{1.97}$$

In such a way the classic ordinary differential equation describing the water surface profile for a steady gradually varied flow is obtained. However, one should remember that it holds only for prismatic channel with $Q = \text{const}$.

1.6.2 Derivation of Governing Equation from the System of Saint-Venant Equations

The steady gradually varied flow can be considered as a particular case of the unsteady gradually varied flow. For this reason it should be possible to derive the governing equations from the system of Saint Venant equations (1.87) and (1.88), previously discussed. This manner of derivation seems to be more instructive, since it clearly shows the direct relations between the derived equations which govern the steady flow and the fundamental principles of mass and momentum conservation used in open channel flow modeling. In addition, such approach allows us to derive the most general form of the steady flow equations.

The Saint Venant equations (1.87) and (1.88) can be rewritten for a channel with variable cross-sections and with non-uniform velocity distribution over the cross-sections as follows:

$$\frac{\partial A}{\partial t} + \frac{\partial Q}{\partial x} = q, \tag{1.98}$$

$$\frac{\partial Q}{\partial t} + \frac{\partial}{\partial x}\left(\frac{\beta \cdot Q^2}{A}\right) + g \cdot A\frac{\partial h}{\partial x} = -g \cdot A \cdot S \tag{1.99}$$

where:

t – time,
x – longitudinal distance,
h – water surface elevation,

q – lateral inflow,
β – momentum correction factor
S – friction slope given by Eq. (1.22).

For steady flow one obtains:

$$\frac{\partial A}{\partial t} = 0, \tag{1.100}$$

$$\frac{\partial Q}{\partial t} = 0. \tag{1.101}$$

Therefore all the functions depend on the variable x only and consequently Eqs. (1.98) and (1.99) become ordinary differential equations. They take the following forms:

$$\frac{dQ}{dx} = q \tag{1.102}$$

$$\frac{d}{dx}\left(\frac{\beta \cdot Q^2}{A}\right) + g \cdot A\frac{dh}{dx} = -g \cdot A \cdot S \tag{1.103}$$

Equation (1.103) can be rearranged to a form more suitable for integration. To this end both its sides are divided by $g{\cdot}A$. This yields:

$$\frac{1}{gA}\frac{d}{dx}\left(\frac{\beta \cdot Q^2}{A}\right) + \frac{dh}{dx} = -S. \tag{1.104}$$

Taking into account the definition of the average flow velocity in the channel cross-section (1.9), the first term of Eq. (1.104) is differentiated as:

$$\frac{1}{gA}\;\frac{d}{dx}\left(\frac{\beta Q^2}{A}\right) = \frac{1}{g}\;\frac{\beta}{A}\;\frac{d}{dx}(Q \cdot U) = \frac{1}{g}\;\frac{\beta}{A}\left(U\frac{dQ}{dx} + Q\frac{dU}{dx}\right) \tag{1.105}$$

Consequently Eq. (1.104) becomes:

$$\frac{\beta}{g}\;\frac{U}{A}\frac{dQ}{dx} + \frac{\beta}{g}\;\frac{Q}{A}\;\frac{dU}{dx} + \frac{dh}{dx} = -S \tag{1.106}$$

Substitution of Eqs. (1.9) and (1.102) into Eq. (1.106) leads to:

$$\frac{d}{dx}\left(h + \frac{\beta \cdot Q^2}{2g \cdot A^2}\right) = -S - \frac{\beta \cdot Q}{g \cdot A^2}q \tag{1.107}$$

Therefore, in the case of steady gradually varied flow the water surface profile is described by the following equations:

$$\frac{dQ}{dx} = q, \tag{1.108}$$

$$\frac{dE}{dx} = -S - \frac{\beta \cdot Q}{g \cdot A^2} q \tag{1.109}$$

where

$$E = h + \frac{\beta \cdot Q^2}{2g \cdot A^2}. \tag{1.110}$$

Note that E represents the total energy of the flowing stream related to the assumed datum. A comment is needed to explain the problem of the correction of the conservative quantity represented by the governing differential equation. This correction is introduced when the locally variable velocity over a cross-section is replaced by the average velocity. If the momentum conservation principle is applied as it is done while deriving the dynamic equation in the Saint Venant system, then the correction parameter β should rather be used. On the other hand, if the energy conservation principle is applied, then the correction parameter α should be used. However, during the derivation of the steady flow equation from the system of Saint Venant equations, the dynamic equation (1.99) is transformed into Eq. (1.109) representing the conservation of energy. Therefore, from now on in the governing equation of the steady flow the correction parameter α will be used. Note that when Eq. (1.92) is derived directly from the mechanical energy conservation principle, the parameter α arises in this equation in natural way.

Since Eqs. (1.108) and (1.109) were derived from the system of Saint Venant equations, they should be applicable for all types of open channels. Indeed, as it will be shown later, numerical integration of Eq. (1.109) by the implicit trapezoidal rule leads to the well known step method, commonly applied in practice. Usually the step method is derived directly from the principle of energy conservation applied for two neighboring cross-sections lying along channel axis, i.e. for discrete system.

Equation (1.109) can be recast into Eq. (1.97), which has been previously derived for rectangular channel from the energy principle and is more familiar in open channel hydraulics. To this order using Eq. (1.108) the spatial derivative of the mechanical energy is transformed to the form:

$$\frac{dE}{dx} = \frac{d}{dx}\left(h + \frac{\alpha \cdot Q^2}{2g \cdot A^2}\right) = \frac{dh}{dx} + \frac{\alpha \cdot Q}{g \cdot A^2} q - \frac{\alpha \cdot Q^2}{g \cdot A^3}\frac{dA}{dx} \tag{1.111}$$

As previously, one can assume that the flow direction coincides with the positive direction of x axis, so that the modulus in the friction slope (Eq. 1.82) is omitted. Substitution of Eqs. (1.82) and (1.111) with Eq. (1.1) in Eq. (1.109) yields:

$$\frac{dh}{dx} = -\frac{\dfrac{n_M^2 \cdot Q^2}{R^{4/3} \cdot A^2} + \dfrac{\alpha \cdot Q^2 \cdot B}{g \cdot A^3}\dfrac{dZ}{dx} + \dfrac{2\alpha \cdot Q}{g \cdot A^2} q}{1 - \dfrac{\alpha \cdot Q^2 \cdot B}{g \cdot A^3}} \tag{1.112}$$

For constant flow discharge along the channel axis ($Q = \text{const.}$), i.e. when the lateral inflow is neglected ($q = 0$), one obtains a simpler version of Eq. (1.112):

$$\frac{dh}{dx} = -\frac{\dfrac{n_M^2 \cdot Q^2}{R^{4/3} \cdot A^2} + \dfrac{\alpha \cdot Q^2 \cdot B}{g \cdot A^3}\dfrac{dZ}{dx}}{1 - \dfrac{\alpha \cdot Q^2 \cdot B}{g \cdot A^3}} \tag{1.113}$$

For a rectangular channel with longitudinal bed slope s, instead of the water stage h the flow depth H can be introduced. To this end the water level h expressed by Eq. (1.1) is substituted in Eq. (1.113) resulting in the following equation:

$$\frac{d}{dx}(H + Z) = -\frac{\dfrac{n_M^2 \cdot Q^2}{R^{4/3} \cdot A^2} + \dfrac{\alpha \cdot Q^2 \cdot B}{g \cdot A^3}\dfrac{dZ}{dx}}{1 - \dfrac{\alpha \cdot Q^2 \cdot B}{g \cdot A^3}} \tag{1.114}$$

Now, introducing the definition of the longitudinal bed slope (1.2) and the Froude number (1.17) into Eq. (1.114) one obtains:

$$\frac{dH}{dx} = \frac{s - S}{1 - \alpha \cdot F_r^2}. \tag{1.115}$$

Note that for $\alpha = 1$ Eq. (1.115) coincides with Eq. (1.99). In such a way the classic form of the governing equation for steady gradually varied flow in prismatic open channel was derived directly from the system of Saint Venant equations.

1.7 Storage Equation

In hydraulic practice one has often to consider a river flowing through reservoir behind a dam. In such a case the dimensions of wetted cross-sections increase significantly, whereas the average flow velocities decrease. Location of a reservoir along the river course influences propagation of flood waves, reducing their peaks flows and volumes. For flood routing purposes it is sometimes reasonable to neglect spatial variability of the flow parameters and to reduce formally the reservoir to a point characterized by time dependent capacity of retained water only. Such assumption leads to the so-called storage equation.

To derive the storage equation let us consider the situation shown in Fig. 1.10. The capacity of the reservoir behind the dam depends on the topography of the valley and on the water stage kept in the reservoir $h(t)$.

If the width and depth of the reservoir are not too large, meaning that the flow process is still one dimensional, the flood wave propagating through reservoir can be modeled using the system of Saint Venant equations. However, if we are not interested in considering the internal distribution of the velocity and pressure fields

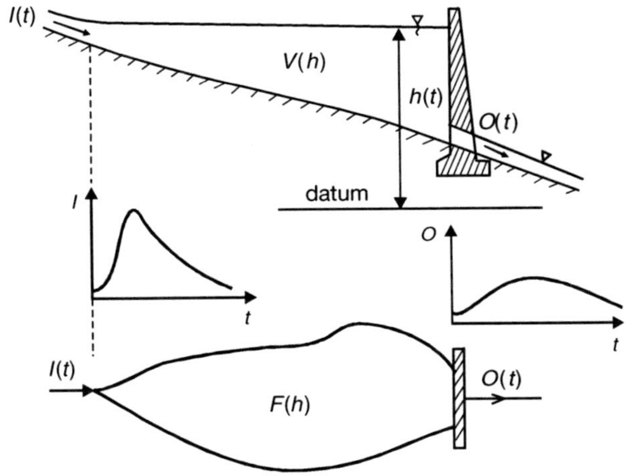

Fig. 1.10 Scheme of a retention reservoir

in the reservoir, the problem of reservoir's dynamic can be reduced to the problem of mass conservation only, considered at the global scale. In such a case we neglect the flow velocities and the behavior of the reservoir will be described by one dependent variable $h(t)$ only.

To obtain the governing equation let us apply the mass conservation principle to a control volume constituted by the entire capacity of the reservoir, limited by an appropriate control surface. The mass conservation principle means that the time variation of the mass contained in reservoir is caused by the net flow through the control surface only. This statement is expressed as follows:

$$\frac{d}{dt}\iiint\limits_V \rho \cdot dV + \iint\limits_\sigma \rho \cdot \mathbf{U} \cdot \mathbf{n} \cdot dA = 0 \tag{1.116}$$

where:

t – time,
ρ – water density,
V – control volume,
σ – control surface,
$\mathbf{U}$ – vector of flow velocity,
$\mathbf{n}$ – unit vector normal to surface σ and oriented towards outside of volume.

If we assume constant water density ($\rho = \text{const}$), then Eq. (1.116) will express the conservation of the water volume in reservoir:

$$\frac{d}{dt}\iiint\limits_V dV + \iint\limits_\sigma \mathbf{U} \cdot \mathbf{n} \cdot d\sigma = 0 \tag{1.117}$$

The first term of equation represents the time variation of volume since:

$$\iiint_V dV = V \tag{1.118}$$

The second term represents the net exchange of the water in reservoir with its surrounding. Taking into account the main ways of exchange, this term can be expressed as a sum the following components:

$$\iint_\sigma \mathbf{U} \cdot \mathbf{n} \cdot d\sigma = -I(t) + O(t) + E(t) - P(t) + G(t) \tag{1.119}$$

where:

$I(t)$ – river inflow,
$O(t)$ – river outflow,
$E(t)$ – evaporation from reservoir surface,
$P(t)$ – rainfall on reservoir surface,
$G(t)$ – infiltration through reservoir bottom.

Substitution of Eqs. (1.118) and (1.119) in Eq. (1.117) yields:

$$\frac{dV}{dt} = I(t) - O(t) - E(t) + P(t) - G(t) \tag{1.120}$$

This equation, known as the storage equation, represents the volume balance of water in the reservoir. Note that this approach leads to the equations involving the derivative with regard to time only. Equation (1.120) is called the volume balance equation (Singh 1996).

For practical reasons the storage equation is more often expressed in terms of the water level instead of the volume. The following relation:

$$dV = F\,(h)\,dh \tag{1.121}$$

where $F(h)$ is area of reservoir at the water level, allows us to rewrite Eq. (1.120) as follows:

$$\frac{dh}{dt} = \frac{1}{F(h)}(I(t) - O(t) - E(t) + P(t) - G(t)). \tag{1.122}$$

Solving this equation one obtains the function $h(t)$ representing time variation of the water level in reservoir caused by the components of balance taken into account in Eq. (1.122).

Equation (1.122) can be derived in other way, directly from the differential continuity equation (1.80). This manner of derivation will be used in Chapter 9 while discussing the lumped flood routing models.

1.8 Equation of Mass Transport

1.8.1 Mass Transport in Flowing Water

The rivers are recipients of many substances originating either from natural geophysical processes or from human activity. For instance during the water run-off from a catchment small mineral particles are rinsed, which together with the river water form a mixture having the features of solution. In a similar way various chemical substances applied in agriculture are delivered to the rivers and streams, including fertilizers, herbicides and pesticides. An important natural constituent of water is the dissolved oxygen, playing fundamental role in aquatic life. All these substances are readily dissolved in water but they do not react with water.

The substances present in the flowing water, which for the sake of shortness will be referred to as pollutants, can be divided into active and passive. Active pollutants have significant effect on the density and viscosity of water, thus changing the hydrodynamic conditions of flow. In such case, the equations describing pollutant transport are coupled with the flow equations and should be solved simultaneously. On the other hand, the passive pollutants have negligible influence on flow conditions, and consequently allow for splitting and separate solution of the hydrodynamic and transport problem.

The pollutants can be also classified with respect to their behaviour in water, into conservative and non-conservative ones. A conservative pollutant does not change its overall mass in time, whereas the non-conservative does. An example of conservative pollutant is the chloride ion. The changes of mass of pollutant may result from its natural properties (for instance radioactive materials), or from chemical and biological processes in which the pollutant is involved (e.g. dissolved oxygen). In the following part we will focus on the transport of passive pollutants, both conservative and non-conservative.

Let us introduce some basic definitions. The quantity of constituent dissolved in water can be expressed by:

- the concentration being the ratio of its mass to the volume of water in which it is dissolved

$$c = \frac{M_p}{V}, \tag{1.123}$$

- the total load in the considered volume of water

$$M_p = c \cdot V \tag{1.124}$$

where:

c – concentration, i.e. mass of considered pollutant per unit volume of water,
M_p – mass of considered pollution,
V – volume of water.

The concentration is usually expressed in kg·m^{-3}, mg·dm^{-3}, or g·m^{-3}. The load of pollutant is expressed in mass units, i.e. kg and its derivatives. In the stream charcterized by the discharge Q, the flow rate of dissolved constituent ϕ is used, instead of the load M_p. It is defined as follows:

$$\phi = c \cdot Q \tag{1.125}$$

where ϕ represents the mass of pollutant flowing through the wetted cross-sectional area normal to the channel axis per unit time.

The intensity of mass transport is defined by the flux, which expresses the mass of pollutants flowing through the unit area of a channel cross-section normal to the velocity vector and per unit time. Then its units are kg·m^{-2}·s^{-1} and their derivatives.

The transport of any pollutant dissolved in the flowing water is considered as a superposition of two fundamental physical processes: advection and diffusion. Advection is the movement of the constituent dissolved in water due to the motion of water itself. Accordingly to Eq. (1.125), the advective flux is expressed as follows:

$$\phi_{ad} = c \cdot U \tag{1.126}$$

where:

ϕ_{ad} – mass flux due to advection,
U – flow velocity or advective velocity.

For constant flow velocity advection does not disturb the initial distribution of the concentration over the channel axis (Fig. 1.11). The concentration distribution moves with the average velocity of the stream, so that after time t_1 it takes the position which is shifted with regard to initial one by the value of $L = U \cdot t_1$. The shape of initial distribution will change for variable flow velocity.

Advection is a reversible process. It means that for the opposite flow conditions imposed at $t = t_1$ the distribution of concentration $c(x, t = t_1)$ will reach the shape and position of initial distribution $c(x, t = 0)$.

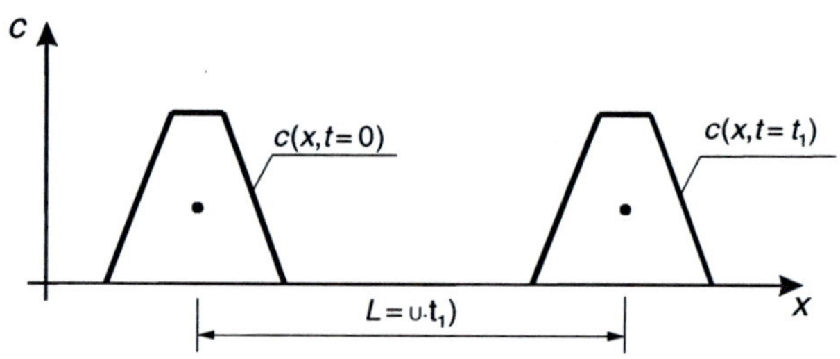

Fig. 1.11 Advective transport of initially imposed distribution of concentration with constant flow velocity U

The second fundamental transport process is the mass diffusion, which causes the mass of dissolved constituent to move from places of higher concentration towards diminishing concentration. This phenomenon is governed by the Fick's Ist law of diffusion (Baehr and Stephan 2006). Accordingly to this law the diffusive flux of mass, for instance in x direction, is expressed as:

$$\phi_x = -D^M \frac{\partial c}{\partial x}, \tag{1.127}$$

where:

ϕ_x – the diffusive flux of mass in x direction,
$\partial c/\partial x$ – gradient of concentration,
D^M – coefficient of molecular diffusion.

The negative sign at the right hand side of Eq. (1.127) ensures a positive mass flux in the direction of decreasing concentration c.

The molecular diffusion is an important transport process when water is at rest or during laminar flow. The value of the diffusion coefficient depends on the physical properties of the solution and its temperature. Typical values of the diffusion coefficient are 10^{-10} to 10^{-9} $m^2 \cdot s^{-1}$ in liquids (Baehr and Stephan 2006). Conversely to advection, the process of diffusion is irreversible. It means that there is no possibility to obtain the initial concentration distribution by reversing flow conditions.

As we will see later, an equation similar to the Fick's law can be applied to describe other processes in flowing stream, which in physical sense have nothing in common with the molecular diffusion. However, the effects of these processes can be considered as similar to the effects of diffusion. Such a case is known as the diffusive analogy, although the coefficient of proportionality does not depend on the physical properties of the considered solution.

1.8.2 Derivation of the Mass Transport Equation

In the control volume V limited by the control surface σ, the mass of constituent dissolved in the water body can vary in time due to the flux through the surface σ or due to the chemical processes taking place in water, which increase or decrease the mass of constituent. This statement is written as:

$$\frac{\partial}{\partial t}\int_V c \cdot dV + \int_\sigma \mathbf{q} \cdot \mathbf{n} \cdot d\sigma - \int_V \delta \cdot dV = 0, \tag{1.128}$$

where:

c – concentration,
V – control volume,

σ – control surface,
q – mass flux through the surface σ,
n – unit vector normal to the surface σ and oriented out of the control volume,
δ – density of internal source/sink, which represents the intensity of generation or decay of dissolved constituent.

Using the Gauss-Ostrogradski theorem (Korn and Korn 1968, McQuarrie 2003), Eq. (1.128) is rewritten as:

$$\frac{\partial c}{\partial t} + \operatorname{div} \mathbf{q} + \delta = 0. \tag{1.129}$$

Next, one can substitute Eqs. (1.126) and (1.127) describing the advection and diffusion fluxes into (1.129). Using a more general vector notation one obtains:

$$\frac{\partial c}{\partial t} + \operatorname{div}(\mathbf{U} \cdot c - \mathbf{D}^M \cdot \mathbf{grad}\, c) + \delta = 0, \tag{1.130}$$

where:

$\mathbf{U} = (u, v, w)^T$ – velocity vector,
$\mathbf{D}^M$ – tensor of molecular diffusion having non-zero elements only on main diagonal.

In scalar form Eq. (1.130) becomes:

$$\begin{aligned} &\frac{\partial c}{\partial t} + \frac{\partial}{\partial x}(u \cdot c) + \frac{\partial}{\partial y}(v \cdot c) + \frac{\partial}{\partial z}(w \cdot c) + \\ &- \frac{\partial}{\partial x}\left(D_x^M \frac{\partial c}{\partial x}\right) - \frac{\partial}{\partial y}\left(D_y^M \frac{\partial c}{\partial y}\right) + - \frac{\partial}{\partial z}\left(D_z^M \frac{\partial c}{\partial z}\right) + \delta = 0, \end{aligned} \tag{1.131}$$

In such a way the 3D advection-diffusion transport equation was derived. It holds for laminar flow only, since the molecular diffusion was taken into consideration. As mentioned earlier, every large-scale flow of surface waters is turbulent (Eagleson 1970), which means that random fluctuations of the velocities and concentration arise. Similarly to the Navier–Stokes equations considered in preceding Section 1.2, the Reynolds approach has to be applied. Therefore it is assumed that the instantaneous values of velocities and concentrations can be expressed as sums of their time-averaged values and fluctuations. The instantaneous value of concentration can be expressed by a formula:

$$c = \bar{c} + c', \tag{1.132}$$

where $\bar{c}$ is the time averaged value of concentration and c' is the fluctuation of concentration from its averaged value. Substituting these relations in Eq. (1.131) and taking into account the rules of operation on the averaged values one obtains:

$$\begin{aligned}&\frac{\partial \overline{c}}{\partial t}+\frac{\partial\left(\overline{u}\cdot\overline{c}\right)}{\partial x}+\frac{\partial\left(\overline{v}\cdot\overline{c}\right)}{\partial y}+\frac{\partial\left(\overline{w}\cdot\overline{c}\right)}{\partial z}+\\&-\frac{\partial}{\partial x}\left(D_x^M\frac{\partial \overline{c}}{\partial x}\right)-\frac{\partial}{\partial y}\left(D_y^M\frac{\partial \overline{c}}{\partial y}\right)-\frac{\partial}{\partial z}\left(D_z^M\frac{\partial \overline{c}}{\partial z}\right)\\&+\overline{u'\cdot c'}+\overline{v'\cdot c'}+\overline{w'\cdot c'}+\delta=0\end{aligned}\tag{1.133}$$

In Eq. (1.133) the bars designate the averaged values, whereas the terms $\overline{u'\cdot c'}, \overline{v'\cdot c'}, \overline{w'\cdot c'}$ containing the products of fluctuations of both velocity and concentration represent the effect of averaging of the advection transport. It is an additional effect of the turbulent transport. Similarly to the turbulent transport of momentum in the Navier–Stokes equations these terms can be expressed using the afore-mentioned diffusive analogy. It implies that the turbulent advective transport can be described with formulas analogous to the Fick's law (1.127):

$$\overline{u'\cdot c'}=-D_x^T\frac{\partial \overline{c}}{\partial x},\tag{1.134a}$$

$$\overline{v'\cdot c'}=-D_y^T\frac{\partial \overline{c}}{\partial y},\tag{1.134b}$$

$$\overline{w'\cdot c'}=-D_z^T\frac{\partial \overline{c}}{\partial z}\tag{1.134c}$$

in which the coefficients D_x^T, D_y^T, D_z^T are called the coefficients of turbulent mass diffusion. Eq. (1.133) combined with Eqs. (1.134a), (1.134b) and (1.134c) takes the following form:

$$\begin{aligned}&\frac{\partial \overline{c}}{\partial t}+\frac{\partial\left(\overline{u}\cdot\overline{c}\right)}{\partial x}+\frac{\partial\left(\overline{v}\cdot\overline{c}\right)}{\partial y}+\frac{\partial\left(\overline{w}\cdot\overline{c}\right)}{\partial z}+\\&-\frac{\partial}{\partial x}\left(\left(D_x^M+D_x^T\right)\frac{\partial \overline{c}}{\partial x}\right)-\frac{\partial}{\partial y}\left(\left(D_y^M+D_y^T\right)\frac{\partial \overline{c}}{\partial y}\right)\\&-\frac{\partial}{\partial z}\left(\left(D_z^M+D_z^T\right)\frac{\partial \overline{c}}{\partial z}\right)+\delta=0\end{aligned}\tag{1.135}$$

The experiments show that the turbulent transport and diffusivity is much greater than the molecular ones (Martin and McCutcheon 1999). Since $D^T \gg D^M$ therefore the molecular diffusion can be neglected. In the following the bars denoting the time-averaged values will be omitted. However, from this moment it should be remembered that the velocities and concentration must be interpreted as averaged values. Consequently Eq. (1.135) is rewritten in the following final form:

$$\begin{aligned}&\frac{\partial c}{\partial t}+\frac{\partial\left(u\cdot c\right)}{\partial x}+\frac{\partial\left(v\cdot c\right)}{\partial y}+\frac{\partial\left(w\cdot c\right)}{\partial z}+\\&-\frac{\partial}{\partial x}\left(D_x^T\frac{\partial c}{\partial x}\right)-\frac{\partial}{\partial y}\left(D_y^T\frac{\partial c}{\partial y}\right)-\frac{\partial}{\partial z}\left(D_z^T\frac{\partial c}{\partial z}\right)+\delta=0\end{aligned}\tag{1.136}$$

Equation (1.136) is 3D advection-diffusion transport equation which is valid for the turbulent flow. Then, if we know the 3D velocity field and we are able to estimate the coefficients of turbulent diffusion, this equation can be solved. For the imposed initial-boundary conditions one obtains the function $c(x, y, z, t)$. Although such solution is possible, it is not a trivial problem. In open channel hydraulic such approach is rarely applied. Instead, one makes use of the particular properties of the open channel flow to simplify Eq. (1.136).

Relatively small depths of rivers compared to widths and lengths cause that mixing process in the vertical direction is more rapid than for the other directions normal to the channel axis. Consequently, the vertical distribution of concentration becomes relatively uniform not far away from the source of pollution. It allows us to eliminate vertical dimension from Eq. (1.136). This elimination is carried out by its integration over depth.

Let us integrate Eq. (1.136) with regard to z changing from the channel bottom to the water stage:

$$\int_{Z}^{h}\left[\frac{\partial c}{\partial t}+\frac{\partial}{\partial x}(u\cdot c)+\frac{\partial}{\partial y}(\mathrm{v}\cdot c)+\frac{\partial}{\partial z}(w\cdot c)+\right.\\ \left.-\frac{\partial}{\partial x}\left(D_x^T\frac{\partial c}{\partial x}\right)-\frac{\partial}{\partial y}\left(D_y^T\frac{\partial c}{\partial y}\right)-\frac{\partial}{\partial z}\left(D_z^T\frac{\partial c}{\partial z}\right)+\delta\right]dz=0. \tag{1.137}$$

It is assumed that any function in Eq. (1.137) is averaged vertically using the formula applied previously for the components of flow velocity. Then we will apply the following:

$$F(x,y,t) = \frac{1}{H}\int_{Z}^{h} f(x,y,z,t)\,dz. \tag{1.138}$$

$$f = F + f'', \tag{1.139}$$

$$\int_{Z}^{h} f''\cdot dz = 0, \tag{1.140}$$

where f'' is the deviation of the actual value of function f from its averaged value F.

All functions are replaced accordingly to the formula (1.139) and afterwards the subsequent integrals in Eq. (1.137) are calculated as in the case of the Reynolds equations. Finally, the following 2D advection-diffusion transport equation is obtained:

$$\frac{\partial(H\cdot C)}{\partial t}+\frac{\partial}{\partial x}\left(H\cdot U\cdot C-H\cdot D_x^T\frac{\partial C}{\partial x}+H\cdot\overline{u''\cdot c''}\right)+\\ +\frac{\partial}{\partial y}\left(H\cdot V\cdot C-H\cdot D_y^T\frac{\partial C}{\partial x}+H\cdot\overline{v''\cdot c''}\right)+H\cdot\delta=0 \tag{1.141}$$

where:

C – vertically averaged concentration,
U, V – vertically averaged the velocity vector components in x and y direction, respectively.

Equation (1.141) has been derived with the assumption that no fluxes through the limits of the water layer exist:

$$D_z^T \left.\frac{\partial c}{\partial z}\right|_h = 0, \tag{1.142}$$

$$D_z^T \left.\frac{\partial c}{\partial z}\right|_Z = 0. \tag{1.143}$$

In the case of mass transport in open channels such assumption is admissible.

In Eq. (1.141) two additional terms, $\overline{u'' \cdot c''}$ and $\overline{v'' \cdot c''}$occurred. These terms represent additional mass fluxes, resulting from the vertical averaging of the variable velocity and concentration functions. They are expressed, in a similar manner to the previously considered turbulent transport, with formulas analogous to the Fick's law:

$$\overline{u'' \cdot c''} = -D_x^D \frac{\partial C}{\partial x}, \tag{1.144a}$$

$$\overline{v'' \cdot c''} = -D_y^D \frac{\partial C}{\partial y}. \tag{1.144b}$$

where D_x^D and D_y^D are called coefficients of dispersion in x and y direction, respectively. Substitution of Eqs. (1.144a) and (1.144b) in Eq. (1.142) yields:

$$\begin{aligned} &\frac{\partial}{\partial t}(H \cdot C) + \frac{\partial}{\partial x}(H \cdot U \cdot C) + \frac{\partial}{\partial y}(H \cdot V \cdot C) + \\ &-\frac{\partial}{\partial x}\left(H(D_x^T + D_x^D)\frac{\partial C}{\partial x}\right) - \frac{\partial}{\partial y}\left(H(D_y^T + D_y^D)\frac{\partial C}{\partial y}\right) + H \cdot \delta = 0 \end{aligned} \tag{1.145}$$

The coefficients of dispersion D_x^D and D_y^D represent the effect of vertical averaging of velocity and concentration. To better explain the concept of their introduction let us follow the imagination of flow and mass transport processes in open channel basing on idea proposed by Cunge et al. (1980) and presented in Fig. 1.12.

If the velocity distribution is non-uniform in the vertical direction and the turbulent diffusion does not exist ($D^T = 0$), then the initial distribution of concentration is traveling along x axis with shape deformation cased by the variable velocity only. Since the particles in the vicinity of the bottom move slower than those in the upper part of the stream, the initial distribution becomes deformed in space as it is shown by the dashed lines in Fig. 1.12. When the flow is turbulent, the rectangular distribution is additionally subjected to increased dispersion caused by random fluctuations

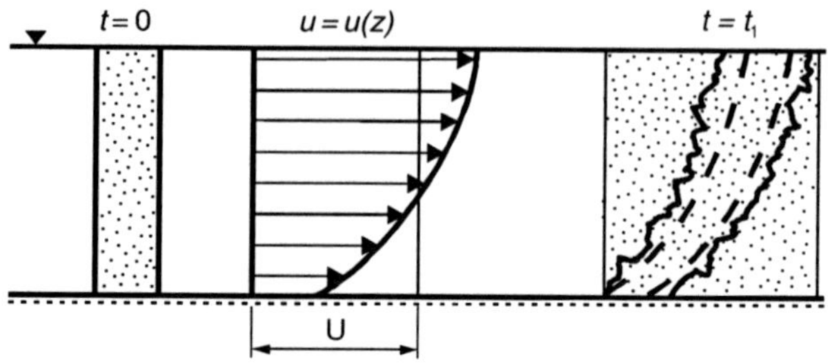

Fig. 1.12 Illustration of dispersion concept in open channel

of velocities. So the deformation of initial distribution occurs in two ways simultaneously. Because both velocity and concentration distributions are replaced by uniform ones, the observed dispersion of particles should be reproduced by an additional term in the transport equation. This is the reason of introducing of the dispersive terms into Eq. (1.145). The introduced term has diffusive character. However, the coefficient D^D is obviously not related to the physical properties of the considered solution but to the degree of non-uniformity of the velocity and concentration distributions.

As results from experiments, the values of the longitudinal dispersion are significantly larger than those of the turbulent diffusion $D^D \gg D^T$. It appears that D^D can reach the value even of the order of 10^2 $m^2{\cdot}s^{-1}$ (Cunge et al. 1980). For this reason sometimes the turbulent diffusion in Eq. (1.145) is neglected.

Considering the pollutant transport in open channels, one can distinguish two different zones in which the mathematical modeling should be carried out using different approaches. If we are dealing with a point source of pollution as in Fig. 1.13, then the pollutant transport in the zones near to the source must be analyzed as 2D (or even 3D) problem.

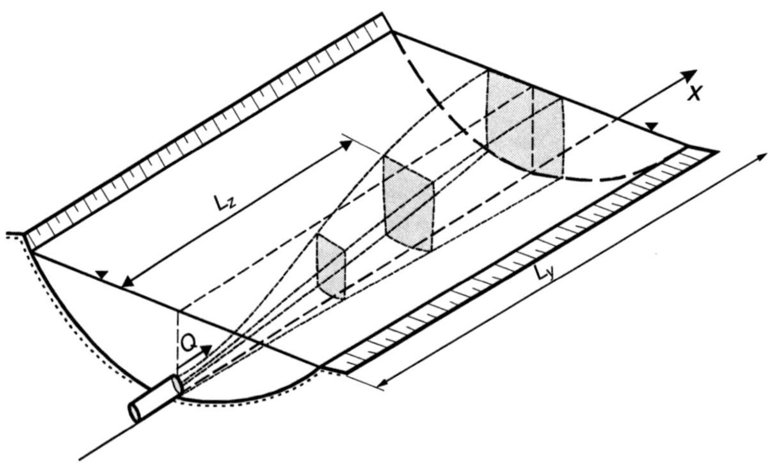

Fig. 1.13 Distribution of concentration near the source of pollutant

In this zone of length L_z, not far away from the source, we can expect the uniform distribution of concentration in the vertical direction only. Assumption of the uniform distribution over the whole wetted cross-sectional area is not admissible. Thus, for such a case Eq. (1.145) is suitable model. 1D model, preferable for open channels, can be applied only when the pollutants are well mixed over whole cross-section of a channel, which sometimes happens at quite large distant (L_y in Fig. 1.13) from the point of release of the pollutant.

When the condition of good mixing is satisfied, Eq. (1.145) can be simplified by elimination of the second dimension normal to the channel axis, i.e. by elimination of the independent variable y. To this order Eq. (1.145) must be integrated over the width of the channel from 0 to B (Fig. 1.14).

Since we assume that the water does not flow in y direction, i.e. $V = 0$, and the depths at the limits of integrations are: $H(y = 0)$, $H(y = B) = 0$ then the integral reduces to:

$$\int_0^B \left(\frac{\partial}{\partial t}(H \cdot C) + \frac{\partial}{\partial x}(H \cdot U \cdot C) - \frac{\partial}{\partial x}\left(H(D_x^T + D_x^D)\frac{\partial C}{\partial x} \right) + H \cdot \delta \right) dy = 0 \tag{1.146}$$

The integrals of subsequent terms of Eq. (1.146) are following:

$$\int_0^B \frac{\partial}{\partial t}(H \cdot C)\, dy = \frac{\partial}{\partial t}\int_0^B H \cdot C \cdot dy = \frac{\partial}{\partial t}\left(A \cdot \overline{C}\right) \tag{1.147}$$

$$\int_0^B \frac{\partial}{\partial x}(H \cdot U \cdot C)\, dy = \frac{\partial}{\partial x}\int_0^B (H \cdot U \cdot C)\, dy = \frac{\partial}{\partial x}\left(Q \cdot \overline{C}\right) \tag{1.148}$$

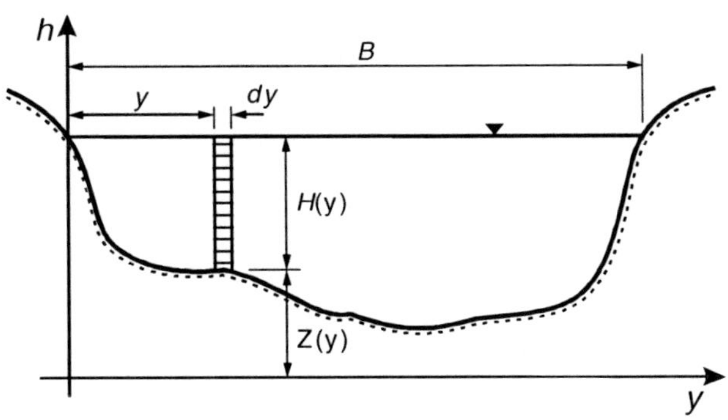

Fig. 1.14 Sketch of cross-section for integration of Eq. (1.145) in y direction

$$\int_0^B \frac{\partial}{\partial x}\left(H(D_x^T + D_x^D)\frac{\partial C}{\partial x}\right)dy = \frac{\partial}{\partial x}\int_0^B \left(H(D_x^T + D_x^D)\frac{\partial C}{\partial x}\right)dy$$
$$= \frac{\partial}{\partial x}\left(A(D_x^T + D_x^D)\frac{\partial \overline{C}}{\partial x}\right) \tag{1.149}$$

$$\int_0^B H\cdot\delta\cdot dy = \delta\int_0^B H\cdot dy = \delta\cdot A \tag{1.150}$$

Substitution of Eqs. (1.147), (1.148), (1.149) and (1.150) in Eq. (1.146) yields

$$\frac{\partial}{\partial t}(A\cdot\overline{C}) + \frac{\partial}{\partial x}(Q\cdot\overline{C}) - \frac{\partial}{\partial x}\left(A\ (D_x^T + D_x^D)\ \frac{\partial\cdot}{\partial x}\right) + A\cdot\delta = 0. \tag{1.151}$$

Consequently 1D advection-diffusion transport equation was obtained. In this equation the bar denotes the averaged concentration over cross-sectional area:

$$\overline{C}(x,t) = \frac{1}{A}\int_0^B\left(\int_Z^h c(x,y,z,t)dz\right)dy \tag{1.152}$$

To simplify notation the bar will be omitted, however, in 1D transport equation considered further, the concentration should be interpreted as averaged over channel cross-section. In addition, in the sum of the diffusion coefficients the turbulent diffusion coefficient is neglected as much smaller than the dispersion coefficient and the subscripts x and the superscript D are omitted, since only one coefficient is used. Consequently Eq. (1.151) becomes:

$$\frac{\partial}{\partial t}(A\cdot C) + \frac{\partial}{\partial x}(Q\cdot C) - \frac{\partial}{\partial x}\left(A\cdot D\frac{\partial C}{\partial x}\right) + A\cdot\delta = 0. \tag{1.153}$$

Equation (1.153) describes 1D advection-diffusion transport of passive pollutant dissolved in water for non-prismatic open channel. In particular cases it can be subjected to additional simplifications.

The products in Eq. (1.153) can be differentiated:

$$C\frac{\partial A}{\partial t} + A\frac{\partial C}{\partial t} + C\frac{\partial Q}{\partial x} + Q\frac{\partial C}{\partial x} - \frac{\partial}{\partial x}\left(A\cdot D\frac{\partial C}{\partial x}\right) + A\cdot\delta = 0. \tag{1.154}$$

If the considered channel is fed by a lateral inflow, then the continuity equation (1.80) is valid. Substitution of this equation as well as of Eq. (1.9) yields:

$$\frac{\partial C}{\partial t} + U\frac{\partial C}{\partial x} - \frac{1}{A}\frac{\partial}{\partial x}\left(A\cdot D\frac{\partial C}{\partial x}\right) + \frac{q\cdot C}{A} + \delta = 0. \tag{1.155}$$

Equation (1.155) can be simplified further. Assuming a constant coefficient of dispersion D and differentiating the diffusive term leads to the following:

$$\frac{\partial C}{\partial t}+\left(U-\frac{D}{A}\frac{\partial A}{\partial x}\right)\frac{\partial C}{\partial x}-D\frac{\partial^2 C}{\partial x^2}+\frac{q\cdot C}{A}+\delta\ =0. \tag{1.156}$$

As results from Eq. (1.156), the variability of the wetted cross-sectional area generates an effect of additional advection. Depending on the sign of the derivative $\partial A/\partial x$ an increase or decrease of the advective transport may occur.

For steady and uniform flow, when $A=$ const, $U=$ const and $q=0$, Eq. (1.156) becomes an advection-diffusion equation with constant parameters:

$$\frac{\partial C}{\partial t}+U\frac{\partial C}{\partial x}-D\frac{\partial^2 C}{\partial x^2}+\delta\ =0. \tag{1.157}$$

It is the simplest form of 1D transport equation. For some initial and boundary conditions Eq. (1.157) can be solved analytically. For this reason it is often used to check the accuracy of the numerical methods applied to solve more realistic forms of transport equations.

1.9 Thermal Energy Transport Equation

Apart from dissolved constituents, the flowing stream also transports heat. This phenomenon has several implications. First of all, the heat transport determines the seasonal changes of open channel hydraulic. The most important change occurs at the beginning of winter in temperate and cold climates, when intensive heat exchange through the water–air interface leads to water cooling. When the water temperature reaches 0°C, ice can occur in the stream. After some time, in suitable circumstances the ice layer at the water surface appears. In such a case the hydraulic conditions of flow change radically, from those typical to a free-surface flow to the ones characteristic for closed-conduit flow. The presence of the ice disturbs the navigation and operation of the hydraulic structures. On the other hand the temperature of water determines the possible concentration of the gases dissolved in water influencing in this way the chemical and biological processes. Finally it determines the self-purification and living conditions for aquatic fauna. Therefore the knowledge of the spatial and temporal evolution of the water temperature has essential meaning from the viewpoint both hydraulic and water-quality related studies.

Variation of the water temperature is described by the head transport equation, which is derived from the energy conservation principle. In this case the conservative quality is the total energy i.e. kinematic and internal (thermal) energy per unit volume of water. Note that for an incompressible liquid like water, the internal energy is defined through its temperature.

In the case of turbulent flow the energy conservation principle applied for the control volume V of the water limited by the control surface σ leads to the following equation:

$$\frac{\partial \theta}{\partial t} + \mathrm{div}(\mathbf{u} \cdot \theta - (\boldsymbol{\alpha} + \boldsymbol{\alpha}^T)\ \mathbf{grad}\ \theta) = 0. \tag{1.158}$$

where:

t – time,
θ – time averaged temperature of water,
$\mathbf{u}$ – time averaged vector of flow velocity,
$\boldsymbol{\alpha}$ – tensor of molecular heat diffusion
$\boldsymbol{\alpha}^T$ – tensor of turbulent heat diffusion

The tensors of heat diffusion are given by formulas:

$$\boldsymbol{\alpha} = \frac{1}{\rho \cdot c_w}\boldsymbol{\lambda}, \quad \boldsymbol{\alpha}^T = \frac{1}{\rho \cdot c_w}\boldsymbol{\lambda}^T \tag{1.159}$$

where:

$\boldsymbol{\lambda}$ – tensor of molecular heat conduction,
$\boldsymbol{\lambda}^T$ – tensor of turbulent heat conduction,
c_w – specific heat of water,
ρ – density of water.

Both tensors have only 3 non-zero elements located on the main diagonal and related with the respective spatial directions. These components are given as:

$$\alpha_x = \frac{\lambda_x}{\rho \cdot c_w},. \tag{1.160a}$$

$$\alpha_y = \frac{\lambda_y}{\rho \cdot c_w}, \tag{1.160b}$$

$$\alpha_z = \frac{\lambda_z}{\rho \cdot c_w} \tag{1.160c}$$

$$\alpha_x^T = \frac{\lambda_x^T}{\rho \cdot c_w},. \tag{1.161a}$$

$$\alpha_y^T = \frac{\lambda_y^T}{\rho \cdot c_w}, \tag{1.161b}$$

$$\alpha_z^T = \frac{\lambda_z^T}{\rho \cdot c_w} \tag{1.161c}$$

where λ_x, λ_y, λ_z are the coefficients of molecular heat conductivity in x, y and z direction, respectively, whereas λ_x^T, λ_y^T, λ_z^T are the coefficients of turbulent heat conductivity in the same directions. In SI system the components of tensor of heat diffusion are expressed in the same units as the coefficient of mass diffusion, i.e. in $m^2 \cdot s^{-1}$. Since the components of turbulent diffusivity tensor are much greater than the coefficients of molecular heat diffusion $(\boldsymbol{\alpha} \ll \boldsymbol{\alpha}^T)$, then $\boldsymbol{\alpha}$ can be neglected. With this assumption Eq. (1.158) can be rewritten in scalar notation as follows:

$$\begin{aligned} &\frac{\partial \theta}{\partial t} + \frac{\partial}{\partial x}(u \cdot \theta) + \frac{\partial}{\partial y}(v \cdot \theta) + \frac{\partial}{\partial z}(w \cdot \theta) + \\ &\quad - \frac{\partial}{\partial x}\left(a_x^T \frac{\partial \theta}{\partial x}\right) - \frac{\partial}{\partial y}\left(a_y^T \frac{\partial \theta}{\partial y}\right) - \frac{\partial}{\partial z}\left(a_z^T \frac{\partial \theta}{\partial z}\right) = 0 \end{aligned} \tag{1.162}$$

As for the mass transport equation, it is well justified to reduce the number of dimensions also in this case. Equation (1.162) can be simplified by integration in the vertical direction from the bottom of water layer to its surface. The integration is carried out with identical assumptions as applied for the mass transport. It means that any function in Eq. (1.162) is averaged in the vertical using the formulas (1.138), (1.139) and (1.140), whereas the velocity vector component in z direction is eliminated ($w = 0$). Since in this case the integration is slightly different, let us carry it out.

Equation (1.162) is integrated over depth:

$$\begin{aligned} \int_Z^h \Bigg[&\frac{\partial}{\partial t}(T + \theta'') + \frac{\partial}{\partial x}((U + u'')(T + \theta'')) + \\ &\frac{\partial}{\partial y}((V + v'')(T + \theta'')) + - \frac{\partial}{\partial x}\left(a_x^T \frac{\partial}{\partial x}(T + \theta'')\right) - \\ &\frac{\partial}{\partial y}\left(a_y^T \frac{\partial}{\partial y}(T + \theta'')\right) - \frac{\partial}{\partial z}\left(a_z^T \frac{\partial \theta}{\partial z}\right) \Bigg] dz = 0, \end{aligned} \tag{1.163}$$

where:

T – vertically averaged water temperature,
U – vertically averaged horizontal flow velocity in x direction,
V – vertically averaged horizontal flow velocity in y direction,
u'' – the deviation of actual value of u from its averaged value U,
θ'' – the deviation of actual value of θ from its averaged value T.

The respective terms of Eq. (1.163) are integrated as follows:

$$I_1 = \int_Z^h \frac{\partial}{\partial t}(T+\theta'')\,dz = \frac{\partial}{\partial t}\int_Z^h (T+\theta'')\,dz = \frac{\partial}{\partial t}(T\cdot H), \tag{1.164a}$$

$$\begin{aligned} I_2 = \int_Z^h \frac{\partial}{\partial x}((U+u'')(T+\theta''))\,dz = &\frac{\partial}{\partial x}\int_Z^h (U\cdot T + u''\cdot T + U\cdot\theta'' + u''\cdot\theta'')\,dz = \\ = &\frac{\partial}{\partial x}(H\cdot U\cdot T + H\cdot\overline{u''\cdot\theta''}), \end{aligned} \tag{1.164b}$$

$$\begin{aligned} I_3 = \int_Z^h \frac{\partial}{\partial y}((V+v'')(T+\theta''))\,dz = &\frac{\partial}{\partial y}\int_Z^h (V\cdot T + v''\cdot T + U\cdot\theta'' + v''\cdot\theta'')\,dz = \\ = &\frac{\partial}{\partial y}(H\cdot V\cdot T + H\cdot\overline{v''\cdot\theta''}), \end{aligned} \tag{1.164c}$$

$$\begin{aligned} I_4 = \int_Z^h \frac{\partial}{\partial x}\left(\alpha_x^T\frac{\partial}{\partial x}(T+\theta'')\right)dz = &\frac{\partial}{\partial x}\alpha_x^T\left(\int_Z^h \frac{\partial}{\partial x}(T+\theta'')\,dz\right) = \\ = &\frac{\partial}{\partial x}\left(\alpha_x^T\cdot H\frac{\partial T}{\partial x}\right), \end{aligned} \tag{1.164d}$$

$$\begin{aligned} I_5 = \int_Z^h \frac{\partial}{\partial y}\left(\alpha_y^T\frac{\partial}{\partial y}(T+\theta'')\right)dz = &\frac{\partial}{\partial y}\alpha_y^T\left(\int_Z^h \frac{\partial}{\partial y}(T+\theta'')\,dz\right) = \\ = &\frac{\partial}{\partial y}\left(\alpha_y^T\cdot H\frac{\partial T}{\partial y}\right), \end{aligned} \tag{1.164e}$$

$$I_6 = \int_Z^h \frac{\partial}{\partial z}\left(\alpha_z^T\frac{\partial\theta}{\partial z}\right)dz = \alpha_z^T\frac{\partial\theta}{\partial z}\bigg|_h - \alpha_z^T\frac{\partial\theta}{\partial z}\bigg|_Z. \tag{1.164f}$$

The integral of the last term of Eq. (1.163) I_6 represents the difference between the fluxes through the water surface at $z = h$ and the channel bottom at $z = Z$. The energy exchange across the bottom is not important, except for the flow under ice cover. Thus one can assume:

$$\alpha_z^T\frac{\partial\theta}{\partial z}\bigg|_Z = 0, \tag{1.165}$$

where ${\alpha_z}^T$ is the coefficient of turbulent conductivity in z direction. On the other hand, the flux through the water surface has essential importance for heat transport since this is the main way of energy exchange, which determines the water temperature. One can assume that:

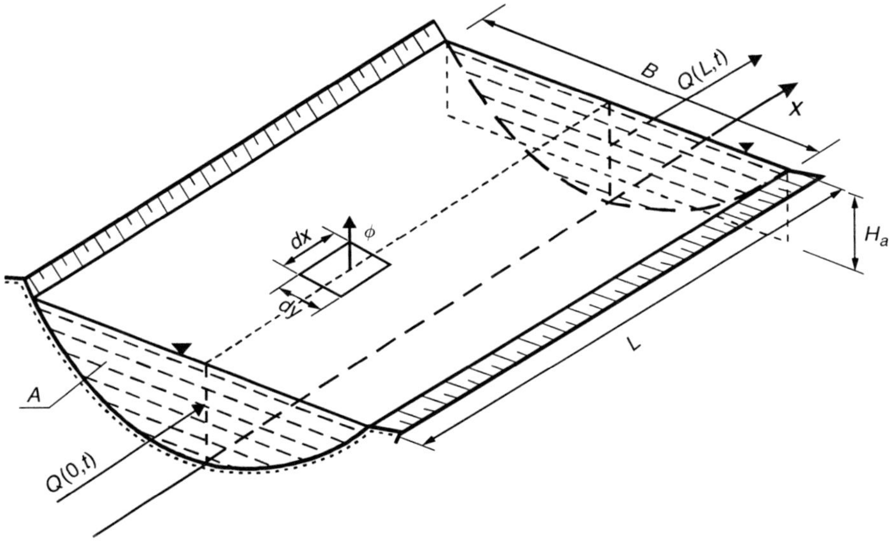

Fig. 1.15 Scheme of a channel reach transporting and exchanging heat

$$-\alpha_z^T \frac{\partial \theta}{\partial z}\bigg|_h = \phi, \tag{1.166}$$

where ϕ is the net rate of heat transport between the water and the air in both ways. It is caused by many physical processes which take place at the air–water interface (Fig. 1.15).

Substitution of all calculated integrals in Eq. (1.163) yields:

$$\begin{aligned}\frac{\partial}{\partial t}(T \cdot H) + \frac{\partial}{\partial x}\left(T \cdot H \cdot U - \alpha^T \cdot H\frac{\partial T}{\partial x} + H \cdot \overline{u'' \cdot \theta''}\right) + \\ + \frac{\partial}{\partial y}\left(T \cdot H \cdot V - \alpha^T \cdot H\frac{\partial T}{\partial y} + H \cdot \overline{v'' \cdot \theta''}\right) - \frac{\phi}{\rho \cdot c_w} = 0\end{aligned}. \tag{1.167}$$

The terms $\overline{u'' \cdot \theta''}$ *and* $\overline{v'' \cdot \theta''}$ and represent the already well-known additional transport effect, this time due to the fluctuations of the velocity and temperature. As previously, it can be expressed using a formula similar to the Fourier law (Baehr and Stephan 2006):

$$\overline{u'' \cdot \theta''} = -\frac{1}{\rho \cdot c_w}\lambda_x^D \frac{\partial T}{\partial x} = -\alpha_x^D \frac{\partial T}{\partial x}, \quad \overline{v'' \cdot \theta''} = -\frac{1}{\rho \cdot c_w}\lambda_y^D \frac{\partial T}{\partial y} = -\alpha_y^D \frac{\partial T}{\partial y}, \tag{1.168}$$

in which α_x^D, α_y^D are the coefficients of longitudinal heat dispersion. Substitution of Eq. (1.168) in Eq. (1.167) yields:

$$\frac{\partial}{\partial t}(H \cdot T) + \frac{\partial}{\partial x}(H \cdot T \cdot U) + \frac{\partial}{\partial y}(H \cdot T \cdot V) + \\ -\frac{\partial}{\partial x}\left(\left(\alpha_x^T + \alpha_x^D\right) H \frac{\partial T}{\partial x}\right) - \frac{\partial}{\partial y}\left(\left(\alpha_y^T + \alpha_y^D\right) H \frac{\partial T}{\partial y}\right) - \frac{\phi}{\rho \cdot c_W} = 0 \tag{1.169}$$

Since the order of magnitude of the turbulent diffusion coefficients is much less than the order of magnitude of the dispersion coefficients, the former can be neglected. It allows us to rewrite Eq. (1.169) as follows:

$$\frac{\partial}{\partial t}(H \cdot T) + \frac{\partial}{\partial x}(H \cdot T \cdot U) + \frac{\partial}{\partial y}(H \cdot T \cdot V) + \\ -\frac{\partial}{\partial x}\left(\alpha_x^D \cdot H \frac{\partial T}{\partial x}\right) - \frac{\partial}{\partial y}\left(\alpha_y^D \cdot H \frac{\partial T}{\partial y}\right) - \frac{\phi}{\rho \cdot c_W} = 0 \tag{1.170}$$

Equation (1.170) describes 2D unsteady heat transport. In open channel hydraulics this equation is applied locally, in the area where the water is not well mixed over the cross-section and two velocity vector components must be taken into consideration.

To obtain 1D equation for a channel reach, in which the water is well mixed in cross-section, Eq. (1.170) should be integrated over the channel width B, i.e. in y direction (Fig. 1.14). Since in y direction the water does not move, $V = 0$. In addition, the depths at the limits of integrations are: $H(y = 0)$, $H(y = B) = 0$. If there is one flow direction only, then the indices connected with the dispersion coefficient D can be omitted. Finally, the integration is carried out for the reduced form of Eq. (1.170):

$$\int_0^B \left(\frac{\partial}{\partial t}(H \cdot T) + \frac{\partial}{\partial x}(H \cdot U \cdot T) - \frac{\partial}{\partial x}\left(H \cdot \alpha \frac{\partial T}{\partial x}\right) - \frac{\phi}{\rho \cdot c_W}\right) dy = 0 \tag{1.171}$$

The integration carried out in the same way as for the mass transport equation, gives:

$$\frac{\partial}{\partial t}(A \cdot T) + \frac{\partial}{\partial x}(Q \cdot T) - \frac{\partial}{\partial x}\left(\alpha \frac{\partial T}{\partial x}\right) - \frac{B \cdot \phi}{\rho \cdot c_W} = 0, \tag{1.172}$$

In this equation:

- $T(x,t) = \frac{1}{A}\int_0^B \left(\int_Z^h \theta \cdot dz\right) dy$ is the temperature averaged over cross-section,
- $Q(x,t) = \int_0^B \left(\int_Z^h u \cdot dz\right) dy$ is the flow rate,
- $A(x,t) = \int_0^B \left(\int_Z^h dz\right) dy$ is the wetted cross-sectional area,

- B(x,t) = $\int_0^B dy$ is the channel width at water surface.

The net flux of heat through the water surface ϕ resulting from the energy balance at the interface has negative sign if the water emits heat towards atmosphere and it has positive sign when heat is absorbed by the water from atmosphere. The term ϕ takes into account all important processes connected with the heat transfer such as:

- short wave solar radiation,
- long wave water surface radiation,
- long wave atmosphere radiation,
- sensible heat conducted between the water and atmosphere,
- latent heat of vaporization or condensation.

Since some of the above listed processes are determined by the water temperature then the net flux is a function of temperature as well: $\phi = \phi(T)$.

For a channel reach, which is not fed by the lateral inflow and for a constant coefficient of longitudinal dispersion α, Eq. (1.172) can be further simplified:

$$\frac{\partial T}{\partial t} + U\frac{\partial T}{\partial x} - \frac{\alpha}{A}\frac{\partial}{\partial x}\left(A\frac{\partial T}{\partial x}\right) - \frac{\phi}{\rho \cdot c_W \cdot H_a} = 0, \tag{1.173}$$

where:

$U = Q/A$ – averaged velocity,
$H_a = A/B$ – hydraulic depth.

If the flow is steady then $A =$ const. and $U =$ const. In such a case Eq. (1.173) is reduced to the simplest form of the advection-diffusion head transport equation:

$$\frac{\partial T}{\partial t} + U\frac{\partial T}{\partial x} - \alpha\frac{\partial^2 T}{\partial x^2} - \frac{\phi}{\rho \cdot c_W \cdot H_a} = 0. \tag{1.174}$$

Assuming constant parameters this equation can be solved analytically for some initial-boundary conditions.

1.10 Types of Equations Applied in Open Channel Hydraulics

In this chapter the basic equations of open channel hydraulics were introduced. They represent a variety of types depending of the considered kind of flow. For instance, equations defining the critical depth (Eq. 1.16) and the normal depth for steady uniform flow (Eq. 1.20) are nonlinear algebraic equations, with respect to the unknown value of H_C or H_n, respectively. In other cases the unknown quantity is a function of a single independent variable, either x or t, and is specified by an ordinary differential equation (ODE), which contains derivatives of the function with respect to the considered variable. An example of ODE is given by the storage equation (1.122), while steady gradually varied flow is described by a system of two coupled ODEs (Eqs. 1.108 and 1.109). Finally, if the unknown function depends on both space and time, it is described by a partial differential equation (PDE), which contains derivatives with respect to both x and t. This is the case of mass and energy transport equations (Eq. 1.153 and 1.172), while the Saint Venant equations (1.87) and (1.88) represent a system of two coupled PDEs.

A common feature of all these equations is that in principle they can be solved only approximately, using numerical methods. Depending on the type of equation to be solved, numerical approaches differ in complexity and may include multiple stages, each of them constituting a separate numerical problem. For instance, ODEs and PDEs are transformed in a process called discretization into systems of linear or nonlinear algebraic equations, which are then solved to obtain the values of the unknown functions at specified points in time-space domain.

In Chapter 2 we discuss the numerical techniques applied to solve single nonlinear algebraic equations and systems of linear and nonlinear algebraic equations, which are the basic building blocks for more complex numerical algorithms. Numerical solution of ODEs and their systems is the subject of Chapters 3. The methods presented in these two chapters are then applied to solve steady flow problems in Chapters 4. Partial differential equations, which are most challenging type of equations from the numerical point, are discussed from Chapters 5 to 8.

References

Abbott MB (1979) Computational hydraulics-elements of the theory of free surface flow. Pitman, London

Abbott MB, Basco DR (1989) Computational fluid dynamics. Longman Scientific and Technical, New York

Akan AO (2006) Open channel hydraulics. Butterworth-Heinemann, Oxford

Baehr HD, Stephan K (2006) Heat and mass transfer, 2nd edn. Springer, Berlin/Heidelberg

Bird RB, Steward WE, Lightfoot EN (1960) Transport phenomena. Wiley, New York

Chanson H (2004) The hydraulics of open channel flow: An introduction, 2nd edn. Elsevier, Oxford

Chow VT (1959) Open channel hydraulics. McGraw-Hill, New York

Cunge J, Holly FM, Verwey A (1980) Practical aspects of computational river hydraulics. Pitman Publishing, London

Eagleson PS (1970) Dynamic hydrology. McGraw-Hill, New York

French RH (1985) Open channel hydraulics. McGraw-Hill, New York

Henderson FM (1966) Open channel flow. Macmillan Company, New York

Hervouet JM (2007) Free surface flow – modelling with the finite element method. Wiley, Chichester

Jain SC (2000) Open channel flow. Wiley, Chichester

Korn GA, Korn TM (1968) Mathematical handbook for scientists and engineers, 2nd edn. McGraw-Hill, New York

Liggett JA (1975) Basic equations of unsteady flow. In: Mahmood K, Yevjevich V (eds.) Unsteady flow in open channels. Water Resources Publications, Fort Collins, CO

Martin JL, McCutcheon SC (1999) Hydrodynamics and transport for water quality modeling. CRC Press, Boca Raton, FL

McQuarrie DA (2003) Mathematical methods for scientists and engineers. University Science Books, Sausalito, CA

Singh VP (1996) Kinematic wave modelling in water resources: Surface water hydrology. Wiley, New York.

Chapter 2
Methods for Solving Algebraic Equations and Their Systems

2.1 Solution of Non-linear Algebraic Equations

2.1.1 Introduction

In this section we consider the solution of the following equation:

$$f(x) = 0, \tag{2.1}$$

where x is real variable and $f(x)$ is real and continuous. The roots of Eq. (2.1) represent the values of $\bar{x}$ for which $f(x) = 0$. For our purpose we are interested in determination of the real roots only.

The determination of the roots is carried out in two stages. At first the searched roots must be located, i.e. we have to know their number and their approximate positions. For this purpose the plot of $f(x)$ is very useful and it should be done. On the other hand, usually some reasonable approximation of the searched roots is available from engineering analysis. For example the flow depth cannot be negative number, so all negative roots are eliminated immediately as they are not the subject of interest.

Each root must be separated from the others, so that in the specified interval designed by two endpoints only a single root exists. Practically this is carried out by examination of the signs of function $f(x)$ at two selected values of x, let say a and b. If $f(x)$ has opposite sign at both ends of interval $\langle a, b\rangle$ then within this interval at least one root exists (Fig. 2.1). Unfortunately this statement says nothing about the number of roots between a and b. To answer this question an examination of the derivative of function $f(x)$ is required. If in the interval (a, b) the derivative of function $f(x)$ exists and it does not change its sign, i.e. $f'(x) > 0$ or $f'(x) < 0$ for $a < x < b$, then there is a single root in $\langle a, b\rangle$ (Fig. 2.2).

When an approximate position of the root is found, one can pass to the second stage of calculations. A suitable method for determination of a single root with required level of precision must be applied. In further discussions we assume that the interval in which a single root exists is already known. Some methods use both endpoints of interval containing a root, whereas other ones need one point from

R. Szymkiewicz, *Numerical Modeling in Open Channel Hydraulics*, Water Science and Technology Library 83, DOI 10.1007/978-90-481-3674-2_2,

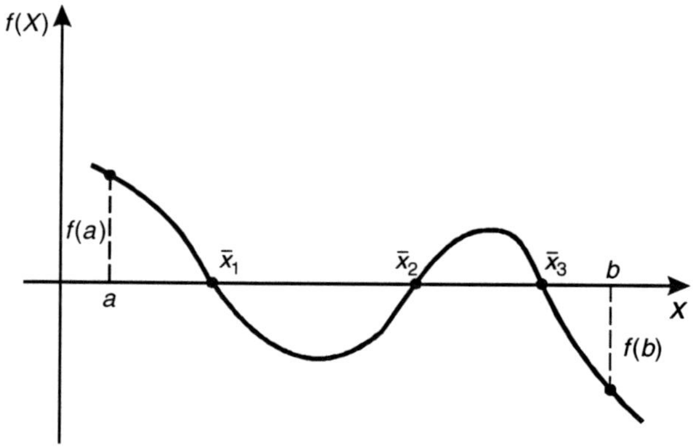

Fig. 2.1 Roots of equation $f(x) = 0$ in the interval $\langle a, b \rangle$

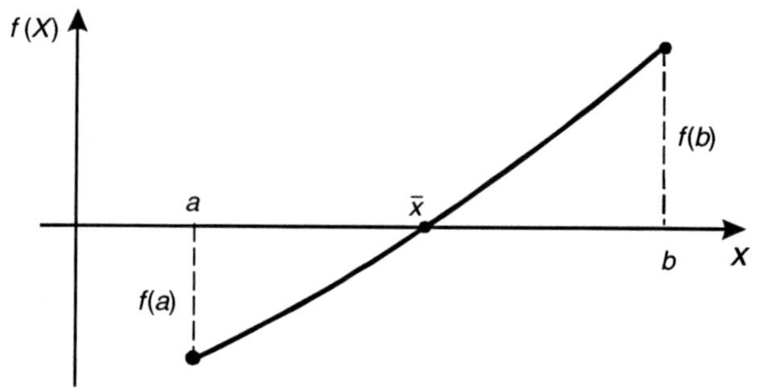

Fig. 2.2 Single root of function $f(x)$ in the interval $\langle a, b \rangle$

this interval only. The former are called bracketing methods and the latter – open methods.

2.1.2 Bisection Method

The concept of bisection method is to divide the interval $\langle a, b \rangle$ in two halves, and to choose for further search the half which contains the root. This procedure is repeated until the required accuracy criterion is satisfied.

The algorithm of determination the approximated value of the root of Eq. (2.1) in interval $\langle a, b \rangle$ by bisection method with the assumed tolerance ε, can be listed as follows:

(1) Compute $c = (a + b)/2$,
(2) Compute $f(c)$,
(3) Verify: if $|f(c)| \leq \varepsilon$ then c is searched value of the root and problem is solved. Otherwise:

(4) Choose from two subintervals $\langle a \cdot c \rangle$ and $\langle c, b \rangle$ this one, in which is the root and determine endpoints of new interval: if $f(a) \cdot f(c) < 0$, then set $b = c$, otherwise set $a = c$;

(5) Go to point 1.

After ith iteration the length of initial interval $\langle a, b \rangle$ is reduced to the following one:

$$b_i - a_i = \frac{1}{2^i}(b - a); \ (i = 1,2,\ldots). \tag{2.2}$$

In general case it is very difficult to estimate the true error of the calculated root. In practice two types of accuracy criteria are commonly used. The first one assumes that x_i is a good approximation of searched root of equation $f(x) = 0$, if $|f(x_i)|$ is a small number and conversely, if $|f(x_i)|$ is large number then x_i is considered as unsatisfying approximation of the root. The iterations are stopped when:

$$|f(x_i)| \leq \varepsilon \tag{2.3}$$

where ε is positive number representing the assumed tolerance. However, sometimes this approach can lead to false results, as illustrated in Fig. 2.3.

Even for small value of ε one can obtain an inaccurate solution of Eq. (2.1) if $f(x)$ varies only insignificantly in neighborhood of the root (Fig. 2.3a).

The second criterion is based on the examination of the length of currently reduced interval. The iterations are stopped, when $b_i - a_i \leq \varepsilon'$ where ε' is specified tolerance however related to the length of interval, not to the value of function $f(x)$, as previously.

Comparing with other methods, the bisection method converges to the exact solution rather slowly. However it has two important advantages. First of all, it is always convergent, what means that it always leads to the solution. Moreover, it does not

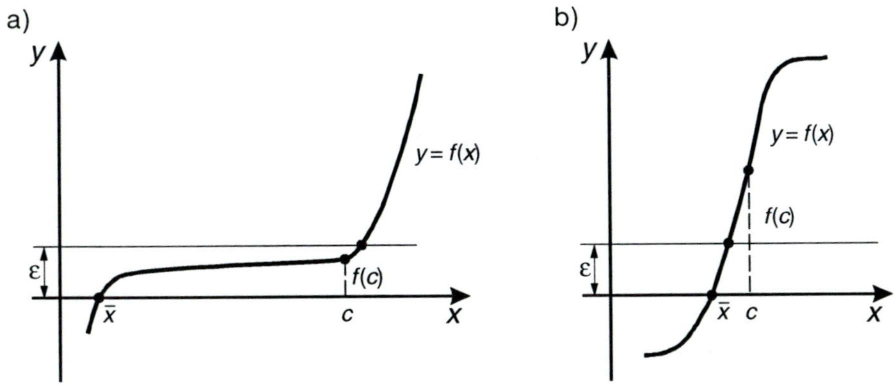

Fig. 2.3 Various forms of $f(x)$ close to the root: slowly (**a**) and rapidly (**b**) varied

require more assumptions on $f(x)$ and its derivative, apart from the continuity of $f(x)$ in $\langle a, b\rangle$.

2.1.3 False Position Method

The false position method is one of the oldest methods of solution of the non-linear equations. It is similar in concept to the bisection method, but the interval containing root is divided in two parts proportionally to the values of $f(x)$ at the endpoints. Consequently, it is usually more efficient than the bisection method.

Consider again the interval $\langle a, b\rangle$ containing a single root (Fig. 2.4).

The points $A(a, f(a))$ and $B(b, f(b))$ we join by a straight line. It intersects the x axis at point c. Its position is calculated using the formula derived from the similarity of triangles (Fig. 2.4):

$$c = \frac{a \cdot f(b) - b \cdot f(a)}{f(b) - f(a)}. \tag{2.4}$$

Close to the root numbers of the same order may appear in both numerator and denominator of Eq. (2.4), so significant round-off errors can be generated. For this reason a rearranged formula is recommended:

$$c = a + f(a)\frac{a - b}{f(b) - f(a)}, \tag{2.5}$$

The second term of Eq. (2.5) can be considered as the correction of a.

According to Eq. (2.5) one or both endpoints of the considered interval can move during the iterations. It depends on the shape of function $f(x)$ within the interval $\langle a, b\rangle$. For monotonic function, when its second derivative $f''(x)$ does not change the sign in $\langle a, b\rangle$, one end will be fixed. For instance, if $f(a) > 0$ and $f''(x) > 0$ for $a \leq x \leq b$, then the end a is fixed (Fig. 2.5).

In such a case Eq. (2.5) may be reformed as follows:

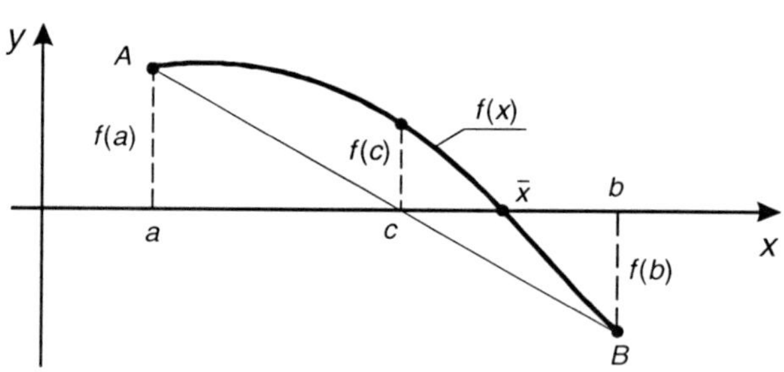

Fig. 2.4 Sketch of false position method

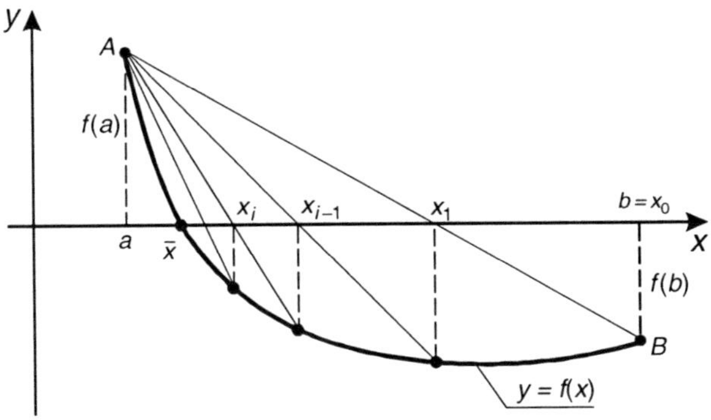

Fig. 2.5 False position method with one fixed endpoint

$$\begin{aligned} x_0 &= b, \\ x_{i+1} &= x_i - \frac{f(x_i)}{f(x_i) - f(a)} \quad (x_i - a) \;\text{ for } i = 1, 2, \ldots \end{aligned} \tag{2.6}$$

Similar formula can be written in opposite situation as well, when the endpoint b is fixed (Fig. 2.6).

Note that as the termination criterion rather relation (2.3) should be used. When one end of the interval is fixed, then we approach the root from side of second end and consequently, it is impossible to check the interval's length. This can be seen in Figs. 2.5 and 2.6.

The algorithm for solution of the non-linear equation using the false position method is similar to the one for bisection method. The only difference is the way of calculation of the point c dividing the interval $\langle a, b\rangle$.

Example 2.1 A trapezoidal channel with $b = 2.5$ m, $n_M = 0.025$, $m = 1.5$, $s = 0.0005$ carries a discharge $Q = 8$ m^3/s. Determine the normal flow depth using the false position method. Assuming the same termination criterion, compare the number of needed iterations with those for the bisection method.

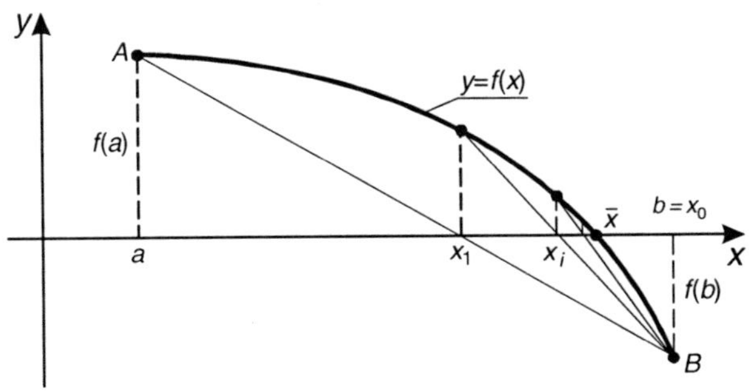

Fig. 2.6 False position method with fixed endpoint b

The normal depth satisfies the steady uniform discharge flow equation expressed using the Manning formula (1.20)

$$Q = \frac{1}{n_M} R^{2/3} \cdot s^{1/2} \cdot A \tag{2.7}$$

For the trapezoidal shape of a channel as presented in Fig. 1.3, the cross-sectional parameters are given by Eqs. (1.5) and (1.6), which lead to the following equation:

$$f(H) = Q - \frac{1}{n}\left(\frac{b \cdot H + m \cdot H^2}{b + 2H\sqrt{1+m^2}}\right)^{5/3} s^{1/2} = 0 \tag{2.8}$$

The root of Eq. (2.8) must be searched in the domain of the positive real numbers. Note that $f(H = 0) = Q$ then $f(H)$ is expressed in units of the flow rate, i.e. in [m^3/s]. This allows us to choose easily the value of the parameter ε representing the tolerance for convergence criterion. The root of Eq. (2.8) was searched in the interval $\langle 0, 2.0 \text{ m}\rangle$.

To obtain the normal depth $H = 1.546$ m with assumed tolerance $\varepsilon = 0.001$ m^3/s, the false position method required 6 iterations, whereas the bisection method needed 11 iterations to reach the identical result.

2.1.4 Newton Method

The non-linear equation in standard form $f(x) = 0$ is solved. Assume that approximated value of its root is x_i and next approximated value x_{i+1} is located from preceding one of distance h, so we have:

$$x_{i+1} = x_i + \delta x, \tag{2.9}$$

where δx is a small value. Assuming that $f(x)$ is continuous and differentiated its value at point x_{i+1} can be expressed using the Taylor series expansion:

$$f(x_i + \delta x) = f(x_i) + \delta x \cdot f'(x_i) + \frac{(\delta x)^2}{2} f''(x_i) + \ldots . \tag{2.10}$$

The Newton method uses Eq. (2.10) in which the terms with 2nd and higher order of derivatives are neglected. On the other hand, it is expected that $f(x_i + \delta x) = 0$. Then Eq. (2.10) is reduced to the following:

$$f(x_i) + \delta x \cdot f'(x_i) = 0 \tag{2.11}$$

and the distance between two consecutive approximations of the root is equal to:

$$\delta x = -\frac{f(x_i)}{f'(x_i)}. \tag{2.12}$$

Substitution of Eq. (2.12) in Eq. (2.9) yields:

$$x_{i+1} = x_i - \frac{f(x_i)}{f'(x_i)}. \tag{2.13}$$

Interpretation of formula (2.13) is given in Fig. 2.7. The new approximation of the root x_{i+1} is determined by the point of intersection of x axis and a straight line tangential to the function $f(x)$ at the old approximation of the root x_i.

The same relation can be derived from the equation of the line tangential to the function $y = f(x)$ at point x_i.

An important problem in the Newton method is the first estimation of root. If $f(a) \cdot f(b) < 0, f'(x) \neq 0$ for $a \leq x \leq b$ and $f''(x)$ does not change the sign, then any value x_0 from the interval $\langle a, b\rangle$ can be taken as the starting point. However even all those conditions are satisfied, the Newton method may not be successful. One can conclude that before using the Newton method, it is worth to examine the derivative of function $f(x)$ and to begin the iteration at the part of interval $\langle a, b\rangle$ in which $f'(x)$ has greater values.

From Eq. (2.13) results that correction of the root δx decreases the value of derivative increases. It means that the Newton method is more efficient for steep function $f(x)$ in surrounding of root. If $f(x)$ has local extreme points, it can be impossible to reach the solution at all.

To stop iterations the criterion (2.3)

$$|f(x_i)| \leq \varepsilon \tag{2.14}$$

is usually applied.

The algorithm of solution of a non-linear equation using the Newton method can be written as follows:

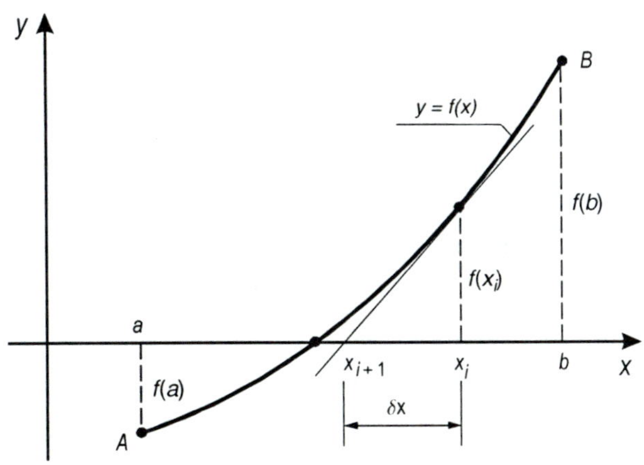

Fig. 2.7 Interpretation of the Newton method

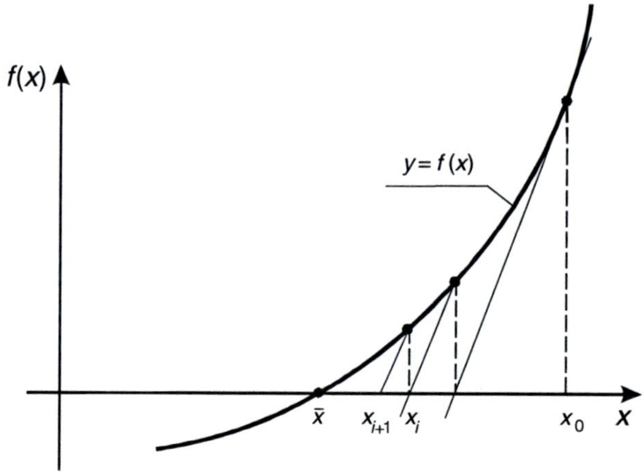

Fig. 2.8 Newton method with constant tangent

(1) Assume the starting point taking x_1 from the interval $\langle a, b\rangle$ containing a single root,
(2) Compute $f(x_1)$,
(3) Compute $f'(x_1)$,
(4) Compute $x_2 = x_1 - f(x_1)/f'(x_1)$
(5) Verify: if $|f(x_2)| \leq \varepsilon$ then x_2 is searched value of the root and problem is solved. Otherwise:
(6) Set $x_1 = x_2$ and go to point 2.

Note that in contrast to the bisection and false position methods the Newton method does not refine the interval containing the searched root, i.e. it belongs to the open methods.

Sometimes the function $f(x)$ is difficult to differentiate. In such a case the Newton method can be modified in two ways. In the first case a constant value of derivative is used, corresponding to the derivative calculated at the starting point. Then the Newton formula becomes the following one:

$$x_{i+1} = x_i - \frac{f(x_i)}{f'(x_0)}, \quad i = 0, 1, 2, \ldots. \tag{2.15}$$

It means that in the consecutive approximations of the root, the lines parallel to the tangent at x_0 are applied instead of the tangents (Fig. 2.8).

The second way of modification of the Newton method is based on the numerical approximation of the derivative $f'(x)$. Using the backward difference formula resulting from the Taylor series expansion of the function $f(x)$, its derivative is expressed as:

$$f'(x_i) \approx \frac{f(x_i) - f(x_{i-1})}{x_i - x_{i-1}}. \tag{2.16}$$

Then Eq. (2.11) takes the following form:

$$x_{i+1} = x_i - \frac{x_i - x_{i-1}}{f(x_i) - f(x_{i-1})} \cdot f(x_i). \tag{2.17}$$

One can notice that to calculate x_{i+1} two preceding approximations are needed. This version of the Newton method is known as the secant method. Note that Eq. (2.17) is very similar to Eq. (2.6).

Example 2.2 A trapezoidal channel with $b = 2.5$ m, $m = 1.5$, carries a discharge $Q = 8$ m^3/s. Determine the normal flow depth using the Newton method with the assumed tolerance $\varepsilon = 0.001$ m.

The critical depth satisfies the formula (1.16):

$$\frac{\alpha \cdot Q^2}{g} = \frac{A^3}{B} \tag{2.18}$$

where:

Q – flow rate,
α – Coriolis coefficient,
g – acceleration due to gravity,
A – wetted cross-section area,
B – channel width at the surface level.

For our purposes Eq. (2.18) is rearranged to the form:

$$B = \frac{g \cdot A^3}{\alpha \cdot Q^2} \tag{2.19}$$

For the trapezoidal shape of a channel, the cross-sectional parameters are given by Eqs. (2.3) and (2.4). Substituting these formulas in Eq. (2.70) yields:

$$f(H) = (b + 2m \cdot H) - \frac{g}{\alpha \cdot Q^2}\left(b \cdot H + m \cdot H^2\right)^3 \tag{2.20}$$

Its derivative with regard to H is following:

$$f'(H) = 2m - \frac{3g}{\alpha \cdot Q^2}\left(b \cdot H + m \cdot H^2\right)^2 (b + 2m \cdot H) \tag{2.21}$$

The root of Eq. (2.20) is located in the interval $\langle 0, 2\ \text{m}\rangle$. At its endpoints the function (2.20) takes the following values: $f(H = 0) = 2.5$ m and $f(H = 2\ \text{m}) = -327.259$ m.

Note, that $f(H)$ is expressed in units of the length i.e. in [m]. This allows us to choose the value of the parameter ε representing the tolerance. It was assumed to be equal $\varepsilon = 0.001$ m. As the first estimation of root end of the interval was assumed.

It means that $H_0 = 2$ m. For the assumed set of data, the critical depth $H = 0.727$ m was obtained after 7 iterations.

2.1.5 *Simple Fixed-Point Iteration*

Similarly to the Newton method, the fixed-point iteration (or successive substitution) belongs the open methods. In this approach the considered non-linear equation $f(x) = 0$, where $f(x)$ is continuous function, is transformed to the equivalent form:

$$x = \varphi(x) \tag{2.22}$$

Assume that the first estimate of the root x_0 is known. The formula (2.22) can be used to predict a new value of x: $x_1 = \varphi(x_0)$. In next step x_1 is substituted in Eq. (2.22) to calculate $x_2 = \varphi(x_1)$. Repeating this substitution one obtains a sequence of approximations of the searched root:

$$\begin{aligned} &x_0, \\ &x_1 = \varphi(x_0), \\ &x_2 = \varphi(x_1), \\ &\vdots \\ &x_i = \varphi(x_{i-1}). \end{aligned} \tag{2.23}$$

where i is the iteration index.

The iteration process (2.23) implies two essential questions:

- How to find the function $\varphi(x)$?
- When is the sequence x_i convergent?

The transformation of equation $f(x) = 0$ into equation $x = \varphi(x)$ can be accomplished by algebraic manipulation leading to the isolation of x at the left-hand side of the equality, by adding x to both sides of solved equation $f(x) = 0$ or by simultaneous application of both mentioned techniques.

As far as the convergence is considered, the following condition ensures that the iterations converge to the root (Chapra and Canale 2006):

$$|\varphi'(x)| < 1 \tag{2.24}$$

If relation (2.24) is satisfied, then the sequence x_i converges for any starting point x_0 taken from the interval $\langle a, b \rangle$. The limit of this sequence is the root. Then the essential problem of solution of the non-linear equation $f(x) = 0$ using the fixed-point iteration, is to transform it into the equation $x = \varphi(x)$ ensuring that the condition (2.24) is satisfied.

The properties of the convergence process for fixed-point iteration can be illustrated graphically. To this order let us reform Eq. (2.22) and express it as a set of two following equations: $y_1 = x$ and $y_2 = \varphi(x)$. These functions, plotted in x–y plane, intersect at the point corresponding to the root of solved equation $f(x) = 0$ (Fig. 2.9). Starting from the first estimate x_0 we tend to the root through the consecutive estimates, depending on the sign of the derivative $\varphi'(x)$.

The root can be approached from one side (monotone pattern – Fig. 2.9) or from both sides alternatively (oscillating pattern – Fig. 2.10).

In Fig. 2.11 an opposite situation is presented. When the condition (2.24) is not satisfied, the iterations diverge. After starting from x_0 the consecutive estimates tend in the opposite direction with regard to the root.

Transformation of $f(x)$ into $\varphi(x)$ has essential meaning for successful application of the fixed-point iteration. The proper choice of function $\varphi(x)$ should ensure that the solution procedure is convergent, and that the convergence is as fast as possible (see Bjorck and Dahlquist (1974) for more details).

The algorithm of solution of a non-linear equation using fixed-point iteration can be written as follows:

(1) Assume the starting point taking x_1 from the interval $\langle a, b \rangle$ containing a single root,
(2) Compute $x_2 = \varphi(x_1)$,
(3) Compute $f(x_2)$,
(4) Verify: if $|f(x_2)| \leq \varepsilon$ then x_2 is the searched value of the root and problem is solved.
Otherwise:
(5) Set $x_1 = x_2$ and go to point 2.

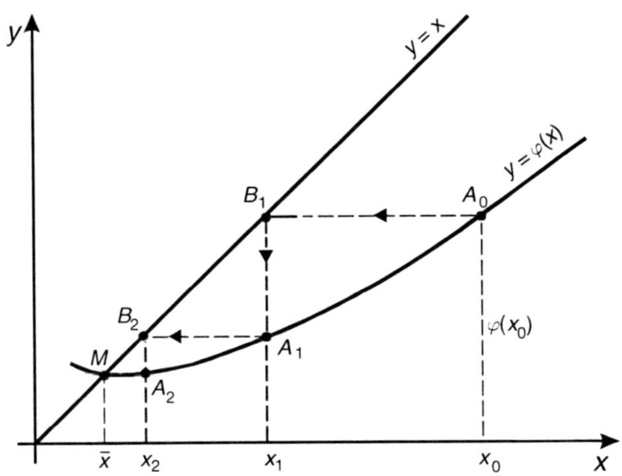

Fig. 2.9 Fixed-point iteration for $1 > \varphi'(x) > 0$

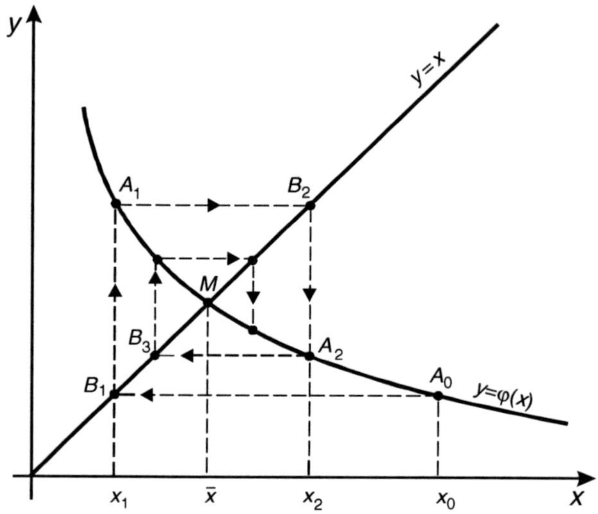

Fig. 2.10 Fixed-point iteration for $0 > \varphi'(x) > -1$

Fig. 2.11 Example of divergent iterations for $\phi'(x) < -1$

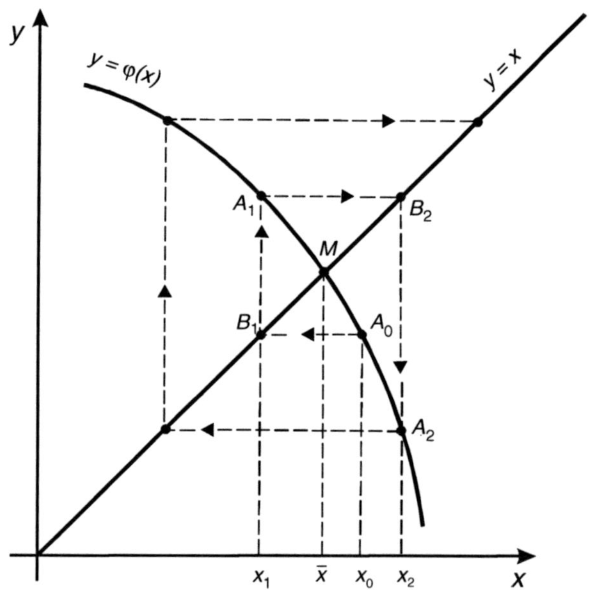

Example 2.3 A trapezoidal channel with $b = 2.5$ m, $m = 1.5$, carries a discharge $Q = 8\ \text{m}^3/\text{s}$. Determine the critical flow depth using the fixed-point iteration with the assumed tolerance $\varepsilon = 0.001$ m.

The formula (2.18) describing the critical flow is transformed as follows:

$$f(H) = \frac{\alpha \cdot Q^2}{g} - \frac{A^3}{B} = 0 \tag{2.25}$$

The notations applied in Eq. (2.18) are valid. For the trapezoidal shape of a channel the cross-sectional parameters A and B are given by Eqs. (1.5) and (1.7). Substitution of these equations causes that the critical depth being the root of Eq. (2.25) must be calculated by iteration.

The fixed-point iteration requires that the function $f(H)$ is rearranged to the form:

$$H = \varphi(H) \tag{2.26}$$

In this case the transformation is a delicate issue, since practically all forms obtained by simple algebraic manipulations on Eq. (2.25) do not ensure convergence. Good results are obtained if Eq. (2.25) is rearranged to the following form:

$$-1 + \frac{1}{A}\left(\frac{\alpha \cdot Q^2 \cdot B}{g}\right)^{1/3} = 0 \tag{2.27}$$

and next, to its both sides the critical depth H is added:

$$H = H - 1 + \frac{1}{A}\left(\frac{\alpha \cdot Q^2 \cdot B}{g}\right)^{1/3} \tag{2.28}$$

Substitution of Eqs. (1.5) and (1.7) in Eq. (2.28) yields the final equation:

$$H = H - 1 + \frac{1}{\left(b \cdot H + m \cdot H^2\right)}\left(\frac{\alpha \cdot Q^2\,(b + 2m \cdot H)}{g}\right)^{1/3} \tag{2.29}$$

Then the formula for fixed-point iteration is:

$$H_{i+1} = H_i - 1 + \frac{1}{\left(b \cdot H_i + m \cdot H_i^2\right)}\left(\frac{\alpha \cdot Q^2\,(b + 2m \cdot H_i)}{g}\right)^{1/3} \tag{2.30}$$

where i is the index of iteration.

As the termination criterion the following relation involving the results of two consecutive iterations is assumed:

$$|H_{i+1} - H_i| \le \varepsilon \tag{2.31}$$

where the parameter ε represents accepted tolerance.

The root of Eq. (2.25) is located, as in Example 2.2, in the interval $\langle 0, 2\ \text{m}\rangle$. At the interval endpoints the function (2.25) takes the following values: $f(H = 0) = 6.524$ m and $f(H = 2\ \text{m}) = -222.739$ m.

As the starting point of iteration the end of interval was assumed, i.e. $H_0 = 2$ m. The critical depth $H = 0.727$ m was obtained after 11 iterations. This is the same result as given by the Newton method in Example 2.2, however to obtain it more iterations were needed.

2.1.6 Hybrid Methods

The list of available methods to solve nonlinear algebraic equations is much longer (see e.g. Bjorck and Dahlquist 1974, Press et al. 1992) and includes some hybrid approaches, which are modifications or combinations of the four basic ones presented above. Two such methods will be presented here: the Ridders method and the Steffensen method.

The Ridders method is a combination of the false position and bisection methods (Press et al. 1992). Assume that a single root of non-linear equation (2.1): $f(x) = 0$ is located in the interval $\langle a, b\rangle$. The iterations are performed according to the following algorithm:

1. Calculate $f(x)$ at midpoint of specified interval: $c = (a + b)/2$;
2. Assume, that $f(x)$ is approximated in $\langle x_1, x_2\rangle$ by an exponential function build on the values of $f(x)$ in known points a, b, c, in the following form

$$f(a) - 2f(c)e^{\alpha} + f(b)e^{2\alpha} = 0. \tag{2.32}$$

It is a quadratic equation with regard to e^{α}. Its solution is following:

$$e^{\alpha} = \frac{f(c) + \mathrm{sign}(f(b))\sqrt{f(c)^2 - f(x_1)f(b)}}{f(b)}, \tag{2.33}$$

Apply the false position method, however not for $f(a), f(c), f(b)$ but for $f(a), f(c)$ e^{α}, $f(b)$ $e^{2\alpha}$. This allows us to calculate new approximation of the root d. With Eq. (2.33) one obtains:

$$d = c + (b - a)\frac{\mathrm{sign}(f(a) - f(a))f(c)}{\sqrt{f(c)^2 - f(a)f(b)}}. \tag{2.34}$$

3. Reduce the dimensions of the interval by checking of the signs in a, b, c and d. It is carry out as follows:

If $f(c) \cdot f(d) < 0$, then $a = c$ and $b = d$,
else
if $f(c) \cdot f(d) > 0$ then if $f(b) \cdot f(d) > 0$ then $b = d$ else $a = d$

Note that the method uses the function $f(x)$ only. Its derivative is not needed. To terminate the iteration the following criterion, similar to this applied in the Newton method, can be used: $|f(d)| \le \varepsilon$, where ε is assumed positive number representing the tolerance.

The Steffenson method is a modified version of the false position method (Bjorck and Dahlquist 1974), where the iteration convergence is accelerated. The following iterative formula is applied:

$$x_{i+1} = x_i - \frac{f(x_i)}{g(x_i)}. \ (i = 0,1,2,\ldots), \tag{2.35}$$

where:

$$g(x_i) = \frac{f(x_i + f(x_i)) - f(x_i)}{f(x_i)}.$$

In Example 2.4 the efficiency of the hybrid methods is compared with the efficiency of those of the standard methods which do not require the derivative of $f(x)$.

Example 2.4 In a rectangular channel of width $b = 50$ m crossed by the sharp-crested weir of height $p = 1.0$ m which spans entire channel, the flow discharge is $Q = 7.20$ m (Fig. 2.12). Determine the depth of the water head h taking into account the velocity head and assuming that the flow coefficient is dependent on the water depth.

Including the velocity head the discharge equation is given as:

$$Q = K \cdot b\sqrt{2g}\left[\left(h + \frac{v_0^2}{2g}\right)^{3/2} - \left(\frac{v_0^2}{2g}\right)^{3/2}\right], \tag{2.36}$$

where:

h – head,
b – weir width,

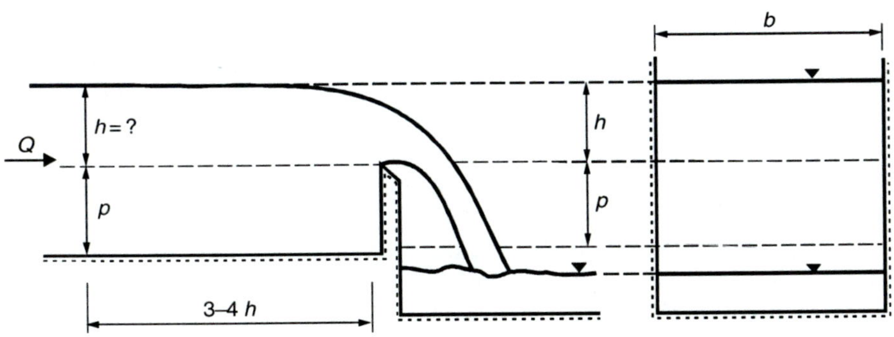

Fig. 2.12 Sharp-crested weir which spans entire width

v_0 – average flow velocity,
g – acceleration due to gravity,
K – flow coefficient.

The average flow velocity behind a weir is given as:

$$v_0 = \frac{Q}{b(p+h)} \tag{2.37}$$

The flow coefficient for the considered type of weir is determined by the following formula (Roberson et al. 1998):

$$K = 0.40 + 0.05\frac{h}{p}. \tag{2.38}$$

This formula is valid up to an h/p value of about 10.
Equation (2.36) is transformed to the typical form:

$$f(h) = Q - K \cdot b\sqrt{2g}\left[\left(h + \frac{v_0^2}{2g}\right)^{3/2} - \left(\frac{v_0^2}{2g}\right)^{3/2}\right] = 0. \tag{2.39}$$

With Eqs. (2.37) and (2.38) this equation is non-linear with regard to head h. For the fixed-point iteration Eq. (2.39) is transformed to the following one:

$$h = \left(\frac{Q}{K \cdot b\sqrt{2g}} + \left(\frac{v_0^2}{2g}\right)^{3/2}\right)^{2/3} - \frac{v_0^2}{2g} \tag{2.40}$$

The single root of Eq. (2.39) is located in the interval $\langle 0,\ 1.5\ \text{m}\rangle$. In the termination criterion, taken in the form of relation $|f(h_i)| \le \varepsilon$, the tolerance parameter $\varepsilon = 0.00001\ \text{m}^3/\text{s}$ is assumed.

The results of consecutive iterations for all applied methods are listed in Table 2.1. Note that the methods differ significantly in efficiency. The total number of iterations needed to reach a solution of the same accuracy varies from 3 to 18. In this case the most effective are the fixed point method and the Ridders method, which require 3 iterations only. Other methods needed much more iterations to reach the solution with the same accuracy. This fact should be taken into considerations, while developing the computer codes for complex problems, in which the non-linear equations are solved many times.

Table 2.1 Comparison of convergence for selected methods applied to solve Eq. (2.39)

	Method				
Iteration	Fixed point	Bisection	False position	Steffensen	Ridders
1	0.78958	0.75000	0.55015	0.15936	0.83379
2	0.78934	1.12500	0.73239	0.31090	0.78680
3	0.78934	0.93750	0.77665	0.45216	0.78935
4	–	0.84375	0.78656	0.57844	–
5	–	0.79688	0.78874	0.68184	–
6	–	0.77344	0.78922	0.75202	–
7	–	0.78516	0.78932	0.78346	–
8	–	0.79102	0.78934	0.78918	–
9	–	0.78809	0.78935	0.78935	–
10	–	0.78955	0.78935	–	–
11	–	0.78882	–	–	–
12	–	0.78918	–	–	–
13	–	0.78937	–	–	–
14	–	0.78928	–	–	–
15	–	0.78932	–	–	–
16	–	0.78934	–	–	–
17	–	0.78936	–	–	–
18	–	0.78935	–	–	–

2.2 Solution of Systems of the Linear Algebraic Equations

2.2.1 *Introduction*

Solution of systems of linear equations is one of the most basic problems of numerical analysis, which arises in multiple applications, for example when solving partial differential equations with finite difference or finite element method. While many ready-made numerical subroutines for linear systems are available, it seems worthwhile to present in this section some basic algorithms, which can be conveniently used in the framework of numerical modeling of open channel flow.

Let us consider a system of N linear equations with N unknowns:

$$\sum_{j=1}^{N} a_{ij}x_j = b_i, \qquad (i = 1, 2, \ldots, N). \tag{2.41}$$

In matrix notation this system is rewritten as follows:

$$\mathbf{AX} = \mathbf{B}, \tag{2.42}$$

where:

$$\mathbf{A} = \begin{bmatrix} a_{11} & a_{12} & \dots & a_{1N} \\ a_{21} & a_{22} & \dots & a_{2N} \\ \vdots & \vdots & & \vdots \\ a_{N1} & a_{N2} & \dots & a_{NN} \end{bmatrix}, \quad \mathbf{X} = \begin{Bmatrix} x_1 \\ x_2 \\ \vdots \\ x_N \end{Bmatrix}, \quad \mathbf{B} = \begin{Bmatrix} b_1 \\ b_2 \\ \vdots \\ b_N \end{Bmatrix}$$

$\mathbf{A} = (a_{ij})$ $(i, j = 1, 2, \dots, N)$ is real square matrix of system coefficients,
$\mathbf{X} = (x_i)$ $(i = 1, 2, \dots, N)$ is column vector of unknowns,
$\mathbf{B} = (b_i)$ $(i = 1, 2, \dots, N)$ is the column vector of constants,
N is dimension of system.

It is assumed that the matrix **A** is nonsingular so that the system (2.42) has a solution. In some cases the matrix **A** assumes one of the following particular forms:

- diagonal matrix, in which the non-zero elements are laying on its main diagonal only, i.e.:

$$a_{ij} \begin{cases} \neq 0 & \text{for} \quad i = j \\ = 0 & \text{for} \quad i \neq j \end{cases} \qquad (i, j = 1,2,\dots, N) \tag{2.43}$$

- identity matrix, which is a diagonal one with all elements on the main diagonal equal to 1:

$$a_{ij} = \begin{cases} 1 \text{ for } i = j, \\ 0 \text{ for } i \neq j, \end{cases} \qquad (i, j = 1, 2, \dots, N) \tag{2.44}$$

Usually the symbol **I** is reserved for this matrix.

- lower triangular matrix having non zero elements on the main diagonal and below:

$$\mathbf{L} = \begin{bmatrix} l_{11} & 0 & 0 & \dots & 0 \\ l_{21} & l_{22} & 0 & \dots & 0 \\ l_{31} & l_{32} & l_{33} & \dots & 0 \\ \vdots & \vdots & \vdots & \ddots & \\ l_{N1} & l_{N2} & l_{N3} & \dots & l_{NN} \end{bmatrix},$$

$$l_{ij} \begin{cases} \neq 0 & \text{for} \quad i > j \\ = 0 & \text{for} \quad i \leq j \end{cases} \qquad (i, j = 1,2,\dots, N) \tag{2.45}$$

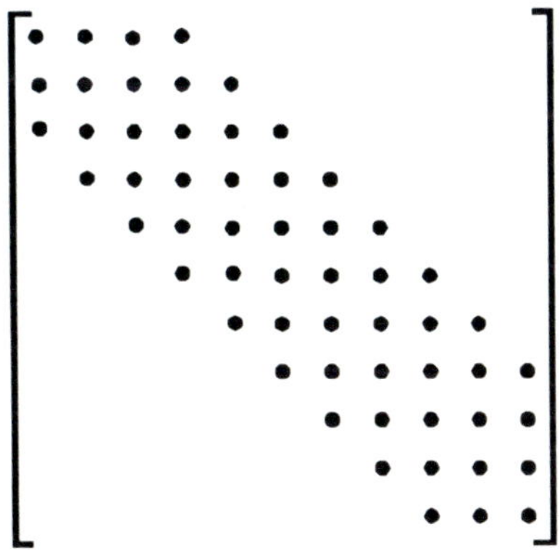

Fig. 2.13 Banded matrix having bandwidth equal to 6

- upper triangle matrix having non zero elements on the main diagonal and above:

$$\mathbf{U} = \begin{bmatrix} u_{11} & u_{12} & u_{13} & \dots & u_{1N} \\ & u_{22} & u_{23} & \dots & u_{2N} \\ & & u_{33} & \dots & u_{3N} \\ & & & \ddots & \vdots \\ & & & & u_{NN} \end{bmatrix};$$

$$u_{ij} \begin{cases} \neq 0 & \text{for} \quad i \leq j \\ = 0 & \text{for} \quad i > j \end{cases} \qquad (i, j = 1,2,..., N) \tag{2.46}$$

- banded matrix, which has all non zero elements concentrated around the main diagonal within a band of specified bandwidth (Fig. 2.13);
- sparse matrix, which has a large number of elements equal to zero – banded matrices can be sparse if the band is narrow or if there are many zeros within the band;
- symmetric matrix for which $a_{ij} = a_{ji}$.

For the cases listed above one can take advantage of the matrix structure to make the solution algorithm much more efficient.

The methods for solving the systems of algebraic equations are divided into two groups:

– direct methods,
– iterative methods.

Direct methods are capable to provide an exact solution after executing a finite number of mathematical operations (it is true on condition that the round-off errors do not exist). Iterative methods start from an initial guess of the solution and provide a series of its better approximations. If this series is convergent, i.e. if it tends to a limit, this limit is a solution of the considered system (2.42). Therefore the iterative methods are capable to provide the approximate solution only, within the assumed tolerance.

The choice of a suitable method of solution depends on the dimension of considered system of equations. Generally it is assumed that direct methods are efficient for the systems having rather small dimensions, say 60 by 60. For larger systems, especially with very sparse matrices, iterative methods are recommended as more effective. In such a case application of the direct methods is a waste of computer memory and time of computation. However in the case of open channel flow modeling situation is particular. As results from Chapter 1, all derived equations or their systems are one dimensional. Consequently, the final systems of algebraic equations obtained from discretization of the ordinary or partial differential equations, even those containing hundreds of unknowns, have banded matrices with very narrow bandwidth. For instance, as we will see in Chapters 6, 7 and 8, for 1D transport equation solved by the implicit methods one obtains systems with tri-diagonal matrices, while for the system of Saint Venant equations the bandwidth is equal to 5 or 7, according to the applied method of solution. The matrices' bandwidths become wider for flow in open channel network. However at the same time they become very sparse, since the total number of the non-zero coefficients in each matrix row is still very small. Consequently, in the case of the equations applied in open channel hydraulics it can be reasonable to apply direct methods to solve even very large systems of equations. However, this can be done on condition that the implemented algorithms for these methods will take into account the untypical properties of matrix. For instance, very efficient solver basing on the Gauss elimination method can be obtained on condition that it works on the non-zero elements only. For this reason this section is limited to direct methods.

2.2.2 Gauss Elimination Method

Let us begin with the solution of a system of linear algebraic equations with upper triangular matrix:

$$\mathbf{UX} = \mathbf{B} \tag{2.47}$$

or in scalar form:

$$\begin{aligned} u_{11}x_1 + \cdots + u_{1,N-1}x_{N-1} + u_{1,N}x_N &= b_1, \\ &\vdots \\ u_{N-1,N-1}x_{N-1} + u_{N-1,N}x_N &= b_{N1}, \\ u_{N,N}x_N &= b_N. \end{aligned} \tag{2.48}$$

Assuming that $u_{ii} \neq 0$ ($i = 1, 2, \ldots, N$), the unknowns can be calculated in the inverse order: $x_N, x_{N-1}, \ldots, x_1$ using the following formulas:

$$
\begin{aligned}
x_N &= \frac{b_N}{u_{N,N}}, \\
x_{N-1} &= \frac{b_{N-1} - u_{N-1,N} \cdot x_N}{u_{N-1,N-1}}, \\
&\dots \\
x_1 &= \frac{b_1 - u_{1,2} \cdot x_2 - u_{1,3} \cdot x_3 - \cdots - u_{1,N-1} \cdot x_{N-1} - u_{1,N} \cdot x_N}{u_{1,1}},
\end{aligned}
\tag{2.49}
$$

Equations (2.49) can be rewritten shorter as:

$$x_N = \frac{b_N}{u_{N,N}} \tag{2.50a}$$

$$x_i = \frac{b_i - \sum\limits_{k=i+1}^{N} u_{i,k} \cdot x_k}{u_{i,i}} \qquad (i = N-1,\ \dots,\ 1). \tag{2.50b}$$

Since the unknowns are calculated from last one to the first one, this algorithm is known as back substitution. Similarly easily one can solve the system of equations with lower triangular matrix. In such a case the computational process is called forward substitution (Bjorck and Dahlquist 1974).

The Gauss elimination method is the most important direct method to solve any system of linear algebraic equations. It bases on such a way of elimination of the subsequent unknowns, which finally reduces the solved system to the one with upper triangular matrix as Eq. (2.48). Afterwards, the obtained system is solved using previously presented algorithm.

Now consider the system (2.42) rewritten follows:

$$
\begin{cases}
a_{11}x_1 + a_{12}x_2 + a_{13}x_3 + \ldots + a_{1N}x_N = b_1, \\
a_{21}x_1 + a_{22}x_2 + a_{23}x_3 + \ldots + a_{2N}x_N = b_2, \\
a_{31}x_1 + a_{32}x_2 + a_{33}x_3 + \ldots + a_{3N}x_N = b_3, \\
\qquad \ldots \\
a_{N1}x_1 + a_{N2}x_2 + a_{N3}x_3 + \ldots + a_{NN}x_N = b_N.
\end{cases}
\tag{2.51}
$$

The elimination is performed in multiple sweeps through the whole system and in each sweep the coefficients are modified. Thus, we introduce additional superscript to distinguish coefficients obtained in subsequent sweeps. The superscript equal to 1 is designated to original matrix, i.e.: $a_{ij}^{(1)} = a_{ij}$ and $b_i^{(1)} = b_i$. Using this convention the system (2.51) is rewritten as:

$$\begin{cases} a_{11}^{(1)}x_1 + a_{12}^{(1)}x_2 + a_{13}^{(1)}x_3 + \ldots + a_{1N}^{(1)}x_N = b_1^{(1)}, \\ a_{21}^{(1)}x_1 + a_{22}^{(1)}x_2 + a_{23}^{(1)}x_3 + \ldots + a_{2N}^{(1)}x_N = b_2^{(1)}, \\ a_{31}^{(1)}x_1 + a_{32}^{(1)}x_2 + a_{33}^{(1)}x_3 + \ldots + a_{3N}^{(1)}x_N = b_3^{(1)}, \\ \ldots \\ a_{N1}^{(1)}x_1 + a_{N2}^{(1)}x_2 + a_{N3}^{(1)}x_3 + \ldots + a_{NN}^{(1)}x_N = b_N^{(1)}. \end{cases} \tag{2.52}$$

Assume that the coefficient matrix is not singular and $a_{11}^{(1)} \neq 0$. Otherwise, the system (2.52) must be rearranged. Let us carry out the following sequence of operations for $i = 2, 3, \ldots, N$:

- define the multiplier $a_{i1}^{(1)} / a_{11}^{(1)}$,
- multiply the first equation by this multiplier,
- subtract the first equation from ith equation.

In such a way one obtains a reduced system, which has dimension $(N-1) \times (N-1)$. Its coefficients are $a_{ij}^{(2)}$. This system written together with the first equation of original system (2.51) is as follows:

$$\begin{aligned} a_{11}^{(1)}x_1 + a_{12}^{(1)}x_2 + a_{13}^{(1)}x_3 + \ldots + a_{1N}^{(1)}x_N &= b_1^{(1)}, \\ a_{22}^{(2)}x_2 + a_{23}^{(2)}x_3 + \ldots + a_{2N}^{(2)}x_N &= b_2^{(2)}, \\ a_{32}^{(2)}x_2 + a_{33}^{(2)}x_3 + \ldots + a_{3N}^{(2)}x_N &= b_3^{(2)}, \\ &\ldots \\ a_{N2}^{(2)}x_2 + a_{N3}^{(2)}x_3 + \ldots + a_{NN}^{(2)}x_N &= b_N^{(2)}, \end{aligned} \tag{2.53}$$

Its coefficients are given by the formulas:

$$a_{ij}^{(2)} = a_{ij}^{(1)} - \frac{a_{i1}^{(1)}}{a_{11}^{(1)}} a_{1j}^{(1)}, \; b_i^{(2)} = b_i^{(1)} - \frac{a_{i1}^{(1)}}{a_{11}^{(1)}} b_1^{(1)} \quad \text{for } i = 2,3,\ldots,N;\; j = 2,3,\ldots,N.$$

Similarly to the preceding step, assume that $a_{22}^{(2)} \neq 0$ and repeat the same sequence of operations but for the reduced system (2.53). We obtain the second reduced system:

$$\begin{aligned} a_{11}^{(1)}x_1 + a_{12}^{(1)}x_2 + a_{13}^{(1)}x_3 + \ldots + a_{1N}^{(1)}x_N &= b_1^{(1)}, \\ a_{22}^{(2)}x_2 + a_{23}^{(2)}x_3 + \ldots + a_{2N}^{(2)}x_N &= b_2^{(2)}, \\ a_{33}^{(3)}x_3 + \ldots + a_{3N}^{(3)}x_N &= b_3^{(3)}, \\ &\ldots \\ a_{N3}^{(3)}x_3 + \ldots + a_{NN}^{(3)}x_N &= b_N^{(3)}. \end{aligned} \tag{2.54}$$

In Eq. (2.54) new coefficients $a_{ij}^{(3)}$ and $b_i^{(3)}$ are given by:

$$a_{ij}^{(3)} = a_{ij}^{(2)} - \frac{a_{i2}^{(2)}}{a_{22}^{(2)}} a_{2j}^{(2)},\ b_i^{(3)} = b_i^{(2)} - \frac{a_{i1}^{(2)}}{a_{22}^{(2)}} b_2^{(2)},\ \text{for } i = 3,4,\ldots,N;\ J = 3,4,\ldots,N.$$

Continuing this way of elimination, after $N - 1$ steps we obtain the following reduced system of equations:

$$\begin{aligned} a_{11}^{(1)}x_1 + a_{12}^{(1)}x_2 + a_{13}^{(1)}x_3 + \ldots + a_{1N}^{(1)}x_N &= b_1^{(1)}, \\ a_{22}^{(2)}x_2 + a_{23}^{(2)}x_3 + \ldots + a_{2N}^{(2)}x_N &= b_2^{(2)}, \\ a_{33}^{(3)}x_3 + \ldots + a_{3N}^{(3)}x_N &= b_3^{(3)}, \\ \ldots \\ a_{NN}^{(N)}x_N &= b_N^{(N)}. \end{aligned} \tag{2.55}$$

The coefficients of the above system were calculated by the following formulas:

$$a_{ij}^{(k+1)} = a_{ij}^{(k)} - \frac{a_{ik}^{(k)}}{a_{kk}^{(k)}} a_{kj}^{(k)}, \tag{2.56}$$

where:

$k = 1, 2, \ldots, N - 1,$
$j = k + 1, k + 2, \ldots, N,$
$i = k + 1, k + 2, \ldots, N.$

Thus, the final result of elimination is a system with upper triangular matrix (2.55). Now the determination of the unknowns can be done easily, using the formulas (2.50a) and (2.50b).

The elements $a_{11}^{(1)}, a_{22}^{(2)}, a_{33}^{(3)}, \ldots, a_{NN}^{(n)}$, determining the elimination process, are called the pivot elements. The elimination was possible since we arbitrary assumed that their values differ to zero. If in kth step of elimination the pivot element appears to be equal to zero, it is necessary to rearrange the matrix by changing of the rows in such a way that at position (k, k) non zero element must appear (Fig. 2.14). Therefore, row k is swapped with row l for which $a_{lk}^{(k)} \neq 0$ (Fig. 2.14).

It is recommended to choose such row for which:

$$\left| a_{lk}^{(k)} \right| = \max_{k \le i \le N} \left| a_{ik}^{(k)} \right|, \tag{2.57}$$

where i is the row index.

There are two cases when the pivot elements certainly differ from zero:

- when matrix **A** has dominating main diagonal, i.e. when:

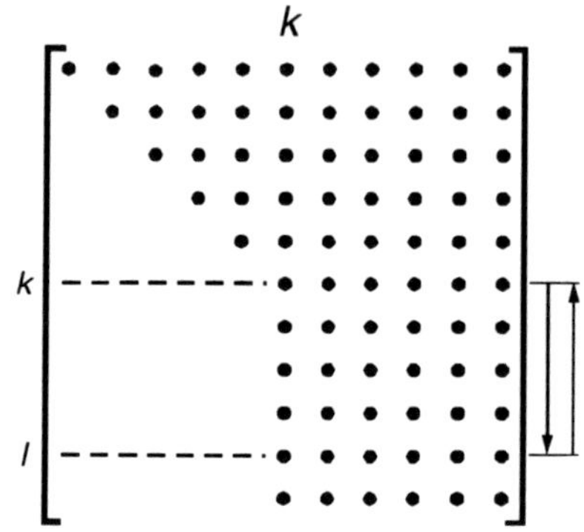

Fig. 2.14 Matrix of coefficients after k steps of elimination

$$|a_{ii}| \geq \sum_{\substack{j=1 \\ j \neq i}}^{N} \left|a_{ij}\right| \quad (i = 1, 2, \ \ldots, N), \tag{2.58}$$

- when matrix **A** is symmetrical and positively defined, i.e. when it satisfies the following conditions:

$$\mathbf{A} = \mathbf{A}^T (\text{symmetrical}) \tag{2.59}$$

and

$$\mathbf{X}^T \mathbf{A} \mathbf{X} > 0 \text{ for each } \mathbf{X} \neq 0 \, (\text{positively defined}) \, . \tag{2.60}$$

For any matrix the condition (2.58) may be verified easily, whereas the condition (2.60) is more difficult to check. More information on the problem is given by Bjorck and Dahlquist (1974).

2.2.3 LU Decomposition Method

Assume that it is possible to decompose the matrix **A** of the system (2.42) and to present it in the form of product of two matrices: lower triangular matrix and upper triangular matrix:

$$\mathbf{A} = \mathbf{LU} \tag{2.61}$$

Then the considered system (2.42) is equivalent to the following one:

$$\mathbf{L}\,\mathbf{U}\,\mathbf{X} = \mathbf{B}. \tag{2.62}$$

The system (2.62) can be split into two systems:

$$\mathbf{L}\,\mathbf{Y} = \mathbf{B} \tag{2.63}$$

$$\mathbf{U}\,\mathbf{X} = \mathbf{Y}, \tag{2.64}$$

which should be solved subsequently.

There is a theorem which states that in fact, each matrix **A** may be rearranged in such a way, that the decomposition on the triangle matrices becomes possible (Bjorck and Dahlquist 1974). This fact results from the equivalency of the Gauss elimination method and **LU** decomposition. Actually, the matrix obtained as the result of elimination is the upper triangular matrix. Accordingly, the elements of the lower matrix are $l_{ik} = a_{ik}^{(k)} / a_{kk}^{(k)}$, which occur in Eq. (2.56) for $i = 2, 3, \ldots, N$; $k = 1, 2, \ldots, i-1$ as well as by the elements at the main diagonal $l_{ii} = 1$ for $i = 1, 2, \ldots, N$. Therefore one can write:

$$\begin{bmatrix} a_{11} & a_{12} & \cdots & a_{1,N-1} & a_{1N} \\ a_{21} & a_{22} & \cdots & a_{2,N-1} & a_{2N} \\ \vdots & \vdots & & \vdots & \vdots \\ a_{N1} & a_{N2} & \cdots & a_{N,N-1} & a_{NN} \end{bmatrix} = \begin{bmatrix} 1 & & & & \\ l_{21} & 1 & & & \\ \vdots & \vdots & & & \\ l_{N1} & l_{N2} & \cdots & l_{N,N-1} & 1 \end{bmatrix} \times \begin{bmatrix} a_{11}^{(1)} & a_{12}^{(1)} & \cdots & a_{1,N-1}^{(1)} & a_{1N}^{(1)} \\ & a_{22}^{(2)} & \cdots & a_{2,N-1}^{(2)} & a_{2N}^{(2)} \\ & & & & \vdots \\ & & & & a_{NN}^{(N)} \end{bmatrix}. \tag{2.65}$$

Note that in the decomposition of **A** using the Gauss elimination the vector of right side hand **B** was not involved. This is an advantage of **LU** decomposition technique, since one can decompose **A** only once and after then to use **L** and **U** for various vectors of right side hand. Such situation occurs very often.

Discretization of 1D partial differential equation using implicit method usually leads to systems of linear algebraic equation (2.42) with tri-diagonal matrices:

$$\mathbf{A} = \begin{bmatrix} \beta_1 & \gamma_1 & & & & \\ \alpha_2 & \beta_2 & \gamma_2 & & & \\ & \alpha_3 & \beta_3 & \gamma_3 & & \\ & & & \ddots & & \\ & & & \alpha_{N-1} & \beta_{N-1} & \gamma_{N-1} \\ & & & & \alpha_N & \beta_N \end{bmatrix}. \tag{2.66}$$

Such a system can be solved using a very simple and effective algorithm, based on the **LU** decomposition. Matrix (2.66) is decomposed in following triangle matrices:

$$\mathbf{L} = \begin{bmatrix} 1 & & & & & \\ l_2 & 1 & & & & \\ & l_3 & 1 & & & \\ & & & \ddots & & \\ & & & l_{N-1} & 1 & \\ & & & & l_N & 1 \end{bmatrix}, \tag{2.67a}$$

$$\mathbf{U} = \begin{bmatrix} u_1 & \gamma_1 & & & & \\ & u_2 & \gamma_2 & & & \\ & & u_3 & \gamma_3 & & \\ & & & \ddots & \ddots & \\ & & & & u_{N-1} & \gamma_{N-1} \\ & & & & & u_N \end{bmatrix}. \tag{2.67b}$$

The elements l_2, l_3, ..., l_N and u_1, u_2, u_3, ..., u_N are computed using the following formulas:

$$u_1 = \beta_1 \tag{2.68a}$$

$$\left.\begin{aligned} l_i &= \frac{\alpha_i}{u_{i-1}} \\ u_i &= \beta_i - l_i \cdot \gamma_{i-1} \end{aligned}\right\} \quad \text{for } i = 2,3,\ldots,N. \tag{2.68b}$$

Knowing the triangular matrices one can determine the vector **X** by solving the two systems (2.63) and (2.64). The computations are carried out as follows:

$$y_1 = b_1, \tag{2.69a}$$

$$y_i = b_i - l_i y_{i-1} \text{ for } i = 2,3,\ldots,N \tag{2.69b}$$

and

$$x_N = y_N/u_N, \tag{2.70a}$$

$$x_i = (y_i - \gamma_i x_{i+1})/u_i \text{ for } i = N-1, N-2, \ldots, 1. \tag{2.70b}$$

If the system of equations is solved only once for given matrix **A** and the right side hand **B**, then it is not necessary to remember the elements l_i, and the algorithm becomes simpler:

$$u_1 = \beta_1; y_1 = b_1, \tag{2.71a}$$

$$\left.\begin{aligned} l &= \alpha_i/u_{i-1}, \\ u_i &= \beta_i - l \cdot \gamma_{i-1}, \\ y_i &= b_i - l \cdot y_{i-1}, \end{aligned}\right\} \quad \text{for } i = 2, 3, \cdots, N \tag{2.71b}$$

and

$$x_N = y_N / u_N \tag{2.72a}$$

$$x_i = (y_i - \gamma_i x_{i+1}) / u_i \quad for\ i = N-1, N-2, 1. \tag{2.72b}$$

The presented technique of solution is called the Thomas method or "double sweep" method (Fletcher 1991). This algorithm is very useful and can be applied for many open channel flow problems.

The Gauss elimination method or **LU** decomposition are especially suitable to solve effectively the systems with banded matrices. After elimination the matrices **L** and **U** are also banded. The only disadvantage is that during elimination the zeros within the band are replaced by non zero elements.

2.3 Solution of Non-linear System of Equations

2.3.1 Introduction

Previously, in Section2.2 we considered systems of linear algebraic equations of the form:

$$\mathbf{AX} = \mathbf{B}, \tag{2.73}$$

where: **A** is square matrix of the coefficients, **X** is vector of unknowns and **B** is vector of right hand side defined as follows:

$$\mathbf{A} = \begin{bmatrix} a_{11} & a_{12} & \ldots & a_{1N} \\ a_{21} & a_{22} & \ldots & a_{2N} \\ \vdots & \vdots & & \vdots \\ a_{N1} & a_{N2} & \ldots & a_{NN} \end{bmatrix} \mathbf{X} = \begin{Bmatrix} x_1 \\ x_2 \\ \vdots \\ x_N \end{Bmatrix}, \quad \mathbf{B} = \begin{Bmatrix} b_1 \\ b_2 \\ \vdots \\ b_N \end{Bmatrix}$$

where a_{ij} and b_i were constant numbers. However, in many engineering applications one obtains systems of algebraic equations in which the coefficients are functions of the unknowns, i.e. $a_{ij} = a_{ij}\ (x_1, x_2, \ldots, x_N)$ or/and $b_{ij} = b_{ij}\ (x_1, x_2, \ldots, x_N)$. Such systems are called non-linear. In open channel hydraulics the systems of non-linear equations usually arise as a part of more complex algorithms. For instance, some numerical methods of solution applied for the ordinary and partial differential equations or their systems lead to the systems of algebraic non-linear equations. In the next chapters we will see numerous examples of such systems.

Usually, instead of the form (2.94), the following notation for the non-linear systems is used:

$$\mathbf{F}\,(\mathbf{X}) = \mathbf{0}, \tag{2.74}$$

where:

$\mathbf{X} = (x_1, x_2, x_3, \ldots, x_N)^T$,
$\mathbf{F}(\mathbf{X}) = (f_1(\mathbf{X}), f_2(\mathbf{X}), \ldots, f_N(\mathbf{X}))^T$,
T – transposition symbol,
N – dimension of system.

Of course, both forms are equivalent, since we have:

$$\mathbf{F}(\mathbf{X}) = \mathbf{AX} - \mathbf{B} \tag{2.75}$$

and both will be used further. In scalar notation Eq. (2.74) is written as:

$$\begin{aligned} f_1\left(x_1, x_2, \ldots x_N\right) &= 0, \\ f_2\left(x_1, x_2 \ldots x_N\right) &= 0, \\ &\vdots \\ f_N\left(x_1, x_2 \ldots x_N\right) &= 0. \end{aligned} \tag{2.76}$$

The vector **X**, which satisfies Eq. (2.76) is its solution.

All numerical methods of solution of the non-linear systems are iterative ones. In the next sections we present the Newton method and the Picard method, which are standard methods commonly applied in open channel hydraulics.

2.3.2 Newton Method

Assume that approximate solution of Eq. (2.74) is known. Then one can write:

$$\mathbf{X} = \mathbf{X}^{(k)} + \Delta\mathbf{X}^{(k)}. \tag{2.77}$$

where:

$\mathbf{X}$ – exact solution of Eq. (2.74),
$\mathbf{X}^{(k)}$ – solution approximation *i* iteration k,
$\mathbf{\Delta X}^{(k)}$ – vector of differences between exact and estimate,
k – index of iteration.

It is assumed that the functions being the components of system are continuous and differentiated with regard to vector **X**. Since form definition results that $\mathbf{F}(\mathbf{X}) = 0$, then using the Taylor series expansion around $\mathbf{X}^{(k)}$, one can write:

$$\mathbf{F}(\mathbf{X}^{(k)} + \Delta\mathbf{X}^{(k)}) \approx \mathbf{F}(\mathbf{X}^{(k)}) + \frac{\partial\mathbf{F}(\mathbf{X}^{(k)})}{\partial\mathbf{X}}\Delta\mathbf{X}^{(k)} \approx 0. \tag{2.78}$$

It means that the Taylor series was truncated and all terms with the derivatives of order higher than one were neglected. The correction vector can be expressed as follows:

$$\Delta \mathbf{X}^{(k)} = \mathbf{X}^{(k+1)} - \mathbf{X}^{(k)}, \tag{2.79}$$

since $\mathbf{X}^{(k+1)}$ estimates of exact solution $\mathbf{X}$. Therefore one can write:

$$\frac{\partial \mathbf{F}(\mathbf{X}^{(k)})}{\partial \mathbf{X}}(\mathbf{X}^{(k+1)} - \mathbf{X}^{(k)}) = -\mathbf{F}(\mathbf{X}^{(k)}), \tag{2.80}$$

where:

$$\frac{\partial \mathbf{F}(\mathbf{X}^{(k)})}{\partial \mathbf{X}} = \mathbf{J}^{(k)} = \begin{bmatrix} \frac{\partial f_1\left(x_1^{(k)}, x_2^{(k)}, \ldots x_N^{(k)}\right)}{\partial x_1} & \cdots & \frac{\partial f_1\left(x_1^{(k)}, x_2^{(k)}, \ldots x_N^{(k)}\right)}{\partial x_N} \\ \vdots & \cdots & \vdots \\ \frac{\partial f_n\left(x_1^{(k)}, x_2^{(k)}, \ldots x_N^{(k)}\right)}{\partial x_1} & \cdots & \frac{\partial f_n\left(x_1^{(k)}, x_2^{(k)}, \ldots x_N^{(k)}\right)}{\partial x_N} \end{bmatrix}$$

is the Jacobian matrix of system (2.74). Then the final form of the Newton method is following:

$$\mathbf{X}^{(k+1)} = \mathbf{X}^{(k)} - \left(\mathbf{J}^{(k)}\right)^{-1} \mathbf{F}\left(\mathbf{X}^{(k)}\right). \tag{2.81}$$

One can see, that the method applied in the form of Eq. (2.81) is time and memory consuming. In each iteration it requires to build the Jacobian matrix of dimension $N \times N$, and next to inverse it. For these reasons the Newton method is applied rather in the form of Eq. (2.80), rewritten as follows:

$$\mathbf{J}^{(k)} \Delta \mathbf{X}^{(k)} = -\mathbf{F}^{(k)}. \tag{2.82}$$

In such a way, instead of inversing of the matrix **J**, a system of linear equations must be solved with regard to the correction vector $\mathbf{\Delta X}^{(k)}$. Afterwards the new estimate of **X** is calculated:

$$\mathbf{X}^{(k+1)} = \mathbf{X}^{(k)} + \Delta \mathbf{X}^{(k)}. \tag{2.83}$$

This approach is more efficient especially when **J** is banded. Note, that the inversion of a banded matrix produces a matrix, which looses banded form and has all non zero elements.

In order to reduce the computational time sometimes the original Newton method is modified, similarly to the modification of this method for scalar equation. Instead of computing the Jacobian matrix $\mathbf{J}$ in each iteration, the Jacobian from the first estimate is applied is all subsequent iterations:

$$\mathbf{J}^{(0)}\Delta\mathbf{X}^{(k)} = -\mathbf{F}^{(k)}. \tag{2.84}$$

The convergence of this iteration is slower, but the Jacobian matrix does not have to be calculated in each iteration what is important when the derivatives must be calculated numerically. The criteria of convergence will be discussed in next section.

2.3.3 Picard Method

The Picard method, also known as the consecutive substitution, is a fixed-point iteration developed for a system of non-linear equations. From the subsequent equations of the system (2.76) one unknown is isolated at left side of equality sign. Then one obtains:

$$\begin{aligned} x_1 &= g_1\left(x_1,x_2\ldots x_N\right), \\ x_2 &= g_2\left(x_1,x_2\ldots x_N\right), \\ &\vdots \\ x_N &= g_N\left(x_1,x_2\ldots x_N\right), \end{aligned} \tag{2.85}$$

or using the matrix notation:

$$\mathbf{X} = \mathbf{G}\left(\mathbf{X}\right) \tag{2.86}$$

where:

$$\begin{aligned} &\mathbf{X} = (x_1, x_2, x_3, \ldots, x_N)^T, \\ &\mathbf{G}(\mathbf{X}) = (g_1(\mathbf{X}), g_2(\mathbf{X}), \ldots, g_N(\mathbf{X}))^T, \end{aligned}$$

Given the first estimate $\mathbf{X}^{(0)}$, one can generate the following sequence of approximations:

$$\mathbf{X}^{(k+1)} = \mathbf{G}(\mathbf{X}^{(k)}), \tag{2.87}$$

where k is iteration index. If this sequence tends to the limit $\mathbf{X}$ for $k \to \infty$, then this limit is the solution of system (2.74).

Modified variants of the Picard method are often used. For instance the Picard method can be applied directly to solve the non-linear system in form of Eq. (2.73). In this case, it is given as:

$$\mathbf{A}^{(k)} \cdot \mathbf{X}^{(k+1)} = \mathbf{B} \tag{2.88}$$

Another version of the method is obtained by transformation of Eq. (2.88) to a formula similar to the Newton one. To this end the term $\mathbf{A}^{(k)} \cdot \mathbf{X}^{(k)}$ is subtracted from both sides of Eq. (2.88):

$$\mathbf{A}^{(k)} \cdot \mathbf{X}^{(k+1)} - \mathbf{A}^{(k)} \cdot \mathbf{X}^{(k)} = -\mathbf{A}^{(k)} \cdot \mathbf{X}^{(k)} + \mathbf{B} \tag{2.89}$$

Substitution of Eqs. (2.79) and (2.75) in Eq. (2.89) yields

$$\mathbf{A}^{(k)} \boldsymbol{\Delta}\mathbf{X}^{(k)} = -\mathbf{F}^{(k)} \tag{2.90}$$

Equation (2.90) differs from the Newton method in the matrix of coefficients only. Since in Eq. (2.90) an original matrix of the solved system is involved then it is simpler. As we will see in Chpater 4, this method can be quite effective.

The iterations carried out with the Newton or Picard methods are stopped when the criterion for convergence is satisfied. In both methods one can apply the following tests:

$$\left|x_i^{(k+1)} - x_i^{(k)}\right| = \left|\Delta x_i^{(k+1)}\right| \le \varepsilon \text{ for } i = 1, 2, \mathrm{N}, \tag{2.91}$$

where:

k– iteration index,
ε– positive number being a specified tolerance.

In this condition the differences between two consecutive iterations are examined. The above criterion is suitable for the cases when the convergence processes is relatively fast. Checking of the solution accuracy for slowly convergent iterations should be carried out thoroughly. Although the consecutive iterations vary insignificantly, the difference between the exact solution and its estimate in kth iteration may be appreciable. Such situation is illustrated in Fig. 2.15.

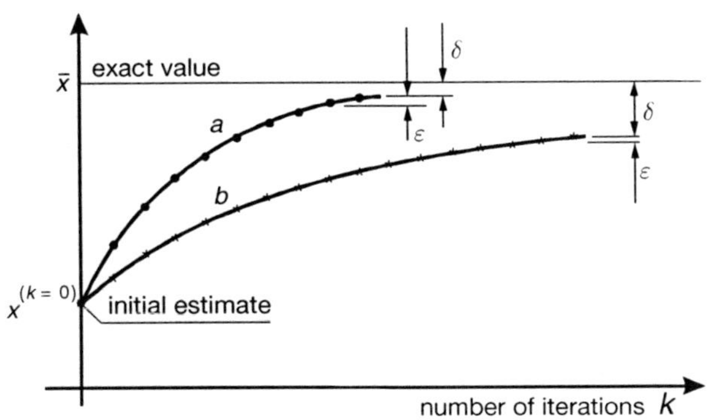

Fig. 2.15 Influence of the criterion (2.91) on final error of solution for: (**a**) quickly convergent iterations; (**b**) slowly convergent iterations

To stop the iterations another criterion, in the form of the Euclidean norm, may be employed. This norm, which serves to quantify the size of vector, this case is defined as norm of residual vector:

$$\|\mathbf{\Delta X}\| = \left(\Delta x_1^2 + \Delta x_2^2 + \Delta x_3^2 + \ldots + \Delta x_N^2\right)^{1/2}. \qquad (2.92)$$

The iterations are terminated when the following relation is satisfied:

$$\left\| \Delta^{(k+1)} \right\| \leq \varepsilon', \qquad (2.93)$$

where $\varepsilon' > 0$ determines assumed tolerance.

Experience shows, that the Newton method converges very quickly on condition that the first estimate is close to the solution. Otherwise it can suffer from possible divergence. Generally it is said, that the Newton method has poor global convergence (Press et al. 1992). As far as the Picard method is considered, it is slower, but converges even if the initial guess is far from the solution. These properties will be confirmed in Chapter 4 while solving some open channel flow problems.

References

Bjorck A, Dahlquist G (1974) Numerical methods. Prentice-Hall, Englewood Cliffs, NJ

Chapra SC, Canale RP (2006) Numerical methods for engineers. McGraw-Hill, New York

Fletcher CAJ (1991) Computational techniques for fluid dynamics, vol. 1. Springer-Verlag, Berlin

Press WH, Teukolsky SA, Veterling WT, Flannery BP (1992) Numerical recipes in C. Cambridge University Press, Cambridge

Roberson JA, Cassidy JJ, Chaudhry MH (1998) Hydraulic engineering, 2nd edn. Wiley, New York

Chapter 3
Numerical Solution of Ordinary Differential Equations

3.1 Initial-Value Problem

3.1.1 Introduction

Ordinary differential equations (ODEs) contain derivatives of an unknown function with respect to a single independent variable. Such equations describe for example steady flow in open channel (Eqs. 1.97, 1.108 and 1.109). In this case the independent variable is the spatial coordinate x. Another example is the storage equation (1.122), which is an ODE with respect to time. More time-dependent ODEs will be introduced in Chapter 9. Moreover ODEs with regard to time appear when unsteady flow and transport equations are solved using the finite element method. Their solution is one of the steps in the computational process.

Consider the following equation:

$$y' = \frac{dy}{dx} = f(\mathrm{x,y}), \tag{3.1}$$

where x is the independent variable, whereas y is the dependent variable. Integration of Eq. (3.1) yields:

$$y(x) = \int f(x,y) \cdot dx + C \tag{3.2}$$

where C is the integration constant. Thus, the obtained solution is not unique. Indeed, we have an infinite number of solutions satisfying Eq. (3.1), corresponding to an infinite number of different possible values of the constant C. However, in practice we are interested rather in a unique solution for the specified range of x from the interval $\langle a, b \rangle$. To choose the proper solution, an auxiliary condition must be formulated. In the case of Eq. (3.1) it is called initial condition. Then the problem can be formulated as follows: determine the function $y(x)$,which satisfies both

R. Szymkiewicz, *Numerical Modeling in Open Channel Hydraulics*, Water Science and Technology Library 83, DOI 10.1007/978-90-481-3674-2_3,

the ordinary differential equation (3.1) in the interval ⟨a, b⟩ and the specified initial condition:

$$y(a) = y_0, \tag{3.3}$$

where a and y_0 are specified values.

The problem of solution of Eq. (3.1) formulated in such a way is called an initial-value problem for the ordinary differential equation. It is assumed that the initial-value problem has a unique solution, i.e. there is only one function $y(x)$ which simultaneously satisfies Eq. (3.1) and the initial condition (3.3). The problem of existence and uniqueness of solution is discussed for instance by Ascher and Petzold (1998).

In majority of engineering applications the initial-value problem for ordinary differential equations is solved using numerical methods. This approach allows us to compute the estimates $y_1, y_2, \ldots, y_N$, of the exact solution $y(x_1), y(x_2), \ldots, y(x_N)$ at selected points $x_1, x_2, \ldots, x_N$ within the interval $\langle a, b\rangle$. Starting from the point (a, y_0) given by the specified initial condition, we have to move over the interval $\langle a, b\rangle$ from one node to the next, calculating the approximated values of the exact solution $y(x_i)$ until the endpoint $x_N = b$ is reached. It is expected that the applied method will generate the approximations $y_1, y_2, \ldots y_i \ldots$ which reproduce the true shape of the function $y(x)$ with sufficient accuracy (Fig. 3.1).

The accuracy of approximation depends on the following two factors:

- selection of the points $x_1, x_2, \ldots,$
- method of calculation of the estimates $y_1, y_2, \ldots$.

The distance between two neighboring nodes, denoted by Δx_i is called the step size:

$$\Delta x_i = x_{i+1} - x_i \qquad (i = 1,2,\ldots). \tag{3.4}$$

A constant step size is often applied $\Delta x_i = \Delta x = \text{const}$ and many numerical methods proposed for integration of the ordinary differential equations were derived for constant Δx. In principle, constant stepsize can be easily applied for

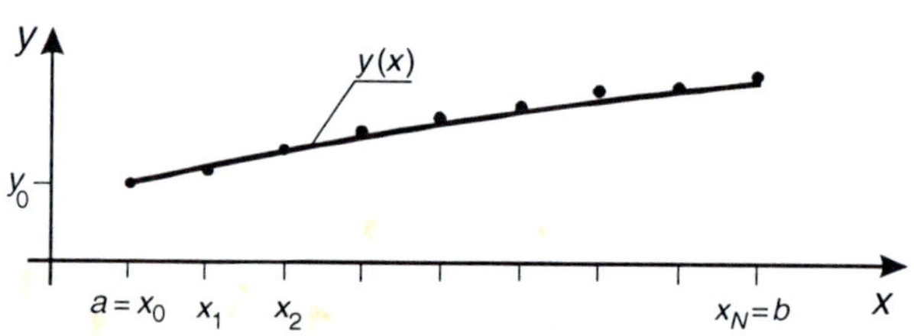

Fig. 3.1 Function $y(x)$ and its estimate calculated using the numerical method

time-dependent ODEs, although even in this case variable stepsizes may be recommended for the sake of efficiency. As far as the space-dependent ODEs are considered, constant Δx can be used for flow in artificial channels having a constant bed slope and constant shape of the cross-sections. In such a case it is easy to determine the parameters of cross-section as wetted area, wetted perimeter and surface width, as a function of position, i.e. of variable x. Conversely, integration of the same equations for natural channels must be carried out with variable step size. The shape of the rivers' cross-sections is irregular and varies continuously along their axes. The computational nodes are chosen in such a way that they correspond to the cross-sections for which data obtained from field measurements are available. Therefore, we must apply non-uniform point spacing along x axis.

The methods of numerical integration of ODEs can be classified according to different criteria. In general, numerical integration schemes lead to algebraic equations, where the unknown represents the value of the function $y(x)$ at the next integration point y_{i+1}. If it can be computed using information only from the last node i, then the method is called one-step. The one-step methods are also called self starting methods. It means that the initial condition is sufficient to perform the computing process by these methods. If the information from more than one preceding step is required (y_i, y_{i-1}, etc) then the method is called multi-step. To initiate multistep methods, apart from imposed the initial condition, additional values of y at the first few points, i.e. $y_1, y_2, \ldots$ must be calculated using a one-step method.

If the resulting algebraic equation is linear, i.e. y_{i+1} can be isolated at one side of the equation:

$$y_{i+1} = G\left(x, y_i, y_{i-1}, y_{i-2}, \ldots\right) \tag{3.5}$$

then the method is called explicit. In the opposite case y_{i+1} cannot be extracted from the algebraic formula, and the following nonlinear relation arises:

$$y_{i+1} = G\left(x, y_{i+1}, y_i, y_{i-1}, y_{i-2}, \ldots\right) \tag{3.6}$$

Such a scheme is called implicit. The nonlinear equation has to be solved with regard to y_{i+1} in each step, using an iterative method. Usually the fixed-point method is applied. Its implementation for Eq. (3.6) is called the "predictor-corrector" method. The prediction of y_{i+1} is performed using the explicit method, whereas the formula (3.6) serves to its correction.

There are a variety of methods which belong to the mentioned types. Due to exceptional properties of the rivers only some of them are applicable in open channel hydraulic and therefore only some of them will be presented in this chapter.

3.1.2 Simple Integration Schemes

Let us begin with a short presentation of the finite differences. The finite difference, being an approximation of the derivative, can be used to derive directly some methods of solution for both the initial-value and the boundary-value problems. Appropriate difference formula comes from the Taylor series expansion of a function around considered point x. For the function depending on one independent variable only, the Taylor series can be written as follows:

$$y(x_{i+1}) = y(x_i) + \sum_{m=1}^{\infty} \frac{\Delta x^m}{m!} \left. \frac{d^m y}{d x^m} \right|_{x_i} . \tag{3.7}$$

As this series has infinite number of terms, taking into account any finite number of its terms introduces a truncation error. Consequently instead of the exact value of function its approximation is obtained. Of course, the error of approximation depends on the number of terms taken into consideration. Assuming that only the first three terms are taken into account, an estimate of $y(x_{i+1})$ can be expressed using Eq. (3.7), rewritten in the following form:

$$y_{i+1} = y_i + \Delta x \left. \frac{dy}{dx} \right|_i + \frac{\Delta x^2}{2} \left. \frac{d^2 y}{dx^2} \right|_i + O(\Delta x^3). \tag{3.8}$$

where:

y_i, y_{i+1} – the values of $y(x)$ at node i and i+1 respectively,
Δx – distance between the nodes i and i+1

The term $O(\Delta x^3)$ indicates that the truncation error is of the order Δx^3. It means that the error varies proportionally to the step size Δx raised to the 3rd power. If we take a half of step size then the estimated error of y_{i+1} is reduced 8 times.

Directly from Eq. (3.8) one can find an estimate of the derivative of 1st order:

$$\left. \frac{dy}{dx} \right|_i = \frac{y_{i+1} - y_i}{\Delta x} - \frac{\Delta x}{2} \left. \frac{d^2 y}{dx^2} \right|_i + \cdots \tag{3.9}$$

One can notice, that the following finite difference

$$\left. \frac{dy}{dx} \right|_i \approx \frac{y_{i+1} - y_i}{\Delta x} \tag{3.10}$$

approximates the derivative of $y(x)$ at node i with accuracy of first order, i.e. $O(\Delta x)$. This formula is called the forward difference.

Similar approach can be applied to derive the backward difference formula. To this order the Taylor series expansion is performed to provide with the estimate of $y(x_{i-1})$. It takes the following form:

$$y_{i-1} = y_i - \Delta x \left. \frac{dy}{dx} \right|_i + \frac{\Delta x^2}{2} \left. \frac{d^2 y}{dx^2} \right|_i + O(\Delta x^3). \tag{3.11}$$

This equation yields the formula:

$$\left.\frac{dy}{dx}\right|_i \approx \frac{y_i - y_{i-1}}{\Delta x} \tag{3.12}$$

The approximation by backward difference introduces a truncation error of order $O(\Delta x)$, similarly to the forward difference (3.10).

Subtracting Eq. (3.8) from Eq. (3.11) gives:

$$y_{i+1} - y_{i-1} = 2\Delta x \left.\frac{dy}{dx}\right|_i + \frac{\Delta x^3}{3} \left.\frac{d^3y}{dx^3}\right|_i + \cdots. \tag{3.13}$$

from which one obtains the approximating formula known as the centered difference:

$$\left.\frac{dy}{dx}\right|_i \approx \frac{y_{i+1} - y_{i-1}}{2\Delta x} \tag{3.14}$$

In contrast to the previous formulas this one ensures accuracy of 2nd order, i.e. the truncation error is $O(\Delta x^2)$.

All the presented formulas approximating the derivative of the 1st order can be used to derive certain basic methods for numerical solution of the initial-value-problem. To this end let us approximate Eq. (3.1) at node i (Fig. 3.2) using the previously derived finite differences. Subsequent substitution of the formulas (3.10), (3.12) and (3.14) in Eq. (3.1) yields:

- For the forward difference:

$$\frac{y_{i+1} - y_i}{\Delta x} = f(x_i,y_i). \tag{3.15}$$

which gives the explicit or forward Euler method:

$$y_{i+1} = y_i + \Delta x \cdot f(x_i,y_i). \tag{3.16}$$

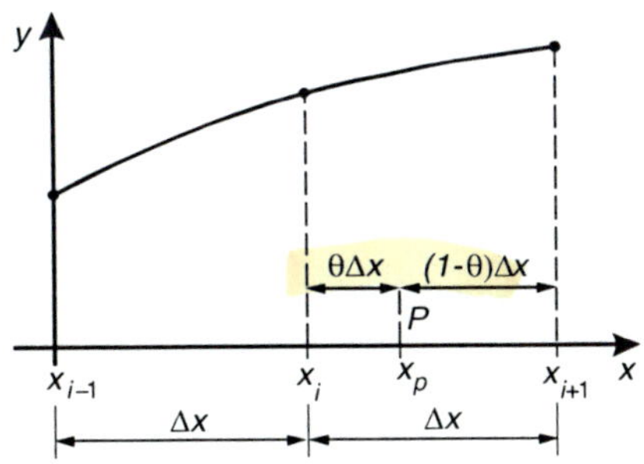

Fig. 3.2 Sketch for approximation of Eq. (3.1)

- For the backward difference:

$$\frac{y_i - y_{i-1}}{\Delta x} = f(x_i, y_i). \tag{3.17}$$

which gives the implicit or backward Euler method:

$$y_i = y_{i-1} + \Delta x \cdot f(x_i, y_i). \tag{3.18}$$

- For the centered difference:

$$\frac{y_{i+1} - y_{i-1}}{2\Delta x} = f(x_i, y_i). \tag{3.19}$$

from which results the Nystrom method:

$$y_{i+1} = y_{i-1} + 2\Delta x \cdot f(x_i, y_i). \tag{3.20}$$

In the same way further methods can be derived. Assume that the approximation of the derivative is performed at midpoint of the interval $\langle x_i, x_{i+1} \rangle$, in which the value of $f(x, y)$ is taken as arithmetic average from both nodes. Then Eq. (3.1) becomes:

$$\frac{y_{i+1} - y_i}{\Delta x} = \frac{1}{2}\left(f(x_i, y_i) + f(x_{i+1}, y_{i+1})\right). \tag{3.21}$$

This is known as the implicit trapezoidal method:

$$y_{i+1} = y_i + \frac{\Delta x}{2}\left(f(x_i, y_i) + f(x_{i+1}, y_{i+1})\right). \tag{3.22}$$

Both Euler methods and the trapezoidal method can be rewritten as a single unified formula. To obtain this formula let us approximate the differential equation (3.1) at point P (Fig. 3.2).

The position of point P, which may move within the interval $\langle x_i, x_{i+1} \rangle$, is determined by the weighting parameter θ defined as follows:

$$\theta = \frac{x_P - x_i}{\Delta x}. \tag{3.23}$$

Since $x_i \le x_P \le x_{i+1}$ then θ ranges from 0 to 1. Approximation of Eq. (3.1) gives:

$$\frac{y_{i+1} - y_i}{\Delta x} = f_P. \tag{3.24}$$

The value of derivative $f(x, y)$ at point P is calculated by linear interpolation between the endpoints of the interval:

$$f_P = (1 - \theta) f(x_i, y_i) + \theta \cdot f(x_{i+1}, y_{i+1}) \tag{3.25}$$

Substitution of Eq. (3.25) in Eq. (3.24) yields:

$$y_{i+1} = y_i + \Delta x\,((1-\theta)f\,(x_i, y_i) + \theta \cdot f(x_{i+1}, y_{i+1})) \tag{3.26}$$

This scheme is equivalent to:

- the forward Euler method for $\theta = 0$,
- the implicit trapezoidal rule for $\theta = 1/2$,
- the Galerkin method for $\theta = 2/3$,
- the backward Euler method for $\theta = 1$.

An alternative way of derivation of theses formulas is to consider the integration of ordinary differential equation (3.1) over the domain defined by the boundaries of the interval:

$$\int_{y_i}^{y_{i+1}} dy = \int_{x_i}^{x_{i+1}} f(x,y)dx. \tag{3.27}$$

From Eq. (3.27) results:

$$y_{i+1} = y_i + \int_{x_i}^{x_{i+1}} f(x,y)dx. \tag{3.28}$$

The presented methods differ in the way of approximating of integral, i.e. the quadrature rule used in its calculation.

Let us consider the forward Euler method (3.16). This formula may be derived from Eq. (3.28) as well. For this purpose the integral in Eq. (3.28) must be calculated using the quadrature in the form of rectangular formula. If the area under the curve in the interval is approximated by the area of the rectangle having dimensions Δx by $f(x_i, y_i)$ (Fig 3.3a) one obtains:

$$\int_{x_i}^{x_{i+1}} f(x,y)dx \approx \Delta x \cdot f(x_i, y_i). \tag{3.29}$$

Introducing this result in Eq. (3.28) one obtains the required forward Euler formula:

$$y_{i+1} = y_i + \Delta x \cdot f(x_i, y_i). \tag{3.30}$$

The forward Euler method has a very simple geometrical interpretation shown in Fig. 3.3b. The function $y(x)$ in the interval $\langle x_i, x_{i+1}\rangle$ is replaced by its tangent at point

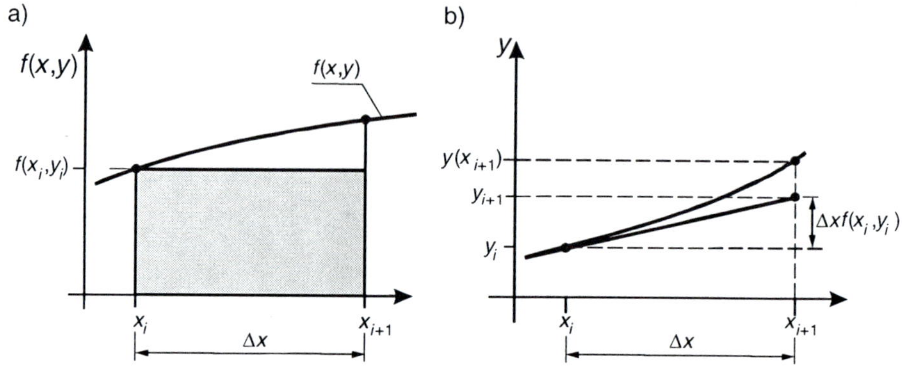

Fig. 3.3 Interpretation of the explicit Euler method

x_i. From this figure results that the value of the step size Δx determines the accuracy of numerical solution. The difference $y(x_i)-y_i$ increases with increasing Δx.

Conversely to the forward Euler method, the backward one is obtained if in Eq. (3.28) the integral is approximated by the area of rectangle having dimensions Δx by $f(x_{i+1}, y_{i+1})$ (Fig. 3.4):

$$\int_{x_i}^{x_{i+1}} f(x,y)dx \approx \Delta x \cdot f(x_{i+1},y_{i+1}). \tag{3.31}$$

Consequently one obtains:

$$y_{i+1} = y_i + \Delta x \cdot f(x_{i+1},y_{i+1}). \tag{3.32}$$

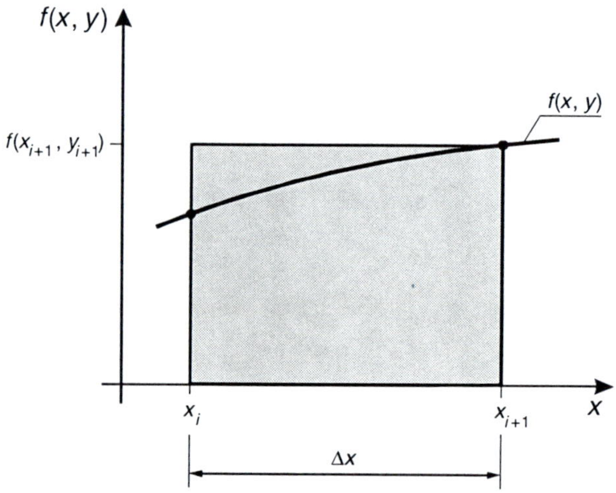

Fig. 3.4 Interpretation of implicit Euler method

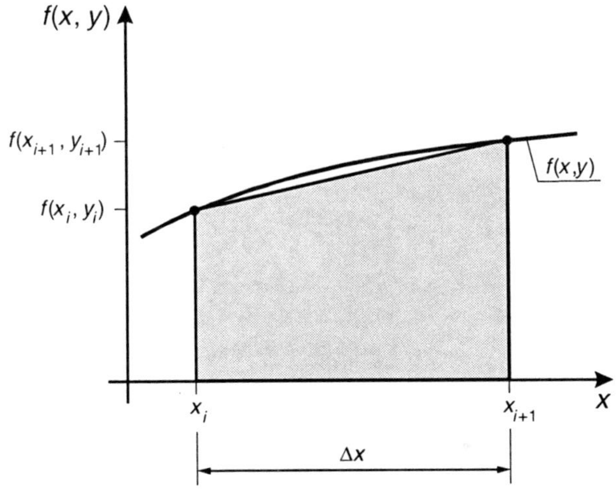

Fig. 3.5 Interpretation of the implicit trapezoidal method

In other words, the function $y(x)$ in the interval $\langle x_i, x_{i+1}\rangle$ is replaced by its tangent at x_{i+1}.

If the trapezoidal quadrature formula is applied one can write (Fig. 3.5):

$$\int_{x_i}^{x_{i+1}} f\,(x,y)\,d\,x \approx \frac{1}{2}\;(f\,(x_i, y_i) + f\,(x_{i+1}, y_{i+1}))\;\;\Delta x \tag{3.33}$$

Substitution of this result in Eq. (3.28) gives the trapezoidal formula (3.22):

$$y_{i+1} = y_i + \frac{\Delta x}{2}\;(f(x_i,y_i) + f(x_{i+1},y_{i+1}))\,. \tag{3.34}$$

One can say that in this case the function $y(x)$ is replaced in the interval $\langle x_i, x_{i+1}\rangle$ by a straight line with the slope equal to the arithmetic average of the slopes at both endpoints of the interval.

3.1.3 Runge–Kutta Methods

More accurate methods can be obtained by increasing the number of intermediate points between x_i and x_{i+1}. For example, the widely used Runge–Kutta methods can be expressed by the following general formulas:

$$y_{i+1} = y_i + \Delta x \sum_{j=1}^{m} w_j \cdot k_j \qquad \text{for} \qquad i = 0,\, 1,\, 2, \ldots,, \tag{3.35a}$$

where

$$k_1 = f(x_i, y_i), \tag{3.35b}$$

$$k_j = f\left(x_i + c_j \cdot \Delta x,\ y_i + \Delta x \sum_{l=1}^{j-1} a_{jl} \cdot k_l\right) \qquad 2 \le j \le m. \tag{3.35c}$$

In these formulas $m \geq 1$, w_j, c_j, a_{jl} are numerical coefficients. If at node x_i an estimate y_i is known, then y_{i+1} at the node $x_{i+1} = x_i + \Delta x$ is obtained from Eq. (3.35a), which uses the previously calculated coefficients $k_1, k_2, \ldots, k_m$. This is repeated in each step of integration: $i = 0, 1, 2, \ldots, N$ where N is number of steps in the interval $\langle a, b\rangle$. If the number of components in Eq. (3.35a) is m, then the corresponding Runge–Kutta method is called m-stage.

Comparison of Eqs. (3.35a) and (3.28) indicates that in fact the Runge–Kutta methods use some specific estimation of the integral. One can write that:

$$\int_{x_i}^{x_{i+1}} f\,(x,y)dx = \Delta x \sum_{j=1}^{m} w_j \, .k_j = \Delta x \, .\bar{f}\,(x, y)\,, \tag{3.36}$$

Where $\bar{f}(x,y)$ is a certain representative average value of the derivative dy/dx in the interval $\langle x_i, x_i + \Delta x\rangle$. One can assume that as a quadrature the rectangular method is applied. The area under the curve $f(x, y)$ is approximated by the area of rectangular having a base equal to Δx and height defined as an average value calculated with the weighting coefficients w_j.

Let us consider the simplest, one stage Runge–Kutta method with $m = 1$. Since in this case

$$k_1 = f\,(x_i, y_i)\,, \tag{3.37}$$

then the condition $w_1 = 1$ must be satisfied and consequently:

$$\int_{x_i}^{x_{i+1}} f(x,y)\,dx \approx \Delta x \cdot f(x_i,y_i) \tag{3.38}$$

Then general formula (3.35a) in this case takes the following particular form:

$$y_{i+1} = y_i + \Delta x f\left(x_i, y_i\right) \tag{3.39}$$

which is equivalent to the forward Euler formula presented earlier (3.16) (Legras 1971).

Improvement of the solution accuracy is possible by increasing the number of stages. For $m = 2$ one obtains two versions of the Runge–Kutta method. They are:

- the improved Euler method for which:

$$k_1 = f(x_i, y_i). \tag{3.40a}$$

$$k_2 = f\left(x_i + \frac{1}{2}\Delta x,\ y_i + \frac{1}{2}\Delta x \cdot k_1\right), \tag{3.40b}$$

$$y_{i+1} = y_i + \Delta x \cdot k_2, \tag{3.40c}$$

- the Euler–Cauchy method for which:

$$k_1 = f(x_i, y_i), \tag{3.41a}$$

$$k_2 = f(x_i + \Delta x,\ y_i + \Delta x \cdot k_1), \tag{3.41b}$$

$$y_{i+1} = y_i + \Delta x \frac{1}{2}(k_1 + k_2), \tag{3.41c}$$

Both versions have simple geometrical interpretations shown in Figs. 3.6 and 3.7 respectively.

In the improved Euler method the function $y(x)$ is represented in the interval $\langle x_i, x_{i+1}\rangle$ by a straight line having slope calculated at midpoint, i.e. at $x_i+\Delta x/2$ (Fig. 3.6). The Euler–Cauchy method uses the representative value of the derivative $\bar{f}(x,y)$

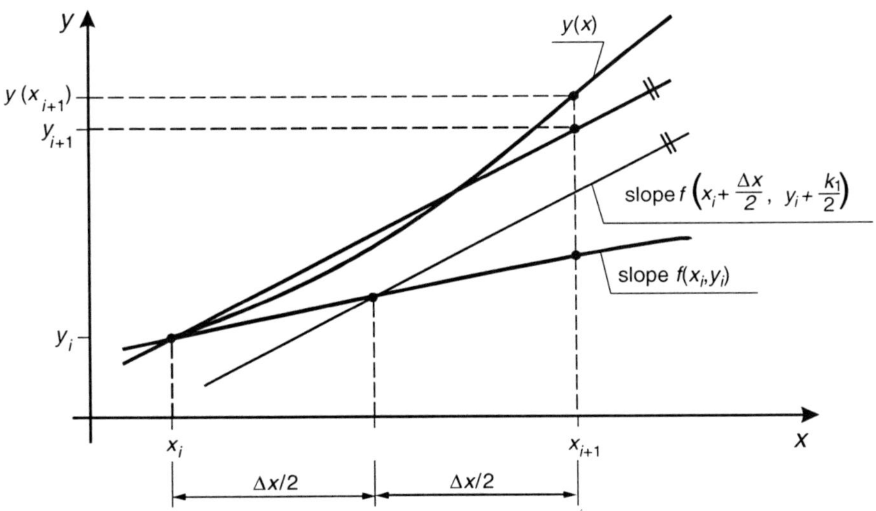

Fig 3.6 Interpretation of the improved Euler method

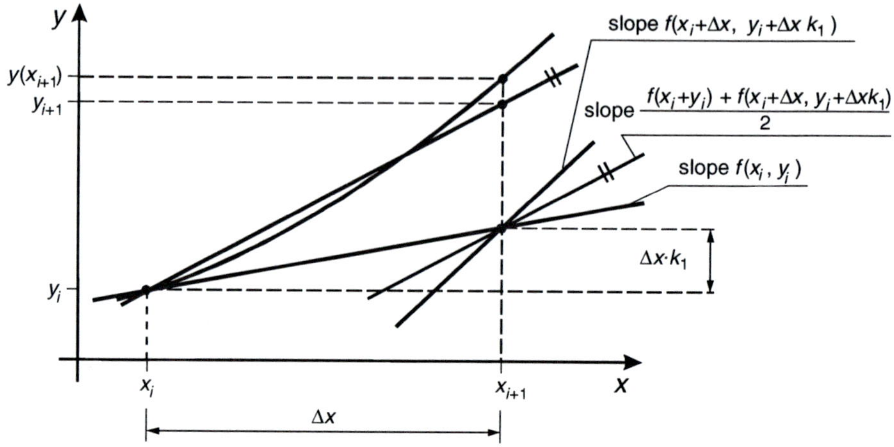

Fig. 3.7 Interpretation of the Euler–Cauchy method

equal to the arithmetic average value taken for both endpoints of the interval: x_i and x_{i+1} (Fig. 3.7).

The most popular and widely applied version of the Runge–Kutta method is the 4-stage one. For $m = 4$ one obtains the following formulas (Legras 1971):

$$k_1 = f(x_i, y_i), \tag{3.42a}$$

$$k_2 = f\left(x_i + \frac{\Delta x}{2},\ y_i + \Delta x \frac{k_1}{2}\right), \tag{3.42b}$$

$$k_3 = f\left(x_i + \frac{\Delta x}{2},\ y_i + \Delta x \frac{k_2}{2}\right), \tag{3.42c}$$

$$k_4 = f(x_i + \Delta x,\ y_i + \Delta x \cdot k_3), \tag{3.42d}$$

$$y_{i+1} = y_i + \frac{\Delta x}{6}(k_1 + 2k_2 + 2k_3 + k_4). \tag{3.42e}$$

This version is very often applied in open channel hydraulics, however rather to integrate the ordinary differential equations with time as the independent variable. There are the Runge–Kutta formulas derived up to $m = 8$ (Legras 1971). However, formulas with more than 4 stages are seldom used in engineering practice. The same is true for the multistep methods, such as Adams–Bashford, Adams–Moulton or Backward Differentiation Formula (BDF) (see Ascher and Petzold (1998) for more details on these schemes).

Example 3.1 A flood wave of known form is passing through a reservoir closed by a dam with a weir. Knowing that reservoir works accordingly to Eq. (1.122) compute the outflow from the reservoir $O(t)$ through a weir. The storage equation (1.122) is applied in a simplified form:

$$\frac{dh}{dt} = \frac{1}{F(h)}(I(t) - O(t)) \tag{3.43}$$

where:

t – time,
$h(t)$ – water level above assumed datum,
$F(h)$ – area of reservoir at the level of surface,
$I(t)$ – inflow into reservoir,
$O(t)$ – outflow from reservoir.

The flood wave is described by the formula:

$$I(t) = q_0 + (q_m - q_0)\left(\frac{t}{t_m}\right)\exp\left(1 - \left(\frac{t}{t_m}\right)\right) \tag{3.44}$$

where:

q_0 – baseflow discharge of the inflow (initial inflow),
q_m – peak discharge of the inflow,
t_m – time of the peak flow.

The outflow $O(t)$ from the reservoir takes place through a weir (Fig. 3.8)

Since the weir is contracted the flow is governed by the following discharge equation (Roberson et al. 1998):

$$O(t) = K \cdot (L - 0.20H)\sqrt{2\,g} \cdot H^{3/2}, \tag{3.45}$$

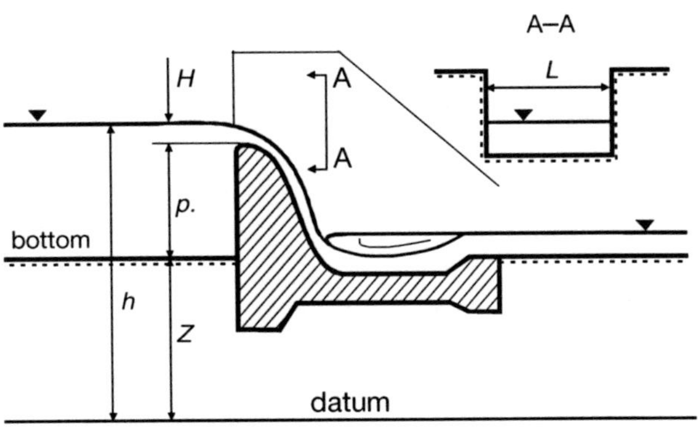

Fig. 3.8 Weir of a dam

where:

H – head,
L – weir width,
p – height of the dam,
Z – bottom elevation above the assumed datum,
g – acceleration due to gravity,
K – flow coefficient given as $K = 0.40 + 0.05H/p$ (Roberson et al. 1997):

Replacing in Eq. (3.45) the head H with the water level h one obtains:

$$O(t) = K\sqrt{2g}\ (L - 0.20\ (h(t) - Z - p))\ (h(t) - Z - p)^{3/2} \tag{3.46}$$

The area of the water surface F varies with the water level elevation h as showed in Fig. 3.9:

Substitution of Eq. (3.162) in Eq. (1.160) yields:

$$\frac{dh}{dt} = \frac{I(t) - K\sqrt{2g}\ (L - 0.20\ (h(t) - Z - p))\ (h(t) - Z - p)^{3/2}}{F(h(t))}. \tag{3.47}$$

Solving this storage equation with appropriate initial condition one obtains the function $h(t)$ representing the evolution in time of the water level in reservoir. The initial condition $h(t{=}0) = h_0$ can be found from the discharge equation (3.45). To this order one can assume that at the beginning the flow is steady with a constant discharge equal to baseflow q_0. The corresponding head at the weir is given by as:

$$H(t = 0) = \left(\frac{q_0}{K\sqrt{2g}\ (L - 0.20\ (h(t) - Z - p))}\right)^{2/3}. \tag{3.48}$$

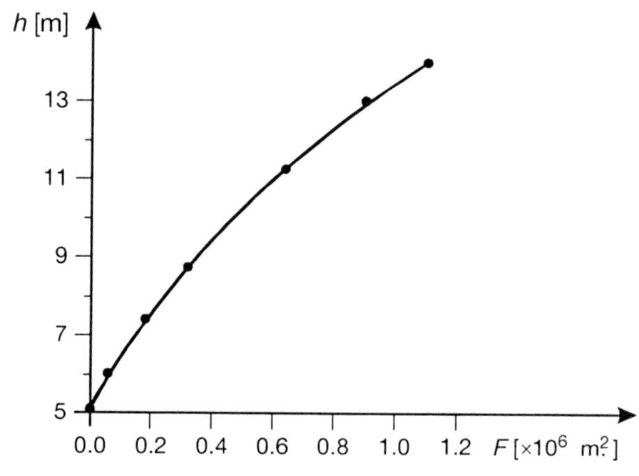

Fig. 3.9 Area of reservoir's water surface versus elevation of the water level above the assumed datum

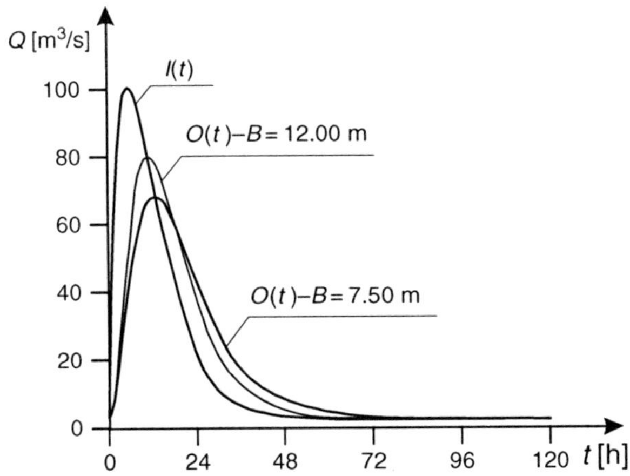

Fig. 3.10 Transformation of the flood wave $I(t)$ flowing through reservoir for two different lengths of weir

Therefore the required initial condition takes the following form:

$$h(t=0) = h_0 = Z + p + \left(\frac{q_0}{K\sqrt{2g}\ (L - 0.20\ (h(t) - Z - p))} \right)^{2/3}. \tag{3.49}$$

Knowing this condition one can use the formulas (3.42) for computation the approximated values of $h(i{\cdot}\Delta t)$ ($i = 1, 2, 3, \ldots$ where Δt is time step and i is index of time step).

The following data were assumed: $q_0 = 2.50$ m^3/s, $q_m = 100$ m^3/s, $t_m = 6$ h and $p = 5$ m. Computations were performed for two values of weir's length: $L = 12.0$ m and $L = 7.50$ m. This parameter mostly influences the discharge through weir so it determines the flood wave transformation process by reservoir. The results of calculations provided by the Runge–Kutta method are shown in Fig. 3.10.

The changes of time step value in the range from 60 to 360 s only insignificantly affected the final results.

3.1.4 Accuracy and Stability

Each numerical method of integration of the initial-value problem gives an approximate value of the exact solution. If we assume that at node x_i the exact value of solution is known, i.e. $y(x_i) = y_i$, then the error at node x_{i+1} is considered as generated in a single step. Obviously, the difference:

$$r(\Delta x) = y(x_i + \Delta x) - y_{i+1}, \tag{3.50}$$

being the error of approximation of the exact solution by the numerical method, should be as small as possible. Smaller value of $r(\Delta x)$ means more accurate method of solution. If instead of y_{i+1}, a formula of applied method, which depends on some coefficients and on the values of the derivative $f(x,y)$, is substituted in Eq. (3.50), then one can assume, that the error r is a function of the step size Δx only. Therefore it can be expanded in the Taylor series with regard to Δx around zero (Ascher and Petzold 1998):

$$r(\Delta \mathrm{x}) = r(O) + \Delta x \left.\frac{dr}{d\Delta x}\right|_0 + \frac{\Delta x^2}{2!}\left.\frac{d^2 r}{d\Delta x^2}\right|_0 + \frac{\Delta x^3}{3!}\left.\frac{d^3 r}{d\Delta x^3}\right|_0 + \cdots \tag{3.51}$$

If the difference (3.50) is equal to zero, all the terms of the Taylor series are equal to zero as well. However the error of approximation exists always, although its magnitude depends on the applied solution scheme. The error of approximation is defined by so called the order of integration method. The considered method is said to be of order p, if for each initial-value problem the first p terms of series (3.51) are equal to zero. Therefore, if a method is of order p, then the error of approximation generated in a single step is:

$$r(\Delta \mathrm{x}) = \frac{\Delta x^{p+1}}{(p+1)!}\left.\frac{d^{p+1} r}{d\Delta x^{p+1}}\right|_0. \tag{3.52}$$

This definition is valid for all numerical methods of integration of the initial-value problem.

Let us examine accuracy of the methods described by Eq. (3.26):

$$y_{i+1} = y_i + \Delta x\left((1-\theta)f_i + \theta \cdot f_{i+1}\right) \tag{3.53}$$

The linear multistep methods frequently applied to solve numerically initial value problems for both scalar ordinary differential equation and system of such equations, derived by means of an interpolating polynomial, have a common general form (Ascher and Petzold 1998):

$$\sum_{j=0}^{k} \alpha_j \cdot y_{i-j} = \Delta x \sum_{j=0}^{k} \beta_j \cdot f_{i-j} \tag{3.54}$$

where:

a_j, β_j – coefficients of the method,
y_i – nodal values of function $y(x)$,
f_i – nodal values of derivative,
i – node index,
k – number of past integration steps,
Δx – step size.

Note that the method (3.53) belongs to this family as well. In this case we have $k = 1$. The corresponding coefficients are given as follows: $\alpha_0 = 1, \alpha_1 = -1, \beta_0 = \theta, \beta_1 = 1-\theta$.

Assume that $y(x)$ is the exact solution of the considered ordinary differential equation (3.1) and it possesses the required derivatives. Introducing this function into formula (3.54) one obtains a local error of solution r_i:

$$\sum_{j=0}^{k} \alpha_j \cdot y_{i-j} - \Delta x \sum_{j=0}^{k} \beta_j \cdot f_{i-j} = r_i \tag{3.55}$$

As it was stated above, the order of the method can be found by the analysis of the local truncation error. Thus, the nodal values of function $y(x)$ and its derivative $f(x)$ in Eq. (3.55) are expanded in Taylor series around the node i. Consequently, the error r_i will be also expressed in the form of Taylor series:

$$r_i = \sum_{p=0}^{\infty} C_p \cdot \Delta x \cdot \left.\frac{d^p y}{dx^p}\right|_i \tag{3.56}$$

with the following coefficients C_p:

$$C_0 = \sum_{j=0}^{k} \alpha_j \tag{3.57a}$$

$$C_p = (-1)^p \left(\frac{1}{p!} \sum_{j=1}^{k} j^p \cdot \alpha_j + \frac{1}{(p-1)!} \sum_{j=0}^{k} j^{p-1} \cdot \beta_j \right) \quad (\text{for } p \geq 1) \tag{3.57b}$$

Calculation of these coefficients allows us to find the order of linear multistep method. If $C_0, C_1, \ldots, C_p = 0$, $C_{p+1} \neq 0$ then considered method is of order p (Ascher and Petzold 1998). For Eq. (3.53) we obtain:

$$C_0 = 1 - 1 = 0, \tag{3.58a}$$

$$C_1 = 1 - \theta - (1 - \theta) = 0, \tag{3.58b}$$

$$C_2 = -\frac{1}{2} + (1 - \theta) = \frac{1}{2} - \theta, \tag{3.58c}$$

$$C_3 = -\frac{1}{6}(-1) - \frac{1}{2}(1 - \theta) = -\frac{1}{3} + \frac{\theta}{2} \tag{3.58d}$$

One can notice that the method (3.53) is of 1st order for $\theta \neq 1/2$ and it is of 2nd order for $\theta = 1/2$. Therefore one can say that:

- both explicit ($\theta = 0$) and implicit ($\theta = 1$) Euler methods are of 1st order since $C_0 = 0$, $C_1 = 0$ and $C_2 \neq 0$,
- the trapezoidal implicit method ($\theta = 1/2$) is of 2nd order since $C_0 = 0$, $C_1 = 0$, $C_2 = 0$ and $C_3 \neq 0$.

The first non-zero coefficient C_{p+1} is called the error constant of the method. Note that it is equal to 1/2 or –1/2 for the Euler methods and –1/12 for the implicit trapezoidal rule.

The second important property of each method of integration is the numerical stability. Roughly speaking, stability means that if the initial condition is slightly changed, then the solution at following points will be also only slightly changed with respect to the original solution. Unstable methods tend to produce oscillatory solutions with unlimited growth of numerical error. To examine this question let us express the linear multistep method in terms of the so-called characteristic polynomials (Ascher and Petzold 1998):

$$\rho(Z) = \sum_{j=0}^{k} \alpha_j \cdot Z^{k-j}, \tag{3.59a}$$

$$\sigma(Z) = \sum_{j=0}^{k} \beta_j \cdot Z^{k-j} \tag{3.59b}$$

where Z is a complex number.

For considered Eq. (3.53) these polynomials take the following forms:

$$\rho(Z) = Z - 1, \tag{3.60a}$$

$$\sigma(Z) = \theta \cdot Z + (1 - \theta) \tag{3.60b}$$

The region of absolute stability of any method is given by the ratio of both polynomials

$$z = \frac{\rho(Z)}{\sigma(Z)} = \frac{\rho(e^{i \cdot \varphi})}{\sigma(e^{i \cdot \varphi})} \tag{3.61}$$

where i is imaginary unit, whereas φ ranges in interval $\langle 0, 2\pi \rangle$. A picture of the stability region is obtained by plotting of boundary given by Eq. (3.61) on the complex plane. Substituting Eq. (3.60) in Eq. (3.61) yields:

$$z = \frac{Z - 1}{\theta \cdot Z + (1 - \theta)} = \frac{e^{i \cdot \varphi} - 1}{\theta \cdot e^{i \cdot \varphi} + (1 - \theta)} \tag{3.62}$$

Using the following well known relation (McQuarrie 2003):

$$e^{i\,\varphi} = \cos(\varphi) + i \cdot \sin(\varphi)$$

Equation (3.62) is rewritten as follows:

$$z = \frac{\cos(\varphi) + i \cdot \sin(\varphi) - 1}{\theta\,(\cos(\varphi) + i \cdot \sin(\varphi)) + (1 - \theta)} = \frac{(\cos(\varphi) - 1) + i \cdot \sin(\varphi)}{(\theta \cdot \cos(\varphi) + (1 - \theta)) + i \cdot \theta \cdot \sin(\varphi)} \tag{3.63}$$

Division of the complex numbers gives:

$$
\begin{aligned}
z = & \frac{(\cos(\varphi) - 1)(\theta \cdot \cos(\varphi) + (1-\theta)) + \theta \cdot \sin^2(\varphi)}{(\theta \cdot \cos(\varphi) + (1-\theta))^2 + \theta^2 \cdot \sin^2(\varphi)} + \\
& + i \frac{(\theta \cdot \cos(\varphi) + (1-\theta))\sin(\varphi) - \theta \cdot (\cos(\varphi) - 1)\sin(\varphi)}{(\theta \cdot \cos(\varphi) + (1-\theta))^2 + \theta^2 \cdot \sin^2(\varphi)}
\end{aligned}
\tag{3.64}
$$

One can see that the extent of the region of absolute stability depends on the weighting parameter θ. In Fig. 3.11 are shown the regions corresponding to the selected values of θ. Note that for $\theta = 0$ (explicit Euler method) the region of absolute stability is very small, having the form of unit circle at left part of the complex plane. Increasing of θ increases the stability region, which for $\theta = 1/2$ (the implicit trapezoidal rule) contains the whole left half of the complex plane (Fig. 3.11b). Further increasing of θ includes a fraction of the real part of the complex plane to the region of absolute stability as well (Fig. 3.11a). The largest region of stability is obtained for the implicit Euler method corresponding to $\theta = 1$. It contains entire complex plane except the shadowed unit circle (Fig. 3.11a).

The theory of the numerical methods for initial value problem in ODEs distinguishes a very important family kind of the so called A-stable methods. This term should not be mistaken with the absolute stability. The difference method is considered as A-stable if its region of absolute stability covers the whole left part of complex plane. This type of methods is recommended for solving so called stiff systems of the ordinary differential equations (Ascher and Petzold 1998).

In the case of Eq. (3.53) the real part of the complex number defining the region of stability (3.64) can be reduced to the following form:

$$
\mathrm{Re}(z) = \frac{(1 - \cos(\varphi))(2\theta - 1)}{(\theta \cdot \cos(\varphi) + (1-\theta))^2 + \theta^2 \cdot \sin^2(\varphi)} \tag{3.65}
$$

Since $0 \leq \varphi \leq 2\pi$ then from Eq. (3.65) results that the region of absolute stability of the method (3.53) will be kept at right half of the complex plane if only:

$$
1/2 \leq \theta \leq 1 \tag{3.66}
$$

This condition confirms the results presented in Fig. 3.11. The method (3.53) will be A-stable as long as its weighting parameter θ satisfies the condition (3.66).

From practical viewpoint the most interesting methods for solving the ordinary differential equations are those which have large region of absolute stability. This explains the reasons of popularity of such methods as the implicit trapezoidal rule and backward Euler method. As we will see in later, these methods are of great importance in open channel hydraulic as well.

More information on the problems of accuracy and stability analysis is given for example by Ascher and Petzold (1998), Bjorck and Dahlquist (1974), LeVeque (2007), Ralston (1965).

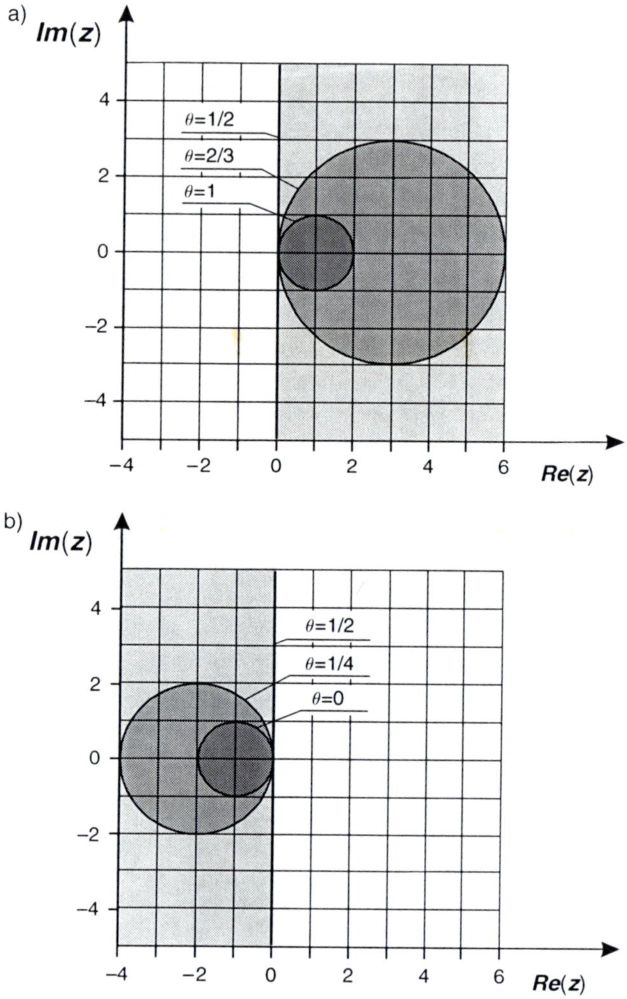

Fig. 3.11 Absolute stability regions of the method (3.53) for various values of weighting parameter θ: (**a**) out of *shadow*; (**b**) inside of *shadow*

3.2 Initial Value Problem for a System of Ordinary Differential Equations

The initial-value problem, discussed previously for a scalar ordinary differential equation, can be also formulated for a system of ordinary differential equations of the following general form:

$$\frac{dy_1}{dx} = f_1(x, y_1, y_2, \ldots, y_N), \tag{3.67a}$$

$$\frac{dy_2}{dx} = f_2(x,y_1,y_2,,\ldots,_N), \tag{3.67b}$$

$$\vdots \qquad \vdots$$

$$\frac{dy_N}{dx} = f_N(x,y_1,y_2,\ldots,y_N) \tag{3.67c}$$

The goal is to find such set of the functions $y_1(x),y_2(x),\ldots,y_N(x)$, which satisfy simultaneously the system of equations (3.67) in the considered interval $\langle a,\, b\rangle$ and the following initial conditions:

$$y_1(a) = y_{1,0}, \quad y_2(a) = y_{2,0}, \ldots, \quad y_N(a) = y_{N,0},$$

where $y_{1,0},\, y_{2,0}, \ldots, y_{N,0}$ are given values.

To solve the initial-value problem one can apply any of the methods derived for the scalar equation. These formulas remain the same. The only difference is that vectors are used instead of scalars.

Using the vector notation the system (3.67) is rewritten shorter, as follows:

$$\frac{d\,\mathbf{y}(x)}{dx} = \mathbf{f}(x,\mathbf{y}) \quad \text{for} \quad a < x \leq b \quad \text{with} \quad \mathbf{y}(a) = \mathbf{y}_0 \tag{3.68}$$

where

$$\mathbf{y}(x) = \begin{Bmatrix} y_1(x) \\ y_2(x) \\ \vdots \\ y_N(x) \end{Bmatrix}, \quad \mathbf{f}\,(x,\mathbf{y}) = \begin{Bmatrix} f_1(x,\mathbf{y}) \\ f_2(x,\mathbf{y}) \\ \vdots \\ f_N(x,\mathbf{y}) \end{Bmatrix}, \quad \mathbf{y}_o = \begin{Bmatrix} y_{1,0} \\ y_{2,0} \\ \vdots \\ y_{N,0} \end{Bmatrix}.$$

If for instance the implicit trapezoidal rule (3.139) is applied to solve the system (3.68), then one obtains:

$$\mathbf{y}_{i+1} = \mathbf{y}_i + \frac{\Delta x}{2}(\mathbf{f}_i + \mathbf{f}_{i+1}). \tag{3.69}$$

where i is index of node and Δx is step size. Other formulas can be implemented in a similar manner.

In many cases, for instance in the numerical solution of 1D partial differential equation using the finite element method, one obtains a systems of ordinary differential equations written in the following form:

$$\mathbf{A}\frac{d\,\mathbf{y}}{d\,t} + \mathbf{B}\cdot\mathbf{y} = \mathbf{0}, \tag{3.70}$$

where $\mathbf{A}$ is a constant matrix, whereas $\mathbf{B}$ is a variable matrix dependent on time t and sometimes, in addition on unknown vector $\mathbf{y}$. Obviously, the system (3.70) can be reduced to the form of Eq. (3.68). However such a transformation is not recommended, since both $\mathbf{B}$ and $\mathbf{A}$ are tri-diagonal matrices. Diagonalization of Eq. (3.70) requires that the matrix $\mathbf{A}$ is inverted. However, it should be remembered that the inversion of a banded matrix gives a matrix $\mathbf{A}^{-1}$ which has all non-zero elements. In such a way the advantage of the tri-diagonal matrix in Eq. (3.70) is lost.

As it was mentioned in the preceding section, the most suitable methods for solution of the systems of ODEs are the A-stable methods. They are recommended to solve the stiff systems, in which the component equations describe the processes having different time scale. It means that there are some processes which occur much more rapidly than others. After Ascher and Petzold (1998), "An initial value problem is stiff in some interval if the step size needed to maintain stability is much smaller than the step size required to represent the solution accuracy". However, it appears that among the standard methods only some of them can be classified as A-stable ones. It is proved that only the implicit methods of order not higher than 2, i.e. for $p \leq 2$, are A-stable (Ascher and Petzold 1998, Bjorck and Dahlquist 1974). Among all standard methods of solution of the initial-value problem only implicit Euler and implicit trapezoidal methods satisfy both mentioned conditions. The previously introduced unified formula (3.53) can be rewritten in vector notation as:

$$\mathbf{y}_{i+1} = \mathbf{y}_i + \Delta \mathrm{t}\,((1-\theta)\,\mathbf{y}'_i + \theta \cdot \mathbf{y}'_{i+1}), \tag{3.71}$$

with θ ranging from 0 to 1. From the preceding section it is known that for $\theta = 1$ Eq. (3.71) becomes the backward Euler method, whereas for $\theta = 1/2$ it is the implicit trapezoidal one. Remember that this formula is A-stable for $1/2 \leq \theta \leq 1$

To solve the system (3.70) the vector of derivatives:

$$\mathbf{y}' = \mathbf{A}^{-1}(-\mathbf{B} \cdot \mathbf{y}), \tag{3.72}$$

is substituted in Eq. (3.71). This gives:

$$\mathbf{y}_{i+1} = \mathbf{y}_i + \Delta t(-(1-\theta)\ \mathbf{A}^{-1} \cdot \mathbf{B}_i \cdot \mathbf{y}_i - \theta\ \cdot \mathbf{A}^{-1} \cdot \mathbf{B}_{i+1} \cdot \mathbf{y}_{i+1}) \tag{3.73}$$

Multiplying both sides by $\mathbf{A}$ and grouping similar terms, yields the following system of algebraic equations:

$$(\mathbf{A} + \Delta t \cdot \theta\ \cdot \mathbf{B}_{i+1})\,\mathbf{y}_{i+1} = (\mathbf{A} - \Delta t(1-\theta)\mathbf{B}_i)\,\mathbf{y}_i, \tag{3.74}$$

which can be rewritten as:

$$\mathbf{R}_{i+1} \cdot \mathbf{y}_{i+1} = \mathbf{F}_i, \tag{3.75}$$

where:

$$\mathbf{R}_{i+1} = \mathbf{A} + \Delta t \cdot \theta\ \cdot \mathbf{B}_{i+1} \tag{3.76a}$$

$$\mathbf{F}_i = (\mathbf{A} - \Delta t(1-\theta)\ \mathbf{B}_i)\ \mathbf{y}_i \tag{3.76b}$$

In such a way the solution of the initial-value problem for a system of ordinary differential equations was reduced to the solution of the system of algebraic equations in each time step. This system may be linear or non-linear one according to the solved problem. In the latter case, an iterative method must be applied. Note that the matrix **R** is banded as the matrices **A** and **B** in the system (3.70).

3.3 Boundary Value Problem

The system of the ordinary differential equations considered in preceding section:

$$\frac{d\mathbf{y}}{dx} = \mathbf{f}(x,\mathbf{y}). \tag{3.77}$$

requires some auxiliary information to be solved. If all needed data are specified at the same value of the independent variable x, then the considered problem is the initial-value one. This problem was discussed in previous section. In contrast to the initial-value problem one can formulate another one, in which the auxiliary conditions are specified at different values of the independent variable. Since usually they are imposed at two different points, corresponding to the ends of the interval $\langle a, b\rangle$, then the problem formulated in such a way is called two-point boundary-value problem or simply the boundary-value problem. Obviously, more than one additional condition is required only by a system of ordinary differential equations or by an equation of higher order than 1st, which can be reduced to the equivalent system of equations. In open channel hydraulics the ordinary differential equations usually appear in the form of a single equation of 1st order. Thus formally it is impossible to formulate a boundary-value problem directly, in the way mentioned above. However, such type of problem can be formulated if we assume that one of the parameters of the ODE is unknown, but constant in the interval $\langle a, b\rangle$, in which the considered equation is integrated (Ascher and Petzold 1998, Press et al. 1992). Therefore we have:

$$\frac{dy}{dx} = f(x,y,\lambda) \tag{3.78}$$

where λ is a constant parameter. Then one more ordinary differential equation is added to the system:

$$\frac{d\lambda}{dx} = 0 \tag{3.79}$$

This means that the constant λ is considered as a function over the interval $\langle a, b\rangle$, which is independent of x.

In such a way we have the system of equations (3.78) and (3.79), for which one can formulate the boundary-value problem. This problem is formulated as follows:

determine the function $y(x)$ over the interval $\langle a, b\rangle$ which satisfies within this interval Eqs. (3.78) and (3.79) and moreover, which satisfies the following boundary conditions:

$$y(a) = y_a \tag{3.80a}$$

and

$$y(b) = y_b \tag{3.80b}$$

where y_a and y_b are specified values of y at both endpoints of the interval.

To solve numerically the above formulated problem two following approaches can be applied (Bjorck and Dahlquist 1974, Stoer and Bulirsch 1980):

- the finite difference method,
- the shooting method.

Both methods are briefly described below, while more details on their application are provided in Chapter 4.

In the finite difference method the interval of integration $\langle a, b\rangle$ is divided using N nodes x_i into the segments of length Δx. Equation (3.79) is approximated at midpoint $x_i + \Delta x/2$ of each segment using the implicit trapezoidal formula (3.22).

$$\frac{y_{i+1} - y_i}{\Delta x} = \frac{1}{2}(f_i + f_{i+1}) \quad \text{for } i = 1, 2, \cdots, N-1 \tag{3.81}$$

In such a way one obtains $N - 1$ algebraic equations containing $N + 1$ unknowns. There are N values of the function y_i at each node and the constant parameter λ. To this system two equations should be added, which correspond to the boundary conditions imposed at the endpoints of the interval. Consequently the closed system may be written as follows:

$$\mathbf{AX} = \mathbf{B} \tag{3.82}$$

where:

A – sparse matrix having dimensions $(N + 1)\times(N + 1)$,
$\mathbf{X} = (y_1, y_2, \ldots, y_{N-1}, y_N, \lambda)^T$ – vector of unknowns,
B – vector of right side hand.

Solving this system one obtains the nodal values of $y(x)$ and the value of parameter λ.

The shooting method is based on converting the boundary-value problem into an equivalent initial-value problem (Chapra and Canale 2006). The solution is obtained

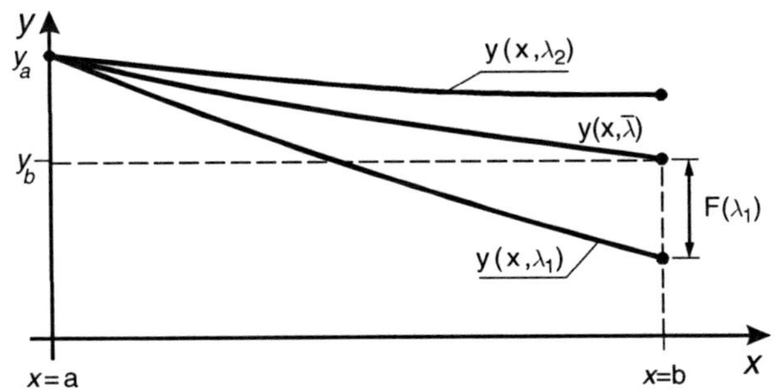

Fig. 3.12 Conversion of the boundary-value problem into a sequence of the initial-value problems

indirectly. A sequence of solutions of the initial-value problems is carried out for various values of the parameter λ until the condition imposed at $x = b$ is satisfied with specified tolerance. Then the following initial-value problem:

$$\frac{dy}{dx} = f(x,y,\lambda) \text{ with } y(a) = y_a \tag{3.83}$$

is solved. Since $y = y(x, \lambda)$ then the solution depends on the assumed value of λ. Solution of the boundary-value problem will be such a solution of the initial-value problem for which the second boundary condition:

$$y(b, \bar{\lambda}) = y_b \tag{3.84}$$

is satisfied (Fig. 3.12).

If we introduce the following function:

$$F(\lambda) = y(b, \lambda) - y_b \tag{3.85}$$

which defines the difference between the value of $y(b)$ computed for given value of λ and imposed value of y as the boundary condition y_b, then solution of problem is reduced to determination of the root of function $F(\lambda)$. To find the root, at first the interval in which it is located must be done. To this order we should specify such two values of λ, say $\lambda^{(1)}$ and $\lambda^{(2)}$, for which the following relation is valid:

$$F(\lambda^{(1)}) \cdot F(\lambda^{(2)}) < 0 \tag{3.86}$$

Afterwards the bisection or false position method can be used to determine the root with required accuracy i.e. the approximated value of $\bar{\lambda}$.

References

Ascher UM, Petzold LR (1998) Computer methods for ordinary differential equations and differential – algebraic equations. SIAM, Philadelphia, PA

Bjorck A, Dahlquist G (1974) Numerical methods. Prentice-Hall, Englewood Cliffs, NJ

Chapra SC, Canale RP (2006) Numerical methods for engineers. McGraw-Hill, New York

Legras J (1971) Méthodes et techniques de l'analyse numérique. Dunod, Paris

LeVeque RJ (2007) Finite difference methods for ordinary and partial differential equations: Steady-state and time dependent problems. SIAM, Philadelphia, PA

McQuarrie DA (2003) Mathematical methods for scientists and engineers. University Science Books, Sausalito, CA

Press WH, Teukolsky SA, Veterling WT, Flannery BP (1992) Numerical recipes in C. Cambridge University Press, Cambridge

Ralston A (1965) A fisrt course in numerical analysis. McGraw-Hill, New York

Roberson JA, Cassidy JJ, Chandhry MH (1998) Hydraulic engineering, 2nd edn. Wiley, New York

Stoer J, Bulirsch R (1980) Introduction to numerical analysis. Springer-Verlag, New York

Chapter 4
Steady Gradually Varied Flow in Open Channels

4.1 Introduction

4.1.1 Governing Equations

In many practical applications the flow in open channels can be considered as steady and gradually varied. Such a flow can take place either in a single channel or in a channel network. The solution of steady flow equation can constitute a separate engineering problem or it can be a part of a more complex problem. The first case arises, for example, while computing the flow distribution and the flow profiles in a watering channel network. The second case can occur in unsteady flow modeling, when steady gradually varied flow in a channel network is assumed as the initial condition for unsteady flow equations.

Sometimes the analysis of the steady gradually varied flow in channel network is carried out using the unsteady flow equations, i.e. the system of Saint Venant equations. Starting from the hydrostatic state, these equations are solved numerically with the boundary conditions corresponding to the required final steady state. Consequently the water stages and discharges asymptotically tend to their steady state values. However, such an indirect approach seems to be somewhat artificial. It is better to apply a direct approach, i.e. to use the appropriate equations describing steady flow and to solve them with a suitable numerical method. Such a concept is presented below.

As it was shown in Section 1.6, the governing equations describing steady gradually varied flow can be obtained from the system of Saint Venant equations:

$$\frac{dQ}{dx} = q, \tag{4.1}$$

$$\frac{dE}{dx} = \frac{d}{dx}\left(h + \frac{\alpha \cdot Q^2}{2g \cdot A^2}\right) = -S - \frac{\alpha \cdot Q}{g \cdot A^2} q, \tag{4.2}$$

where:

h – water level above the accepted datum,
Q – discharge,

R. Szymkiewicz, *Numerical Modeling in Open Channel Hydraulics*, Water Science and Technology Library 83, DOI 10.1007/978-90-481-3674-2_4,

A – wetted cross-sectional area,
x – longitudinal distance,
q – lateral inflow,
α – energy correction factor
S – friction slope given by equation:

$$S = \frac{(n_M)^2 \cdot Q^2}{R^{4/3} \cdot A^2} \tag{4.3}$$

where:

n_M – Manning coefficient,
R – hydraulic radius.

Since Eqs. (4.1) and (4.2) are derived from the Saint Venant equations, they should be applicable for all types of open channels. Indeed, as will be shown later, numerical integration of Eq. (4.2) by the implicit trapezoidal rule leads to the well known step method, commonly applied in practice. Usually the step method is derived directly from the principle of energy conservation applied for the discrete system of two neighboring cross-sections along channel axis.

4.1.2 Determination of the Water Surface Profiles for Prismatic and Natural Channel

For a rectangular channel and $Q = \text{const}$. Equations (4.1) and (4.2) with $\alpha = 1$ may be rearranged to the form of Eq. (1.97):

$$\frac{dH}{dx} = \frac{s - S}{1 - F_r^2}. \tag{4.4}$$

where:

H – flow depth,
s – channel bed slope,
F_r – Froude number.

Equation (4.4) is the base for theoretical analysis of the water surface profiles arising for various flow conditions in open channels. The profiles can be deduced by taking into account the relation between the bed slope s, critical slope s_c and the friction slope S, as well as the relations between the flow depth H, normal depth H_n and critical depth H_c. This rather theoretical analysis, presented in detail for example by Chow (1959), French (1985), Singh (1996) and others, provides some qualitative conclusions. Typical water profiles which can occur during the steady gradually varied flow in open channel are presented in Fig. 4.1.

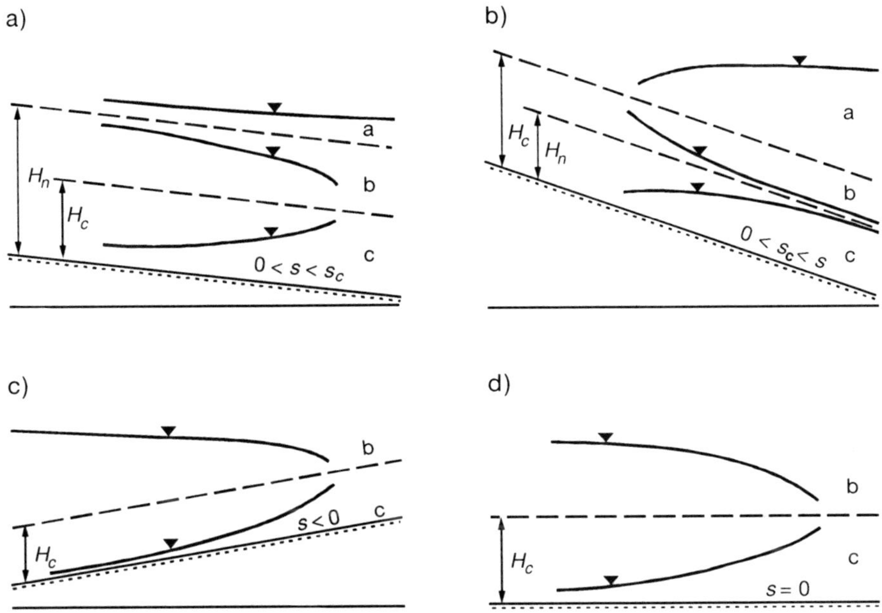

Fig. 4.1 Typical forms of the water flow profiles resulting from the analysis of Eq. (4.4) for channel with mild slope, steep slope, inverse slope and horizontal bed

Some of the presented profiles are important for engineering practice, such as the backwater curve presented in Fig. 4.2.

Others have rather theoretical significance, although they can occur locally in particular circumstances, as it is shown in Figs. 4.3 and 4.4.

Note that the profiles shown in Fig. 4.1 were deduced from the analysis of the relations between the variables involved in Eq. (4.4) but not from its direct solution. To summarize, Eq. (4.4) is rather not applied in modern hydraulics to solve practical problems, although years ago it was adapted to solve some particular cases (Chow 1959).

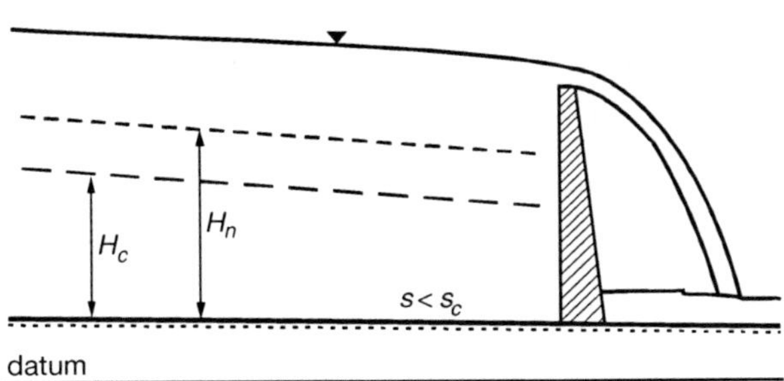

Fig. 4.2 Typical flow profile behind a dam

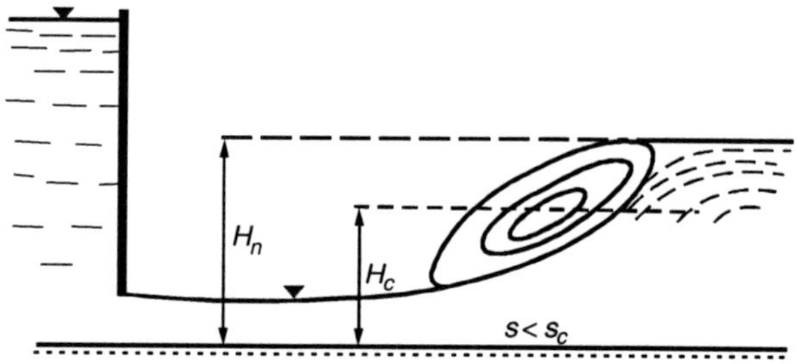

Fig. 4.3 Typical flow profiles at downstream of gate

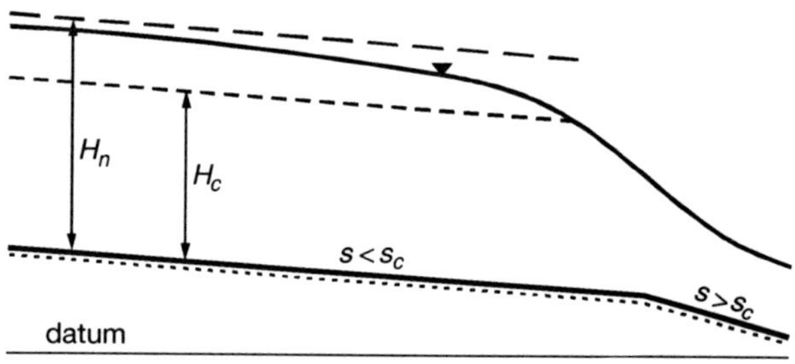

Fig. 4.4 Typical flow profile occurring between channels of mild slope and channel of steep slope

For practical application, when natural channels are considered, instead of Eq. (4.4) the discrete energy equation leading to the so-called step method is commonly used. Let us recall this approach.

A channel reach enclosed by two sections 1 and 2, as presented in Fig. 4.5, is taken into consideration. Neglecting the lateral inflow, i.e. for $q = 0$ and $Q_1 = Q_2$, one can write the following equation representing the energy conservation principle:

$$h_1 + \frac{\alpha \cdot Q^2}{2g \cdot A_1^2} = h_2 + \frac{\alpha \cdot Q^2}{2g \cdot A_2^2} + \Delta H_f \qquad (4.5)$$

where ΔH_f is the friction loss in the channel reach.

Sometimes in Eq. (4.5) an additional term representing the so-called eddy losses is introduced as well (French 1985). It takes into account the effects of expansion or contraction caused by sudden change of the cross-sectional area, which can be appreciable in non – prismatic channels. However, it seems that if these losses are neglected in the Saint Venant equations, then they can be omitted in the case of steady flow as well. If the cross-section area changes suddenly at any point in the considered channel, it should be treated as a point of discontinuity of $h(x)$, which divides the channel into two segments. At this point special equations connecting both segments must be introduced. They result from the mass and energy conservation principles. This problem is discussed in Section 4.2.4.

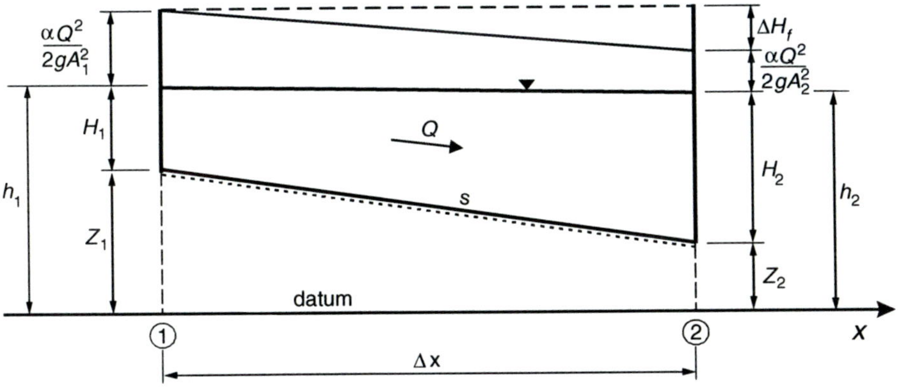

Fig. 4.5 Channel reach definition for the step method

The term ΔH_f is usually approximated using the arithmetic average of the friction slope over the considered channel reach. Then the following formula can be written:

$$\Delta H_f = \Delta x \cdot \overline{S} = \Delta x \frac{S_1 + S_2}{2} \tag{4.6}$$

where S_1 and S_2 are the friction slopes given by formula (4.3). Therefore they are calculated as follows:

$$S_1 = \frac{(n_M)_1^2 \cdot Q^2}{R_1^{4/3} \cdot A_1^2}, \tag{4.7a}$$

$$S_2 = \frac{(n_M)_2^2 \cdot Q^2}{R_2^{4/3} \cdot A_2^2} \tag{4.7b}$$

Other methods to estimate the average friction slope for a channel reach exist (French 1985). However, they will not be considered here. As it will be shown later, the use of arithmetic average is consistent with the discrete form of the differential equation (4.2) obtained using the implicit trapezoidal method.

Substitution of Eqs. (4.7a) and (4.7b) in Eq. (4.5) yields:

$$h_1 + \frac{\alpha \cdot Q^2}{2g \cdot A_1^2} = h_2 + \frac{\alpha \cdot Q^2}{2g \cdot A_2^2} + \frac{\Delta x}{2}\left(\frac{(n_M)_1^2 \cdot Q^2}{R_1^{4/3} \cdot A_1^2} + \frac{(n_M)_2^2 \cdot Q^2}{R_2^{4/3} \cdot A_2^2}\right) \tag{4.8}$$

It should be remembered, that the calculation of the flow profile proceeds always towards the direction for which the water surface asymptotically tends to the final level corresponding to the normal depth. Depending on the node numeration, it can be the direction of either increasing or decreasing indices. If for example the water level in cross-section number 2, is known, then h_1 can be easily calculated. Since Eq. (4.8) is a non-linear algebraic equation with regard to h_1, then to solve it one of the well-known methods as for example the bisection, secant or Newton method should be applied. If h_1 is calculated, one can pass to the next reach enclosed by the next pair of sections. This procedure, called the step method (Chow 1959, French 1985), is repeated until the criterion for convergence to the normal water level is satisfied.

While comparing the two presented approaches to solve the steady gradually varied flow, i.e. by the solution of the ordinary differential equation (4.4) and by the step method, one can notice some kind of inconsistency. For a prismatic channel the general approach, based on the unsteady flow equations, allowed to derive the equations for steady flow, whereas for a non-prismatic channel a different approach had to be used. This makes impression that the steady flow in open channel must be analyzed with different methods to reach the same goal, depending on the geometry of the channel. This fact has been noticed by other authors as well. For instance Roberson et al. (1998) explicitly distinguish the two possible approaches: "Modern methods for computing the water surface profile may be divided into two categories: methods based on the solution of the energy equation between different channel sections and methods based on the solution of the differential equation describing the rate of change of depth with distance". However, as the matter of fact, both approaches are equivalent. This coincidence will be shown in the next section.

4.1.3 Formulation of the Initial and Boundary Value Problems for Steady Flow Equations

The water profile along channel axis in steady gradually varied flow can be formally obtained by solving one of the previously presented ordinary differential equations. To this end a proper problem of their solution must be formulated. For practical applications it is more convenient to write the ordinary differential equations with regard to the water stage $h(x)$ than with regard to the depth $H(x)$. For example, given the flow discharge Q the function $h(x)$ can be obtained via the solution of the so-called initial value problem for ordinary differential equation. Such situation is illustrated in Fig. 4.6.

In a channel, which carries the discharge Q, the water level was raised owing to a dam from the level h_n, corresponding to the normal depth, to the assumed level h_0. Then behind a dam the backwater curve occurs (Fig. 4.6). This curve can be calculated on condition that its value h_0 at control $x = 0$ is known. In such a way it is possible to reproduce all the typical water profiles showed in Fig. 4.1, which can occur in open channel.

Sometimes the flow profile must be determined in another way. If at both ends of open channel the water levels are imposed and the flow discharge in channel is unknown, the flow profile can be obtained via the solution of the so-called boundary problem (or two points boundary value problem) for ordinary differential equations.

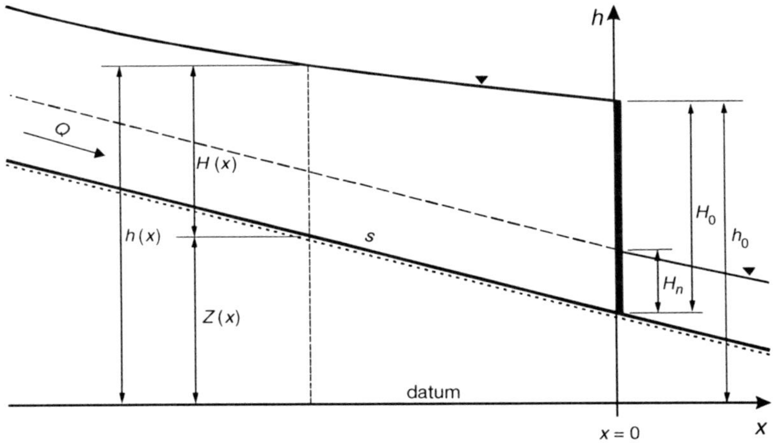

Fig. 4.6 Backwater curve behind a dam – an example of the initial value problem for ordinary differential equation

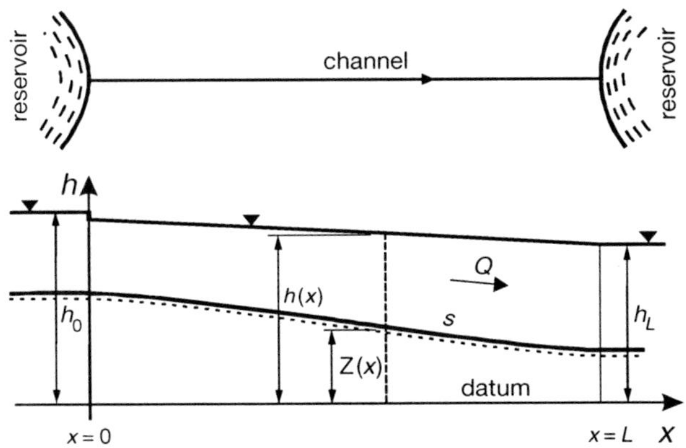

Fig. 4.7 Channel connecting two reservoirs having constant water levels

This problem has been already presented in Section 3.3. Such a situation takes place when a channel of length L connects two reservoirs, as it is shown in Fig. 4.7. If the water levels in reservoirs are time – invariant, the curve $h(x)$ must satisfy the steady gradually varied flow equations in the domain $0 \leq x \leq L$ and two additional conditions imposed at both ends of the channel.

Another particular set of problems occur for steady gradually varied flow in open channel network. These questions will be taken up in Section 4.4.

4.2 Numerical Solution of the Initial Value Problem for Steady Gradually Varied Flow Equation in a Single Channel

4.2.1 Numerical Integration of the Ordinary Differential Equations

The steady flow profile can be obtained using a number of relatively simple graphical or analytical approaches, which do not involve numerical computations. However, these methods are limited to prismatic channels, and thus they will not be considered here. Their detailed presentation can be found in French (1985). In this chapter we shall focus on a more general approach, suitable for channels with arbitrary cross-section geometry.

A general mathematical model of steady flow is given by the previously presented system of the ordinary differential equations (4.1) and (4.2), which with Eq. (4.3) is rewritten as follows:

$$\frac{dQ}{dx} = q, \tag{4.9}$$

$$\frac{d}{dx}\left(h + \frac{\alpha \cdot Q^2}{2g \cdot A^2}\right) = -\frac{n_M^2 \cdot Q^2}{R^{4/3} \cdot A^2} - \frac{\alpha \cdot Q}{g \cdot A^2}q \tag{4.10}$$

These equations describe the flow profile $h(x)$ and the discharge $Q(x)$ along the channel axis. To calculate these functions for a single channel an appropriate problem of solution for Eqs. (4.9) and (4.10) should be formulated.

Let us assume a channel of length L, divided by N nodes (at which the channel cross-sections are known), into intervals of length $\Delta x_i = x_{i+1} - x_i$. When the discharge Q at any channel cross-section is given and the distribution of lateral inflow along channel is known, Eq. (4.9) can be solved immediately. Its integration over the interval Δx_i yields:

$$Q_{i+1} = Q_i + \int_{x_i}^{x_{i+1}} q(x) \cdot dx \tag{4.11}$$

This formula allows us to determine the discharges at all nodes along the considered channel reach. To this order the integral in Eq. (4.11) is calculated numerically using any quadrature formula, such as the rectangular or trapezoidal rule. Knowing the discharges Q_i ($i = 1, 2, \ldots, N$) one can solve Eq. (4.10). For this purpose the following initial value problem is formulated: calculate the water flow profile $h(x)$, which satisfies Eq. (4.10) and the initial condition $h(x = x_0) = h_0$ imposed at control end of the considered channel reach.

There exist numerous methods to solve the initial value problem for the ordinary differential equations (Ascher and Petzold 1998, Le Veque 2007, Press et al. 1992, Stoer and Bulirsch 1980). Some of them were presented in Chapter 3. As it was explained, the cross sections of an open channel obtained by field measurements are distributed along channel in a non-uniform way, so the lengths of intervals Δx_i vary. Consequently it is better to avoid such methods of solution, which need an interpolation between neighboring nodes as for instance the methods of Runge–Kutta. Such type of methods can be applied rather for prismatic channels in which interpolation is possible. Similarly, multi-step methods should be avoided, as they are suitable rather for uniformly spaced grid points. For natural rivers with variable cross-sections only the Adams–Moulton method of the lowest order should be applied, since it uses the cross-sectional parameters at the grid points only. For these reasons the implicit trapezoidal rule (3.22) seems to be the best choice:

$$y_{i+1} = y_i + \frac{\Delta x}{2}(y'_i + y'_{i+1}). \tag{4.12}$$

where:

y_i, y_{i+1} are the values of the unknown function at nodes i and $i + 1$ respectively,
y'_i, y'_{i+1} are the values of the derivatives of $y(x)$ at nodes i and $i + 1$ respectively,
Δx is the integration step.

The trapezoidal rule, being an implicit method, has two important advantages:

- it ensures approximation of 2nd order of accuracy,
- it requires data from cross – sections i and $i + 1$ only, thus it is especially suitable for integration with variable step size as it occurs in open channel.

Application of the formula (4.12) to solve Eq. (4.10) yields:

$$\left(h_{i+1} + \frac{\alpha \cdot Q_{i+1}^2}{2g \cdot A_{i+1}^2}\right) = \left(h_i + \frac{\alpha \cdot Q_i^2}{2g \cdot A_i^2}\right) + \\ - \frac{\Delta x_i}{2}\left(\frac{(n_M)_i^2 \cdot Q_i^2}{R_i^{4/3} \cdot A_i^2} + \frac{\alpha \cdot Q_i}{g \cdot A_i^2} q_i + \frac{(n_M)_{i+1}^2 \cdot Q_{i+1}^2}{R_{i+1}^{4/3} \cdot A_{i+1}^2} + \frac{\alpha \cdot Q_{i+1}}{g \cdot A_{i+1}^2} q_{i+1}\right) \tag{4.13}$$

where:

i – index of cross-section,
Δx_i – length of interval number i.

Note that if the index i increase as x decreases, then in Eq. (4.13) the step size Δx_i should have negative sign.

Sometimes the lateral inflow $q(x)$ can be neglected. With $q_i = q_{i+1} = 0$ the discharge is $Q_i = Q_{i+1} =$ const. and Eq. (4.13) can be simplified:

$$\left(h_{i+1} + \frac{\alpha \cdot Q^2}{2g \cdot A_{i+1}^2}\right) = \left(h_i + \frac{\alpha \cdot Q^2}{2g \cdot A_i^2}\right) - \frac{\Delta x_i}{2}\left(\frac{(n_M)_i^2 \cdot Q^2}{R_i^{4/3} \cdot A_i^2} + \frac{(n_M)_{i+1}^2 \cdot Q^2}{R_{i+1}^{4/3} \cdot A_{i+1}^2}\right) \tag{4.14}$$

Let us compare Eq. (4.14) with Eq. (4.8). It is easy to find out that they are identical. It means that the ordinary differential equation (4.10) solved by the trapezoidal rule coincides with the discrete energy equation (4.8), usually applied to calculate the water profile in natural channels. Note that Eq. (4.14) has been derived directly from the most general mathematical model of 1D unsteady flow in open channel, i.e. the Saint Venent equations. Therefore any special treatment of the steady gradually varied flow in natural channels is not needed. On the other hand, any special procedures for numerical integration of Eq. (4.4) are not needed as well.

Equation (4.14) should be solved either for prismatic or non-prismatic channel by one of the methods previously listed for Eq. (4.8). In such a way it is possible to obtain the flow profiles corresponding to all zones distinguished by the theoretical analysis of Eq. (4.4) and presented in Fig. 4.1.

4.2.2 Solution of the Non-linear Algebraic Equation Furnished by the Method of Integration

The goal of the applied approach is to determine the water flow profile for specified depth imposed at the downstream end. Since all variables in Eq. (4.14) corresponding to the node *i* are known, this algebraic equation contains only one unknown h_{i+1}. However as h_{i+1} is involved in other parametrs, it cannot be expressed explicitly. To solve the problem Eq. (4.14) is reformed as follows:

$$f(h_{i+1}) = \left(h_{i+1} + \frac{\alpha \cdot Q^2}{2g \cdot A_{i+1}^2}\right) - \left(h_i + \frac{\alpha \cdot Q^2}{2g \cdot A_i^2}\right) + \frac{\Delta x_i}{2}\left(\frac{(n_M)_i^2 \cdot Q^2}{R_i^{4/3} \cdot A_i^2} + \frac{(n_M)_{i+1}^2 \cdot Q^2}{R_{i+1}^{4/3} \cdot A_{i+1}^2}\right) \tag{4.15}$$

The value of h_{i+1} that makes $f(h_{i+1}) = 0$ is the root of Eq. (4.15) and is the solution to the problem. To obtain it one of the numerical methods for determining roots of the non-linear algebraic equations described in Section 2.1 must be applied.

Numerical experiments provide us with interesting insights into the properties of Eq. (4.15). Apparently no difficulties occur for the standard problem of backwater curve behind a dam, where the intervals Δx are of the order of hundreds meters. However, for other type of flow profiles presented in Fig. 4.1 some computational problems can arise. For instance it appears that starting from the same initial condition $H_0 = H_c$, i.e. when the critical depth is imposed at $x = 0$, two different physically admissible solutions can be obtained. On the other hand, sometimes the correct solution cannot be obtained at all. The possible existence of many solutions results from the existence of multiple roots of the function $f(h_{i+1})$.

In order to investigate these effects in a more detailed way, let us consider steady gradually varied flow in a trapezoidal channel with $b = 3.5$ m, $n_M = 0.015$, $m = 1.5$, $s = 0.001$ that carries a discharge $Q = 4$ m^3/s. To simplify interpretation of the results of calculations let us change the independent variable. Let us use the depth H_{i+1} instead of the water level h_{i+1}. The objective is to determine the water flow profiles for various characteristic depths imposed at the downstream end as initial condition.

For the assumed data the normal depth satisfying the Manning formula:

$$Q = \frac{1}{n_M} R^{2/3} \cdot s^{1/2} \cdot A \tag{4.16}$$

solved by the false position method (see Section 2.1.3) is equal to $H_n = 0.664$ m, whereas the critical depth satisfying the equation:

$$\frac{Q^2}{g} = \frac{A^3}{B} \tag{4.17}$$

and solved by the same method, is equal to $H_c = 0.490$ m.

The flow profiles are calculated for the following initial conditions imposed at $x = 0$:

- $H_0 = 1.00$ m ($H_0 > H_n$ and $H_0 > H_c$)
- $H_0 = 0.55$ m ($H_0 < H_n$ and $H_0 > H_c$)
- $H_0 = 0.45$ m ($H_0 < H_n$ and $H_0 < H_c$).

An analysis of the function $f(H_{i+1})$ concerns the first reach of channel, which is bounded by the cross-sections 0 and 1. So it will deal with the function $f(H_1)$ only.

It is well known that while solving any non-linear algebraic equation, at first the searched root must be separated. In considered Eq. (4.15) all possible roots of function $f(H_1)$ exist in the domain of positive real numbers. In the interval defined by two endpoints, only one root may exist. Afterwards a suitable method for determination a single root is applied. To check the number of roots and their positions the graph of $f(H_1)$ should be drawn.

The plots of function $f(H_1)$ obtained for the assumed data are presented in Figs. 4.8, 4.10 and 4.11.

It appears that for the imposed initial value of H_0 the shape of the function $f(H_1)$ strongly depends on the value of the stepsize Δx. One can notice, that depending on the assumed values of H_0 and Δx there are one, two or three roots of the function (4.15) in the domain of real positive numbers. Therefore, while solving Eq. (4.15)

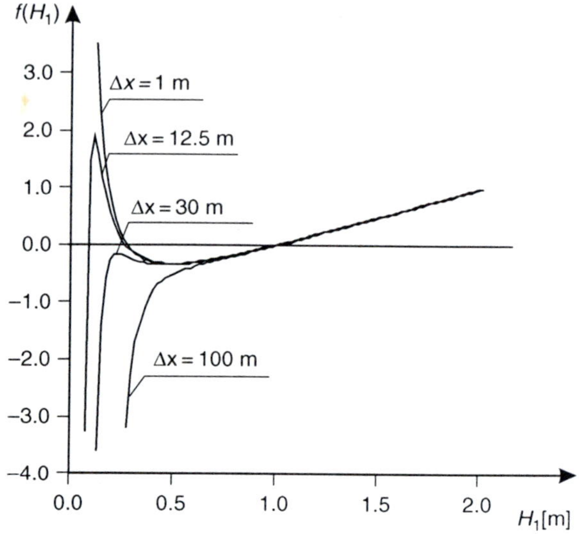

Fig. 4.8 Function $f(H_1)$ for various values of stepsize Δx with $s = 0.001$ and $H_0 = 1.00$ m

the separation of the roots must be done very thoroughly. The single root should be properly selected and then its value must be approximated with sufficient accuracy. Otherwise one can find a false root and consequently obtain an unphysical solution.

The presented plots of $f(H_1)$ show why only some attempts of computations can be successful, while others can fail or to deliver false solutions. The crucial point during the computational process is to choose the proper value of stepsize Δx, since as results form the presented figures, it determines both the shape of function $f(H_1)$ and the number of its roots. The significance of this parameter is due to the fact that one of the most important terms of Eq. (4.7) representing the losses of energy is governed by the stepsize Δx. For example, Fig. 4.8 shows that in the case of a large stepsize applied for the subcritical flow ($H_0 = 1.00$ m $> H_c$) only one root exists. Then the solution will be always reached without any difficulties. On the other hand, for small values of Δx one or even two additional roots occur. However both of them are physically inadmissible. The appropriate root can be identified taking into account that the extreme minimum value of $f(H_{i+1})$ roughly corresponds to the critical depth, which in this case is equal to $H_c = 0.49$ m. This is clearly shown in Fig. 4.9.

Similar results to the ones presented in Fig. 4.8 are also obtained for the following initial condition: $H_c = 0.49$ m $< H(x = 0) = H_0 = 0.55$ m $< H_n = 0.664$ m. They are shown in Fig. 4.10.

One should remember that while the calculations proceed from the cross section i to the cross section $i + 1$ ($i = 1, 2, 3, \ldots$), the depth H_{i+1} varies. Therefore at the same time the form of the function $f(H_{i+1})$ varies as well. For this reason in every cross-section, a separation of the interval containing the searched root must de done very thoroughly.

Fig. 4.9 Function $f(H_1)$ for step size $\Delta x = 1.0$ m with $s = 0.001$ and $H_0 = 1.0$ m has its minimum point at $H_1 \approx H_c = 0.49$ m

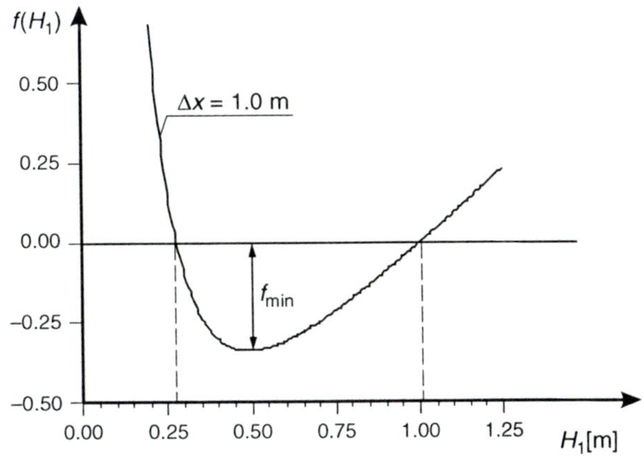

Fig. 4.10 Function $f(H_1)$ for various values of step size Δx with $s = 0.001$ and $H_0 = 0.55$ m

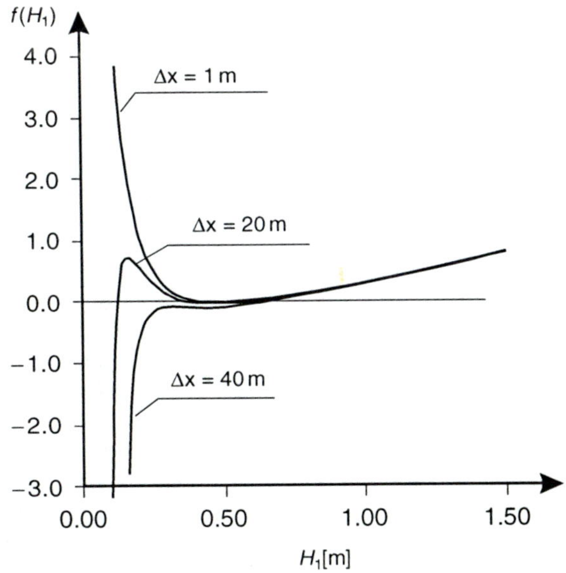

If the imposed depth H_0 is greater than critical depth, as in the considered case, the root at the right-hand side of the extreme point should be chosen as the proper one. If we choose one of the roots from the left-hand side of the minimum, the calculated solution will have unphysical character.

Figure 4.11 shows a situation opposite to the ones presented in Figs. 4.8 and 4.10. One can see that with too large step size applied for the supercritical flow ($H_0 = 0.45$ m $< H_c$) it is impossible to obtain a physically reasonable solution. On the other hand, for small values of Δx two roots occur. However, only one of them is physically admissible. To choose the suitable root one can use previously presented information on the minimum point of $f(H_{i+1})$, which corresponds to the

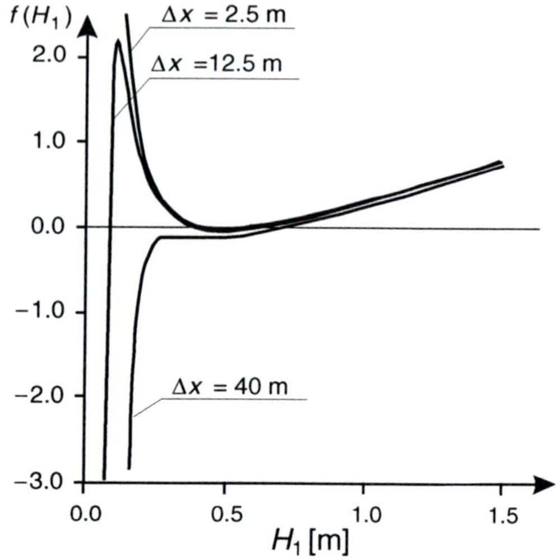

Fig. 4.11 Function $f(H_1)$ for various values of step size Δx with $s = 0.001$ and $H_0 = 0.45$ m

critical depth. Therefore if the imposed depth H_0 is less than critical one, as in this case, the root at left side of extreme point should be chosen. Conversely, if H_0 is greater than H_c we have to choose the one from the right-hand side.

The influence of the stepsize on the number of roots is a very important conclusion, since in the numerical analysis it is usually considered that the step of integration influences only the accuracy and stability of the numerical solution. The presented results suggest that in some cases the correct solution can be obtained only if additional conditions on the stepsize are satisfied.

From the above discussed properties of the function $f(H_{i+1})$ results that it is possible to calculate numerically all types of the flow profiles, which can occur during steady varied flow in open channel. The calculations can be carried out for both artificial and natural channels, since the approach based on the ordinary differential energy equation is capable of predicting the depth of flow at arbitrary longitudinal distances. This feature is an important advantage as in rivers the distances between neighboring cross-sections are variable. However the numerical solution of the non-linear equations should be performed especially thoroughly. The function $f(H_{i+1})$ given by Eq. (4.15) can have one, two or three roots, which in addition can move along axis. Some of them are physically inadmissible. For these reasons the separation of the interval containing the single root must be carried out very precisely. To determine the root within an interval the bracketing methods should be used, rather than the open ones. Thus, the bisection and false position methods will be more reliable than the Newton and simple fixed-point iteration methods. The latter methods can be applied on condition that the interval containing the correct root is sufficiently narrow, so that $f(H_{i+1})$ varies monotonically over the interval.

4.2.3 Examples of Numerical Solutions of the Initial Value problem

Example 4.1 A trapezoidal channel with $b = 3.5$ m, $n_M = 0.015$, $m = 1.5$, $s = 0.001$ carries a discharge $Q = 4$ m^3/s. Determine the water flow profiles for various characteristic depths imposed at the downstream end.

For the assumed data the normal depth satisfying the Manning formula is equal to $H_n = 0.664$ m, whereas the critical depth satisfying the Equation (4.16) is equal to $H_c = 0.490$ m. The flow profiles were computed for $h(x = 0)$ corresponding to the depths $H_0 = 1.0$ m $> H_n$, $H_0 = 0.664$ m $= H_n$ and $H_0 = 0.40$ m $= H_c$. They are shown in Fig. 4.12. As it could be expected, the profiles obtained by numerical solution have typical forms resulting from the theoretical analysis of Eq. (4.4) and presented in Fig. 4.1. The plots correspond to case A – zones a and b. The backwater curve calculated for zone c, i.e. for supercritical flow, is discussed later.

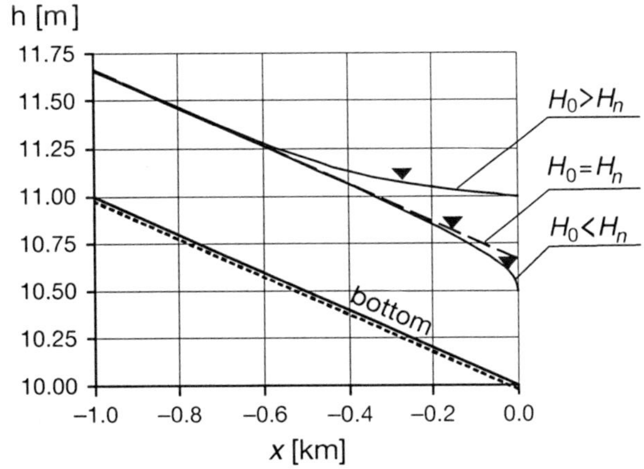

Fig. 4.12 Steady gradually varied flow profiles in a channel with uniform longitudinal slope for various depths H_0 imposed at $x = 0$

Example 4.2 A trapezoidal channel with $b = 3.5$ m, $n_M = 0.015$, $m = 1.5$ and horizontal bed ($s = 0.0$) carries a discharge $Q = 4$ m^3/s. Determine the water flow profiles for various depths imposed at the downstream end.

The calculation performed for $H_0 = H_c$, corresponding to the free overfall at the channel end gave the flow profile in the form of a typical drawdown curve obtained for the critical depth imposed at downstream end (Fig. 4.13). Similar, but more gentle drawdown curve results for a greater value of H_0 imposed as the initial condition at $x = 0$.

Example 4.3 A trapezoidal channel with $b = 3.5$ m, $n_M = 0.015$, $m = 1.5$ and having adverse bed slope $s = -0.001$ carries a discharge $Q = 4$ m^3/s. Determine the water flow profile assuming that the channel terminates in a free overfall, i.e. the critical depth is imposed at the downstream end.

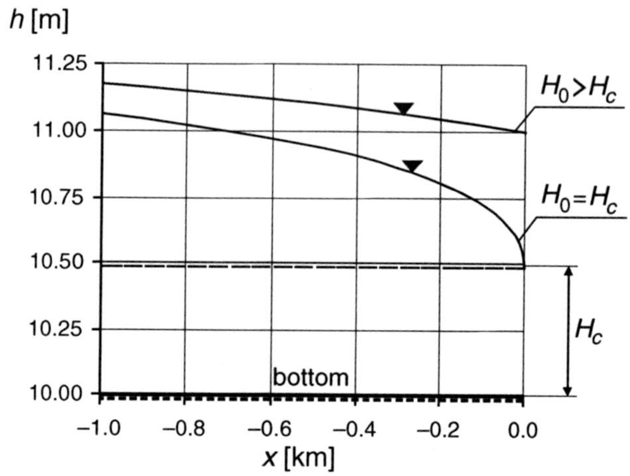

Fig. 4.13 Steady gradually varied flow profiles in a channel with horizontal bed for various depth H_0 imposed at $x = 0$

Fig. 4.14 Steady gradually varied flow profiles in a channel with adverse slope for $H_0 = H_c$ imposed at $x = 0$

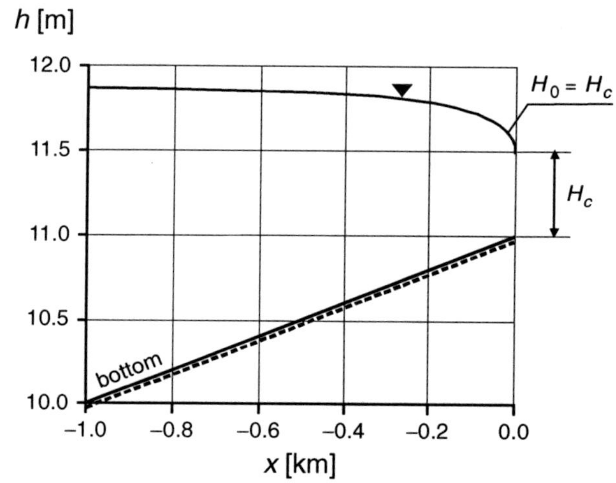

In Fig. 4.14 one can see that computed drawdown curve corresponds to the one presented in Fig. 4.3 (case C, zone b).

Example 4.4 A trapezoidal channel with $b = 3.5$ m, $n_M = 0.010$, $m = 1.5$ and a steep bed slope $s = 0.005$ carries a discharge $Q = 4$ m^3/s. Determine the water flow profile for the following initial conditions, imposed at $x = 0$:

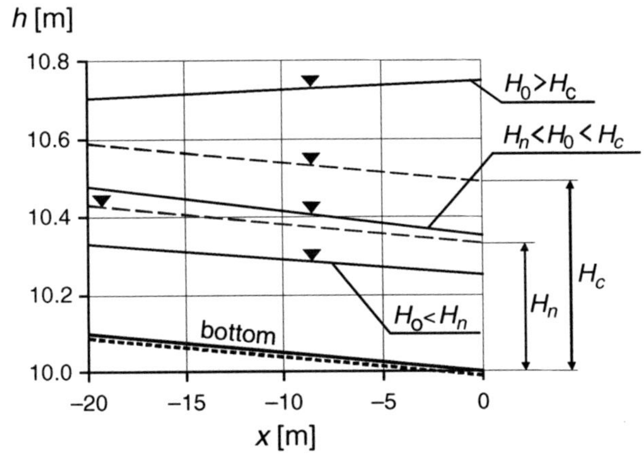

Fig. 4.15 Steady gradually varied flow profile in channel with steep slope for various depths imposed at $x = 0$

- $H_0 = 0.75$ m $> H_c$,
- $H_n < H_0 = 0.35$ m $< H_c$,
- $H_0 = 0.25$ m $< H_n$.

The considered type of flow corresponds to case B in Fig. 4.1. To obtain the results displayed in Fig. 4.15 small stepsize equal to $\Delta x = 1.0$ m had to be applied. Note that the flow profiles for all three zones a, b, and c can be reproduced using the same approach.

Example 4.5 An interesting case arises when the critical depth is imposed as the initial condition. To illustrate this problem let us consider steady flow in the channel from Example 4.4. It is a channel with mild slope $s = 0.001 < s_c$ for which the critical depth is equal to $H_c = 0.49$ m. This value is imposed at $x = 0$ as the initial condition $H_0 = H_c$. Let us plot the function $f(H_1)$ given by Eq. (4.15). Figure 4.16 presents the values of the function for the stepsize $\Delta x = 0.25$ m. One can see that there are two roots, both located in the vicinity of the minimum point of $f(H_1)$, which is at $H_1 = H_c$. Therefore a question arises, which root should be taken into consideration. The choice of the root depends on the type of curve which is required. If the curve in zone b (Fig. 4.1, case A) is searched, then the root at the right-hand side of the minimum point must be chosen. Conversely, if the drawdown curve from zone c is computed, the other root should be taken. Summarizing, the initial condition $H_0 = H_c$ can generate the backwater (zone c) or drawdown (zone c) curve depending on the chosen root. The two possible profiles are presented in Fig. 4.17.

The presented examples show that practically all possible forms of the flow profile displayed in Fig. 4.1 can be reproduced by the numerical solution of the governing equation for the steady gradually varied flow (4.2).

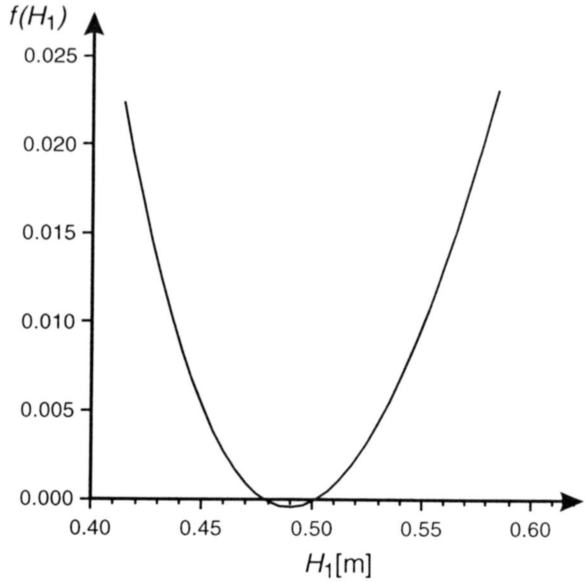

Fig. 4.16 Plot of the function $f(H_1)$ for initial condition $H_0 = H_c = 0.49$ m and $\Delta x = 0.25$ m

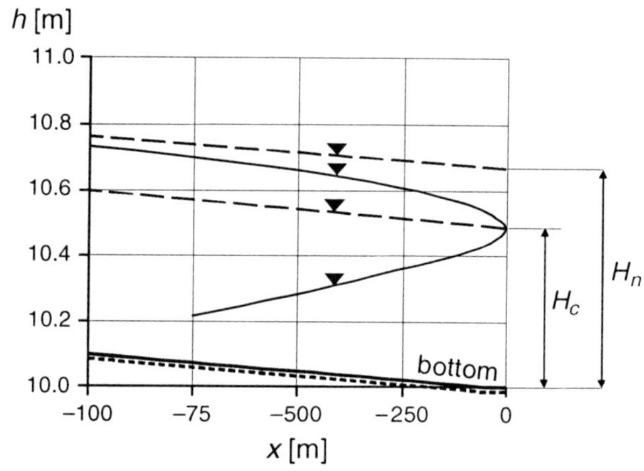

Fig. 4.17 Two different flow profiles computed with $\Delta x =$ 0.25 m for the same initial condition $H_0 = H_c$ at $x = 0$

4.2.4 Flow Profile in a Channel with Sudden Change of Cross-Section

Sometimes the backwater curve should be computed for a channel composed of two reaches with a sharp discontinuity in the cross-section geometry. In such a case a significant local change of the water level can occur. The considered situation is presented in Fig. 4.18.

Let us assume that the computation of the backwater curve proceeds step by step from the downstream end, where the initial condition is imposed, towards the upstream end. However sometimes it is impossible to continue the computations

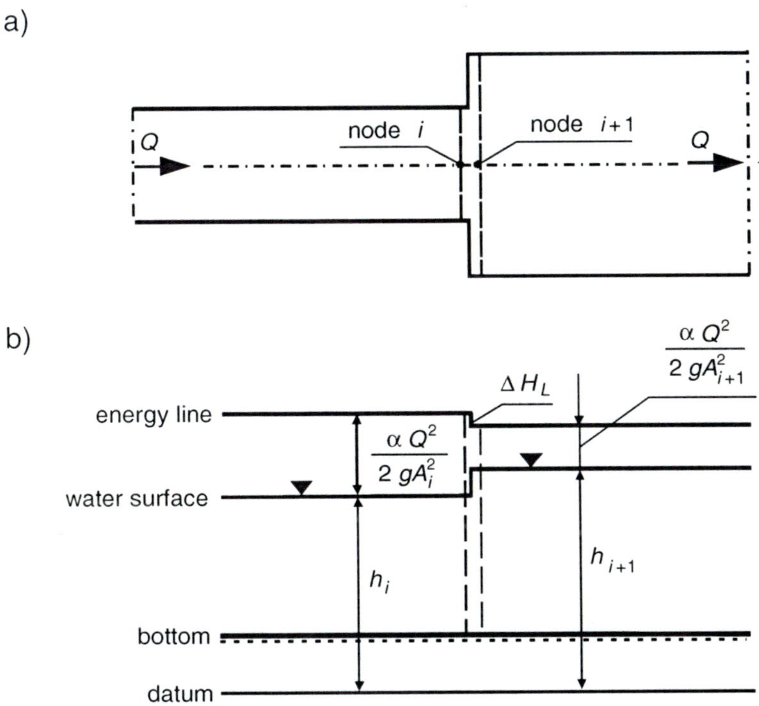

Fig. 4.18 Channel changing a cross-section area

using the same algorithm. It can occur in the junction of two segments having significantly different cross-sections. To overcome the problem, one has to introduce two computational cross-sections, located very close to each other at the two sides of the discontinuity (Fig. 4.18).

The equation relating water stages and discharges in both segments results from the energy conservation principle written for the two cross-sections. It has the following form:

$$\left(h_i + \frac{\alpha \cdot Q^2}{2g \cdot A_i^2}\right) = \left(h_{i+1} + \frac{\alpha \cdot Q^2}{2g \cdot A_{i+1}^2}\right) + \Delta H_L \tag{4.18}$$

where

ΔH_L – local energy loss,
i – index of cross-section at left side of junction,
$i+1$ – index of cross-section at right side of junction.

In open channels the local losses are usually neglected, so Eq. (4.18) becomes:

$$\left(h_i + \frac{\alpha \cdot Q^2}{2g \cdot A_i^2}\right) = \left(h_{i+1} + \frac{\alpha \cdot Q^2}{2g \cdot A_{i+1}^2}\right) \tag{4.19}$$

Note that Eq. (4.19) is equivalent to Eq. (4.14) applied in the step method, if in the latter $\Delta x_i = 0$ is assumed. Thus, in order to compute the flow profile for a channel with sudden change in cross-section parameters, only a slight modification of the algorithm presented in Section 4.2.2 is required.

Example 4.6 A channel with constant bed slope $s = 0.001$ and carrying a discharge $Q = 4$ m³/s is composed of two segments of lengths L_1 and L_2. Each part has a trapezoidal cross-section characterized by different parameters:

– part 1: $L_I = 550$ m, $b_I = 3.5$ m, $n_{M1} = 0.015$, $m_1 = 1.5$,
– part 2: $L_2 = 450$ m, $b_2 = 2.0$ m, $n_{M2} = 0.015$, $m_2 = 1.4$.

Determine the water flow profile for the water level $h(x = 0) = h_0$ imposed at the downstream end.

The normal depths satisfying the Manning formula are $H_{n1} = 0.664$ m for the first part of channel and $H_{n2} = 0.857$ m for the second one. The imposed $h(x = 0)$ corresponds to the depth $H_0 = 1.50$ m $> H_{n1}$. The results obtained for $\Delta x = 50$ m = const. are shown in Fig. 4.19.

As it could be expected, in the first segment of the channel, a typical backwater curve was obtained. In the node $x = -550$ m a discontinuity of the water surface occurred, resulting from the sudden change of the channel parameters. Since the velocity head in the first part is smaller than in the second one, the water surface falls while passing between channel segments. Afterwards the flow profile tends asymptotically to H_{n2}, which is the normal depth in the second segment of channel.

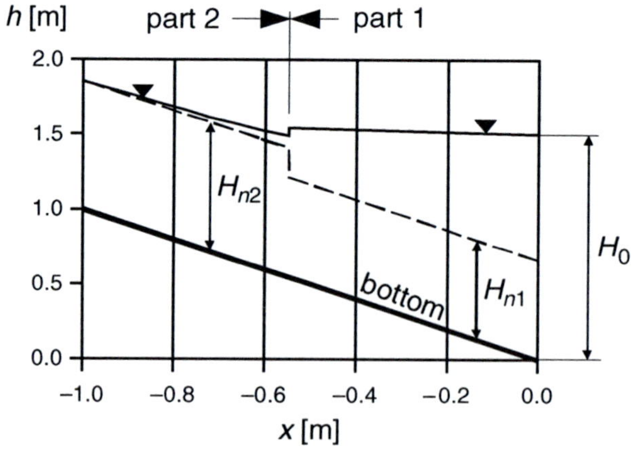

Fig. 4.19 Steady gradually varied flow profile in channel with varying cross-sections for depth $H_0 = 1.5$ m $> H_{n1}$ imposed at $x = 0$

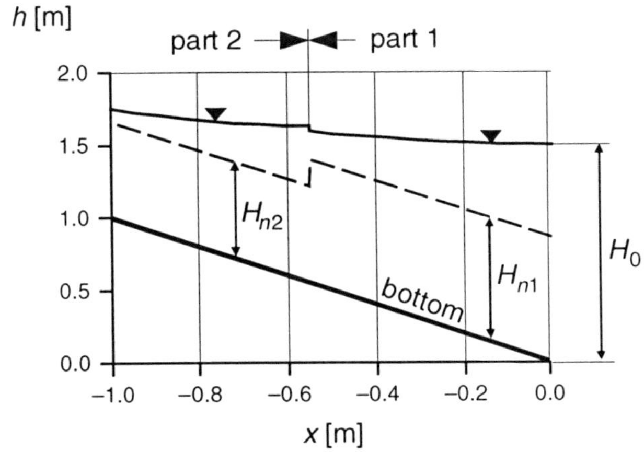

Fig. 4.20 Steady gradually varied flow profile in channel with varying cross-sections for depth $H_0 = 1.5$ m $> H_{n1}$ imposed at $x = 0$

In next example the same data as in the previous case is used, but the parameters b, n_M and m of the two segments are swapped.

The results of computation are presented in Fig. 4.20. This time the velocity head in the first part is larger than in the second one, so that when moving towards the upstream end one can observe a significant rising of the water surface at the junction of the segments.

4.2.5 Flow Profile in Ice-Covered Channel

In cold climate regions with long seasons of low temperature the rivers can be covered by ice layer. The appearance of the ice cover radically changes the flow regime. Flow with free surface becomes flow in closed conduit. Consequently the velocity distribution over a cross-section becomes similar to the one in pipes and the energy losses increase remarkably. Additionally, variable coefficient of roughness occurs along the wetted perimeter. Due to the ice cover the flow profile is modified. Comparing with the free surface flow the same discharge Q causes increase of the water levels especially in the upstream part of channel, so that the backwater curve becomes much longer. The knowledge of the new position of the water surface is of great importance for the design of flood banks.

The flow profile in an ice covered channel can be computed using the same equations for the steady gradually varied flow and a similar solution algorithm. The modifications concern the calculation of the cross-sectional parameters and the equivalent Manning roughness coefficient, which takes into account the additional friction coming from the ice cover.

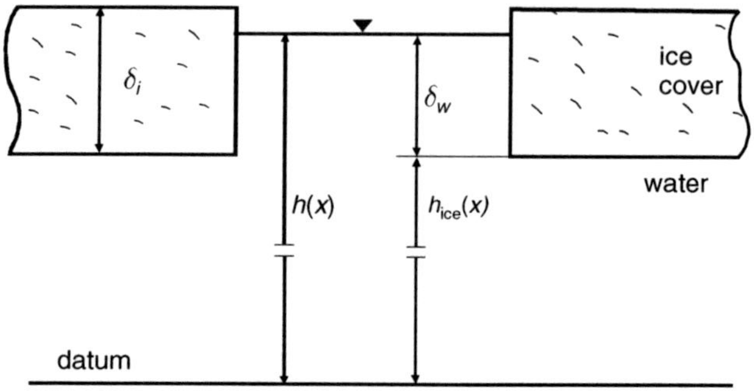

Fig. 4.21 View of the cross-section of ice layer

Let us assume that the thickness of ice layer at the channel surface is equal to δ_{ice}. In a hole made in the ice layer, as in Fig. 4.21, the surface of water is below the ice surface.

The difference is caused by different specific weights of ice and water. Their ratio is $\gamma_{\text{ice}}/\gamma_w = 0.92$. Therefore, the bottom of the ice layer is located in water at the depth:

$$\delta_w = 0.92 \cdot \delta_{\text{ice}} \tag{4.20}$$

whereas the elevation of the bottom of the ice layer above the accepted datum is:

$$h_{\text{ice}}(x) = h(x) - 0.92 \cdot \delta_{\text{ice}} \tag{4.21}$$

where:

δ_{ice}– thickness of ice layer,
δ_w– submersion of ice layer in water,
$h(x)$– elevation of the free water surface above the assumed datum,
$h_{\text{ice}}(x)$ – elevation of the bottom of ice layer above the assumed datum.

While solving Eq. (4.14) the unknown function is $h(x)$, as previously. However, all the cross-sectional parameters have to be calculated for $h_{\text{ice}}(x)$.

The next modification concerns the Manning roughness coefficient. Since one part of the wetted perimeter corresponds to the channel bed and the other one to the ice bottom, then as in other 1D flow models, a single equivalent roughness coefficient must be introduced. To this order the relatively simple Belokon formula is applied (French 1985):

$$(n_M)_e = (n_M)_b \left(1 + \frac{p_{\text{ice}}}{p_b}\left(\frac{(n_M)_{\text{ice}}}{(n_M)_b}\right)^{3/2}\right)^{2/3} \tag{4.22}$$

where:

$(n_M)_e$– equivalent roughness coefficient,
$(n_M)_b$– roughness coefficient associated with the bed,
$(n_M)_{\text{ice}}$– roughness coefficient associated with bottom of ice cover,
p_{ice}– wetted perimeter associated with ice cover,
p_b– wetted perimeter associated with the bed.

Example 4.7 A trapezoidal channel with $b = 3.5$ m, $m = 1.5$, $s = 0.001$ carries a discharge $Q = 4$ m^3/s. The Manning roughness coefficient taken for channel bed is equal to $(n_M)_b = 0.024$. The normal depth satisfying the Manning formula is $H_n = 0.664$ m. Determine the water flow profiles for two different cases:

- flow with free surface,
- flow with ice cover,

assuming in each case the same initial condition imposed at the downstream end, i.e. $h(x = 0)$ corresponding to the depth $H_0 = 2.0$ m $> H_n = 0.664$ m. The thickness of the ice cover is constant, $\delta_{\text{ice}} = 0.25$ m, whereas the Manning coefficient for bottom surface of the ice takes the same value as for channel bed i.e. $n_b = 0.024$. It should be remembered, that the actual values of $(n_M)_{\text{ice}}$ can range in a relatively large interval, from 0.01 even to 0.10 (Gray and Prowse, 1993). The computations were carried out for $\Delta x = 50$ m. The flow profiles are shown in Fig. 4.22.

A significant difference between the two cases can be observed. The appearance of the ice cover at the water surface increases the friction loss and consequently also increases the flow depths. In this case, in which the data were assumed arbitrarily, the increase exceeds 30% in the upstream part of channel.

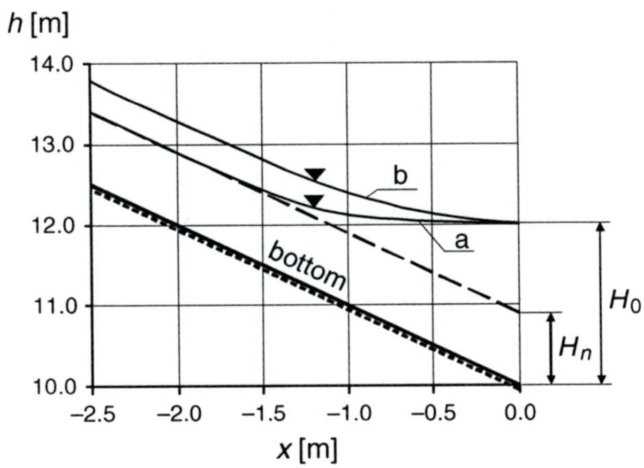

Fig. 4.22 The steady gradually varied flow profile in channel with free surface (**a**) and with ice cover (**b**) for initial depth $H_0 = 2.0$ m imposed at $x = 0$

4.3 Solution of the Boundary Problem for Steady Gradually Varied Flow Equation in Single Channel

4.3.1 Introduction to the Problem

Apart from the initial value problem for steady gradually varied flow equation, the boundary value problem can be formulated as well. The boundary value problem arises when the depths (or water stages) at the two endpoints of the considered channel are known, but the discharge Q is unknown. The solution of the boundary problem consists in the determination of the water flow profile between the two endpoints and the value of the discharge Q. A typical example of such a problem is the flow in a channel connecting two reservoirs with constant water levels (compare Section 4.1.3). A similar problem can be formulated for a river reach bordered by the upstream and downstream ends. If during the steady gradually varied flow the water levels are measured at both ends of considered reach, then the water flow profile $h(x)$ and the discharge Q can be obtained via the solution of the boundary value problem as well.

Another type of the boundary problem arises when the discharge and the depths at the endpoints are known, but the Manning roughness coefficient n_M is unknown. In this case, the solution of the problem results in the flow profile and the value of n_M (on condition that n_M is constant along the channel reach).

In practice the boundary problems can be solved using different approaches. The first one is based on the numerical integration of the unsteady flow equations, i.e. the system of Saint Venant equations. The equations are solved for the boundary conditions which asymptotically tend to the required constant values and after a sufficiently long time the steady state solution is reached. Cunge et al. (1980) showed that this system of equations approximated with the box scheme of the finite difference method in the case of steady flow does not coincide with the discrete Bernoulli equation used for computation of the steady flow in non-prismatic channels. Consequently the differences between both approaches can occur. It should be also remembered that while the discharge Q results directly from the solution of the Saint Venant equations, the coefficient n_M must be estimated by the trial and error method.

The second method of calculation of the flow profile and the values of Q or n_M is based on the solution of the initial value problem for steady flow. It is performed repetitively for the values of Q or n_M which are chosen by trial and error, until the condition imposed at the opposite channel end is satisfied. This approach is known as the shooting method (Bjorck and Dahlquist 1974). A simple variant of this method was applied by Chow (1959) to solve the steady flow in a channel connecting two reservoirs.

The third possible approach is to solve directly the boundary value problem. To this end the steady flow equation is discretized over the considered channel reach, for example using the finite difference method. This results in a system of algebraic equations involving the water depths at the cross-sections within the channel reach and the discharge Q (or the Manning coefficient n_M). The system is nonlinear and should be solved using an iterative method.

Although the boundary value problem for ordinary differential equations is not formulated in open channel hydraulic, however practically, it is solved using the trial and error approach. In certain cases such problem ought to be formulated, since it allows us to apply general well known mathematical methods to calculate some hydraulic problems as for example the steady flow profile and discharge in channel.

As the boundary value problem can be formulated for the ordinary differential equations of higher order than 1 or for the system of differential equations, we have to consider the system of equations (4.1) and (4.2):

$$\frac{dE}{dx} = -S - \frac{\alpha \cdot Q}{g \cdot A^2} q, \tag{4.23}$$

$$\frac{dQ}{dx} = q \tag{4.24}$$

After its solution the water flow and discharge are known. However one can assume another situation. Knowing the discharge Q and the water levels at both channel's ends, one can try to compute the water flow profile and the Manning roughness coefficient n_M as a solution of the boundary value problem as well. To this order a constant value of n_M must be assumed. Owing to this assumption and neglecting the lateral inflow q one can write formally the following system of ordinary differential equations:

$$\frac{dE}{dx} = -S, \tag{4.25}$$

$$\frac{dn_M}{dx} = 0 \tag{4.26}$$

which should be solved instead of Eqs. (4.23) and (4.24). An approach, in which a new variable is introduced and a new equation is added, is very often applied while solving the boundary value problem for the ordinary differential equations (Ascher and Petzold 1998, Press et al. 1992).

4.3.2 Direct Solution Using the Newton Method

To solve the boundary problem for the ordinary differential equations by the finite difference method, the channel of length $\langle 0, L \rangle$ is divided by N nodes into $N-1$ intervals Δx_i. For simplicity let us assume that the lateral inflow does not exist so we have $q = 0$. Equations (4.23) and (4.24) are approximated at the midpoint of each interval $x_i + \Delta x_i/2$ by a centered difference corresponding to the implicit trapezoidal rule, applied previously:

$$\frac{E_{i+1} - E_i}{\Delta x_i} = -\frac{1}{2}\left(S_i + S_{i+1}\right), \tag{4.27}$$

$$\frac{Q_{i+1} - Q_i}{\Delta x_i} = 0 \tag{4.28}$$

where:

i – index of cross section,
Δx_i – length of interval number i.

From Eq. (4.28) results that $Q_i = Q_{i+1} = Q = \text{const.}$ for $i = 1, 2, \ldots, N$, whereas introducing into Eq. (4.25) the energy E and the friction slope S defined by the expressions (4.2) and (4.3) respectively, yields:

$$\left(h_{i+1} + \frac{\alpha \cdot Q^2}{2g \cdot A_{i+1}^2}\right) - \left(h_i + \frac{\alpha \cdot Q^2}{2g \cdot A_i^2}\right) + \frac{\Delta x_i}{2}\left(\frac{n_M^2 \cdot Q^2}{R_i^{4/3} \cdot A_i^2} + \frac{n_M^2 \cdot Q^2}{R_{i+1}^{4/3} \cdot A_{i+1}^2}\right) = 0 \tag{4.29}$$

Note that formally this equation coincides with the well known discrete energy equation usually applied to calculate the water profile in natural channels. However in the case of the initial value problem, in Eq. (4.29) only one unknown exists. This allows us to calculate the water level step by step, marching along channel axis from cross-section to cross-section. When solving the boundary value problem, the water levels described by Eq. (4.29) must be solved simultaneously at all cross-sections, since they have to satisfy two imposed boundary conditions. To this order similar equations should be written for each interval $\Delta x_i (i = 1, 2, \ldots, N - 1)$. In such a way one obtains a system of $N - 1$ algebraic equations with $N + 1$ unknowns. There are N water levels h_i at nodes and flow discharge Q. When flow in a single channel is considered, this system has to be completed by two additional equations resulting from the imposed boundary conditions (Fig. 4.23).

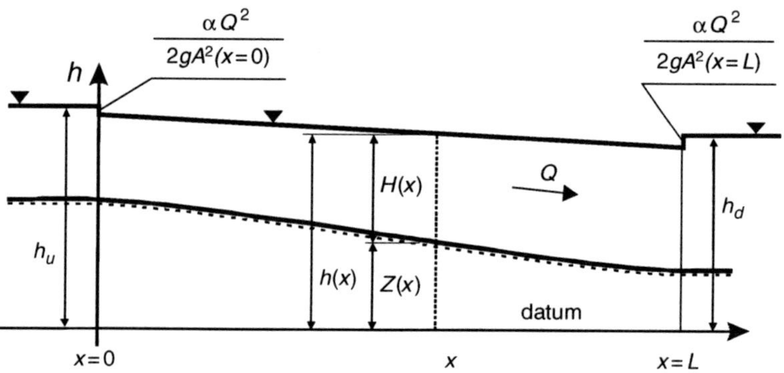

Fig. 4.23 Channel connecting two reservoirs having constant water levels

Assuming subcritical flow in the channel, the following conditions should be imposed at the ends of channel:

$$E(x=0) = h_1 = h_u - \frac{\alpha \cdot Q^2}{2g \cdot A_1^2} \tag{4.30}$$

$$E(x=L) = h_N = h_d - \frac{\alpha \cdot Q^2}{2g \cdot A_N^2} \tag{4.31}$$

where:

h_u, h_d – water levels imposed at upstream and downstream reservoir respectively,
h_1, h_N – water levels at upstream and downstream end of channel respectively.

Since the discharge Q is unknown, the boundary problem formulated above has non–linear boundary conditions.

The final system of equations can be written in matrix form:

$$\mathbf{AX} = \mathbf{B} \tag{4.32}$$

where:

A – matrix of coefficients,
$\mathbf{B} = (h_u,0,\ldots,0,h_d,0)^T$ – vector of right side,
$\mathbf{X} = (h_1,h_2,\ldots,h_{N-1},h_N,Q)^T$ – vector of unknowns,
T – transposition symbol.

The matrix **A** of dimensions $(N+1) \times (N+1)$ is very sparse (Fig. 4.24).
Its non-zero elements are defined as follows:

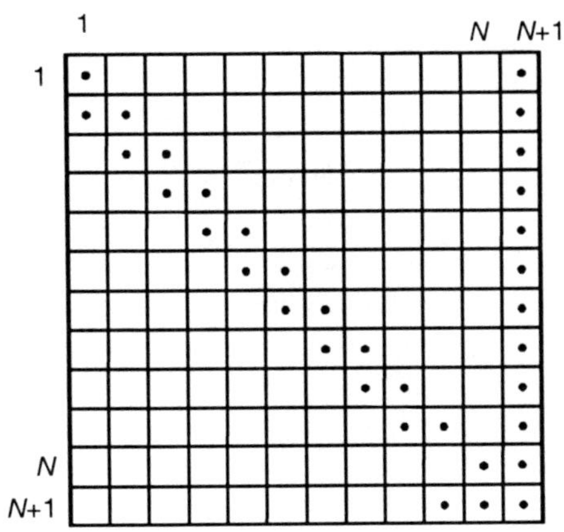

Fig. 4.24 Structure of matrix **A**

$$a_{1,1} = 1, \tag{4.33a}$$

$$a_{1,N+1} = \frac{\alpha \cdot Q}{2g \cdot A_1^2}, \tag{4.33b}$$

$$a_{i,i} = 1, \text{ for } i = 2,3, \cdots, N-1, \tag{4.33c}$$

$$a_{i,i-1} = -1, \text{ for } i = 2,3, \cdots, N-1 \tag{4.33d}$$

$$a_{i,\,N-1} = -\frac{\alpha \cdot Q}{2g \cdot A_{i-1}^2} + \frac{\alpha \cdot Q}{2g \cdot A_i^2} + \frac{\Delta x_{i-1}}{2}\left(\frac{n_M^2 \cdot Q}{R_{i-1}^{4/3} \cdot A_{i-1}^2} + \frac{n_M^2 \cdot Q}{R_i^{4/3} \cdot A_i^2}\right) \tag{4.33e}$$

for $i = 2,3,\cdots, N-1$,

$$a_{N,N} = 1, \tag{4.33f}$$

$$a_{N,N+1} = \frac{\alpha \cdot Q}{2g \cdot A_N^2} \tag{4.33g}$$

$$a_{N+1,\,N-1} = -1; \tag{4.33h}$$

$$a_{N+1,\,N} = 1 \tag{4.33i}$$

$$a_{N+1,N+1} = -\frac{\alpha \cdot Q}{2g \cdot A_{N-1}^2} + \frac{\alpha \cdot Q}{2g \cdot A_N^2} + \frac{\Delta x_{N-1}}{2}\left(\frac{n_M^2 \cdot Q}{R_{N-1}^{4/3} \cdot A_{N-1}^2} + \frac{n_M^2 \cdot Q}{R_N^{4/3} \cdot A_N^2}\right) \tag{4.33j}$$

The system of non-linear equations (4.32) can be solved by the Newton method, which is implemented as follows (see Section 2.3.2):

$$\mathbf{J}^{(k)} \cdot \mathbf{\Delta} X^{(k+1)} = -\mathbf{F}^{(k)} \tag{4.34}$$

where:

k is iteration index,
$\mathbf{\Delta X}^{(k)} = \mathbf{X}^{(k+1)} - \mathbf{X}^{(k)}$ is correction vector,
$\mathbf{F}^{(k)} = \mathbf{A}^{(k)}\,\mathbf{X}^{(k)} - \mathbf{B}$ is vector of residuals in Eq. (4.32),
$\mathbf{J}^{(k)}$ is Jacobian matrix.

After assuming the first estimation of the unknown vector $\mathbf{X}^{(0)}$, the iterative process is continued until two succeeding solutions satisfy the following criterions for convergence:

$$\left| X_i^{(k+1)} - X_i^{(k)} \right| \le \varepsilon_H \quad \text{for } i = 1, \ldots, N \tag{4.35}$$

and

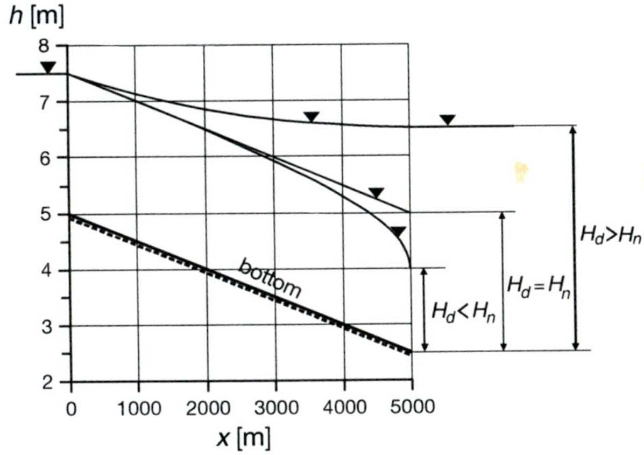

Fig. 4.25 Water flow profiles in channel connecting two reservoirs calculated with Newton method for various water levels in downstream reservoir

$$\left|X_{N+1}^{(k+1)} - X_{N+1}^{(k)}\right| \leq \varepsilon_Q \tag{4.36}$$

where ε_H and ε_Q represent the specified tolerances for water level H_i and discharge Q respectively.

Example 4.8 Let us consider an idealized channel with the following geometrical parameters: $L = 5000$ m, is $s = 0.0005$, $b = 5$ m and $m = 1.4$. The Manning roughness coefficient is $n_M = 0.020$. The channel is divided by $N = 101$ nodes into 100 reaches of the constant length $\Delta x = 50$ m. The bed elevation above the datum changes linearly from $z(x = 0) = 4.0$ m at the upstream end to $z(x = L) = 2.5$ m at the downstream end. The boundary condition imposed at the upstream reservoir was equal to $h_u = 7.5$ m whereas at the downstream one the water level h_d takes the following values: 6.5, 5 and 4 m. The velocity head in boundary conditions (see Eqs. (4.30) and (4.31)) is neglected.

Numerical tests showed that the Newton iterative process was convergent in particular cases only, when the starting value of the discharge Q was close enough to the exact solution. With the initial water level corresponding to the hydrostatic state, rapid convergence is ensured for $Q^{(0)}$ ranging from 5 to 80 m^3/s. For example starting from $Q^{(0)} = 40$ m^3/s only 4 iterations are needed to obtain the solution $Q = 27.594$ m^3/s with $\varepsilon_H = 0.001$ m and $\varepsilon_Q = 0.1$ m^3/s. Unfortunately for values of $Q^{(0)}$ outside this range the iterations failed to converge. Thus one can conclude that the Newton method is very effective only on condition that the starting value of the discharge $Q^{(0)}$ is properly chosen. The calculated flow profiles are shown in Fig. 4.25.

4.3.3 Direct Solution Using the Newton Method with Quasi–Variable Discharge

Schulte and Chaudhry (1987) solved the steady gradually – varied flow in looped network of open channels using both discrete energy and continuity equations. This approach can be applied for a single channel as well. In the proposed algorithm two unknowns, the water level and the discharge are introduced at each grid point. Therefore in the numerical solution the discharge is treated as variable which formally varies along x axis, although from the physical point of view the discharge is constant over the entire channel. This approach can be called the solution with quasi–variable discharge.

Equations (4.27) and (4.28) can be rewritten as follows:

$$\left(h_{i+1}+\frac{\alpha\cdot Q_{i+1}^2}{2g\cdot A_{i+1}^2}\right)-\left(h_i+\frac{\alpha\cdot Q_i^2}{2g\cdot A_i^2}\right)+\frac{\Delta x_i}{2}\left(\frac{n_M^2\cdot Q_i^2}{R_i^{4/3}\cdot A_i^2}+\frac{n_M^2\cdot Q_{i+1}^2}{R_{i+1}^{4/3}\cdot A_{i+1}^2}\right)=0 \tag{4.37}$$

$$-Q_i+Q_{i+1}=0 \text{ for } i=1,2,3,\ldots,N-1 \tag{4.38}$$

They constitute the system of non-linear equations, which can be expressed in matrix form similar to Eq. (4.32). However, this time the vector of unknowns has the following structure:

$$\mathbf{X}=(h_1,Q_1,h_2,Q_2,\ldots,h_{N-1},Q_{N-1},h_N,Q_N)^T \tag{4.39}$$

As one can see, this method leads to a larger system of non-linear algebraic equations, since instead of a system of dimension $(N+1)\times(N+1)$ one obtains a system of dimension $(2N)\times(2N)$. As we have $2(N-1)$ equations in form of Eqs. (4.37) and (4.38) two additional equations are required to complete the system. They come from the imposed boundary conditions. The structure of matrix **A** corresponding to the assumed structure of vector **X** (Eq. 4.39) is shown in Fig. 4.26.

The goal of such approach is to improve the convergence of the Newton iteration process, which in the preceding algorithm can suffer from the lack of convergence.

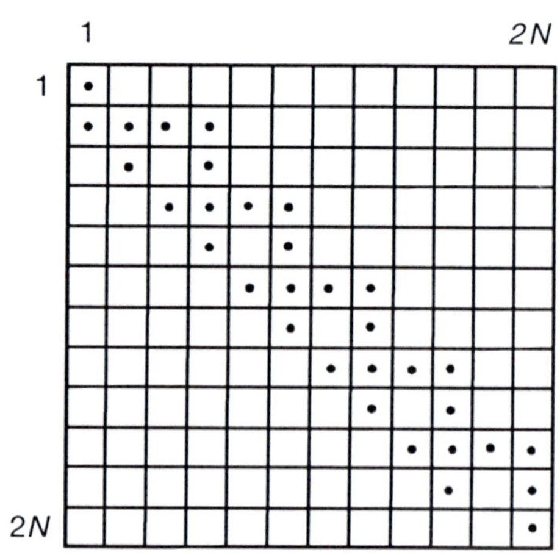

Fig. 4.26 Structure of matrix **A** for the system obtained in Schulte and Chaudhry (1987) method (dots represent non-zero elements)

Compared to the previously described approach, in which only one value of the flow discharge in the channel was computed, solution system of Eqs. (4.37) and (4.38) is more costly in terms of numerical work. However, the iterative process appears less sensitive on the initial guess of the discharge. Practically always a convergent solution is reached. The water profiles computed for the same data as in Example 4.8 are identical to the ones displayed in Fig. 4.25.

4.3.4 Direct Solution Using the Improved Picard Method

To solve the non linear system of equations the Picard method can be applied as well. The Picard iterative scheme for Eq. (4.32) can be written in the following form (compare Section 2.3.3):

$$\mathbf{A}^{(k)} \cdot \mathbf{X}^{(k+1)} = \mathbf{B} \tag{4.40}$$

where k is iteration index. Its application for the single channel considered previously in Example 4.8 showed that regardless of the first evaluation of the discharge Q it was impossible to reach the solution. The typical situation for $Q^{(0)} = 50\ \mathrm{m^3/s}$ is presented in Fig. 4.27 as tooth oscillations. While the water levels relatively quickly tend to the expected values, the discharge $Q^{(k)}$ oscillates with constant amplitude during subsequent iterations. Therefore the standard Picard method cannot be used in the considered case.

For the equations of steady gradually varied flow an improvement of the iteration process that it becomes convergent, is possible. As results from Fig. 4.27 the discharge Q oscillates around its real value in the subsequent iterations. This fact suggests the way of improvement of the Picard method. Instead of Eq. (4.40) the following modified scheme can be used (Szymkiewicz and Szymkiewicz 2004):

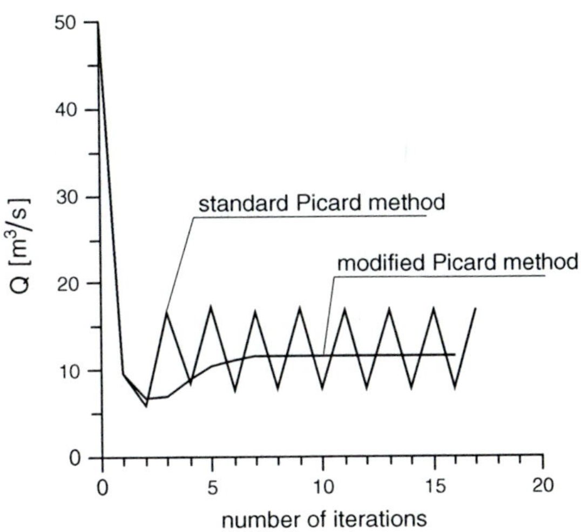

Fig. 4.27 Convergence of the Picard iterative process

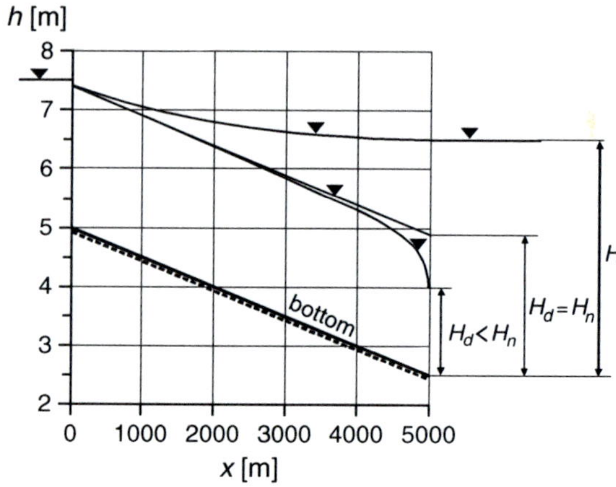

Fig. 4.28 Water flow profile in channel connecting two reservoirs calculated with modified Picard method for various water levels in downstream reservoir

$$\mathbf{A}^* \cdot \mathbf{X}^{(k+1)} = \mathbf{B} \tag{4.41}$$

where $\mathbf{A}^* = \mathbf{A}\left(0.5\left(\mathbf{X}^{(k)} + \mathbf{X}^{(k-1)}\right)\right)$ is the modified matrix of coefficients. It means that to calculate the vector **X** in iteration $k + 1$, the matrix **A** is calculated using the arithmetic average value of **X** from two preceding iterations. Consequently the estimation accepted for the next iteration is closer to its real value. For $k = 1$ $\mathbf{A}^* = \mathbf{A}(\mathbf{X}^{(0)})$ is recommended. The positive results of the proposed improvement are confirmed by the following example.

Example 4.9 The calculations carried out for the same data as previously with the assumed tolerances $\varepsilon_H = 0.001$ m and $\varepsilon_Q = 0.1$ m^3/s were always successful. The first estimation of the water profile corresponded to the hydrostatic state with $h(x) = h_u =$ const. The final discharge $Q = 27.969$ m^3/s was computed for the initial approximations of $Q^{(0)}$ ranging from $-1{,}000$ to $1{,}000$ m^3/s after $12 \div 15$ iterations. The results of improved convergence obtained for $Q^{(0)} = 50$ m^3/s are presented in Fig. 4.28.

One can notice that the iteration process (4.41) relatively quickly converges to the solution. The performed tests show that $Q^{(0)}$ can take any positive or negative value except zero. In the last case the hydrostatic state is obtained. This is because for $Q^{(0)} = 0$ the system of equations is split into a series of independent equations, each with only one unknown. These results prove indirectly that the system (4.32) has a unique solution. This fact was confirmed additionally by the solution of Eq. (4.32) with the following boundary conditions: $h_u = 7.5$ m and $h_d = 4.0$ m, corresponding to the uniform steady state flow. The water flow profile parallel to the channel bed ($h(x) = 2.39$ m = const.) and a discharge $Q = 30.1$ m^3/s satisfying the Manning formula were obtained.

Comparing the Newton and Picard methods one can state that the Newton method ensures rapid convergence on condition that it is convergent. It needs only 4–6 iterations to obtain a solution, whereas the modified Picard method needs 12–15 iterations to obtain the solution with the same tolerances. However the Newton method failed for some sets of data while the Picard one appeared unconditionally convergent. The observed difference in the number of iterations is evident because the Picard method converges linearly while the convergence of the Newton one is quadratic (Bjorck and Dahlquist 1974).

To solve the system of linear algebraic equations (4.41) with sparse matrix of coefficients an iterative method seems suitable. However the attempts to solve it with Gauss–Seidel and SOR methods failed. Finally, Eqs. (4.34) and (4.41) should be rather solved by the Gauss elimination method in its version using non-zero elements of matrix **J** or **A** only.

Let us consider the second possible type of the boundary problem for steady gradually varied flow equations, in which the Manning coefficient n_M is considered as an unknown. This case is described by Eqs. (4.25) and (4.26). The elements of the vectors **X** and **b** as well as the elements of the matrix **A** must be modified. The vector **X** is defined as follows:

$$X = (h_1, h_2, \dots, h_N, n_M)^T, \tag{4.42}$$

whereas the matrix **A** has the following elements:

$$a_{1,1} = a_{N,N} = 1, \tag{4.43a}$$

$$a_{i,i} = 1, \quad a_{i,i-1} = -1 \text{ for } i = 2,3,\dots, N-1, \tag{4.43b}$$

$$a_{i,N+1} = \frac{\Delta x_{i-1}}{2}\left(\frac{n_M \cdot Q^2}{R^{4/3}_{i-1} \cdot A^2_{i-1}} + \frac{n_M \cdot Q^2}{R^{4/3}_{i} \cdot A^2_{i}}\right) \text{ for } i = 2,3,\dots, N-1, \tag{4.43c}$$

$$a_{N,N+1} = \frac{\Delta x_{N-1}}{2}\left(\frac{n_M \cdot Q^2}{R^{4/3}_{N-1} \cdot A^2_{N-1}} + \frac{n_M \cdot Q^2}{R^{4/3}_{N} \cdot A^2_{N}}\right). \tag{4.43d}$$

The vector of the right-hand side $\mathbf{B} = (b_1, b_2, \dots, b_N, b_{N+1})^T$ has the following components:

$$b_1 = h_o, \quad b_N = h_L, \tag{4.44a}$$

$$b_i = \frac{\alpha \cdot Q^2}{2g \cdot A^2_{i-1}} - \frac{\alpha \cdot Q^2}{2g \cdot A^2_{i}} \text{ for } i = 2, \dots, N-1, \tag{4.44b}$$

$$b_{N+1} = \frac{\alpha \cdot Q^2}{2g \cdot A^2_{N-1}} - \frac{\alpha \cdot Q^2}{2g \cdot A^2_{N}}. \tag{4.44c}$$

Numerical tests carried out for the channel described in Example 4.8 show that for $\varepsilon_H = 0.0001$ and $\varepsilon_n = 0.001$ the finite difference method with the Picard iteration ensures a convergent solution for any starting value of ${n_M}^{(0)}$. However, the number of iterations depends on the initial choice of ${n_M}^{(0)}$.

4.3.5 Solution of the Boundary Problem Using the Shooting Method

In this approach the process of solution of the boundary problem for the system of equations (4.23) and (4.24) is carried out as a sequence of solutions of initial value problems. Assuming $q = 0$ the corresponding initial value problem can be written in the following form (see Section 3.3):

$$\frac{dE}{dx} = -S \text{ with } E\,(x = 0) = E_o. \tag{4.45}$$

Since $E = E(x;\, Q)$ the solution depends on the chosen value of Q. Therefore for any assumed value of Q one obtains the corresponding value of $E(x = L,\, Q)$. Solution of the boundary value problem coincides with the solution of the initial value problem with such a value of Q, for which the boundary condition at the downstream end is satisfied, i.e.:

$$E\,(L,\, Q) = E_L. \tag{4.46}$$

Let us introduce the function:

$$F\,(Q) = E\,(L,Q) - E_L, \tag{4.47}$$

which represents the difference between the calculated value of E at $x = L$ for the assumed Q and the imposed boundary condition at $x = L$ equal to E_L (Fig. 4.29).

In such a way the solution of the boundary problem can be considered as finding the root of Eq. (4.47). The value of $F(Q)$ can be calculated for any value of Q. To this end the initial value problem (4.45) should be solved to obtain $E(x = L,\, Q)$. If a couple of values of Q are known, say $Q^{(1)}$ and $Q^{(2)}$, which satisfy the following relation:

$$F\left(Q^{(1)}\right) \cdot F\left(Q^{(2)}\right) < 0, \tag{4.48}$$

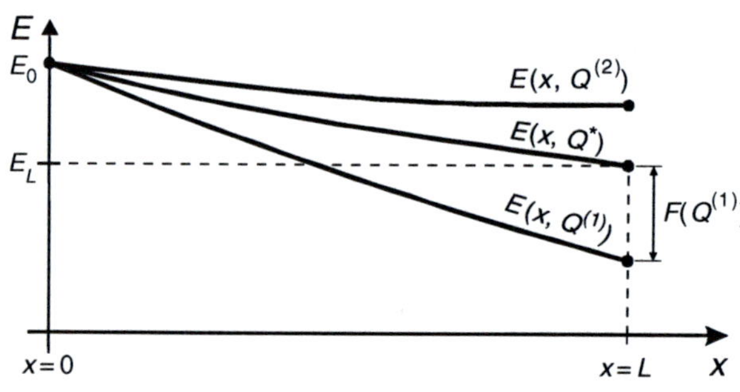

Fig. 4.29 Solution of the boundary problem as a sequence of initial value problems

then the root Q^* of $F(Q)$ is located inside the interval $\langle Q^{(1)}, Q^{(2)} \rangle$. The root can be found using one of the methods of solution of the non linear algebraic equations as the bisection method or the secant method. Note that the Newton method is rather less suitable since it requires the derivative of function $F(Q)$, which in this case can be obtained via numerical approximation only.

Let us apply the method to compute the water flow profile and flow discharge in channel connecting two reservoirs with different but constant in time water levels presented in Fig. 4.23.

Example 4.10 A channel connecting two reservoirs of length L = 5,000 m with bed slope s = 0.0005 has trapezoidal cross section with base width b = 10 m and side slope m = 1. The Manning coefficient is equal to $n_M = 0.030$. The channel is divided into N = 50 intervals having constant length Äx = 100 m. The bed elevation varies linearly from 4.0 m above the datum at the upstream end to 2.5 m at the downstream end. As it was showed by Chow (1959), the shape of the flow profile h(x) depends on the relation between the water levels in reservoirs and the normal depth in the channel H_n.

In the upstream reservoir a constant water level equal to $h_o = 10.0$ m, which corresponds to the depth $H_o = 4.0$ m, is assumed. The velocity head is neglected. Computations were carried out for the following values of the water levels in the downstream reservoir: $h_L = 8.75$, 7.50 and 4.25 m. These values correspond to the depths: $H_L = 4.25$, 4.00 and 3.75 m. The results of calculations are presented in Fig. 4.30.

The shapes of the flow profiles agree with those obtained by Chow (1959). The computed discharge for $H_L = 4.25$ m is equal to $Q = 102.24$ m^3/s, whereas for $H_L = 3.75$ m it is equal to $Q = 123.07$ m^3/s. Further calculations were carried out for the same set of data, but taking into account the velocity head in the boundary condition at the upstream end. The computed velocity heads at the upstream end were: 0.092 m for $h_L = 8.75$ and 0.134 m for $h_L = 4.25$ m. For this

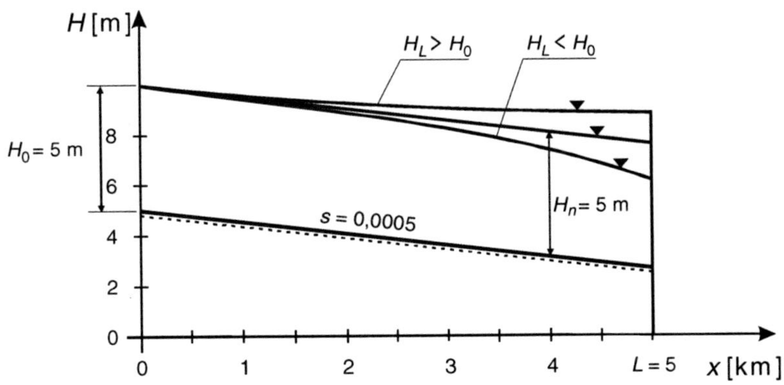

Fig. 4.30 The flow profiles for various water levels in downstream reservoir

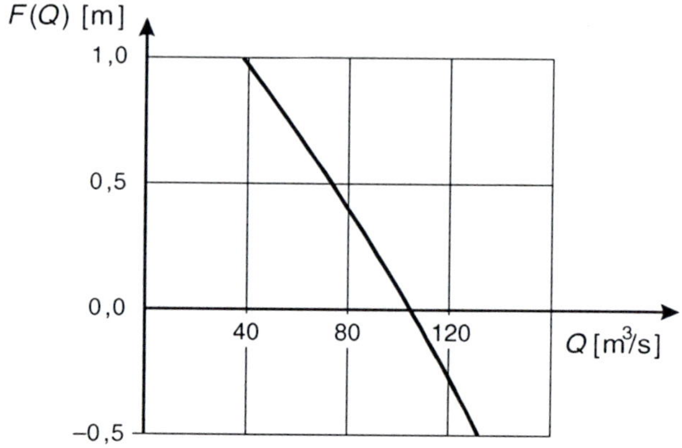

Fig. 4.31 Plot of function $F(Q)$

reason the previously computed discharges were modified to $Q = 98.022\ \text{m}^3/\text{s}$ and $Q = 117\ \text{m}^3/\text{s}$ respectively.

In the presented example the function $F(Q)$ has a regular shape, as presented in Fig. 4.31, and possesses one root. To find it the bisection method has been applied. The number of iterations depends on the dimension of interval in which the root is localized, and on the required tolerance of solution. For $\varepsilon_H = 0.0001$ m the total number of iteration was less than 20.

In contrast to the presented approach Chow (1959) solved the same problem by the trial and error method using in addition the so-called discharge curve $Q(h_L)$. However, formulation of the boundary value problem allows us to calculate both the flow profile $h(x)$ and the discharge Q directly.

The shooting method can be also used to find the value of the Manning coefficient n_M when the discharge Q and the water levels at the endpoints are known. In this case, instead of the root of function $F(Q)$, the root of function $F(n_M)$ should be determined. The function $F(n_M)$ is defined as follows:

$$F\left(n_M\right) = E\left(L, n_M\right) - E_L, \tag{4.49}$$

whereas the criteria for convergence take the following form:

$$\left|h_i^{(k+1)} - h_i^{(k)}\right| \leq \varepsilon_H \text{ for } i = 1,2,\ldots, N \tag{4.50a}$$

and

$$\left|n_M^{(k+1)} - n_M^{(k)}\right| \leq \varepsilon_n \tag{4.50b}$$

where ε_n is the assumed tolerance for the Manning coefficient n_M.

Example 4.11 A rectangular channel with $B = 0.385$ m, and $L = 10$ m has longitudinal slope equal to $s = 0.0005$ (Fig. 4.32). Computations were performed over the length of $L_1 = 8$ m, which was divided into segments of length equal to $\Delta x = 0.5\ \text{m} = \text{const.}$

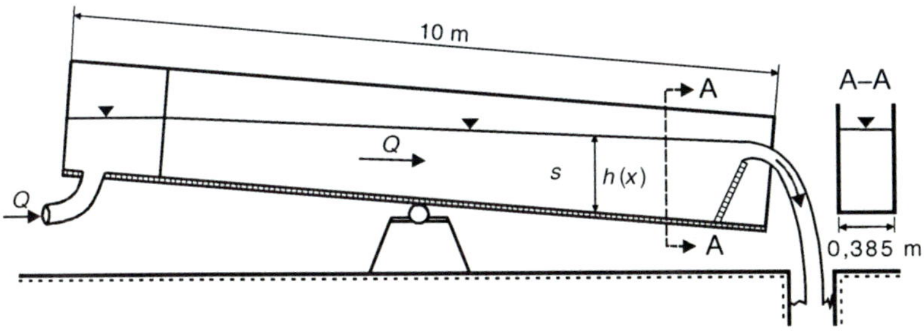

Fig. 4.32 The considered laboratory flume (Szymkiewicz 2000)

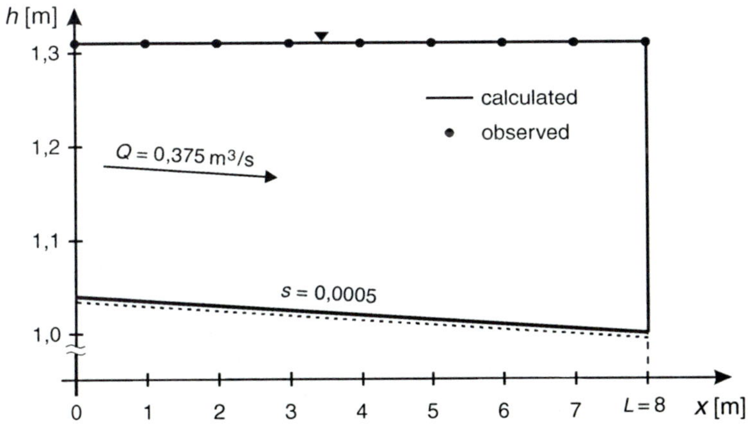

Fig. 4.33 Calculated and measured water profile in channel (Szymkiewicz 2000)

An experiment performed in hydraulic laboratory provided the following data: $Q = 0.0375$ m^3/s, $h_o = 1.314$ m, $h_L = 1.309$ m. Only few iterations are needed to obtain the solution satisfying the following tolerances: $\varepsilon_H = 0.0001$ and $\varepsilon_n = 0.001$.

Regardless of the starting value of the Manning coefficient $n_M^{(0)}$, convergent solutions were reached. This fact confirms that the iterative process converges unconditionally. The differences between the computed and observed water levels did not exceed 0.002 m, whereas the computed Manning coefficient is equal to $n_M = 0.0201$. In Fig. 4.33 computed and observed water levels are compared.

4.4 Steady Gradually Varied Flow in Open Channel Networks

4.4.1 Formulation of the Problem

In practical applications the steady gradually varied flow is considered not only in a single open channel but in open channel networks as well. There are two types of channel networks: tree – type and looped ones (Fig. 4.34).

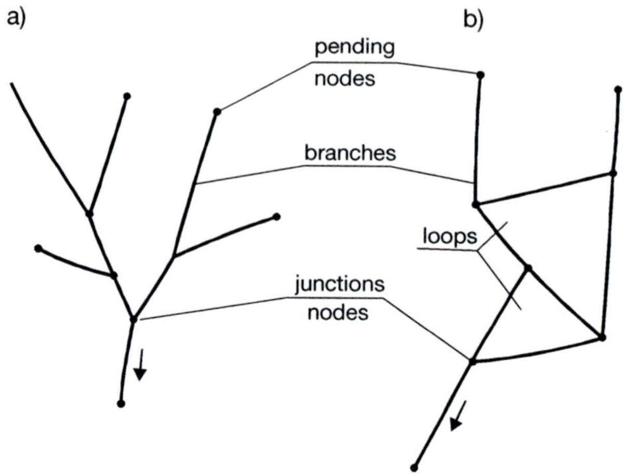

Fig. 4.34 Tree – type (**a**) and looped (**b**) network

A network consists of branches, which join each other in the junction. Usually the junction gathers three branches. If a particle of water starting from an arbitrary point of the network can reach another point via only one possible route, the network is called a tree – type one, whereas if it can travel towards the destination point using multiple alternative ways, the network is called looped. The upper parts of river systems tend to form tree – type networks. In natural conditions the looped network can appear rather in the lowest parts of rivers (deltas).

In an open channel network one can distinguish two kinds of branches. The first one connects two junctions, i.e. its both ends are located at the neighboring junctions. The second kind of branch has one end located at a junction, whereas the opposite end is a pendant one. Note that the additional conditions, which are required to ensure the solution of flow problem, must be imposed at pendant ends.

While for a single channel both initial value and boundary value problems can be formulated, for a channel network in principle only the boundary value problem should be formulated, since the discharges in branches and the flow profiles are the goal of computations. Although the governing equations describing this kind of flow are the same, the solution of flow equations for network becomes more complicated comparing with a single channel.

Various approaches were proposed to solve the boundary value problem for a channel network. Some of them are applicable to tree type networks only. Similarly to a single channel, the final system of nonlinear algebraic equations cannot be efficiently solved by the standard Picard or Newton methods because the iterative process suffer from the lack of convergence. Thus, Schulte and Chaudhry (1987) introduced the discharge Q as an additional unknown at each computational node (see Section 4.3.3), although they do not vary within each branch. This modification allows solving efficiently the resulting system of equations by the Newton method, but the system itself is considerably larger than the system resulting from the approach in which the discharge Q is treated as a single unknown for the whole branch. Consequently, the computational cost of the solution is increased.

Naidu et al. (1997) have proposed an algorithm based on the decomposition of the channel network into small units. The problem is solved in each unit using the IV order Runge–Kutta method and the local solutions are connected to obtain the final solution for whole network using the shooting method. However, the proposed algorithm works for tree-type networks only.

The steady flow problem can be also solved indirectly using the Saint-Venant equations describing the unsteady flow, with the specified, constant in time boundary conditions. After a sufficiently long time the system reaches steady state and the corresponding water surface profiles in the channel network are obtained (Cunge et al. 1980).

All the mentioned methods do not have general character. However, it is possible to propose a very effective general algorithm, which will work for any type of network and for any kind of boundary conditions imposed at pendant channel ends. This approach based on the approximation of the governing equations by the finite difference method appeared a robust tool in the case of single channel and can be successfully adapted for a channel network.

4.4.2 Numerical Solution of Steady Gradually Varied Flow Equations in Channel Network

Let us consider a channel network consisting of M channels (branches). It can be either tree type or looped one. Each channel is divided into N_j ($j = 1, 2, \ldots, M$) intervals of constant or variable length Δx_i. The total number of grid points in whole network is ΣN_j. In addition, let us assume positive direction of flow in each channel corresponding to the direction of increasing of the indexes.

To solve the boundary problem for the considered network the governing ordinary differential equations (4.23) and (4.24) are approximated by the finite difference method. Equation (4.23) is approximated, as in the previous section, by centered differences, coinciding with the implicit trapezoidal rule. Since in the network branches the water can flow in both directions, this fact should be taken into account in the sign of the friction force. Therefore for channel j Eq. (4.23) is approximated by the following formula:

$$\left(h_{i+1} + \frac{\alpha \cdot Q_j^2}{2g \cdot A_{i+1}^2}\right) - \left(h_i + \frac{\alpha \cdot Q_j^2}{2g \cdot A_i^2}\right) + \frac{\Delta x_i}{2}\left(\frac{n_M^2 \cdot Q_j \cdot |Q_j|}{R_i^{4/3} \cdot A_i^2} + \frac{n_M^2 \cdot Q_j \cdot |Q_j|}{R_{i+1}^{4/3} \cdot A_{i+1}^2}\right) = 0 \tag{4.51}$$

Similar equations are written for each interval $\Delta x_i (i = 1,2, \ldots, N_j - 1)$. In this way one obtains a system of $N_j - 1$ algebraic equations with $N_j + 1$ unknowns. There are N_j water levels h_i at the nodes and the flow discharge Q_j.

The system of equations (4.51) can be presented in the matrix form:

$$\mathbf{AX} = \mathbf{B} \tag{4.52}$$

where:

A – matrix of coefficients,
$\mathbf{B} = (0, 0, \ldots,0)^T$ – vector of right hand side,
$\mathbf{X} = \left(h_1, h_2, \ldots ,h_{N_j},Q_j\right)^T$ – vector of unknowns,
T – transposition symbol.

The matrix **A** of dimensions $(N_j + 1) \times (N_j + 1)$ is very sparse. Its structure is similar to the one obtained for a single channel (Fig. 4.24). Its non–zero elements are defined as follows:

$$a_{i,i} = 1, \text{ for } i = 2,3,\ldots,N_j - 1, \tag{4.53a}$$

$$a_{i,i-1} = 1, \text{ for } i = 2,3,\ldots,N_j - 1, \tag{4.53b}$$

$$a_{i,N_j-1} = -\frac{\alpha \cdot Q_j}{2g \cdot A_{i-1}^2} + \frac{\alpha \cdot Q_j}{2g \cdot A_i^2} + \frac{\Delta x_{i-1}}{2}\left(\frac{(n_M)_j^2 \cdot |Q_j|}{R_{i-1}^{4/3} \cdot A_{i-1}^2} + \frac{(n_M)_j^2 \cdot |Q_j|}{R_i^{4/3} \cdot A_i^2}\right)$$

for $i = 2,3,\ldots,N_j - 1,$

$$\tag{4.53c}$$

$$a_{N_j,N_j} = 1, \tag{4.53d}$$

$$a_{N_j,N_j-1} = -1 \tag{4.53e}$$

$$a_{N_j+1,N_j+1} = -\frac{\alpha \cdot Q_j}{2g \cdot A_{N_j-1}^2} + \frac{\alpha \cdot Q_j}{2g \cdot A_{N_j}^2} + \frac{\Delta x_{N-1}}{2}\left(\frac{(n_M)_j^2 \cdot |Q_j|}{R_{N-1}^{4/3} \cdot A_{N-1}^2} + \frac{(n_M)_j^2 \cdot |Q_j|}{R_N^{4/3} \cdot A_N^2}\right) \tag{4.53f}$$

When flow in a single channel was considered this system had to be completed by two additional equations resulting from the imposed boundary conditions. However, in the case of channel network such system of equations is written for each branch separately. To assembly them, apart from the mentioned boundary conditions, which are imposed at pendant nodes only, additional equations must be specified for each junction. For a junction of three channels I, J, K formed by the nodes i, j, k as presented in Fig. 4.35. one can write the continuity equation:

$$Q_I = Q_J + Q_K \tag{4.54}$$

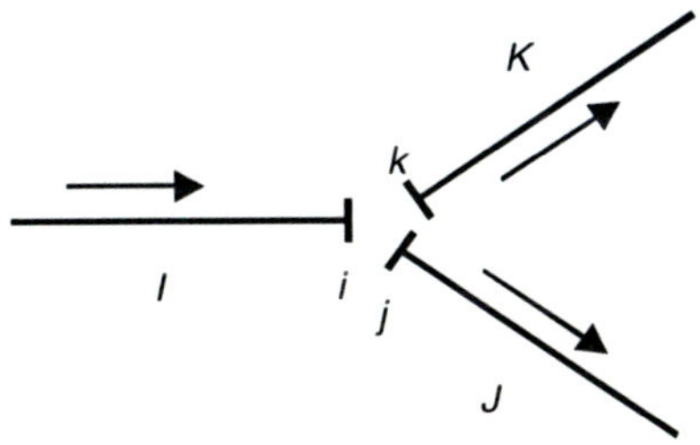

Fig. 4.35 Junction of three channels

and the energy equations:

$$h_i + \frac{\alpha \cdot Q_I^2}{2g \cdot A_i^2} = h_j + \frac{\alpha \cdot Q_J^2}{2g \cdot A_j^2} = h_k + \frac{\alpha \cdot Q_K^2}{2g \cdot A_k^2} \tag{4.55}$$

In the above equations the energy losses are neglected. Sometimes the velocity heads can be also neglected as relatively small.

For each junction of the considered channel network three additional equations in form of Eqs. (4.54) and (4.55) can be written. This enables us to close the global system of equations for the entire network. The matrix of this system contains submatrices corresponding to each channel in form presented in Eq. (4.52), connected by the equations for junctions. The final matrix is banded and very sparse.

To better illustrate the presented procedure of solution let us consider a looped channel network as presented in Fig. 4.36. It consists of seven channels and four junctions. The network has two pendant nodes, at which the boundary conditions are imposed. In all branches of the network subcritical flow is assumed. The discharge is assumed positive if the direction of the flow corresponds to the direction of increasing node indices, as shown by the arrows in Fig. 4.36. Each channel is divided into 10 intervals of constant length. Then the total number of grid points is 77.

For each channel one can write a system of 10 equations in form of Eq. (4.51). Therefore for the entire network one obtains a global system containing 7 subsystems, written for each branch separately. Then one has 70 equations, whereas the

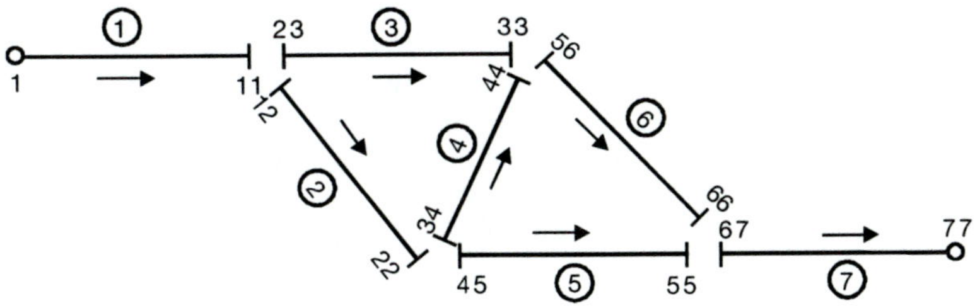

Fig. 4.36 Example of looped channel network

total number of unknowns is equal to 84 (77 water levels in grid point and 7 discharges in branches). To close the system the following additional equations are introduced:

- for the four junctions – 4 continuity equations of type (4.54) and 8 energy equations of type (4.55);
- for the two pendant channel ends (number 1 and 77) – 2 equations given by the prescribed boundary conditions.

Finally one obtains the global system of dimension 84×84 having 84 unknowns. In each row of the global matrix there are only 3 or 4 non-zero elements.

Assuming subcritical flow in channel, the following conditions can be imposed at the pendant channel's ends:

- the water level with velocity head;
- the water level;
- the discharge.

Preliminary numerical tests carried out for the considered network showed that the Newton iterative method applied to solve Eq. (4.52) failed to converge. This conclusion coincides with the well-established opinion of its poor global convergence properties (Press et al. 1992). On the other hand, application of the modified Picard method described in Subsection 4.3.4 is successful. Therefore the non-linear system of equations (4.52) is solved as follows:

$$\mathbf{A}^* \cdot \mathbf{X}^{(k+1)} = \mathbf{B} \tag{4.56}$$

where k is index of iteration and

$$\mathbf{A}^* = \mathbf{A}\left(0.5\left(\mathbf{X}^{(k)} + \mathbf{X}^{(k-1)}\right)\right) \tag{4.57}$$

is modified matrix of coefficients. Numerous tests confirmed that with the presented algorithm it is possible to overcome the problem of poor convergence of the standard Picard method.

Example 4.12 Let us consider a problem of the flow past an island, which occurs when a long island divides the river into two channels, as shown in Fig. 4.37. This is a well-known problem in open channel hydraulics. While some special algorithms were proposed to solve such problem (French 1985), it can be considered as a particular case of the flow in a looped channel network and solved with the method described above.

The system consists of 4 channels and 2 junctions. The network has 2 pendant nodes at which the boundary conditions are imposed. In all branches of the network subcritical flow is assumed. The flow is assumed from the upstream end (node u in Fig. 4.37) towards the downstream end (node d in Fig. 4.37). The positive direction

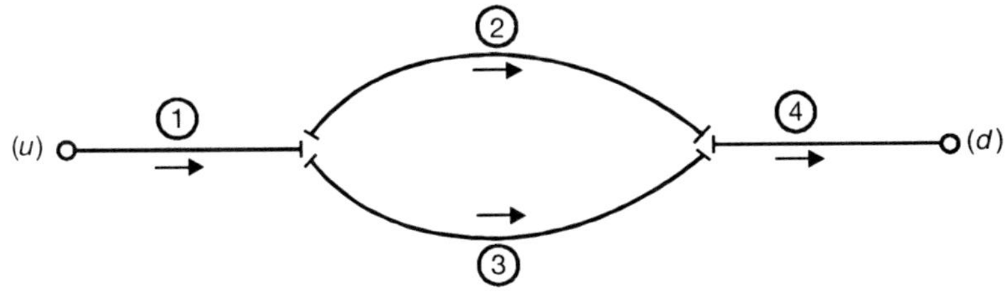

Fig. 4.37 Considered channel network

of flow in each channel is marked with an arrow. The channels have trapezoidal cross-sections. The bed elevation at the upstream end is assumed to be equal to 10.000 m above the datum, whereas at the downstream it is equal to 9.000 m. Each channel is divided into 10 intervals of constant length, which gives the total number of grid points equal to 44. First let us consider a symmetric system, where the two channels surrounding the island have the same parameters (Table 4.1). In such a case the solution must be symmetrical as well.

The boundary conditions are specified in terms of water levels at the upstream end and the downstream end of the network: h_u= 11.500 m, h_d= 10.500 m. The results obtained for the data from Table 4.1 are displayed in Table 4.2.

As it could be expected the flow in channel 1 bifurcated symmetrically into channels 2 and 3. To obtain these results, with the assumed accuracy of ε_h= 0.001 m and ε_Q= 0.001 m^3/s, 18 iterations are needed. The iterations converged for arbitrary initial approximations of the discharges $Q^{(0)}$ in the channels.

The second case concerns an asymmetrical network, in which the channels surrounding the island have different characteristics (Table 4.3). Compared to channel 2, channel 3 is longer, narrower, less smooth and has smaller slope. Consequently, the flow in channel 1 bifurcates into channels 2 and 3 in asymmetrical way. As it is shown in Table 4.4, the discharges in the two channels differ by a factor larger than 3. Also in this case no problem with convergence occurs.

Example 4.13 Let us consider the steady gradually varied flow in a tree – type network taken from Naidu et al. (1997), which is presented in Fig. 4.38. It consists of 41 branches. The total number of nodes is 470.

The parameters of the channels are given in Table 4.5.

Table 4.1 Characteristics of the channels from Fig. 4.37 – the symmetric case

Channel number	Length [m]	Bed width [m]	Side slope	Bed slope	n_M	Δx [m]
1	300	4.0	1.5	0.001	0.025	30
2	400	3.5	1.5	0.001	0.035	40
3	400	3.5	1.5	0.001	0.035	40
4	300	4.0	1.5	0.001	0.025	30

Table 4.2 Results for the network from Fig. 4.37 – the symmetric case

Channel number	Discharge Q [m^3/s]	Upstream water level [m]	Downstream water level [m]
1	14.451	11.500	11.176
2	7.225	11.176	10.810
3	7.225	11.176	10.810
4	14.451	10.810	10.500

Table 4.3 Characteristics of the channels from Fig. 4.37 – the asymmetric case

Channel number	Length [m]	Bed width [m]	Side slope	Bed slope	n_M	Δx [m]
1	300	4.0	1.5	0.001	0.025	30
2	400	3.5	1.5	0.001	0.035	40
3	800	1.5	1.5	0.0005	0.045	80
4	300	4.0	1.5	0.001	0.025	30

Table 4.4 Results for the network from Fig. 4.37 – the asymmetric case

Channel number	Discharge Q [m^3/s]	Upstream water level [m]	Downstream water level [m]
1	12.418	11.500	11.305
2	9.370	11.305	10.743
3	3.048	11.305	10.743
4	12.418	10.743	10.500

In order to simplify the problem the velocity heads at the junctions are neglected. The first solution is carried out for exactly the same data as in the paper by Naidu et al. (1997). Since their method requires that the discharge is known in at least one branch, $Q = 40$ m^3/s is imposed at node number 1, while the water levels are known at all pendant nodes except the node 1 (Table 4.6).

As the result of calculation the water profiles and discharges in all branches, including the water level at node 1, are obtained. The results of calculation by the modified Picard method differ only insignificantly from those obtained by Naidu et al. (1997). The final result with $\varepsilon_H = 0.001$ m and $\varepsilon_Q = 0.1$ m^3/s was obtained after 18 iterations. The greatest difference, equal to 0.007 m, exists at node 1. At other nodes the depths calculated by both approaches are practically the same.

The second solution is obtained assuming that the discharges in all branches, including the first one, are unknown. On the other hand, the water levels are known

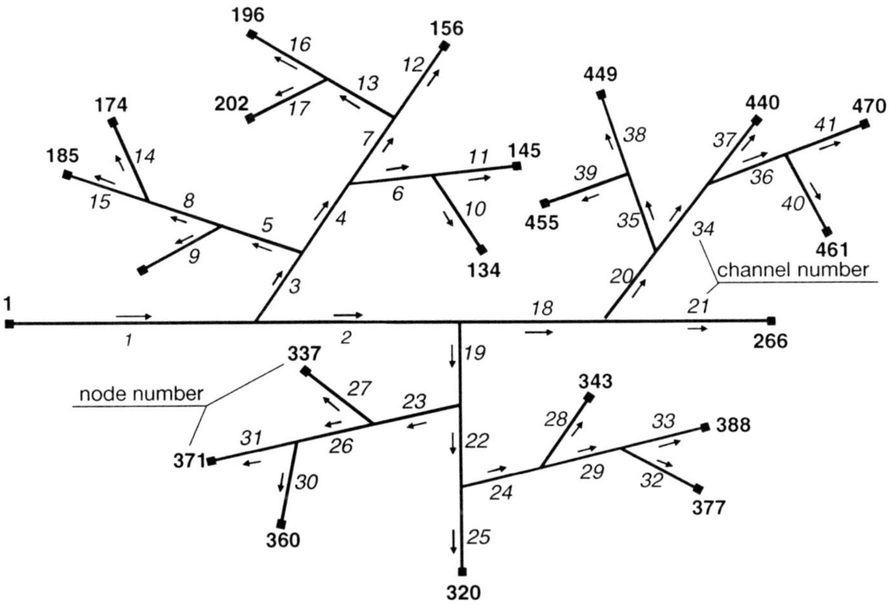

Fig. 4.38 Channel network used in Example 4.13 (after Naidu et al. 1997)

Table 4.5 Characteristics of the network from Fig. 4.38

Channel	Bed width [m]	Side slope	Bed slope	n_M	Length [m]	Number of reaches
1	10.0	2.0	0.000133	0.015	2,500.0	20
2	8.5	2.0	0.00015	0.016	2,000.0	20
3	4.0	2.0	0.00020	0.020	1,500.0	15
4	4.0	2.0	0.00021	0.020	1,400.0	15
5	1.5	2.0	0.00024	0.022	1,000.0	10
6	1.5	2.0	0.00024	0.022	1,100.0	10
7	3.0	2.0	0.00022	0.020	1,200.0	15
8	1.0	1.0	0.00025	0.022	1,000.0	10
9	0.5	1.0	0.00050	0.030	700.0	5
10	0.5	1.0	0.00050	0.030	700.0	5
11	1.0	1.0	0.00025	0.022	1,000.0	10
12	2.0	2.0	0.00024	0.022	1,000.0	10
13	1.5	2.0	0.00024	0.022	1,000.0	10
14	0.5	1.0	0.00050	0.030	700.0	5
15	1.75	2.0	0.00024	0.022	1,000.0	10
16	0.9	0.9	0.00025	0.022	900.0	10
17	0.5	1.0	0.00050	0.030	700.0	5
18	7.0	2.0	0.00016	0.017	1,700.0	15
19	3.5	1.0	0.00025	0.022	1,400.0	15
20	2.5	2.0	0.00022	0.022	1,300.0	15
21	4.0	2.0	0.00017	0.018	1,500.0	15

Table 4.5 (continued)

Channel	Bed width [m]	Side slope	Bed slope	n_M	Length [m]	Number of reaches
22	2.7	1.0	0.00022	0.022	1,200.0	15
23	1.75	2.0	0.00024	0.022	1,200.0	10
24	2.0	2.0	0.00024	0.020	1,200.0	10
25	1.75	2.0	0.00024	0.022	1,000.0	15
26	1.5	2.0	0.00024	0.022	1,100.0	10
27	0.5	1.0	0.00050	0.030	700.0	5
28	0.5	1.0	0.00050	0.030	700.0	5
29	1.75	2.0	0.000024	0.022	1,000.0	10
30	0.5	1.0	0.00050	0.030	700.0	5
31	1.0	1.0	0.00025	0.025	1,000.0	10
32	0.5	1.0	0.00050	0.030	700.0	5
33	1.5	2.0	0.00024	0.022	900.0	10
34	1.5	1.0	0.00025	0.022	1,200.0	15
35	1.5	1.0	0.00025	0.022	900.0	10
36	1.25	2.0	0.00024	0.022	800.0	8
37	1.0	2.0	0.00022	0.022	1,000.0	15
38	1.0	1.0	0.00025	0.022	800.0	8
39	0.5	1.0	0.00050	0.030	700.0	5
40	0.5	1.0	0.00050	0.030	700.0	5
41	0.75	2.0	0.00024	0.022	700.0	8

Table 4.6 Water levels imposed at pendant nodes in Fig. 4.38

Node	Water levels [m]	Node	Water levels [m]
1	12.199	337	9.608
128	10.265	343	9.497
134	10.246	360	9.600
145	10.242	371	9.587
156	10.235	377	9.494
174	10.248	388	9.496
185	10.249	440	9.212
196	10.232	449	9.568
202	10.236	455	9.675
266	9.761	461	9.231
320	9.489	470	9.234

at all pendant nodes, including node 1, where the previously calculated value is applied. Thus one can expect the resulting discharge value Q in the first branch to be close to the value specified in the previous calculation, i.e. $Q = 40$ m^3/s. Indeed, the obtained result is $Q = 40.005$ m^3/s. Other discharges are in similarly good agreement with the previous results (differences being less than 0.4%). The solution was obtained after 17 iterations. Thus, it seems that the approach basing on the finite difference discretization of the steady flow equation combined with improved Picard iterative scheme works efficiently for both tree – type and looped channel network and for any type of boundary conditions specified at its pendant ends.

References

Ascher UM, Petzold LR (1998) Computer methods for ordinary differential equations and differential – algebraic equations. SIAM, Philadelphia

Bjorck A, Dahlquist G (1974) Numerical methods. Prentice-Hall, Englewood Cliffs, NJ

Chow VT (1959) Open channel hydraulics. Mc Graw-Hill, New York

Cunge J, Holly FM, Verwey A (1980) Practical aspects of computational river hydraulics. Pitman Publishing, London

French RH (1985) Open channel hydraulics. McGraw-Hill, New York

Gray DM, Prowse TD (1993) Snow and floating ice. In: Maidment DR (ed.) Handbook of hydrology. McGraw-Hill, New York

Le Veque RJ (2007) Finite difference methods for ordinary and partial differential equations: Steady–state and time dependent problems. SIAM, Philadelphia.

Naidu BJ, Murty Bhallamudi S, Narasimhan S (1997) GVF computation in tree type channel networks. J. Hydr. Engng. ASCE 123 (8):700–708

Press WH, Teukolsky SA, Veterling WT, Flannery BP (1992). Numerical recipes in C, Cambridge University Press

Roberson JA, Cassidy JJ, Chaudhry MH (1998) Hydraulic engineering, 2nd edn. Wiley, New York

Schulte AM, Chaudhry MH (1987) Gradually-varied flow in open channel network. J. Hydr. Res. 25 (3):358–371

Singh VP (1996) Kinematic wave modelling in water resources: Surface water hydrology. John Wiley, New York

Stoer J, Bulirsch R (1980) Introduction to numerical analysis. Springer-Verlag, New York

Szymkiewicz R (2000) Mathematical modeling of river flow. Polish Scientific Publisher PWN, Warsaw (in polish)

Szymkiewicz R, Szymkiewicz A (2004) Method to solve the non-linear systems of equations for steady gradually varied flow in open channel network. Commun. Numer. Methods Engng. 20 (4):299–312

Chapter 5
Partial Differential Equations of Hyperbolic and Parabolic Type

5.1 Types of Partial Differential Equations and Their Properties

5.1.1 Classification of the Partial Differential Equations of 2nd Order with Two Independent Variables

The governing equations for unsteady flow and transport processes presented in Chapter 1 are partial differential equations (PDEs). They result from the application of the laws of conservation of some physical quantities like mass and momentum, and describe the evolution of an unknown function (like the water stage, discharge or concentration) with respect to temporal and spatial variables. Other examples of PDEs, in form of simplified flood routing models, will be introduced later in this book. This chapter discusses in more detail the basic properties of PDEs and numerical techniques applied to solve them.

As we could see in Chapter 1, a PDE may contain derivatives of various orders (first or second order in the presented examples). Thus, any PDE is additionally labeled as being of 1st order, 2nd order, 3rd order and so on. The order of a partial differential equation corresponds to the highest order of derivative which appears in the considered equation. Moreover, PDEs can be classified as hyperbolic, parabolic and elliptic ones. Since the proper designation of the type of considered equation is essential to formulate well-posed solution problem, let us begin with a brief discussion of classification of the partial differential equations.

For an equation of 2nd order with two independent variables there is very simple way of classification (Potter 1973). Such kind of equation can be written in the following general form:

$$a_1 \frac{\partial f}{\partial t^2} + a_2 \frac{\partial^2 f}{\partial t \cdot \partial x} + a_3 \frac{\partial^2 f}{\partial x^2} + a_4 \frac{\partial f}{\partial t} + a_5 \frac{\partial f}{\partial x} + a_6 \cdot f = \delta(x,t), \tag{5.1}$$

where:

t, x – independent variable (time and space position respectively),
$f(x, t)$ – dependent variable,

R. Szymkiewicz, *Numerical Modeling in Open Channel Hydraulics*, Water Science and Technology Library 83, DOI 10.1007/978-90-481-3674-2_5,

$a_1, \ldots, a_6$ – equation's coefficients,
$\delta(x, t)$ – source term,

Depending on the form of coefficients, Eq. (5.1) is considered as:

- linear equation with constant coefficients when $a_1, a_2, a_3, \ldots = \text{const.}$,
- linear equation with variable coefficients when $a_1 = a_1(x, t)$, $a_2 = a_2(x, t)$, $a_3 = a_3(x, t)$, ...,
- quasi-linear equation when $a_1 = a_1(x, t, f)$, $a_2 = a_2(x, t, f)$, $a_3 = a_3(x, t, f)$, ...,
- non-linear equation when $a_1 = a_1(x, t, f, \partial f/\partial t, \partial f/\partial x)$, $a_2 = a_2(x, t, f, \partial f/\partial t, \partial f/\partial x)$, $a_3 = a_3(x, t, f, \partial f/\partial t, \partial f/\partial x)$, ...

As far as the type of equation is considered, Eq. (5.1) is called:

- elliptic, when $\Delta = a_2^2 - a_1 \cdot a_3 < 0$
- parabolic, when $\Delta = a_2^2 - a_1 \cdot a_3 = 0$,
- hyperbolic, when $\Delta = a_2^2 - a_1 \cdot a_3 > 0$.

Note that the value of discriminant Δ determining the type of equation depends only on the values of the coefficients related to the derivatives of the 2nd order. Thus the type of 2nd order equation does not depend on the presence of the terms containing derivatives of 1st order or on the presence of source terms. The system of classification is valid globally only for linear equations with constant coefficients. When the coefficients vary it is possible that Eq. (5.1) can change its type in the considered domain, so the classification is valid only locally.

In the case of open channel flow the partial differential equations of 2nd order are associated with pollutant transport and simplified models of unsteady flow in the form of diffusive wave (see Chapters 1, 7 and 9). They have the form of advection-diffusion equation, i.e.:

$$\frac{\partial f}{\partial t} + U\frac{\partial f}{\partial x} - D\frac{\partial^2 f}{\partial x^2} = 0, \tag{5.2}$$

where:

f – scalar function as pollutant concentration or temperature,
U – cross-sectional average flow velocity,
D – coefficient of diffusion.

Comparing Eqs. (5.2) and (5.1) one obtains:

$$\Delta = a_2^2 - 4a_1 \cdot a_3 = 0 - 4 \cdot 0 \cdot D = 0,$$

which means, that the advection-diffusion equation (5.2) is of parabolic type.

There are some general relations, which enable us to predict the type of PDE even without a detailed analysis. It is known, that the flow problems variable in time (unsteady ones) are described by the equations of hyperbolic or parabolic type. If the considered problem is related to wave propagation, then the equations will be certainly of hyperbolic type. However, if in the equations some processes of dissipation are represented, as for example the stresses resulting from the liquid viscosity, heat conduction or mass diffusion, then one can expect equations of parabolic type. Finally, elliptic equations usually describe steady problems and equilibrium states and are rarely used in open channel flow modeling.

5.1.2 Classification of the Partial Differential Equations via Characteristics

The presented method of classification, while simple and clear, can be applied for the equations of 2nd order only. If we have a single equation of first order or a system of such equations, their type can be defined by the analysis of characteristics. To better explain this approach, let us make a short introduction to the characteristics concept using the simplest type of equation, i.e. the pure advection equation:

$$\frac{\partial f}{\partial t} + U\frac{\partial f}{\partial x} = 0, \tag{5.3}$$

where U is advective velocity. Both sides of Eq. (5.3) are multiplied by dt:

$$\frac{\partial f}{\partial t}dt + U \cdot dt\frac{\partial f}{\partial x} = 0. \tag{5.4}$$

If we introduce the following notation:

$$dx = U \cdot dt \tag{5.5}$$

and substitute it in Eq. (5.4), then we obtain:

$$\frac{\partial f}{\partial t}dt + \frac{\partial f}{\partial x}dx = 0. \tag{5.6}$$

The right hand side of Eq. (5.6) is the total derivative of the function $f(x, t)$, so that Eq. (5.4) can be rewritten as follows:

$$df = 0 \quad \text{for,} \tag{5.7a}$$

$$dx = U \cdot dt \tag{5.7b}$$

or

$$\frac{df}{dt} = 0 \quad \text{for} \quad \frac{dx}{dt} = U. \tag{5.8}$$

Therefore Eq. (5.3), which is a PDE, can be integrated as an ordinary differential equation, along the lines defined by Eq. (5.5) in (x, t) plane. From Eq. (5.8) results that in this case, the time variation of function f is equal to zero along this line. The formula:

$$\frac{dx}{dt} = U \tag{5.9}$$

defines the characteristics of the pure advection equation, or more precisely, it determines the slope of the line tangent to the characteristic lines. In a general case, when $U \neq$ const., the characteristics are curves in (x, t) plane.

The concept of characteristics can be used to classify the partial differential equations of 2nd order like Eq. (5.1). Fletcher (1991) showed, that the discriminant $\Delta = a_2^2 - a_1 \cdot a_3$ determines both type of equation and the nature of its characteristics. He found out that:

- hyperbolic equation has two real characteristics,
- parabolic equation has one real characteristic;
- elliptic equation has two imaginary characteristics.

This is valid for the equations of 1st order as well.

Coming back to the pure advection equation (5.3), one can find out that this equation is of hyperbolic type. It has one characteristic only, which is real. In addition one can notice that with $U =$ const. its characteristics are straight lines as presented in Fig. 5.1.

Since along these lines the function $f(x, t)$ does not vary, the governing equation can be easily integrated as an ordinary differential equation, on condition that the initial data are given along any line in (x, t) plane, which is not a characteristics. These properties form the base of the method of characteristics, one of the fundamental method for solution of PDEs of hyperbolic type (Abbott and Basco 1989). For a long time this method was the only one commonly applied to solve such kind of equations. Even now, the method of characteristics is widely applied to solve the unsteady pipe flow equations, very similar to the Saint Venant equations (Abbott and Basco 1989). In open channel hydraulic, as we shall see later, if the advection diffusion equation is solved by a splitting approach, the method of characteristics is frequently used for solving the advective part.

As an instructive exercise, let us apply this method to solve of the pure advection equation (5.3). From Eq. (5.7a) results that the function $f(x, t)$ does not vary along the characteristic line (5.7b). This fact allows us to find the following formula, representing an exact solution of Eq. (5.3):

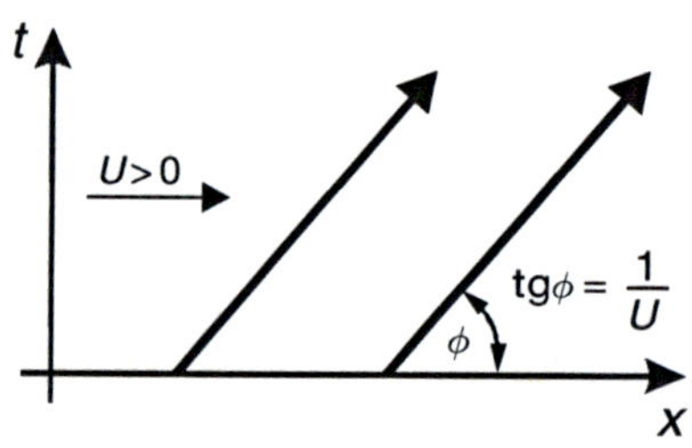

Fig. 5.1 Characteristics of the pure advection equation for $U =$ const. > 0

$$f(x,t) = f(x_0, t_0) = f\left(x - \int_{t_0}^{t} U \cdot dt, t_0\right) \tag{5.10}$$

Therefore a particle, which at current moment t is in the position defined by x, at $t = t_0$ was in the following position:

$$x_0 = x - \int_{t_0}^{t} U \cdot dt \tag{5.11}$$

Since in this case we assumed a constant flow velocity, then the characteristics being a straight line of slope $1/U$, allows us to rewrite Eq. (5.10) as

$$f(x,t) = f(x - (t - t_0)U, t_0) \tag{5.12}$$

From Eq. (5.12) results that the advection equation with constant advective velocity U causes pure translation of the initial distribution of $f(x, t_0)$ along the channel axis. This information is very useful, since it helps us to evaluate the accuracy of numerical methods of solution.

In open channel hydraulics the pure advection equation describes the advective transport of pollutants derived in Chapter 1, and as well as the kinematic wave equation, which is a simplified flood routing model (see Chapter 9). Apart from single equations of 1st order, systems of 1st order PDEs exist, as for example the system of Saint Venant equations. Classification of a system of equations can be also carried out using analysis of characteristics.

In the case of two independent variables t and x the system of two 1st order PDEs with two dependent variables $u(x, t)$ and $v(x, t)$ has the following general form (Fletcher 1991):

$$A_{11}\frac{\partial u}{\partial t} + B_{11}\frac{\partial u}{\partial x} + A_{12}\frac{\partial v}{\partial t} + B_{12}\frac{\partial v}{\partial x} = F_1, \tag{5.13a}$$

$$A_{21}\frac{\partial u}{\partial t} + B_{21}\frac{\partial u}{\partial x} + A_{22}\frac{\partial v}{\partial t} + B_{22}\frac{\partial v}{\partial x} = F_2, \tag{5.13b}$$

In matrix notation it can be written as:

$$\mathbf{A}\frac{\partial \mathbf{\Phi}}{\partial t} + \mathbf{B}\frac{\partial \mathbf{\Phi}}{\partial x} = \mathbf{F}, \tag{5.14}$$

With

$$\mathbf{\Phi} = \begin{Bmatrix} u \\ v \end{Bmatrix}, \quad \mathbf{F} = \begin{Bmatrix} F_1 \\ F_2 \end{Bmatrix}, \quad \mathbf{A} = \begin{bmatrix} A_{11} & A_{12} \\ A_{21} & A_{22} \end{bmatrix}, \quad \mathbf{B} = \begin{bmatrix} B_{11} & B_{12} \\ B_{21} & B_{22} \end{bmatrix}.$$

where:

$\mathbf{\Phi}$ – vector of unknown functions,
A, **B** – matrices of coefficients,
F – vector of right hand side.

From the definition of characteristics results that the total derivative of $\mathbf{\Phi}$ is constant along a characteristic, so that

$$d\mathbf{\Phi} = \mathbf{I}\frac{\partial \mathbf{\Phi}}{\partial t}dt + \mathbf{I}\,\frac{\partial \mathbf{\Phi}}{\partial x}dx = 0, \tag{5.15}$$

where **I** is identity matrix given as follows:

$$\mathbf{I} = \begin{bmatrix} 1 & 0 \\ 0 & 1 \end{bmatrix}.$$

Thus, Eqs. (5.10), (5.11) and (5.12) can be written together as:

$$\begin{bmatrix} \mathbf{A} & \mathbf{B} \\ \mathbf{I}dt & \mathbf{I}dx \end{bmatrix} \begin{bmatrix} \dfrac{\partial \mathbf{\Phi}}{\partial t} \\ \dfrac{\partial \mathbf{\Phi}}{\partial x} \end{bmatrix} = \begin{bmatrix} \mathbf{F} \\ \mathbf{0} \end{bmatrix}. \tag{5.16}$$

From the formal definition, the characteristic are such lines in time-space domain, at which the governing system of equation does not possesses a unique solution. This take place when the determinant of the coefficient matrix is equal to zero. This condition leads to the ordinary differential equation, which defines the characteristics. Equation (5.14) has no solution when the determinant of the matrix of coefficients in Eq. (5.16) is equal to zero, i.e. if $\det(\mathbf{A} \cdot dx - \mathbf{B} \cdot dt) = 0$. This condition can be written as:

$$\begin{aligned}(A_{11}A_{22} - A_{21}A_{12})\left(\frac{dx}{dt}\right)^2 - (A_{11}B_{22} - A_{21}B_{12} + B_{11}A_{22} - B_{22}A_{12})\frac{dx}{dt} \\ + (B_{11}B_{22} - B_{21}B_{12}) = 0.\end{aligned} \tag{5.17}$$

Equation (5.17) has two roots. Their values depend on the value of discriminant Δ, which is given by:

$$\begin{aligned}\Delta = (A_{11}B_{22} - A_{21}B_{12} + A_{22}B_{11} - A_{12}B_{21})^2 + \\ -4(A_{11}A_{22} - A_{21}A_{12})(B_{11}B_{22} - B_{21}B_{12}).\end{aligned} \tag{5.18}$$

The value of discriminant Δ determines the type of the system of equation. The possible cases are following:

- for $\Delta > 0$ Eq. (5.17) has two real roots and the system (5.13) is hyperbolic,
- for $\Delta = 0$ Eq. (5.17) has one real and one imaginary root and the system (5.13) is parabolic,
- for $\Delta < 0$ Eq. (5.17) has two imaginary roots and the system (5.13) is elliptic.

Although we consider classification of the systems of PDEs, the same method of classification is valid for a scalar equation as well. Let us reconsider the pure advection equation (5.3). For this equation relation (5.16) takes the following form:

$$\begin{bmatrix} 1 & U \\ dt & dx \end{bmatrix} \begin{bmatrix} \dfrac{\partial f}{\partial t} \\ \dfrac{\partial f}{\partial x} \end{bmatrix} = \begin{bmatrix} 0 \\ 0 \end{bmatrix}. \tag{5.19}$$

The equation of characteristic is given by the matrix determinant of Eq. (5.19), which must be equal to zero. Then the condition:

$$\det \begin{bmatrix} 1 & U \\ dt & dx \end{bmatrix} = 0 \tag{5.20}$$

gives the following relation:

$$dx = U \cdot dt \tag{5.21}$$

In such a way we obtained previously derived Eq. (5.9), which defines the characteristic of pure advection equation. Since this characteristic is a real one, the advection equation is classified as a hyperbolic PDE.

5.1.3 Classification of the Saint Venant System and Its Characteristics

Let us apply the considered approach to examine the system of Saint Venant equations (1.77) and (1.78):

$$\frac{\partial U}{\partial t} + U\frac{\partial U}{\partial x} + g\frac{\partial H}{\partial x} = g(s - S), \tag{5.22a}$$

$$\frac{\partial H}{\partial t} + H\frac{\partial U}{\partial x} + U\frac{\partial H}{\partial x} = 0. \tag{5.22b}$$

This system is rewritten in the form of Eq. (5.14) with:

$$\boldsymbol{\Phi} = \begin{Bmatrix} U \\ H \end{Bmatrix}, \quad \mathbf{F} = \begin{Bmatrix} g(s-S) \\ 0 \end{Bmatrix}, \quad \mathbf{A} = \begin{bmatrix} 1 & 0 \\ 0 & 1 \end{bmatrix}, \quad \mathbf{B} = \begin{bmatrix} U & g \\ H & U \end{bmatrix}.$$

The equation of characteristics is derived from the condition (5.15), which gives:

$$\det \begin{bmatrix} 1 & 0 & U & g \\ 0 & 1 & H & U \\ dt & 0 & dx & 0 \\ 0 & dt & 0 & dx \end{bmatrix} = 0. \tag{5.23}$$

From Eq. (5.23) one obtains the following equation:

$$\left(\frac{dx}{dt}\right)^2 - 2U\frac{dx}{dt} + (U^2 - g \cdot H) = 0. \tag{5.24}$$

Equation (5.24) has positive discriminant $\Delta = 4\,U^2 - 4(U^2 - g \cdot H) = 4\,g \cdot H$. Its square root is equal to:

$$\sqrt{\Delta} = 2\sqrt{g \cdot H}. \tag{5.25}$$

and the equations of characteristics are:

$$\left.\frac{dx}{dt}\right|_1 = \frac{2\,U + 2\sqrt{g \cdot H}}{2} = U + \sqrt{g \cdot H}, \tag{5.26a}$$

$$\left.\frac{dx}{dt}\right|_2 = \frac{2\,U - 2\sqrt{g \cdot H}}{2} = U - \sqrt{g \cdot H}. \tag{5.26b}$$

Therefore the system of Saint Venant equations, having two real characteristics, is classified as hyperbolic. In a similar manner one can classify the systems containing n equations of first order. We will apply this approach in Section 8.5 to classify the system of three equations.

The physical interpretation of the expression $\sqrt{g \cdot H}$ which appeared in Eq. (5.26) results from the analysis of the simplified form of governing equations (5.22). The first simplification deals with neglecting the source and the advective terms. Equation (5.22) become the following ones:

$$\frac{\partial U}{\partial t} + g\frac{\partial H}{\partial x} = 0, \tag{5.27a}$$

$$\frac{\partial H}{\partial t} + H\frac{\partial U}{\partial x} = 0 \tag{5.27b}$$

Note that omitting the right side of Eq. (5.22a) means that the frictionless flow in horizontal channel is considered. In the next step the linearization of Eq. (5.27b) is performed. Instead of variable coefficient $H(x, t)$ a constant one, equal to the average flow depth $\overline{H}$ is introduced. Therefore we obtain:

$$\frac{\partial U}{\partial t} + g\frac{\partial H}{\partial x} = 0, \tag{5.28a}$$

$$\frac{\partial H}{\partial t} + \overline{H}\frac{\partial U}{\partial x} = 0. \tag{5.28b}$$

Both equations are differentiated with regard to x and to t respectively, yielding:

$$\frac{\partial^2 U}{\partial t \cdot \partial x} + g\frac{\partial^2 H}{\partial x^2} = 0, \tag{5.29a}$$

$$\frac{\partial^2 H}{\partial t^2} + \overline{H}\frac{\partial U}{\partial x \cdot \partial t} = 0 \tag{5.29b}$$

Combining these equations one obtains:

$$\frac{\partial^2 H}{\partial t^2} = g \cdot \overline{H}\frac{\partial^2 H}{\partial x^2} \tag{5.30}$$

In a similar way, inverting the order of differentiation of Eqs. (5.28a) and (5.28b), we arrive to the second possible form of the shallow water wave equation

$$\frac{\partial^2 U}{\partial t^2} = g \cdot \overline{H}\frac{\partial^2 U}{\partial x^2} \tag{5.31}$$

Setting:

$$c = \sqrt{g \cdot \overline{H}} \tag{5.32}$$

Equations (5.30) and (5.31) can be presented in the common form:

$$\frac{\partial^2 f}{\partial t^2} = c^2\frac{\partial^2 f}{\partial x^2}. \tag{5.33}$$

In this equation f represents the flow depth $H(x, t)$ or the flow velocity $U(x, t)$.

This equation has the form of the well-known wave equation, describing a vibrating stretched string, where c represents the wave speed in the string (for details see e.g. Billingham and King (2000)). In the case of shallow water equations c is the speed of wave occurring at the free surface of water. Coming back to the equations of characteristics (5.26) one can find out that they represent the net velocities of propagating small disturbances of H or U arising at the surface of stream flowing at the velocity U. These disturbances travel in space and time depending of the relation between U and c. More information on interpretation of the characteristics is provided by an appropriate transformation of the system of shallow water equations (5.22) (Billingham and King 2000, Hervouet 2007, Tan Weiyan 1992).

Le us consider the system of Eqs. (5.22a) and (5.22b) in its homogeneous version:

$$\frac{\partial U}{\partial t} + U \frac{\partial U}{\partial x} + g \frac{\partial H}{\partial x} = 0, \tag{5.34a}$$

$$\frac{\partial H}{\partial t} + H \frac{\partial U}{\partial x} + U \frac{\partial H}{\partial x} = 0. \tag{5.34b}$$

Using relation (5.32) the flow depth is expressed as:

$$H = \frac{c^2}{g} \tag{5.35}$$

Substituting Eq. (5.32) in the continuity equation (5.34b) yields:

$$\frac{\partial}{\partial t}\left(\frac{c^2}{g}\right) + \frac{c^2}{g}\frac{\partial U}{\partial x} + U\frac{\partial}{\partial x}\left(\frac{c^2}{g}\right) = 0. \tag{5.36}$$

Developing the derivatives in (5.36) and dividing both sides of the obtained equation by the factor $g \cdot c$ gives:

$$\frac{\partial (2c)}{\partial t} + U\frac{\partial (2c)}{\partial x} + c\frac{\partial U}{\partial x} = 0. \tag{5.37}$$

Dynamic equation (5.34a) coupled with Eq. (5.35) yields:

$$\frac{\partial U}{\partial t} + U \frac{\partial U}{\partial x} + c\frac{\partial (2c)}{\partial x} = 0, \tag{5.38}$$

If we now add and subtract Eqs. (5.37) and (5.38) we obtain the following two equations:

$$\frac{\partial (U + 2c)}{\partial t} + (U + c)\frac{\partial (U + 2c)}{\partial x} = 0. \tag{5.39a}$$

$$\frac{\partial (U - 2c)}{\partial t} + (U - c)\frac{\partial (U - 2c)}{\partial x} = 0. \tag{5.39b}$$

In such a way the governing system of shallow water equations is expressed in an equivalent form, in which the characteristic equations are involved. Note that both equations describe advective transport of the quantities $U + 2c$ or $U - 2c$ with advective celerity $U + c$ or $U - c$, respectively. Therefore using Eqs. (5.26a) and (5.26b) they can be rewritten as:

$$\frac{d(U + 2c)}{dt} = 0 \quad \text{for} \quad \left.\frac{dx}{dt}\right|_1 = U + c \tag{5.40a}$$

$$\frac{d\,(U-2c)}{dt}=0 \quad \text{for} \quad \left.\frac{dx}{dt}\right|_2 = U-c \tag{5.40b}$$

These equations show that the quantities $U \pm 2c$ are constant on the characteristic curves determined by $U \pm c$. The transported quantities are called the Riemann invariants of the system (5.34). The characteristic equations are of great importance for numerical solution of the hyperbolic equations since they allow for an appropriate formulation of the required auxiliary conditions. This is because the character of characteristics vary depending on the relation between the flow velocity U and the wave celerity c. On the other hand these variables are related to each other via the Froude number F_r defined by Eq. (1.17). This number, being the ratio of U and c, allows us to determine the character of channel flow. Assuming $U > 0$ the following situations are possible:

- if $U < c$, ($F_r < 1$) the flow is subcritical: the wave celerity exceeds the flow velocity, so any flow disturbance at the considered channel reach travels in both directions, downstream and upstream;
- if $U > c$, ($F_r > 1$) the flow is supercritical: the flow velocity exceeds the wave celerity, so any flow disturbance at the considered channel reach travels in one direction – downstream;
- if $U = c$, ($F_r = 1$) the flow is critical.

The relation between the characteristics of hyperbolic equations and the required auxiliary conditions is the subject of the next section.

5.1.4 Well Posed Problem of Solution of the Hyperbolic and Parabolic Equations

Classification of the partial differential equations has essentials meaning, because it allows us to formulate properly the problem of solution for the considered PDE. PDEs are solved in strictly defined domains and for strictly defined auxiliary conditions, imposed at the limits of the solution domain. In other words, each solution of the equation or the system of equations corresponds to the conditions a priori imposed on this unknown solution. Therefore the function being the solution of the considered problem must satisfy simultaneously both the solved equation in the domain of solution and the additional conditions imposed at its limits. These additional conditions usually are called the limit conditions. They are divided in two kinds: the initial conditions and the boundary conditions. The initial conditions provide information on the function or the functions in the solution domain at the initial moment t_0. They are present in the unsteady problems, when one of the independent variable is time. The boundary conditions, on the other hand, provide information imposed at the physical boundary the considered domain of solution. In open channel flow problems such boundaries are constitute by the upstream end and the downstream end of a channel. If at the boundary of the considered domain the value of the unknown function is given, then it is said that Dirichlet condition is imposed, whereas if the derivative of the function is set, then it is said that Neumann condition is imposed. Apart from the mentioned conditions so-called mixed conditions can be formulated as well. Since in open channel flow they are not applied we do not them discussed here.

After Hadamard (Fletcher 1991) the problem equation and auxiliary conditions of solution, i.e. the limit conditions are mathematically well posed if they ensure:

- existence of solution,
- uniqueness of solution.
- continuous dependence of solution on the auxiliary conditions.

The problem of proving the existence of solution is important from the point of view of mathematical formalism. For the purposes of engineering practice, however, it is reasonable to assume that the model equations based on the principles of conservation laws have solution in most cases.

The uniqueness of solution is related to the auxiliary conditions formulated for the considered type of equation. Generally, if there are too few boundary conditions no unique solution can be obtained, whereas too many conditions imposed at the boundary can generate unphysical solution (Fletcher 1991). In order to properly define the auxiliary conditions one needs to know the type of solved equations. For this reason, the correct classification of considered partial differential equations and the knowledge of their characteristics structure has fundamental meaning for well posedness of solution problem.

For some problems the way of proper formulation of the initial and boundary conditions is well known. For instance for the hyperbolic equations the following rule is valid: at every boundary of the considered solution domain it is necessary to impose as many additional conditions as many characteristics enter the solution domain from this boundary. This rule results from the nature of characteristics discussed in the preceding section. As an example, let us consider the pure advection equation (5.13):

$$\frac{\partial f}{\partial t} + U\frac{\partial f}{\partial x} = 0 \quad \text{for} \quad U = \text{const.} \tag{5.41}$$

The characteristics of Eq. (5.41) on the plane (x,t), given by Eq. (5.9) and dependent on the sign of the flow velocity U, are presented in Fig. 5.2. Then if we want to solve Eq. (5.41) in the domain: $0 \leq x \leq L$ and $t \geq 0$ with $U > 0$, the following auxiliary conditions must be prescribed (Fig. 5.2a):

- initial condition:

at $t = 0$ $f(x,0) = f_i(x)$ for $0 \leq x \leq L$ should be imposed

- boundary condition:

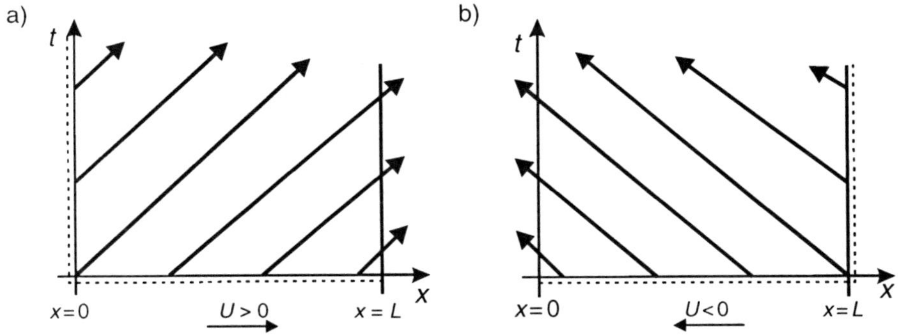

Fig. 5.2 The characteristics for pure advection equation (5.27) for $U > 0$ (**a**) and for $U < 0$ (**b**)

at the upstream end $x=0$ $f(0,t)=f_0(t)$ for $t \geq 0$ should be imposed where $f_i(x)$ and $f_0(t)$ are given.

If the velocity vector has opposite (negative) sign, as in Fig. 5.2b, then the boundary condition must be imposed at the opposite end of the channel reach i.e. at $x = L$. Generally, for the pure advection equation the boundary condition is prescribed at the end, where the water inflows.

As the second example let us consider the system of Saint Venant equations (5.22). In the preceding section it was shown that this system possesses two families of real characteristics given by Eqs. (5.26a) and (5.26b):

$$\left.\frac{dx}{dt}\right|_1 = U + \sqrt{g \cdot H}, \tag{5.42a}$$

$$\left.\frac{dx}{dt}\right|_2 = U - \sqrt{g \cdot H} \tag{5.42b}$$

In this case the structure of the characteristics in the solution domain ($0 \leq x \leq L$ and $t \geq 0$) can vary according to the relation between the flow velocity U and the wave celerity in the shallow water $c = (g \cdot H)^{1/2}$. If $U < (g \cdot H)^{1/2}$, i.e. for the subcritical flow when the Froude number is less than unity ($F_r < 1$), both families of characteristics are inclined in the opposite directions. Their form is presented in Fig. 5.3.

Since two characteristics enter the solution domain through the line $t = 0$, two initial conditions must be prescribed: $U(x,0) = U_i(x)$, $H(x,0) = H_i(x)$ for $0 \leq x \leq L$. As only one characteristic enters the solution domain through each of the boundaries $x = 0$ and $x = L$, then at each channel end only one additional condition must be imposed: $U(0,t) = U_0(t)$ or $H(0,t) = H_0(t)$ for $t \geq 0$ and $U(L,t) = U_L(t)$ or $H(L,t) = H_L(t)$ for $t \geq 0$.

If $U > (g \cdot H)^{1/2}$, i.e. for the supercritical flow in the channel, when the Froude number is greater than unity ($F_r > 1$), both families of characteristics are inclined in the same direction. Their form is presented in Fig. 5.3 by dashed lines. In this

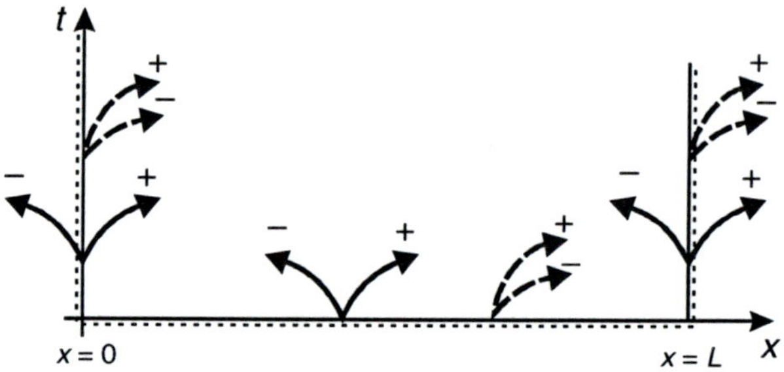

Fig. 5.3 Characteristics for the system of Saint Venant equations with $U < (g \cdot H)^{1/2}$ (*solid lines*) and for $U > (g \cdot H)^{1/2}$ (*dashed lines*)

case the initial conditions are prescribed identically as previously, so both functions H and U must be imposed for $t = 0$ and $0 \leq x \leq L$. The boundary conditions are specified as follows: at the upstream end of the channel $x = 0$, both functions $U(0,t) = U_0\,(t)$ and $H\,(0,t) = H_0\,(t)$ for $t \geq 0$ have to be imposed, whereas at the downstream end $x = L$ no condition is required, since through this boundary no characteristic enters the solution domain (Fig. 5.3).

As results from the presented discussion, the knowledge of the characteristics structure is indispensable in the case of hyperbolic equations because it allows us to prescribe correctly the required auxiliary conditions and consequently to ensure well-posedness of the solution problem.

Quite a different situation appears in the case of the partial differential equations of parabolic type. In such a case the characteristics are not useful since some of them have imaginary character. The problem of proper formulation of the auxiliary conditions for these equations is well known. To ensure the unique solution of the parabolic equation as diffusion equation (5.17):

$$\frac{\partial f}{\partial t} - D\frac{\partial^2 f}{\partial x^2} = 0 \tag{5.43}$$

in the solution domain, i.e. for $0 \leq x \leq L$ and $t \geq 0$, the following initial-boundary conditions should be specified (Fig. 5.4):

- for $t = 0$ the initial condition takes the form: $f(x{,}0) = f_i(x)$ for $0 \leq x \leq L$,
- at the upstream end $x = 0$ the Dirichlet boundary condition $f(0,\ t) = f_0(t)$ for $t \geq 0$, is applied,
- at the downstream end $x = L$ either the Dirichlet boundary condition $f(0,\ t) = f_L(t)$ for $t \geq 0$, or the Neumann condition $\partial f(L,t)/\partial x = \phi_L(t)$ for $t \geq 0$ is applied.

Of course, the Dirichlet and Neumann conditions can be inverted and imposed at opposite ends as listed above. However we have to remember that for the advection-diffusion equation, which is also of parabolic type, the Dirichlet condition must

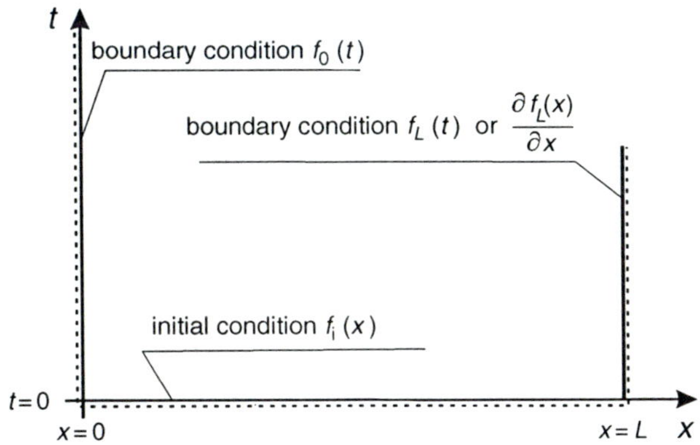

Fig. 5.4 Auxiliary conditions on the solution domain limits for the parabolic equation

be specified at least at one boundary. For a system of parabolic equations a set of additional conditions, similar to the one presented above, have to be imposed for each dependent variable (Fletcher 1991).

The initial and boundary conditions i.e. the functions f_p, f_0 and f_L, one should be consistent. Otherwise, false gradients of the function $f(x, t)$ can occur during the computation. They can give rise to numerical disturbances.

The third condition of well-posedness, i.e. stability, requires that insignificant variations of the initial and boundary conditions cause only insignificant variations of the obtained solution. Usually the additional conditions have approximate character. If they are not fulfilled exactly then the errors generated during computation will increase in uncontrolled manner and consequently these errors can dominate the numerical solution. This problem is especially important for the hyperbolic equations.

By analogy one can formulate similar requirements in relation to the numerical algorithm. It is assumed that the numerical problem is well-posed if:

- the numerical solution exists.
- the numerical solution is unique,
- the numerical solution continuously depends on the approximate auxiliary conditions.

As states Fletcher (1991), the well-posed numerical problem requires not only the well-posedness of the solution problem for the partial differential equations, but it requires also that the numerical algorithm of solution is well-posed (stable). Only in such a case an approximate solution of the well-posed numerical problem will be close to the exact solution of the well-posed problem for the partial differential equation.

5.1.5 Properties of the Hyperbolic and Parabolic Equations

A classical hyperbolic equation, often used as an example to demonstrate the properties of this type of equations, is the linear wave equation (5.33):

$$\frac{\partial^2 f}{\partial t^2} - c^2 \frac{\partial^2 f}{\partial x^2} = 0 \tag{5.44}$$

where:

- t – time,
- x – position,
- c – celerity of the wave propagation (in this case $= (g \cdot \overline{H})^{1/2}$)
- $f(x, t)$ – function satisfying Eq. (5.44), which can be either the flow depth $H(x, t)$ or the flow velocity $U(x, t)$.

Generally Eq. (5.44) describes the propagation of a small disturbance on the surface of shallow body of perfect liquid. Assuming $c = 1$, one obtains:

$$\frac{\partial^2 f}{\partial t^2} - \frac{\partial^2 f}{\partial x^2} = 0 \tag{5.45}$$

Assume that we seek the solution of Eq. (5.45) in the following domain: $0 \leq x \leq 1$ and $t \geq 0$. Since this equation has two real characteristics described by the equations $dx/dt = 1$ and $dx/dt = -1$, respectively. shown in Fig. 5.5 (Fletcher 1991), additional conditions must be given at $t = 0$, $x = 0$ and $x = 1$.

After Fletcher (1991) let us assume the following initial-boundary conditions:

– initial conditions

$$f(x, t = 0) = \sin(\pi \cdot x), \quad \frac{\partial f(x, t = 0)}{\partial t} = 0 \quad \text{for} \quad 0 \leq x \leq 1$$

– boundary conditions

$$f(x = 0, t) = 0 \quad \text{and} \quad f(x = 1, t) = 0 \quad \text{for} \quad t \geq 0$$

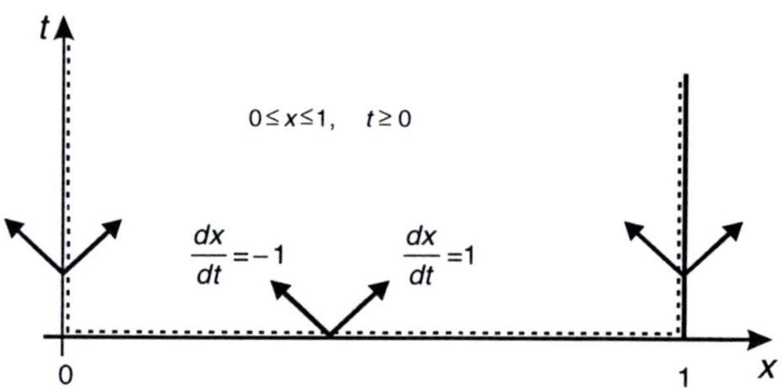

Fig. 5.5 Characteristics of Eq. (5.45)

Table 5.1 The values of function (5.33) for $0 \le x \le 1$

x	$f(x, t)$	x	$f(x, t)$
0.00	0.000	0.60	−0.294
0.10	0.294	0.70	−0.476
0.20	0.476	0.75	−0.500
0.25	0.500	0.80	−0.476
0.30	0.476	0.90	−0.294
0.40	0.294	1.00	0.000
0.50	0.000	–	–

For the above specified conditions, Eq. (5.45) has an exact solution (Fletcher 1991):

$$f(x,t) = \sin(\pi \cdot x) \cdot \cos(\pi \cdot x) \quad 0 \le x \le 1 \tag{5.46}$$

This solution has character of a non-damped oscillation, independent of the time t. The values of function (5.46) are displayed in Table 5.1.

The lack of attenuation of the propagating wave is a typical feature of linear hyperbolic equations. For this reason any discontinuity, introduced into their solution by initial or boundary conditions, will never disappear. Moreover, in the case of the hyperbolic non-linear equations a discontinuity in the solution domain may occur even if the initial and boundary conditions are continuous.

As far as parabolic PDEs are considered, they arise for problems associated with dissipative mechanisms like, for instance, the liquid viscosity or the diffusive mass or heat transport. The most popular example of such type of equation is the diffusion equations describing the transport of mass or energy without the advection. For the coefficient of diffusion equal to unity ($D = 1$ m^2/s) the diffusion equations is as follows:

$$\frac{\partial f}{\partial t} = \frac{\partial^2 f}{\partial x^2}. \tag{5.47}$$

We are looking for its solution in the same domain as it was done previously for the hyperbolic equation, i.e. for $0 \le x \le 1$ and $t \ge 0$, with the following auxiliary conditions:

- initial condition: $f(x, 0) = \sin(\pi \cdot x)$ for $0 \le x \le 1$,
- boundary conditions: $f(0, t) = f(1, t) = 0$ for $t > 0$.

For those conditions Eq. (5.47) has the following analytical solution (Fletcher 1991):

$$f(x,t) = \sin(\pi \cdot x) \exp(-\pi^2 \cdot t). \tag{5.48}$$

From Eq. (5.48) results that the function $f(x, t)$ is attenuated exponentially. This process illustrates the mechanism of dissipation. Note that in the hyperbolic

Table 5.2 The values of function (5.48) for $0 \leq x \leq 1$

x	$f(x, t = 0.0)$	$f(x, t = 0.05)$	$f(x, t = 0.10)$
0.00	0.000	0.000	0.000
0.10	0.309	0.189	0.115
0.20	0.588	0.359	0.219
0.30	0.809	0.494	0.302
0.40	0.951	0.581	0.354
0.50	1.000	0.610	0.373
0.60	0.951	0.581	0.354
0.70	0.809	0.494	0.302
0.80	0.588	0.359	0.219
0.90	0.309	0.189	0.115
1.00	0.000	0.000	0.000

equations the lack of such a mechanism leads to non-attenuated oscillations. As the solution of a parabolic equation advances in time, it is diffused in space. The presence of the dissipation mechanism causes that even if the initial or boundary conditions contain a discontinuity, it disappears after some time, and the solution inside the considered domain will be always smooth. The function $f(x, t)$ for the selected values of time: $t = 0$, $t = 0.05$ and $t = 0.10$ are displayed in Table 5.2. One can see, that the wave is strongly attenuated as time increases.

The solutions of hyperbolic and parabolic equations presented above can be interpreted in a more general way, via analysis of the Fourier representation of a general solution of wave equation. The Fourier representation of the solution of wave equation can be written as follows (Fletcher 1991):

$$f(x, t) = \sum_{m=-\infty}^{\infty} f_m(x, t) \tag{5.49}$$

where $f_m(x, t)$ is mth component of representation. Since both considered equations were assumed to be linear, the components of the Fourier representation can be considered independently. Each of them has the following form:

$$f_m(x, t) = A_m \cdot e^{-p(m)t} \cdot e^{i \cdot m \cdot (x - q(m) \cdot t)} \tag{5.50}$$

where:

A_m – amplitude of the mth component,
m – wave-number related to the components' wave length λ by formula $m = 2\pi/\lambda$,
$p(m)$ – dissipation parameter, which determines how rapidly the amplitude of the wave is attenuated,
$q(m)$ – wave propagation speed,
i – imaginary unit.

Equation (5.50) describes the propagation of a plane wave that is subjected to both dissipation and dispersion. The propagating wave is described by two parameters: amplitude and phase speed. If during propagation the amplitude is attenuated, i.e. decreases in time, it is said that the wave is subjected to dissipation. If during propagation the phase speed varies, it is said that the wave is subjected to dispersion.

Let us assume, that the movement is described by the equations discussed earlier. Introducing successively Eq. (5.50) into these equations one can determine the dissipation parameter $p(m)$ and the wave speed $q(m)$. Consequently one obtains a particular form of Eq. (5.50) corresponding to the considered equations:

Since the wave described by Eq. (5.50) must satisfy the partial differential equation (5.44), it can be substituted into this equation. The derivatives of 2nd order in Eq. (5.44) are given as:

$$\begin{aligned}\frac{\partial^2 f}{\partial t^2} &= \frac{\partial}{\partial t}\left(\frac{\partial}{\partial t}(A_m \cdot \exp(-p(m)t)\exp(-i \cdot m(x - q(m)t))\right) = \\ &= \frac{\partial}{\partial t}(A_m \cdot \exp(-p(m)t)\exp(-i \cdot m(x - q(m)t) \cdot (-p(m) + i \cdot m \cdot q(m))) = \\ &= A_m \cdot \exp(-p(m)t)\exp(-i \cdot m(x - q(m)t) \cdot (-p(m) + i \cdot m \cdot q(m))^2\end{aligned} \tag{5.51}$$

$$\begin{aligned}\frac{\partial^2 f}{\partial x^2} &= \frac{\partial}{\partial x}\left(\frac{\partial}{\partial x}(A_m \cdot \exp(-p(m)t)\ \exp(-i \cdot m(x - q(m)t))\right) = \\ &= \frac{\partial}{\partial x}(A_m \cdot \exp(-p(m)t)\exp(-i \cdot m(x - q(m)t)) \cdot (-i \cdot m)) = \\ &= A_m \cdot \exp(-p(m) \cdot t)\exp(-i \cdot m(x - q(m)t))(-i \cdot m)^2\end{aligned} \tag{5.52}$$

Substitution of Eqs. (5.51) and (5.52) in Eq. (5.44) yields:

$$\begin{aligned}&A_m \cdot \exp(-p(m)t)\exp(-i \cdot m(x - q(m)t))(-p(m) + i \cdot m \cdot q(m))^2 + \\ &-c^2 \cdot A_m \cdot \exp(-p(m)t)\exp(-i \cdot m(x - q(m)t)(-i \cdot m)^2 = 0\end{aligned} \tag{5.53}$$

Dividing both sides of Eq. (5.53) by Eq. (5.50) and rearranging them one obtains the following expression:

$$-p(m) + i \cdot m \cdot q(m) = -c \cdot i \cdot m \tag{5.54}$$

from which results:

$$p(m) = 0, \tag{5.55}$$

$$q(m) = -c \tag{5.56}$$

Then the wave (5.50) satisfying the wave equation (5.44) is expressed as follows:

$$f_m(x, t) = A_m \cdot e^{-i \cdot m(x - c \cdot t)} \tag{5.57}$$

This is a confirmation of the previously obtained result – the plane wave satisfying the wave equation (5.44) moves without any deformation. In other words, such a wave does not dissipate.

A similar approach can be applied to the examine the propagation of a plane wave governed by the diffusion equation:

$$\frac{\partial f}{\partial t} = D\frac{\partial^2 f}{\partial x^2}. \tag{5.58}$$

As the wave described by Eq. (5.50) must satisfy the partial differential equation (5.58), let us substitute them . The derivative of 1st order with regard to time is given as:

$$\begin{aligned}\frac{\partial f}{\partial t} &= \frac{\partial}{\partial t}(A_m \cdot \exp(-p(m)t)\exp(-i \cdot m(x - q(m)t)) = \\ &= A_m \cdot \exp(-p(m)t)\exp(-i \cdot m(x - q(m)t) \cdot (-p(m) + i \cdot m \cdot q(m))\end{aligned} \tag{5.59}$$

On the other hand, the derivative of 2nd order with regard to x is given by Eq. (5.52). Substitution of Eqs. (5.59) and (5.52) in Eq. (5.58) yields:

$$\begin{aligned}&A_m \cdot \exp(-p(m)t)\exp(-i \cdot m(x - q(m)t)) \cdot (-p(m) + i \cdot m \cdot q(m)) = \\ &= D \cdot A_m \cdot \exp(-p(m)t)\exp(-i \cdot m(x - q(m)t)) \cdot (-i \cdot m)^2\end{aligned} \tag{5.60}$$

After dividing both sides of Eq. (5.60) by Eq. (5.50) one obtains:

$$-p(m) + i \cdot m \cdot q(m) = D \cdot (-i \cdot m)^2 \tag{5.61}$$

From this relation results that:

$$p(m) = D \cdot m^2 \quad \text{and} \quad q(m) = 0 \tag{5.62, 5.63}$$

Therefore the plane wave governed by the diffusion equation (5.58) is as follows:

$$f_m(x,t) = A_m \cdot e^{-D \cdot m^2 \cdot t} \cdot e^{i \cdot m \cdot x} \tag{5.64}$$

In this equation one can notice the factor representing the dissipation process which is responsible for attenuation of the wave amplitude with time. Therefore the properties of the parabolic partial differential equation observed in its exact solution (5.48) are confirmed by the Fourier analysis.

5.1.6 Properties of the Advection-Diffusion Transport Equation

The wave equation and the diffusion equation are rather seldom applied in open channel hydraulics. However, both equations are closely related to the advection-diffusion transport equations (1.157) or (1.174) describing the transport of mass or heat in open channels. They can be rewritten in simplified form as:

$$\frac{\partial f}{\partial t} + U\frac{\partial f}{\partial x} - D\frac{\partial^2 f}{\partial x^2} = \delta, \tag{5.65}$$

where:

$f(x, t)$ – function represented transported quantity,
U – cross-sectional average flow velocity,
D – coefficient of diffusion.
δ – source term.

The advection-diffusion transport equation is one of the most challenging equations in mathematical physics, as it represents a superposition of two very different transport processes: advection and diffusion. The first one has hyperbolic character, whereas the second one has parabolic character. In the preceding section we discussed the properties of the hyperbolic and parabolic partial differential equations separately, showing that they are completely different. According to the relative intensity of the advective and diffusive transport, Eq. (5.65) can be dominated by either hyperbolic or parabolic features. Numerical problems arise when the transport process is dominated by advection. Thus, it seems instructive to evaluate in more detail the role of subsequent terms of Eq. (5.65). To this order we follow the way applied by Szymkiewicz and Mitosek (2007).

As we found out previously, the partial differential equations of hyperbolic type are related to the problem of wave propagation without any dissipative process. Therefore in the solution of hyperbolic equation no damping is observed. Consequently any discontinuity introduced into the solution by initial or boundary conditions cannot disappear in time as it happens in case of parabolic equation.

In order to explain the considered process of damping and smoothing of the propagating wave let us begin with the 1D pure advection equation being the simplest example of the hyperbolic equation. The following equation

$$\frac{\partial f}{\partial t} + U\frac{\partial f}{\partial x} = 0 \tag{5.66}$$

where:

f – scalar quantity (temperature or concentration),
U – flow velocity

describes the transport of heat or mass of dissolved substance by flowing stream of fluid. The term $\partial f/\partial t$ represents accumulation for unsteady process whereas the term $U\ \partial f/\partial x$ represents advection. If $U =$ const., then according to Eq. (5.66) the initial distribution of f will be translated along x axis without any deformation. In Fig. 5.6 an exact solution of the advective transport equation with constant velocity $U = 0.5$ m/s is shown. The rectangular distribution of $f(x, t)$ imposed at the boundary $x = 0$ is transported towards the downstream end keeping its initial shape.

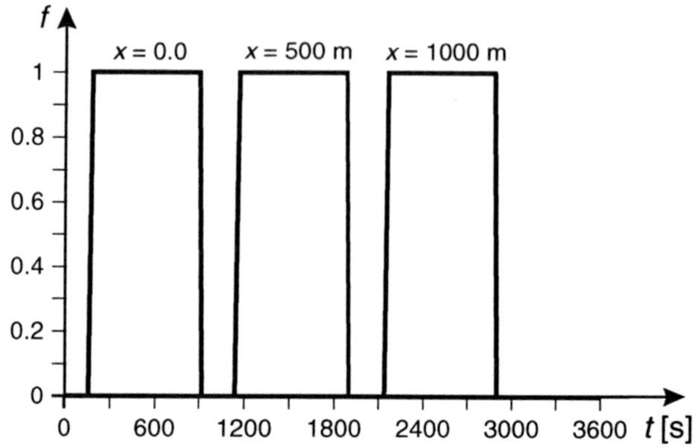

Fig. 5.6 Pure translation of *f* along *x* axis

Sometimes a hyperbolic equation contains a source term, which represents the exchange of transported quantity or describes a chemical process. To show the role of source term let us consider the following equation, similar to Eq. (5.66):

$$\frac{\partial f}{\partial t} + U\frac{\partial f}{\partial x} = \delta \tag{5.67}$$

where δ represents source ($\delta > 0$) or sink ($\delta < 0$) term. Let us assume the simplest form of the source term:

$$\delta = \kappa \cdot f \tag{5.68}$$

where κ is a decay coefficient. Exact solution of Eq. (5.67) for the set of data accepted previously and with $\kappa = 0.0001\ \mathrm{s}^{-1}$ is presented in Fig. 5.7. One can notice that in this case the initially rectangular distribution of *f* is transported along *x* axis without any smoothing. Its height is systematically reduced in time with intensity determined by the term δ acting as a sink. Of course, this effect can be considered as some kind of damping. However it has a non-dissipative character.

It is well known that the dissipative processes are irreversible while advection with source term can be inverted. To this order the signs of *U* and δ should be reversed. Then starting from final distribution of *f* at the downstream end one can get the boundary condition at the upstream end.

To obtain the effect of smoothing of the sharp fronts of *f*, a diffusion term should be introduced into Eq. (5.67). In this way pure advection equation becomes the advection-diffusion equation (5.65):

$$\frac{\partial f}{\partial t} + U\frac{\partial f}{\partial x} - D\frac{\partial^2 f}{\partial x^2} = \delta, \tag{5.69}$$

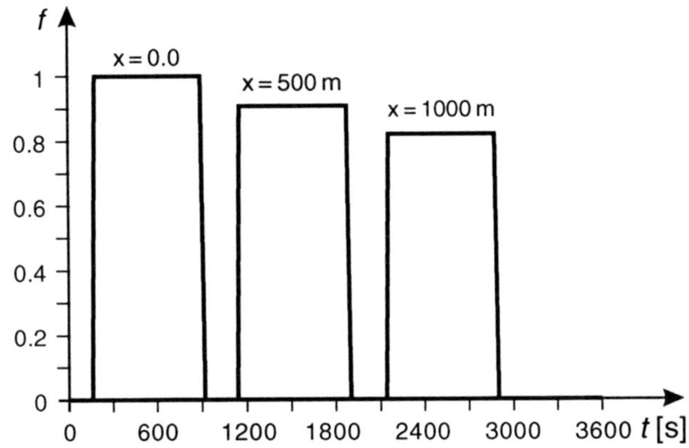

Fig. 5.7 Advective transport with $\delta = 0.0001 \cdot f$

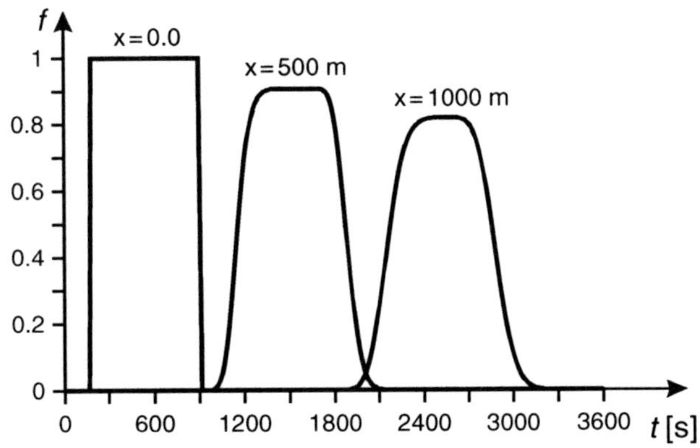

Fig. 5.8 Solution of advective-diffusive transport with $U = 0.5$ m/s, $D = 0.625$ m^2/s, $\delta = 0.0001 \cdot f$

Now instead of a hyperbolic equation we have a parabolic one. Consequently while solving the transport of f having a rectangular distribution at $x = 0$ one obtains simultaneously the following effects:

- translation with flow velocity U,
- reduction of the height of rectangle,
- smoothing.

It means that the transported rectangular distribution looses its initial steep shape. In Fig. 5.8 the combined effect of advection with flow velocity $U = 0.5$ m/s, diffusion with $D = 0.625$ m^2/s and sink with $\delta = 0.0001 \cdot f\,\mathrm{s}^{-1}$ is presented.

The solution of Eqs. (5.66), (5.67) and (5.69) presented above, can be interpreted in a more general way, via analysis of the Fourier representation of a general solution of wave equation (5.50) introduced in the preceding section. First, let us list the functions f and δ as well as the derivatives represented in Eq. (5.69). We have:

$$f(x,t) = A_m \cdot \exp(-p(m)t) \cdot \exp(-i \cdot m(x - q(m) \cdot t) \tag{5.70}$$

$$\delta = \kappa \cdot f = \kappa \cdot A_m \cdot \exp(-p(m)t) \cdot \exp(-i \cdot m(x - q(m) \cdot t) \tag{5.71}$$

$$\frac{\partial f}{\partial t} = A_m \cdot \exp(-p(m) \cdot t) \cdot \exp(-i \cdot m(x - q(m) \cdot t) \cdot (-p(m) + i \cdot k \cdot q(m)) \tag{5.72}$$

$$\frac{\partial f}{\partial x} = A_m \cdot \exp(-p(m) \cdot t) \cdot \exp(-i \cdot m(x - q(m) \cdot t) \cdot (-i \cdot m) \tag{5.73}$$

$$\frac{\partial^2 f}{\partial x^2} = A_m \cdot \exp(-p(m) \cdot t) \exp(-i \cdot m(x - q(m)t))(-i \cdot m)^2 \tag{5.74}$$

Using these relations one obtains the following Fourier components:

- for pure advection equation (Eq. 5.66) we have $p(m) = 0$ and $q(m) = U$, which gives:

$$f_m(x,t) = A_m \cdot e^{i \cdot m \cdot (x - U \cdot t)} \tag{5.75}$$

- for advection equation with source term (Eq. 5.67) we have $p(m) = \kappa$ and $q(m) = U$, which gives:

$$f_m(x,t) = A_m \cdot e^{-\kappa \cdot t} \cdot e^{i \cdot m \cdot (x - U \cdot t)} \tag{5.76}$$

- for advection-diffusion equation (5.69) without the source term ($\delta = 0$) we have $p(m) = D\,m^2$ and $q(m) = U$, which gives:

$$f_m(x,t) = A_m \cdot e^{-D \cdot m^2 \cdot t} \cdot e^{i \cdot m \cdot (x - U \cdot t)} \tag{5.77}$$

- for advection-diffusion equation (5.69) with the source term in the form of Eq. (5.68) we have $p(m) = \kappa + D \cdot m^2$ and $q(m) = U$, which gives:

$$f_m(x,t) = A_m \cdot e^{-(\kappa + D \cdot m^2) \cdot t} \cdot e^{i \cdot m \cdot (x - U \cdot t)} \tag{5.78}$$

It can be seen that all the considered transport equations, i.e. Equations (5.66), (5.67) and (5.69) ensure wave propagation at the same speed regardless of its wavelength λ (or wave number m since $m = 2\pi/\lambda$). In each case all components have the same speed, equal to the advective velocity $q(m) = U$, while the damping process depends on the type of equation.

For the plane wave governed by pure advection equation (Eq. 5.66) its amplitude is not damped. The propagating wave keeps constant amplitude and consequently initial distribution of transported quantity f is not disturbed, as it was shown in Fig. 5.6. Introduction of a source term into the pure advection equation (Eq. 5.67) changes the behavior of the propagating wave. While traveling, the wave amplitude is decreased in every time decrement Δt by exp $(-\kappa \cdot \Delta t)$. Although the general solution (Eq. 5.49) contains an infinite number of the components of the Fourier representation, the process of attenuation is the same for all of them, since it is not dependent on the wave number. All amplitudes are decreased in the same way. Consequently, as the initial distribution of f is decreased proportionally with time, it retains the initially imposed general form, as it was presented in Fig. 5.7.

The solution of the advection-diffusion transport without the source term (Eq. (5.69) with $\delta = 0$) shows that the propagating plane wave is attenuated because of the diffusion. The diffusive term acts with variable intensity, since the damping depends on the wave number m. The wave attenuation is more rapid for short waves than for long ones. Consequently the amplitude of any component of the Fourier representation is attenuated depending on its wavelength. Since a general solution of the propagation wave (Eq. 5.49) contains an infinite number of components, each of them is damped in a different way. The effect of this process is observed as a smoothing of the initial distribution of f, especially significant for steep fronts. If in the considered advection-diffusion equation the source term with a decay coefficient κ is present, its solution contains coupled effect of both diffusive and source terms (Eq. 5.69). The amplitude of propagating wave is decreased by a term partially dependent on the decay coefficient κ and partially on the coefficient of diffusion D multiplied by the wave number m raised to the second power. The combined effect of both processes is seen in Fig. 5.8.

The effects caused by various terms of the advection-diffusion transport equations (5.65) are worth to remember, since they are helpful in the interpretation the results of its numerical solution. Numerical errors often produce effects similar to the results of physical processes, for example numerical diffusion influences the solution in the same way as physical diffusion. Thus it is important to know what kind of solution should result from the physical transport mechanisms included in the considered equation.

5.2 Introduction to the Finite Difference Method

5.2.1 Basic Information

Partial differential equations describing flow and transport in open channels are routinely solved using numerical methods. Analytical solutions have limited applicability for practical problems, which are characterized by irregular and non-uniformly spaced channel cross-sections and variable in time boundary conditions. Thus, analytical solutions are useful mostly as a tool to verify the accuracy of numerical methods, which are then applied to solve real-life problems.

Numerical methods allow to solve the mathematical problems in which both the data and the results of calculations are given in the form of numbers. Operating on numbers is their basic feature. The same feature is characteristic of computers. For this reason there exist a very close relation between the progress in numerical methods and in computer technology. This is particularly true in the domain of solution of the partial differential equations.

It is well known that differential calculus is based on the concept of continuous medium. This idea must be applied to derive these equations. However, the main concept of the numerical methods is a number. Therefore the continuous domain and the partial differential equations must be converted to the form suitable for a treatment by the computational tools, i.e. a PDE or a system of PDEs must be converted into a system of algebraic equations, suitable for numerical solution.

In general, this process of transformation is carried out in two stages (Fletcher 1991). At first the continuous domain of solution of the considered equation is discretized, i.e. it is converted into a discrete domain constituted by a set of points, separated from each other. The idea of such conversion is presented in Fig. 5.9.

The continuous time-space domain C, defined as: $x_u \leq x \leq x_d$ and $t_0 \leq t \leq t_{\max}$, is replaced by a set of points called nodes. The nodes can be uniformly spaced with constant intervals Δx = const. and Δt = const., or they can be located in non-uniform way. The position of any node is defined by its co-ordinates x_j and t_n, with $1 \leq j \leq M$, $1 \leq n \leq N$. The discrete domain D is then constituted by $M \times N$ nodes. The numerical methods allow to calculate the approximate values of the solution of the partial differential equations in these nodes only. To this order the second stage of transformation is necessary, i.e. the conversion of the partial differential equation (or system of K equations) into the system of algebraic equations, i.e.:

$$\left.\begin{array}{l} \dfrac{\partial f_1}{\partial t} + \cdots = 0 \\ \vdots \\ \dfrac{\partial f_K}{\partial t} + \cdots = 0 \end{array}\right\} \rightarrow \mathbf{A} \cdot \mathbf{X} = \mathbf{B}, \tag{5.79}$$

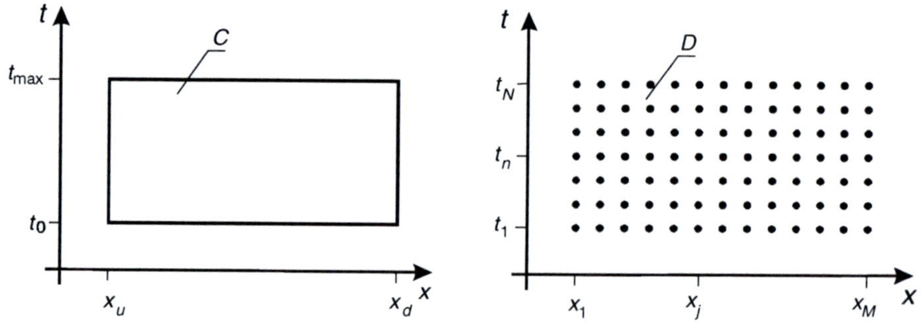

Fig. 5.9 Conversion of continuous domain C into discrete domain D

where:

A – matrix of coefficients of the system of algebraic equations,
X – vector of unknowns,
B – vector of right hand side.

This process is called discretization of the partial differential equations. Its details depend on the applied numerical method. Solving the system of algebraic equations one obtains an approximate solution of the partial differential equation or system of equations.

Sometimes the discretization is made in two steps. At first, a discretization with regard to one of the independent variables, for instance with regard to x, is carried out. In such a way one obtains a system of ordinary differential equations with regard to the second independent variable, which is usually the time t. In the second step this system is integrated numerically using one of the methods presented in Chapter 3. Such an approach is called semi-discretization and will be applied in the following part of the book. Practically the derivatives with regard to time are always discretized by the finite difference technique, whereas the discretization in space is carried out using various possible methods, like finite difference (FD), finite element (FE) or finite volume (FV).

As far as the spatial discretization of the open channel flow equations is considered, the finite difference method dominates. This is due to one-dimensional character of the majority of open channel flow problems. Note that practically all equations derived in Chapter 1 are one dimensional. However, some specific flow problems require other methods. For example, to model the flow with shock waves as caused by a dam break, the finite volume method is recommended. On the other hand the well-known finite element method seems to be best suited for 2D or 3D flow problems, while it is generally believed to be less useful for 1D problems. Later in this book we will show that a modification of the standard FE method makes it a very efficient tool also for 1D flow and transport equations. In this chapter basic FD and FE algorithms will be introduced.

5.2.2 Approximation of the Derivatives

The finite difference method is one of the most frequently used method of solution of the partial differential equations. Its concept is to replace directly the derivatives in the equation by approximate formulas in the form of the difference equations. To this end the continuous domain, which in the case of 1D equations has always a rectangular shape, is covered by the mesh of nodes (grid points) as in Fig. 5.10. This grid is build of two families of straight lines. The first family is parallel to the x axis. The distances between the lines equal to Δx_j, can vary. The second family is parallel to the time axis t. and it is usually distanced with constant time step Δt. If the intervals Δx are constant as well, then the grid is uniform.

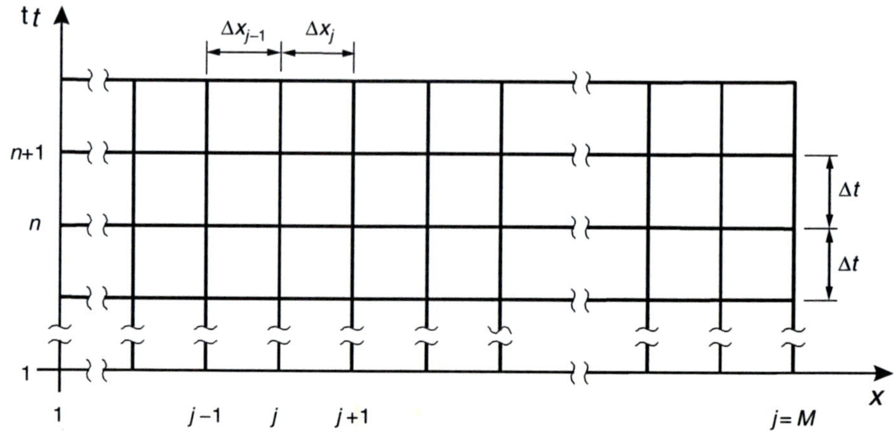

Fig. 5.10 Grid points covering the domain of solution at the plane $x-t$

In the intersection of both families of lines one obtains the nodes, in which the approximate value of solution will be calculated. In a uniform grid the position of any node can be defined by two indices j and n. The first one determines the spatial co-ordinate $x_j = (j-1)\,\Delta x$, whereas the second one – the time $t_n = (n-1)\,\Delta t$.

To convert the partial differential equation into a system of algebraic equations appropriate difference formulas approximating the derivatives must be used. In Chapter 3 some of the simplest expressions resulting directly from the Taylor series expansion were derived. These formulas such as forward (Eq. 3.10), backward (Eq. 3.12) or centered differences (Eq. 3.14) approximating the derivative of 1st order can be applied for discretization the considered equation. There are a variety of approximating formulas, which can be used to transform the partial differential equation. They may be derived using a more general approach which is presented here after Fletcher (1991).

The general form of applied approximation should be selected a priori. Assume that we are looking for a symmetrical approximating formula of 1st order derivative at node j. Its general equation is assumed to be:

$$\left.\frac{\partial f}{\partial x}\right|_j = a \cdot f_{j-1} + b \cdot f_j + c \cdot f_{j+1} + O(\Delta x^m)\,, \tag{5.80}$$

where a, b, c are the coefficients to be determined and $O(\Delta x^m)$ indicates the accuracy of applied approximation. Since for approximation carried out at the node j the values of function f in neighboring nodes $j-1$ and $j+1$ are involved, then the formula (5.80) is said to be symmetrical.

To determine the unknown coefficients in Eq. (5.80) the nodal values of f are expanded in the Taylor series around the node j. For the function $f(x, t)$, depending on two independent variables, the Taylor series is given as:

$$f(x+\Delta x, t+\Delta t) = f(x,t) + \sum_{m=1}^{\infty} \frac{1}{m!}\left(\Delta x \frac{\partial}{\partial x} + \Delta t \frac{\partial}{\partial t}\right)^m f(x,t). \tag{5.81}$$

At any time level n, for the nodes $j+1$ and $j-1$ distanced respectively by Δx and $-\Delta x$ from node j, formula (5.81) becomes:

$$f_{j+1} = f_j + \Delta x \left.\frac{\partial f}{\partial x}\right|_j + \frac{\Delta x^2}{2}\left.\frac{\partial^2 f}{\partial x^2}\right|_j + O\left(\Delta x^3\right). \tag{5.82a}$$

$$f_{j-1} = f_j - \Delta x \left.\frac{\partial f}{\partial x}\right|_j + \frac{\Delta x^2}{2}\left.\frac{\partial^2 f}{\partial x^2}\right|_j + O\left(\Delta x^3\right). \tag{5.82b}$$

Using the Taylor series (5.82) we define the values of f_{j-1} and f_{j+1} (Fig. 5.11). To this end let us consider equally spaced nodes for which $\Delta x_{j-1} = \Delta x_j = \Delta x$.

Next we substitute these expressions into Eq. (5.80). Gathering similar terms one obtains:

$$a \cdot f_{j-1} + b \cdot f_j + c \cdot f_{j+1} = (a+b+c) f_j + (-a+c)\,\Delta x \left.\frac{\partial f}{\partial x}\right|_j +$$

$$+ (a+c)\,\frac{\Delta x^2}{2}\left.\frac{\partial^2 f}{\partial x^2}\right|_j + (-a+c)\,\frac{\Delta x^3}{6}\left.\frac{\partial^3 f}{\partial x^3}\right|_j + \ldots. \tag{5.83}$$

Comparison of the left hand side of Eq. (5.80) and the right side hand of Eq. (5.83) yields:

$$a+b+c=0,\ (-a+c)\ \Delta x = 1.$$

For any value of the parameter c, the above equations give:

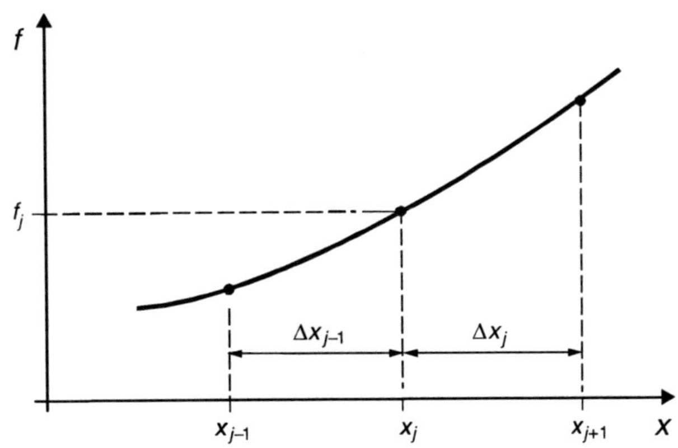

Fig. 5.11 Sketch for approximation of the derivatives of $f(x, t)$ with regard to x

$$a = c - \frac{1}{\Delta x}, \quad b = -2c + \frac{1}{\Delta x}.$$

If we assume such value of c that cancels the third term of the right side hand of Eq. (5.83), then one obtains an approximating formula as accurate as possible for the assumed form of Eq. (5.80). This case corresponds to:

$$c = -a = \frac{1}{2}\Delta x \qquad \text{and} \qquad b = 0.$$

Substitution of these relations in Eq. (5.80) yields:

$$\left.\frac{\partial f}{\partial x}\right|_j = \frac{1}{2\Delta x}(-f_{j-1} + f_{j+1}) - \frac{\Delta x^2}{6}\left.\frac{\partial^3 f}{\partial x^3}\right|_j + \ldots. \tag{5.84}$$

In such a way the centered difference formula (3.14) is derived. It ensures the accuracy of 2nd order i.e. produces a truncation error of order $O(\Delta x^2)$.

Following a similar way one can derive a symmetrical formula for the derivative of 2nd order. Assume that the approximation will be carried out using the formula analogical to Eq. (5.80), i.e.:

$$\left.\frac{\partial^2 f}{\partial x^2}\right|_j = a \cdot f_{j-1} + b \cdot f_j + c \cdot f_{j+1} + O(\Delta x^m), \tag{5.85}$$

then its right side hand is given by Eq. (5.83) as well. Comparison of the left hand side of Eq. (5.85) and the right hand side of Eq. (5.83) yields a system of 3 equations with 3 unknown parameters a, b and c:

$$a + b + c = 0, (-a + c)\,\Delta x = 0, \; (a + c)\,\frac{\Delta x^2}{2} = 1.$$

Its solution is the following:

$$a = \frac{1}{\Delta x^2}, \quad b = \frac{2}{\Delta x^2}, \quad c = \frac{1}{\Delta x^2}.$$

Consequently the approximating formula for the derivative of 2nd order is:

$$\left.\frac{\partial^2 f}{\partial x^2}\right|_J = \frac{f_{j-1} - 2f_j + f_{j+1}}{\Delta x^2} + O(\Delta x^2). \tag{5.86}$$

This well known three-point symmetrical formula approximates the derivative of 2nd order with accuracy of order $O(\Delta x^2)$. Usually Eq. (5.86) is derived directly using the Taylor series expansion.

The main advantages of the presented approach are: the possibility of derivation of symmetrical or asymmetrical formulas, involving various number of nodes depending on the required order of formula and easy application for both uniform and non-uniform grids. For illustration let us derive three-points asymmetrical formula approximating the derivative of first order. Assume that its general form is the following:

$$\left.\frac{\partial f}{\partial x}\right|_j = af_j + bf_{j+1} + cf_{j+2} + O(\Delta x^m). \tag{5.87}$$

Substituting f_{j+1} and f_{j+2} as the Taylor series expansion around the node j and regrouping, one obtains:

$$\left.\frac{\partial f}{\partial x}\right|_j = (a + b + c)f_j^n + (b \cdot \Delta x + c \cdot 2\,\Delta x)\left.\frac{\partial f}{\partial x}\right|_j + \\ + \left(\frac{b \cdot \Delta x^2}{2} + \frac{c(2\Delta x)^2}{2}\right)\left.\frac{\partial^2 f}{\partial x^2}\right|_j + \ldots. \tag{5.88}$$

Comparing the respective sides of Eqs. (5.87) and (5.88) one can find the following conditions:

$$a + b + c = 0,\ b \cdot \Delta x + c\,(2\Delta x) = 1,\ \frac{b \cdot \Delta x^2}{2} + \frac{c\,(2\Delta x)^2}{2} = 0,$$

which ensure the most accurate approximating formula. These conditions give:

$$a = -\frac{1.5}{\Delta x},\quad b = \frac{2}{\Delta x},\quad c = -\frac{0.5}{\Delta x}.$$

Finally the searched asymmetric formula is as follows:

$$\left.\frac{\partial f}{\partial x}\right|_j = \frac{-1.5f_j + 2f_{j+1}^n - 0.5\,f_{j+2}}{\Delta x} - \frac{\Delta x^2}{3}\left.\frac{\partial^3 f}{\partial x^3}\right|_j + \ldots. \tag{5.89}$$

As it can be seen, this formula approximates the first order derivative with accuracy of $O(\Delta x^2)$. In such a way one can derive various formulas approximating the derivatives of various order of the function $f(x)$.

The solution of flow or transport equations for natural open channels is usually performed on non-uniform grids. In such a case the appropriate approximating formulas can be easily derived using the same approach. Assume that the first order derivative is approximated in non-equally spaced nodes at the cross section j as presented in Fig. 5.12.

A general form of the searched formula is given by Eq. (5.80). The nodal values of function f at $j-1$ and $j+1$ are calculated with Taylor series expansion around the node j. They are following:

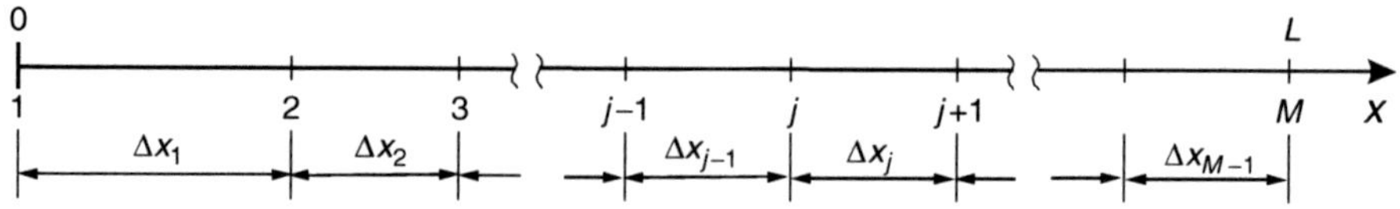

Fig. 5.12 Non-uniformly spaced grid points: $\Delta x_1 \neq \Delta x_2 \neq \Delta x_3 \neq \ldots$

$$f_{j-1} = f_j - \Delta x_{j-1} \left.\frac{\partial f}{\partial x}\right|_j + \frac{\Delta x_{j-1}^2}{2} \left.\frac{\partial^2 f}{\partial x^2}\right|_j + O\left(\Delta x^3\right). \tag{5.90}$$

$$f_{j+1} = f_j + \Delta x_j \left.\frac{\partial f}{\partial x}\right|_j + \frac{\Delta x_j^2}{2} \left.\frac{\partial^2 f}{\partial x^2}\right|_j + O\left(\Delta x^3\right). \tag{5.91}$$

Substitution of Eqs. (5.90) and (5.91) in Eq. (5.80) yields after regrouping:

$$\begin{aligned} \left.\frac{\partial f}{\partial x}\right|_j &= (a + b + c) f_j + (-a \cdot \Delta x_{j-1} + c \cdot \Delta x_j) \left.\frac{\partial f}{\partial x}\right|_j \\ &+ \left(\frac{a \cdot \Delta x_{j-1}^2}{2} + \frac{c \cdot \Delta x_j^2}{2} \right) \left.\frac{\partial^2 f}{\partial x^2}\right|_j + \ldots \end{aligned} \tag{5.92}$$

Comparing the respective sides of Eqs. (5.80) and (5.92) one obtains the following relations:

$$a + b + c = 0, \quad -a \cdot \Delta x_{j-1} + c \cdot \Delta x_j = 1, \; a \cdot \Delta x_{j-1}^2 + c \cdot \Delta x_j^2 = 0.$$

Solution of this system of equations gives:

$$a = -\frac{1}{2\Delta x_{j-1}}, \tag{5.93a}$$

$$b = -\frac{1}{2\Delta x_{j-1}} \left(1 - \frac{1}{\chi^2} \right), \tag{5.93b}$$

$$c = -\frac{1}{2\chi^2 \cdot \Delta x_{j-1}} \tag{5.93c}$$

where χ is the ratio of the neighboring space intervals:

$$\chi = \frac{\Delta x_j}{\Delta x_{j-1}} \tag{5.94}$$

Then the approximating formula is given by:

$$\left.\frac{\partial f}{\partial x}\right|_j = \frac{1}{2\Delta x_{j-1}} \left(-f_{j-1} + \frac{\chi^2 - 1}{\chi^2} f_j + \frac{1}{\chi^2} f_{j+1} \right). \tag{5.95}$$

The approximating formula for the second order derivative is carried out similarly. We assume a general formula in the form of Eq. (5.85), in which the nodal values of f given by Eqs. (5.90) and (5.91) are substituted:

$$\left.\frac{\partial^2 f}{\partial x^2}\right|_j = (a+b+c)f_j + (-a\cdot\Delta x_{j-1} + c\cdot\Delta x_j)\left.\frac{\partial f}{\partial x}\right|_j + \left(\frac{a\cdot\Delta x_{j-1}^2}{2} + \frac{c\cdot\Delta x_j^2}{2}\right)\left.\frac{\partial^2 f}{\partial x^2}\right|_j + \dots \tag{5.96}$$

Comparing the respective sides of Eqs. (5.85) and (5.96) one obtains the following relations:

$$a+b+c=0,\quad -a\cdot\Delta x_{j-1}+c\cdot\Delta x_j=0,\quad \frac{a\cdot\Delta x_{j-1}^2}{2}+\frac{c\cdot\Delta x_j^2}{2}=1.$$

which yields:

$$a = -\frac{2}{\Delta x_{j-1}^2(1+\chi)} \tag{5.97a}$$

$$b = -\frac{2}{\Delta x_{j-1}^2}\frac{1}{1+\chi}\left(1+\frac{1}{\chi}\right) \tag{5.97b}$$

$$c = \frac{2}{\Delta x_{j-1}^2}\frac{1}{\chi+\chi^2} \tag{5.97c}$$

Substitution of Eqs. (5.97a), (5.97b) and (5.97c) in Eq. (5.85) gives:

$$\left.\frac{\partial^2 f}{\partial x^2}\right|_j = \frac{2}{\Delta x_{j-1}^2}\frac{1}{1+\chi}\left(f_{j-1} - \left(1+\frac{1}{\chi}\right)f_j + \frac{1}{\chi}f_{j+1}\right). \tag{5.98}$$

It is obvious that any discretization introduces an error. The exceptions are represented by some particular cases when the exact solution has very simple analytical form. For instance the backward or forward difference ensure exact approximation only when the solution is a linear function of x (Fletcher 1991).

Summarizing the problem of approximation of the derivative of 1st order, let us list the basic formulas discussed previously in Chapter 3 and in current one. We have:

– forward difference

$$\left.\frac{df}{dx}\right|_j = \frac{f_{j+1}-f_j}{\Delta x} + O(\Delta x), \tag{5.99}$$

– backward difference

$$\left.\frac{df}{dx}\right|_j = \frac{f_j-f_{j-1}}{\Delta x} + O(\Delta x), \tag{5.100}$$

– centred difference

$$\left.\frac{df}{dx}\right|_j = \frac{f_{j+1} - f_{j-1}}{2\Delta x} + O(\Delta x^2). \tag{5.101}$$

Using the presented formulas for approximation of the derivatives existing in considered equation, one can combine various versions of the finite difference method. These versions are usually called schemes of the finite difference method. However the numerical schemes constructed using the backward, forward and centred differences for the partial differential hyperbolic equations can generate some negative effects in the solution. These issues will be presented in details in the next sections. Now we can say generally that the obtained numerical solution can contain unphysical oscillations or it can be artificially smoothed. These phenomena are caused by the properties of above mentioned approximating formulas. The wiggles are caused by the centered difference approximating the advective term. This phenomenon arises because the approximation of the first derivative at node j is independent of f_j (Abbott and Basco 1989). In Fig. 5.13 one can see two different functions $f_1(x)$ and $f_2(x)$ having the same value of 1st order derivative approximated by the centered difference.

On the other hand, approximation of the first derivative by the backward difference eliminates the wiggles, but at the same time it gives rise to intensive artificial smoothing. To limit both effects it is sometimes necessary to include the nodal value of the unknown function to approximate its first derivative at the same node. Such formula, which covers simultaneously Eqs. (5.99), (5.100) and (5.101) can be simply derived using the previously applied approach.

The assumed formula for the first derivative (5.80) compared with the result of the Taylor series expansion (5.83) allows us to write the following conditions:

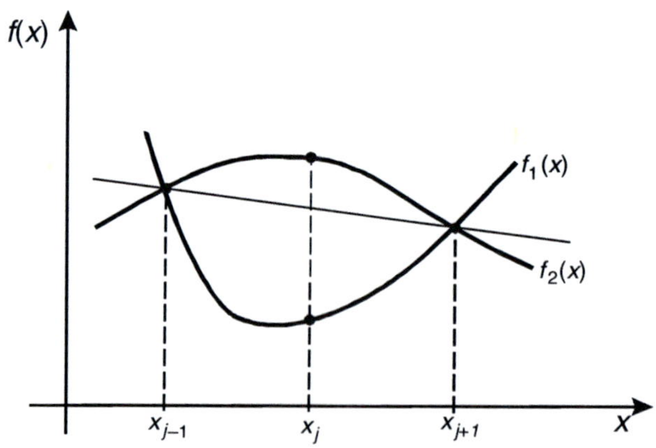

Fig. 5.13 Approximation of the 1st order derivative of $f(x)$ using the centred difference does not involve $f(x_j)$

$$a + b + c = 0, \; (-a + c)\,\Delta x = 1.$$

The coefficients b and c can be expressed in relation to a as follows:

$$c = \frac{1}{\Delta x} + a \tag{5.102a}$$

$$b = -2a - \frac{1}{\Delta x} \tag{5.102b}$$

Substitution of these relations in Eq. (5.80) yields:

$$\left.\frac{\partial f}{\partial x}\right|_j \approx \frac{a \cdot \Delta x \cdot f_{j-1} - (2a \cdot \Delta x + 1) f_j + (1 + a\Delta x) f_{j+1}}{\Delta x} \tag{5.103}$$

Setting $a \cdot \Delta x = -\eta$ one obtains:

$$\left.\frac{\partial f}{\partial x}\right|_j \approx \frac{-\eta \cdot f_{j-1} + (2\eta - 1) f_j + (1 - \eta) f_{j+1}}{\Delta x} \tag{5.104}$$

This formula involves the weighting parameter η. According to its value, Eq. (5.104) describes all the standard approximations of the 1st derivative given by Eqs. (5.99), (5.100) and (5.101). For $\eta = 0$ it gives the forward difference, for $\eta = 0.5$ – the centred difference, whereas for $\eta = 1$ – the backward difference. However the most interesting property of Eq. (5.104) is that, if necessary, it enables a gradual increase of the weight of f_j in the approximation of the derivative.

An in-depth discussion of various aspects of the finite difference method is given by Fletcher (1991). His main suggestions and recommendations, very useful for computational practice, can be summarized as follows:

1. For a smooth function the truncation error of the Taylor series expansion is dominated by the first term of its truncated part and consequently it determines the accuracy of approximation. This is valid for all approximating formulas.
2. The accuracy of approximation of the derivatives increases with the order of approximating formula.
3. The error of approximation depends on the value of mesh dimension Δx. It can be expected that this error is more efficiently reduced by decreasing Δx than by increasing the order of approximating formula. Therefore, while increasing the order of approximation, grid refinement should be applied simultaneously.
4. A negative aspect of increasing of the order of approximation is that more nodal values of the function $f(x)$ are involved in approximating formulas. Moreover, increasing of formula's order only very slowly improves the results of computations for sparse mesh.
5. Increasing of the approximating formula order does not guarantee better accuracy approximation for rapidly varied function, especially when discontinuities

occur. This is because in such a case the truncation error of the Taylor series is not dominated by the first term of the neglected part of the series.

As far as the approximation of the derivative of the first order is considered, Fletcher (1991) concludes that the approximation which should be applied in practice is the one of 2nd order, whereas approximations of higher order should be used in exceptional circumstances only.

Another very important problem concerns the representation of waves on the numerical grid. This has particular meaning in open channel flow modeling as the considered flow processes have the character of wave motion. Discrete character of grid points gives rise to limited possibility of wave representation. It appears that the shortest waves which can be represented on the mesh spaced with Δx have the length equal to double space interval. Then the numerical solution of the partial differential equation at a grid point should be considered as a long-wave representation of its exact solution. As it was shown by Fletcher (1991), all difference formulas better approximate the derivatives in the case of long waves than short ones. This suggests the following conclusion: to improve the results of computations dealing with the wave propagation problems, grid refinement should be applied. Such approach, allowing to take into account the shorter and shorter waves, ought to be used in the case when large gradients are expected in the solution. Comprehensive discussion of this question is presented by Abbott and Basco (1989), Fletcher (1991) and others. We will take up these questions in Chapter 6, while discussing the properties of the applied numerical methods. However at the moment one can find out that except the prismatic channels the grid refinement rather cannot be applied. This is because the data on channel geometry at the grid points are obtained from field measurements.

5.2.3 Example of Solution: Advection Equation

Using the finite difference method let us solve the pure advection equation (5.3), the simplest example of hyperbolic equation:

$$\frac{\partial f}{\partial t} + U\frac{\partial f}{\partial x} = 0 \qquad (5.105)$$

where:

f – scalar function,
U – flow velocity.

Assume that $U = \text{const.} > 0$ and the solution is searched in the domain: $0 \leq x \leq L$ and $t \geq 0$, where L is the length of the channel reach, in which flowing water transports the dissolved pollutant. For positive advection velocity, the following initial-boundary conditions are imposed:

– initial condition: $f(x, t = 0) = f_i(x)$ for $0 \leq x \leq L$
– boundary condition: $f(x = 0, t) = f_0(t)$ for $t \geq 0$

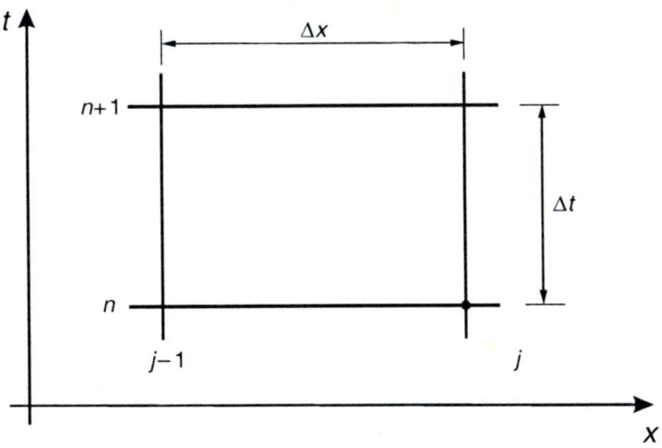

Fig. 5.14 Grid points for the upwind scheme

where $f_i(x)$ and $f_0(t)$ are given functions, satisfying the condition of consistency: $f_i(0) = f_0(0)$.

To solve Eq. (5.105) let us apply the difference scheme known as the upwind scheme. This scheme uses the grid points shown in Fig. 5.14.

The derivatives are approximated at the node (j, n) as follows:

$$\left.\frac{\partial f}{\partial t}\right|_j^n = \frac{f_j^{n+1} - f_j^n}{\Delta t}, \tag{5.106a}$$

$$\left.\frac{\partial f}{\partial x}\right|_j^n = \frac{f_j^n - f_{j-1}^n}{\Delta x} \tag{5.106b}$$

where:

j – index of cross-section,
n – index of time level,
Δx – spatial mesh dimension,
Δt – time step,

Substitution of Eq. (5.106) in advection equation (5.105) yields:

$$\frac{f_j^{n+1} - f_j^n}{\Delta t} + U\frac{f_j^n - f_{j-1}^n}{\Delta x} = 0 \quad \text{for} \quad j = 2, 3, \cdots, M \tag{5.107}$$

where M is the total number of cross-sections. Note that f_1^{n+1} is known from the boundary condition. Therefore in each equation of the system (5.107) only one unknown f_j^{n+1} exists. It can be easily calculated, giving:

$$f_j^{n+1} = C_a f_{j-1}^n + (1 - C_a)\, f_j^n \tag{5.108}$$

where C_a denotes so called advective Courant number, defined as follows:

$$C_a = \frac{U \cdot \Delta t}{\Delta x} \tag{5.109}$$

Setting in Eq. (5.108) $j = 2, 3, \ldots, M$ one can compute the nodal values of the function f in all nodes at the time level t_{n+1}. Note that the computations can be run in any order – not only from the upstream end towards the downstream end. It is impossible to reverse the order of calculations. The numerical schemes, which do not require respecting the imposed order of calculations of f at the new time level, are called explicit schemes. If calculations must be carried out in the required order at the new time level, it is said that an implicit scheme is applied.

The upwind scheme is extremely simple, but it is rather seldom used to solve practical problems. This classical and very instructive finite difference scheme usually serves to explain some questions of numerical solution of the hyperbolic equations. To examine how the upwind scheme works, let us solve the pure advection equation (5.105) for arbitrary assumed data.

Example 5.1 In a prismatic channel of length L the water flows with constant velocity U. Assume that:

- $U = 0.5$ m/s = const.,
- $\Delta x = 100$ m = const.,
- initial condition is $f(x,t = 0) = 0$ for $0 \le x \le L$;
- boundary condition is

$$f(x=0,t) = \begin{cases} F_m \cdot \frac{t}{T_m} & \text{for} \quad 0 \le t \le T_m \\ F_m \cdot \left(2 - \frac{t}{T_m}\right) & \text{for} \quad T_m \le t \le 2T_m \\ 0 & \text{for} \quad t > 2T_m \end{cases}$$

with $F_m = 100$ and $T_m = 1{,}200$ s.

Therefore we will examine advective transport of a scalar quantity represented by the function f, which has the initial distribution in the form of triangle having height of $F_m = 100$ and a base equal to $2T_m$.

Taking into account the analytical solution of the advection equation in the form of expression (5.12), one can predict the form of the exact solution in the case considered here. The triangular distribution forced at the upstream end will travel along channel axis with any shape deformation. Then in each cross-section it must be the same, shifted in time only. For our analysis we take the cross-section located in position distanced 5,000 m from the upstream end.

The results of computations are shown in Fig. 5.15. One can see that the upwind scheme gives exact solution for the Courant number equal to unity ($C_a = 1$) only.

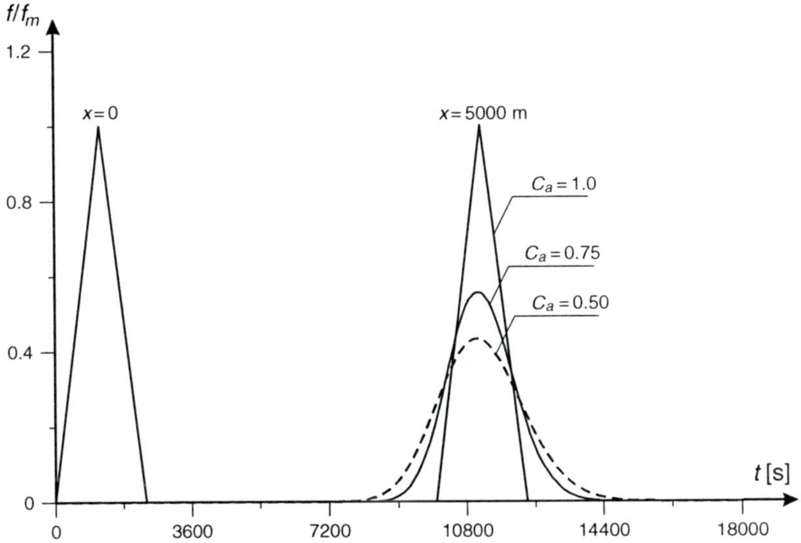

Fig. 5.15 Solution of the pure advection equation using the upwind scheme for various Courant number

For other values it produces results completely different from the exact one. One observe the attenuation of triangular distribution, which increases as the Courant number decreases. On the other hand, for $C_a > 1$ very intensive instabilities in the solution occur (Fig. 5.16).

The results of numerical experiments carried out for various set of Δx and Δt indicate that the accuracy of solution is determined by the assumed values of the numerical parameters. The exact solution is obtained only for a particular set of data, for which the advective Courant number is equal to unity. For other sets one obtains either oscillating solutions or damped solutions. Since the only physical process represented in the solved equation is advection, these observed effects must have numerical roots. A more detailed explanation of such behavior is the subject of Section 5.4.

5.3 Introduction to the Finite Element Method

5.3.1 General Concept of the Finite Element Method

The finite element method is one of the most popular approaches for solving the partial differentia equations. Its advantages are especially appreciable in application to 2D and 3D problems. On the other hand, it is seldom applied for 1D problems, as in this case it does not seem to possess significant benefits compared with the finite difference method, while being more complicated in implementation. However, using the finite element method one can develop very effective algorithms for solution

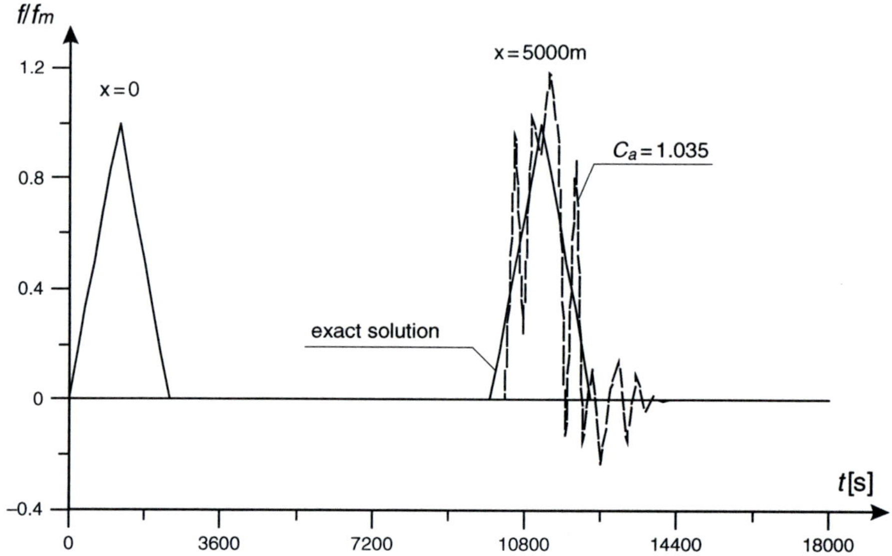

Fig. 5.16 Solution of the pure advection equation using the upwind scheme for $C_a = 1.035$

the open channel flow problems as well. For this reason our presentation of this method will be limited to 1D case. Comprehensive presentations of the finite element method oriented to solve the fluid mechanics problems are given by Gresho and Sani (1998), Fletcher (1991), Oden and Reddy (1976), Zienkiewicz (1972) among others.

Similarly to the finite difference method, the goal of application of the finite element method is to find an approximate solution of a partial differential equation (or a system of such equations), which must be satisfied by unknown function in the solution domain. As in the finite difference method, in FE method the approximate solution is searched in a set of isolated nodes, which constitute discrete domain covering the continuous physical domain. However, the manner of transformation of the solved differential equation into the system of algebraic equations is quite different. The finite element method belongs to the family of the so-called weighted residual methods. To explain briefly the concept of this approach, let us consider the problem of solution of the partial differential equation written in a symbolic form:

$$\Phi(f) = 0, \tag{5.110}$$

with the auxiliary conditions imposed at the limits of the solution domain C

$$B(f) = 0. \tag{5.111}$$

which must be satisfied by the unknown function $f(x, t)$. The weighted residual methods assume the following general form of the approximate solution of Eq. (5.110):

$$f_a(x,t) = \sum_{j=1}^{K} \alpha_j(t) \cdot N_j(x) \tag{5.112}$$

where:

$f_a(x, t)$ – approximation of exact solution $f(x, t)$,
$\alpha_j(t)$ – coefficients, which are to be determined,
$N_j(x)$ – assumed approximating functions.

For an arbitrary approximation Eq. (5.110) will not be satisfied. Substitution of this approximate solution in Eq. (5.110) yields:

$$\Phi(f_a) = R \neq 0. \tag{5.113}$$

where R is the equation's residual. We should try to choose such values of the coefficients $\alpha_j(t)$ that R is minimized over the solution domain, i.e the best approximation of f is ensured. To find these values the following condition is applied:

$$\int_C w_j \cdot R \cdot dC = \int_C w_j \cdot \Phi(f_a) \cdot dC = 0, \tag{5.114}$$

which means that the integral of the weighted residual must be equal to zero. Assuming a set of the weight functions w_j one obtains a system of ordinary differential equations for unsteady problem or a system of algebraic equations for steady problem. There are a couple possibilities to choose the weight functions, which lead to various numerical schemes. If we assume that the approximating functions applied in Eq. (5.112) are used, i.e. as the weight functions:

$$w_j(x) = N_j(x) \tag{5.115}$$

then the Galerkin method is obtained. For j tending to infinity, the approximated solution $f_a(x, t)$ should tend to the exact one $f(x, t)$ (Fletcher 1991).

Remember that in 1D unsteady flow process the unknown function f depends on two independent variables: the spatial co-ordinate x and the time t. In such a case the partial differential equation (5.110) is at first approximated in space only using the finite element method. This leads to a system of the ordinary differential equation with regard to time t. Next, the obtained system must be integrated over time. To this end one can use the finite element method as well. However such approach does not give any advantages. It is better to use standard methods of numerical solution of the initial-value problem for the ordinary differential equations, which were presented in Chapter 3. Application of one of these methods leads to the system of algebraic equations, which can be linear or non-linear depending on the solved differential equation. The obtained system is completed by introducing the imposed boundary conditions and after that it is solved usually using direct methods. Let us remember that approximation of 1D partial differential equation leads to the system with banded tridiagonal matrix. For the system of equations the bandwidth is larger.

In the finite element method the continuous domain is divided into smaller subareas called the finite elements. It is assumed that these elements are joined in a finite number of points lying on the element's circumference. These points are the nodes, in which the approximate solution will be computed. Since in open channel hydraulics we have 1D equations, then the domain of solution has the form of channel reach of length L so that we have $0 \leq x \leq L$. Consequently, in this case the applied finite elements can take the form of linear segments only.

The unknown function f in the solution domain is approximated as follows:

$$f_a\,(x,t) = \mathbf{N}\,(x) \cdot \;\mathbf{f}\,(t)\,, \tag{5.116}$$

where:

f_a – approximation of the function f in C,
$\mathbf{N}(x) = (N_1,\ldots,N_j,N_{j+1},\ldots N_M)$ – matrix of basis or shape functions,
$\mathbf{f}(t) = (f_1,\ldots,f_j, f_{j+1},\ldots,f_M)^T$ – vector of the nodal values of function f, which for time dependent problems has the components being functions of time,
T – transposition symbol,
M – total number of nodes.

The row matrix $\mathbf{N}$ has components dependent on space co-ordinates only. They should be chosen in such a way that Eq. (5.116) is satisfied at the nodes. This means that substitution of the node co-ordinates in Eq. (5.116) ensures that the approximation f_a must be equal to the nodal value of f. In other words, for $x = x_j$ is $f_a = f_j$. The set of functions $N_j(x)$ is called basis or shape functions (Fletcher 1991, Zienkiewicz 1972).

Equation (5.114) determining the best approximation of f, in such a case will take the following form:

$$\int_0^L N_j \cdot \Phi\,(\mathbf{N} \cdot \mathbf{f}) \cdot dx = 0 \quad \text{for} \quad j = 1, 2, ..., M. \tag{5.117}$$

Since the considered channel reach was divided with M nodes into $M - 1$ finite elements of length Δx_j, Eq. (5.117) can be rewritten as follows:

$$\int_0^L \mathbf{N} \cdot \Phi\,(\mathbf{N} \cdot \mathbf{f}) \cdot dx = \sum_{j=1}^{M-1} \int_0^{\Delta x_j} \mathbf{N} \cdot \Phi\,(\mathbf{N} \cdot \mathbf{f}) \cdot dx = 0, \tag{5.118}$$

The main question, which arises while solving any equation with the finite element method, is the choice of elements and the trial functions. As we found out

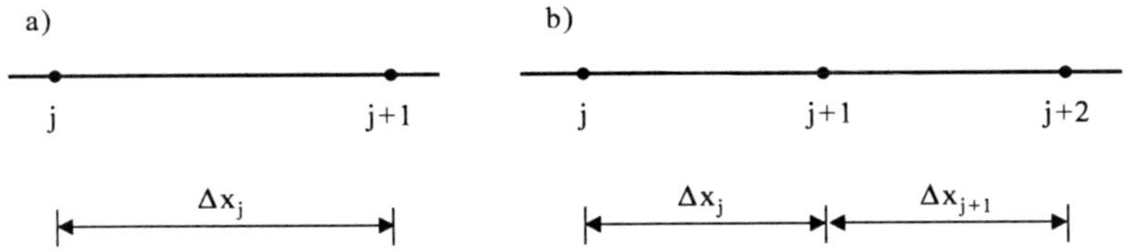

Fig. 5.17 Linear finite elements containing 2 nodes (**a**) and 3 nodes (**b**)

previously, the shape of elements depends on the dimensionality of solved problem. For open channel flow equation this choice is restricted to segments of straight line. The number of nodes which constitute a segment, depends on the degree of polynomial applied for approximation. For our purposes we will consider the simplest element as shown in Fig. 5.17a, which allows us to apply the linear shape functions only.

The chosen shape functions must be continuous over an element and they have to ensure that f_a is continuous between elements.

Let us consider a channel reach of length L, which is divided into the sub-intervals-elements bounded by nodes j and $j+1$. Their lengths are $\Delta x_j = x_{j+1} - x_j$ (Fig. 5.17a). In such element the function f can be approximated using a linear polynomial:

$$f_a(x) = a \cdot x + b \text{ for } x_j \leq x \leq x_{j+1}, \tag{5.119}$$

The coefficients a and b can be determined by the solution of the following system of equations:

$$f_j = a \cdot x_j + b, \tag{5.120a}$$

$$f_{j+1} = a \cdot x_{j+1} + b \tag{5.120b}$$

which gives:

$$a = \frac{f_{j+1} - f_j}{x_{j+1} - x_j}, \tag{5.121a}$$

$$b = -\frac{f_{j+1} - f_j}{x_{j+1} - x_j} x_j + f_j. \tag{5.121b}$$

Substitution of Eq. (5.121) in Eq. (5.119) yields:

$$f_a(x) = \frac{f_{j+1} - f_j}{x_{j+1} - x_j} x - \frac{f_{j+1} - f_j}{x_{j+1} - x_j} x_j + f_j. \tag{5.122}$$

This equation can be rewritten as follows:

$$f_a(x) = N_j(x) \cdot f_j + N_{j+1}(x) \cdot f_{j+1}, \tag{5.123}$$

where:

$$N_j(x) = \frac{x_{j+1} - x}{\Delta x_j}, \text{ for } x_j \le x \le x_{j+1} \tag{5.124a}$$

$$N_{j+1}(x) = \frac{x - x_j}{\Delta x_j}, \text{ for } x_j \le x \le x_{j+1} \tag{5.124b}$$

Therefore for an element containing two nodes the appropriate shape functions have the form of linear Lagrange polynomials. Very similar expression to Eq. (5.123) is derived for the preceding element $j-1$. In the approximaring formula for this element the trial functions $N_{j-1}(x)$ and $N_j(x)$ will occur. Consequently, the function $N_j(x)$ contributes to the elements containing the node j only, i.e. it is non-zero in the elements $j-1$ and j. In others elements this function is equal to zero. Then $N_j(x)$ is defined as follows:

$$N_j(x) = 0 \text{ for } x < x_{j-1}, \tag{5.125a}$$

$$N_j(x) = \frac{x - x_{j-1}}{\Delta x_{j-1}} \text{ for } x_{j-1} \le x \le x_j \text{ (element } j-1), \tag{5.125b}$$

$$N_j(x) = 1 \text{ for } x = x_j, \tag{5.125c}$$

$$N_j(x) = \frac{x_{j+1} - x}{\Delta x_j} \text{ for } x_j \le x \le x_{j+1} \text{ (element } j), \tag{5.125d}$$

$$N_j(x) = 0 \text{ for } x > x_{j+1}. \tag{5.125e}$$

The function $N_j(x)$ shown in Fig. 5.18 has typical form, because of which it is called hat function.

Differentiation of the shape functions with regard to x in element j gives:

$$\frac{dN_j}{dx} = \frac{d}{dx}\left(\frac{x_{j+1} - x}{\Delta x_j}\right) = -\frac{1}{\Delta x_j}, \tag{5.126a}$$

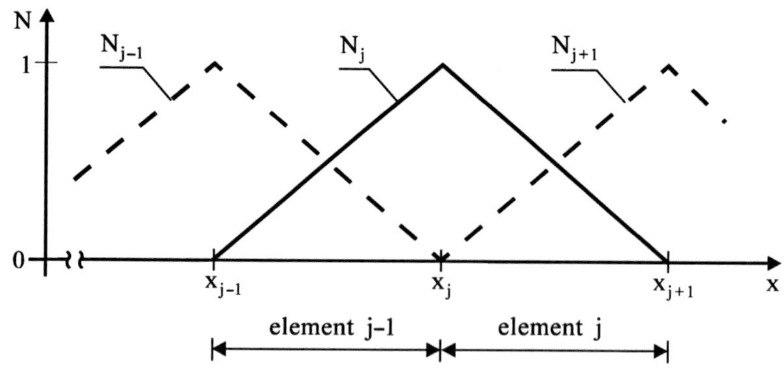

Fig. 5.18 The form of linear trial function $N_j(x)$

$$\frac{dN_{j+1}}{dx} = \frac{d}{dx}\left(\frac{x - x_j}{\Delta x_j}\right) = \frac{1}{\Delta x_j} \tag{5.126b}$$

Then the derivative of the approximating function $f_a(x)$ given by Eq. (5.123) is:

$$\frac{df_a}{dx} = -\frac{1}{\Delta x_j} f_j + \frac{1}{\Delta x_j} f_{j+1} = \frac{-f_j + f_{j+1}}{\Delta x_j}. \tag{5.127}$$

While solving unsteady flow equations the time derivative of unknown function occurs as well. This derivative in element j is given as:

$$\frac{df_a}{dt} = \frac{d}{d\,t}(N_j(x) \cdot f_j(t) + N_{j+1}(x) \cdot f_{j+1}(\mathrm{t})) = N_j(x)\frac{df_j(t)}{d\,t} + N_{j+1}(x)\frac{df_{j+1}(t)}{d\,t}. \tag{5.128}$$

Integration of the shape functions over element j is carried out as follows:

$$\int_{x_j}^{x_{j+1}} N_j(x)dx = \int_{x_j}^{x_{j+1}} \frac{x_{j+1} - x}{\Delta x_j} dx = \frac{1}{2}\Delta\, x_j, \tag{5.129a}$$

$$\int_{x_j}^{x_{j+1}} N_{j+1}(x)dx = \int_{x_j}^{x_{j+1}} \frac{x - x_j}{\Delta x_j} dx = \frac{1}{2}\Delta\, x_j, \tag{5.129b}$$

whereas for integration of a product of shape functions the following formula can be used (Zienkiewicz 1972):

$$\int_{x_j}^{x_{j+1}} N_j^{\alpha} \cdot N_{j+1}^{\beta} \cdot dx = \int_{x_j}^{x_{j+1}} N_j^{\beta} \cdot N_{j+1}^{\alpha} \cdot dx = \frac{\alpha! \cdot \beta!}{(\alpha + \beta + 1)!}\Delta\, x_j, \tag{5.130}$$

where α and β are the exponents of powers of the shape functions.

5.3.2 Example of Solution: Diffusion Equation

Let us consider the diffusion equation with $D = \text{const.}$:

$$\frac{\partial f}{\partial t} - D\frac{\partial^2 f}{\partial x^2} = 0. \tag{5.131}$$

This equation will be solved for the following auxiliary conditions:

– initial condition:

$$f\,(x, t = 0) = f_i\,(x) \;\text{ for }\; 0 \le x \le L(L\,\text{is length of considered channel reach})$$

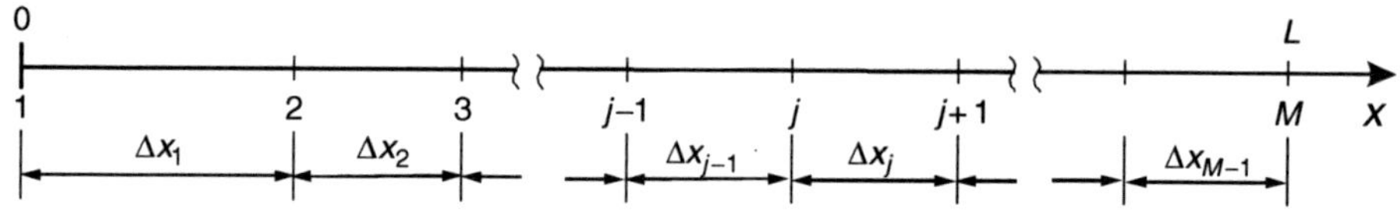

Fig. 5.19 Assumed discrete solution domain for the diffusion equation

– boundary conditions:

$$f\,(x=0,t)=f_0\,(t)\text{ and }f\,(x=L,t)=f_L\,(t)\text{ for }t\geq 0.$$

The functions $f_i(x)$, $f_0(t)$ and $f_L(t)$ are given.

To solve Eq. (5.131) the Galerkin finite element method is applied. Assume that considered segment of a channel is divided with M nodes into $M-1$ elements of length Δx_j ($j = 1,2,\ldots, M-1$). The discretized solution domain is shown in Fig. 5.19.

According to the Galerkin procedure the numerical solution must satisfy the condition (5.118). For Eq. (5.131) this condition takes the following form:

$$\sum_{j=1}^{M-1}\int_{x_j}^{x_{j+1}}\left(\frac{\partial f_a}{\partial t}-D\frac{\partial^2 f_a}{\partial x^2}\right)\mathbf{N}(x)\cdot dx=0, \tag{5.132}$$

where $\mathbf{N}(x)$ is the vector of shape functions having components given by Eq. (5.125). In Eq. (5.132) the subscript a denotes the approximation of function according to formula (5.123).

Let us calculate one integral-component of the above sum, corresponding to the element j. Since in this element only two components of the vector $\mathbf{N}$(x), i.e. $N_j(x)$ and $N_{j+1}(x)$, are non-zero then in Eq. (5.132) only two following non-zero products will occur:

$$I^{(j)}=\int_{x_j}^{x_{j+1}}\left(\frac{\partial f_a}{\partial t}-D\frac{\partial^2 f_a}{\partial x^2}\right)N_j(x)\cdot dx, \tag{5.133}$$

$$I^{(j+1)}=\int_{x_j}^{x_{j+1}}\left(\frac{\partial f_a}{\partial t}-D\frac{\partial^2 f_a}{\partial x^2}\right)N_{j+1}\,(x)\cdot dx, \tag{5.134}$$

Calculation of the integral (5.133) carried out term by term provides consecutively:

– The time variation term

$$I_1^{(j)} = \int_{x_j}^{x_{j+1}} \frac{\partial f_a}{\partial t} dx = \int_{x_j}^{x_{j+1}} \left(N_j(x)\frac{df_j}{dt} + N_{j+1}(x)\frac{df_{j+1}}{dt} \right) N_j(x)dx = \\ = \frac{\Delta x_j}{3}\frac{df_j}{dt} + \frac{\Delta x_j}{6}\frac{df_{j+1}}{dt}. \tag{5.135}$$

– The diffusive term

Calculation of this term is slightly more complicated. This is because linear polynomials were applied for approximation, whereas in this term the derivative of 2nd order occurs, which cannot be calculated directly. To solve this problem, at first one can carry out an integration by parts decreasing the order of derivative. This is performed as follows:

$$I_2^{(j)} = \int_{x_j}^{x_{j+1}} D\frac{\partial^2 f_a}{\partial x^2} N_j(x)dx = D \cdot N_j \left. \frac{\partial f_a}{\partial x} \right|_{x_j}^{x_{j+1}} - D \int_{x_j}^{x_{j+1}} \frac{\partial f_a}{\partial x}\frac{\partial N_j}{\partial x} dx = \\ = -D \left. \frac{df}{dx} \right|_j - D\left(\frac{-f_j + f_{j+1}}{\Delta x_j} \right)\left(-\frac{1}{\Delta x_j} \right) \int_{x_j}^{x_{j+1}} dx = -D \left. \frac{df}{dx} \right|_j + \frac{D}{\Delta x_j}(-f_j + f_{j+1}). \tag{5.136}$$

Finally, the integral (5.133) equal to $I_1^{(j)} - I_2^{(j)}$, is given as:

$$I^{(j)} = \frac{\Delta x_j}{3}\frac{df_j}{dt} + \frac{\Delta x_j}{6}\frac{df_{j+1}}{dt} - \frac{D}{\Delta x_j}(-f_j + f_{j+1}) + D \left. \frac{df_j}{dx} \right|_j \tag{5.137}$$

Calculation of the integral (5.134) is performed similarly. For its consecutive terms one obtains:

$$I_1^{(j+1)} = \int_{x_j}^{x_{j+1}} \frac{\partial f_a}{\partial t} N_{j+1}(x)dx = \int_{x_j}^{x_{j+1}} \left(N_j(x)\frac{df_j}{dt} + N_{j+1}(x)\frac{df_{j+1}}{dt} \right) N_{j+1}(x)dx = \\ = \frac{\Delta x_j}{6}\frac{df_j}{dt} + \frac{\Delta x_j}{3}\frac{df_{j+1}}{dt} \tag{5.138}$$

$$I_2^{(j+1)} = \int_{x_j}^{x_{j+1}} D\frac{\partial^2 f_a}{\partial x^2} N_{j+1}(x)dx = D \cdot N_{j+1} \left. \frac{\partial f_a}{\partial x} \right|_{x_j}^{x_{j+1}} - D \int_{x_j}^{x_{j+1}} \frac{\partial f_a}{\partial x}\frac{\partial N_{j+1}}{\partial x} dx = \\ = D \left. \frac{df}{dx} \right|_{j+1} - D\left(\frac{-f_j + f_{j+1}}{\Delta x_j} \right)\frac{1}{\Delta x_j} \int_{x_j}^{x_{j+1}} dx = D \left. \frac{df}{dx} \right|_j - \frac{D}{\Delta x_j}(-f_j + f_{j+1}). \tag{5.139}$$

Finally, the integral (5.134) equal to $I_1^{(j+1)} - I_2^{(j+1)}$, is given as:

$$I^{(j+1)} = \frac{\Delta x_j}{6} f_j + \frac{\Delta x_j}{3} f_{j+1} + \frac{D}{\Delta x_j}(-f_j + f_{j+1}) - D \left. \frac{df}{dx} \right|_{j+1}. \tag{5.140}$$

The equations similar to (5.137) and (5.140) are obtained for all elements ($j = 1, 2, \ldots, M-1$). According to Eq. (5.132) they should be assembled leading to the following global system of ordinary differential equations:

– for $j = 1$

$$\frac{\Delta x_j}{3}\frac{d f_j}{d t}+\frac{\Delta x_j}{6}\frac{d f_{j+1}}{d t}-\frac{D}{\Delta x_j}(-f_j+f_{j+1})+D\left.\frac{d f_j}{d x}\right|_j=0 \tag{5.141a}$$

– for $j = 2, 3, \ldots, M-1$

$$\begin{aligned}\frac{\Delta x_{j-1}}{6}\frac{d f_{j-1}}{d t}+\left(\frac{\Delta x_{j-1}}{3}+\frac{\Delta x_j}{3}\right)\frac{d f_j}{d t}+\frac{\Delta x_j}{6}\frac{d f_{j+1}}{d t}+\\ +\frac{D}{\Delta x_{j-1}}(-f_{j-1}+f_j)-\frac{D}{\Delta x_j}(-f_j+f_{j+1})=0\end{aligned} \tag{5.141b}$$

– for $j = M$

$$\frac{\Delta x_{j-1}}{6}f_{j-1}+\frac{\Delta x_{j-1}}{3}f_j+\frac{D}{\Delta x_{j-1}}(-f_{j-1}+f_j)\;-D\left.\frac{df}{dx}\right|_j=0. \tag{5.141c}$$

Equation (5.141) can be rewritten in matrix notation as:

$$\mathbf{A}\frac{d\mathbf{f}}{dt}+\mathbf{B}\cdot\mathbf{f}+\mathbf{F}=\mathbf{0} \tag{5.142}$$

where:

$\mathbf{A}$ – constant matrix, symmetrical and tri-diagonal,
$\mathbf{B}$ – constant matrix, symmetrical and tri-diagonal,
$\mathbf{f} = (f_1, f_2, \ldots, f_M)^T$ – vector of unknowns set up from nodal values of f,
$\frac{d\mathbf{f}}{dt} = \left(\frac{df_1}{dt}, \frac{df_2}{dt}, \ldots, \frac{df_M}{dt}\right)^T$ – vector of time derivatives,
$\mathbf{F} = \left(D\frac{df_1}{dt}, 0, \cdots, 0, -D\frac{df_M}{dt}\right)^T$ – vector representing fluxes through the ends,
T – symbol of transposition.

Both matrices $\mathbf{A}$ and $\mathbf{B}$ have dimensions of $(2M) \times (2M)$.

The initial-value problem for the system (5.142) can be solved using the previously presented implicit trapezoidal method (3.22), yielding the following system of algebraic equations:

$$(\mathbf{A}+0.5\Delta t\cdot\mathbf{B})\,\mathbf{f}_{n+1}=(\mathbf{A}-0.5\Delta t\;\cdot\mathbf{B})\,\mathbf{f}_n-0.5\Delta t\cdot\mathbf{F}_{n+1}-0.5\Delta t\;\cdot\mathbf{F}_n, \tag{5.143}$$

where:

n – index of time level,
Δt – time step.

In such a way the Crank–Nicolson finite element method is obtained. After introducing the required boundary conditions this system of linear algebraic equations is solved with double sweep method giving approximate values of f at next time level.

Assume that we consider the solution domain divided into 6 elements. In this case the matrices **A** and **B** have the following structure:

$$\mathbf{A}=\begin{bmatrix} \frac{\Delta x_1}{3} & \frac{\Delta x_1}{6} & 0 & 0 & 0 & 0 & 0 \\ \frac{\Delta x_1}{6} & \frac{\Delta x_1}{3}+\frac{\Delta x_2}{3} & \frac{\Delta x_2}{6} & 0 & 0 & 0 & 0 \\ 0 & \frac{\Delta x_2}{6} & \frac{\Delta x_2}{3}+\frac{\Delta x_3}{3} & \frac{\Delta x_3}{6} & 0 & 0 & 0 \\ 0 & 0 & \frac{\Delta x_3}{6} & \frac{\Delta x_3}{3}+\frac{\Delta x_4}{3} & \frac{\Delta x_4}{6} & 0 & 0 \\ 0 & 0 & 0 & \frac{\Delta x_4}{6} & \frac{\Delta x_4}{3}+\frac{\Delta x_5}{3} & \frac{\Delta x_5}{6} & 0 \\ 0 & 0 & 0 & 0 & \frac{\Delta x_5}{6} & \frac{\Delta x_5}{3}+\frac{\Delta x_6}{3} & \frac{\Delta x_6}{6} \\ 0 & 0 & 0 & 0 & 0 & \frac{\Delta x_6}{6} & \frac{\Delta x_6}{3} \end{bmatrix}$$

$$\mathbf{B}=\begin{bmatrix} \frac{D}{\Delta x_1} & -\frac{D}{\Delta x_1} & 0 & 0 & 0 & 0 & 0 \\ -\frac{D}{\Delta x_1} & \frac{D}{\Delta x_1}+\frac{D}{\Delta x_2} & -\frac{D}{\Delta x_2} & 0 & 0 & 0 & 0 \\ 0 & -\frac{D}{\Delta x_2} & \frac{D}{\Delta x_2}+\frac{D}{\Delta x_3} & -\frac{D}{\Delta x_3} & 0 & 0 & 0 \\ 0 & 0 & -\frac{D}{\Delta x_3} & \frac{D}{\Delta x_3}+\frac{D}{\Delta x_4} & -\frac{D}{\Delta x_4} & 0 & 0 \\ 0 & 0 & 0 & -\frac{D}{\Delta x_4} & \frac{D}{\Delta x_4}+\frac{D}{\Delta x_5} & -\frac{D}{\Delta x_5} & 0 \\ 0 & 0 & 0 & 0 & -\frac{D}{\Delta x_5} & \frac{D}{\Delta x_5}+\frac{D}{\Delta x_6} & -\frac{D}{\Delta x_6} \\ 0 & 0 & 0 & 0 & 0 & -\frac{D}{\Delta x_6} & \frac{D}{\Delta x_6} \end{bmatrix}$$

Vector **F** represents the fluxes of the transported quantity through the boundary. This vector can have only one nonzero component. This takes place when a Neumann boundary condition $D \cdot \partial f / \partial x$ is imposed at $x = 0$ or at $x = L$. Otherwise its all elements disappear.

It is interesting to compare the semi-discrete forms of the FE and FD methods. For any internal node for uniform grid points ($\Delta x = \text{const.}$), the finite element method gives:

$$\frac{1}{6}\frac{df_{j-1}}{dt} + \frac{4}{6}\frac{df_j}{dt} + \frac{1}{6}\frac{df_{j+1}}{dt} - D\frac{f_{j-1} - 2f_j + f_{j+1}}{\Delta x^2} = 0 \tag{5.144}$$

whereas spatial discretization of the diffusion equation by the finite difference method leads to the following formula:

$$\frac{df_j}{dt} - D\frac{f_{j-1} - 2f_j + f_{j+1}}{\Delta x^2} = 0. \tag{5.145}$$

One can see that in this case the only difference between the methods deals with the approximation of the time derivative at node j. In the finite element method it is considered as a weighted averaged taken from three nodes: $j - 1, j$ and $j + 1$. Consequently this method requires to solve the system of equations with tri-diagonal matrix for both explicit and implicit schemes.

Example 5.2 Solve the diffusion equation using the Crank–Nicolson finite element method for the following auxiliary conditions:

- initial condition: $f(x, 0) = 0$ for $0 \le x \le L$.
- boundary conditions: $f(0, t) = 1.0$ and $f(L, t) = 0.0$ for $t > 0$

Note, that in this case the values of f_1^{n+1} and f_M^{n+1} known as imposed boundary conditions must be introduced into the system (5.143).

The results of calculations carried out for $L = 4$ m, $\Delta x = 0.1$, $D = 0.01$ m^2/s are shown in Fig. 5.20. It appears that they are very similar for different values of time step. This fact suggests that conversely to the advection equation solved in Example 5.1 by the finite difference upwind scheme, this time the value of the applied time step is not restricted.

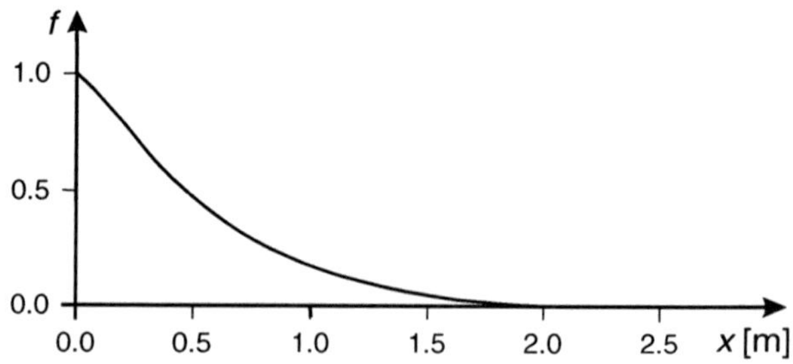

Fig. 5.20 Solutions of the diffusion equation obtained for $t = 50$ s

The Crank–Nicolson scheme in both version i.e. with the finite difference and element methods is known as very effective and reliable tool for solving the diffusion equation. A detailed examination of its numerical properties will be given later.

5.4 Properties of the Numerical Methods for Partial Differential Equations

5.4.1 Convergence

Each numerical method of solution of the partial differential equations is useful on condition that it is convergent. The theory of truncation error says that refinement of the grid provides more accurate numerical solution. Generalizing, one can expect that approximate solution tends to the exact one, while the mesh dimensions Δt and Δx are systematically reduced. More formally, one can say that solution of the system of algebraic equations, which approximates the considered partial differential equation is convergent, if it tends to the exact solution of the differential equation for any value of the independent variable, when the mesh dimensions tend to zero (Abbott and Basco 1989, Fletcher 1991). Then it is required that:

$$f_j^n \to f\left(x_j, t_n\right) \text{ while } \Delta x, \Delta t \to 0,$$

where

f_j^n – approximate value of the function f at the node (j, n),
$f(x_j, t_n)$ – exact value of the function f at the same node (x_j, t_n).

If this requirement is satisfied, the method is said to be convergent. This statement, while obvious, is usually difficult to be proved theoretically.

The difference between the exact solution of the partial differential equation and the exact solution of the system of algebraic equations, called the solution error, is given as:

$$E_j^n = f\left(x_j, t_n\right) - f_j^n. \tag{5.146}$$

where e_j^n is the solution error at the node (j, n). It should be remembered that the exact solution of the system of algebraic equation is still only the approximate solution of the differential equation. We assume that in the solution of the algebraic equations no errors caused by the iteration process as well as by the round-off errors during the computation are present.

The value of solution error E_j^n at the node (j, n) depends on approximation accuracy of the differential equation, i.e. on the values of mesh dimensions Δx and Δt, as well as on the values of higher order derivatives neglected in the numerical approximation. These issues can be illustrated on the examples of the pure advection equation and the pure diffusion equation, which have exact solutions.

In Section 5.2.3 we considered the numerical solution of advection equation (5.105) using the up-wind scheme given by formula (5.108). The results of this application are shown in Figs. 5.15 and 5.16. Now we will continue the numerical experiments with this scheme using the same data as assumed previously in Example 5.1. Let us remember that this equation has an exact solution. Starting with $\Delta x = 100$ m and $\Delta t = 100$ s, which ensures $C_a = 0.5$, the grid is refined in such a way that constant value of the Courant number C_a is kept. The results of calculation carried out for systematically reduced values of Δx and Δt are presented in Fig. 5.21.

Comparing the consecutive solutions with the exact one, we see that the numerical solution approaches the analytical one as the mesh dimensions are reduced. This means that the up-wind scheme provides the solutions which converge to the exact one.

In this case we managed to demonstrate the convergence of the applied numerical scheme directly via numerical tests, since the exact solution of solved equation is known. Unfortunately, in practical cases the exact solutions of equations are unknown. For this reason for some classes of solved problems the convergence may be proved indirectly. To this end the Lax Equivalence Theorem is applied, which reads as follows: "Given a properly posed initial-value problem and a finite difference approximation to it that satisfies the consistency condition, stability is necessary and sufficient condition for convergence" (Abbott and Basco (1983) after Rychtmyer and Morton (1967)). Therefore, if the considered numerical method satisfies both consistency and stability conditions at the same time it satisfies the convergence condition. The advantage of this approach is obvious, since it is relatively easy to verify the consistency and stability.

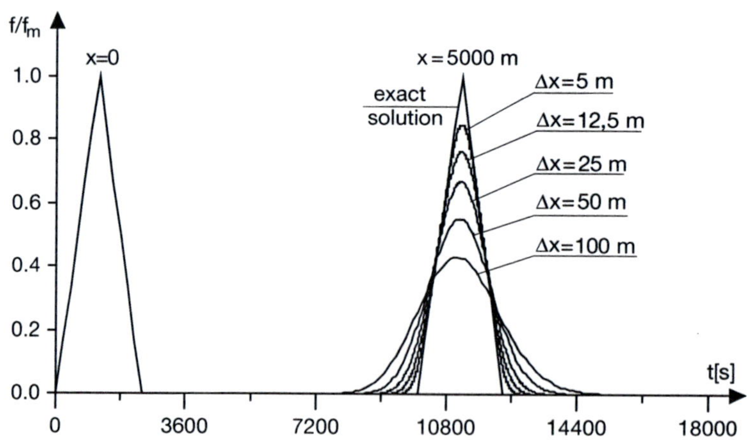

Fig. 5.21 Effect of mesh shrinking for pure advection equation solved by the up-wind difference scheme

5.4.2 Consistency

The system of algebraic equations obtained as an approximation of the partial differentia equation is said to be consistent with this equation if in the limit, when the mesh dimensions tend to zero, it becomes the governing partial differential equation at every node of the grid. This condition is examined by replacing the nodal values of functions by their Taylor series expansion around the considered node. One can say that this is a process inverse to discretization. The equation obtained in such a way is not exactly the same as the original equation, since it contains additional terms. The condition of consistency requires that these extra terms tend to zero as the grid is refined.

Let us verify the consistency condition for the up-wind scheme for advection equation described in the preceding section (Eq. 5.107):

$$\frac{f_j^{n+1}-f_j^n}{\Delta t}+U\frac{f_j^n-f_{j-1}^n}{\Delta x}=0 \tag{5.147}$$

All nodal values of f should be replaced by the Taylor series expansion round the node (j, n) (see Fig. 5.14):

$$f(x+\Delta x,t+\Delta t)=f(x,t)+\sum_{m=1}^{\infty}\frac{1}{m!}\left(\Delta x\frac{\partial}{\partial x}+\Delta t\frac{\partial}{\partial t}\right)^m f(x,t). \tag{5.148}$$

The consecutive nodal values present in Eq. (5.147) are following:

$$f_{j-1}^n=f_j^n-\Delta x\left.\frac{\partial f}{\partial x}\right|_j^n+\frac{\Delta x^2}{2}\left.\frac{\partial^2 f}{\partial x^2}\right|_j^n-\frac{\Delta x^3}{6}\left.\frac{\partial^3 f}{\partial x^3}\right|_j^n+\cdots \tag{5.149}$$

$$f_j^{n+1}=f_j^n+\Delta t\left.\frac{\partial f}{\partial t}\right|_j^{n+1}+\frac{\Delta t^2}{2}\left.\frac{\partial^2 f}{\partial t^2}\right|_j^{n+1}+\frac{\Delta t^3}{6}\left.\frac{\partial^3 f}{\partial t^3}\right|_j^{n+1}+\cdots \tag{5.150}$$

Substitution of Eqs. (5.149) and (5.150) in Eq. (5.147) yields:

$$\frac{\partial f}{\partial t}+U\frac{\partial f}{\partial x}=R(f) \tag{5.151}$$

The right hand side of Eq. (5.151) is equal:

$$R(f)=-\frac{\Delta t}{2}\frac{\partial^2 f}{\partial t^2}-\frac{U\cdot\Delta x}{2}\frac{\partial^2 f}{\partial x^2}+\frac{\Delta t^2}{6}\frac{\partial^3 f}{\partial t^3}-\frac{U\cdot\Delta x^2}{6}\frac{\partial^3 f}{\partial x^3}+\cdots \tag{5.152}$$

In Eqs. (5.151) and (5.152) the indices are omitted, but we have to remember that it is valid for the node (j, n). The extra term $R(f)$ results from the truncation error and it depends on the applied method of solution. We expect that $R(f)$ will disappear for mesh dimensions tending to zero.Indeed, one can notice that for $\Delta x \to 0$

and $\Delta t \to 0$ $R(f) \to 0$ and consequently Eq. (5.151) becomes the original advection equation (5.105). The same approach can be used to prove the consistency of any other numerical scheme.

5.4.3 Stability

As it was fund out previously, a convergent numerical method of solution of the partial differential equation must be not only consistent but stable as well. Stability is a property of a numerical method that ensures damping of any random disturbance which can occur during the computational process. Such disturbance can arise from auxiliary conditions or round-off errors. Typical symptoms of instability are unphysical oscillations occurring in the numerical solution. Let us come back to the results of solution of the pure advection equation using the upwind scheme, which were presented in Figs. 5.15 and 5.16. One can see that as long as the advective Courant number is not greater than unity, the obtained solutions are smooth although they are sometimes inaccurate (Fig. 5.15). However, if the Courant number slightly exceeds unity, the solution is immediately affected by strong oscillations (Fig. 5.16). These oscillations cause that the solution looses its physical sense. The amplitudes of oscillations increase with time and with increasing distance from bounds. It should be mentioned that there are another kinds of oscillations, which can be precluded by stable numerical methods as well. Such oscillations should be distinguished from those generated by instability. In next section this question will be taken up.

The numerical method is considered stable if insignificant variation of input data causes similarly insignificant variation of the final results. Conversely, if the numerical method is not capable to damp such errors, then these errors can increase in an unlimited way and finally they can dominate the solution. Such method is called unstable one. When an unstable method is applied to solve the open channel unsteady flow equations, then usually the computation fails. This is because the oscillating depth takes the negative values, for which the solved equations are not valid.

All numerical methods can be divided into the following ones: absolutely stable, conditionally stable and unstable. For obvious reasons the unstable methods are not interesting. To conclude on the stability of applied scheme, an analysis of its stability must be done. This analysis provides the detailed information of particular conditions, which must be satisfied to ensure a stable solution. As it was shown previously, while discussing the solutions of pure advection and pure diffusion equations, a stable solution required certain relations between the mesh dimensions Δx and Δt, and the physical parameters, being the coefficients of solved equation. The stability analysis can be carried out using the matrix method. The details of this approach are given for instance by Fletcher (1991). However the most commonly applied method of stability analysis is the Neumann approach (Potter 1973, Fletcher 1991, Abbott and Basco 1983). This method is applied below.

The numerical methods applied for solving the partial differential equation provides the solution, which approximates the exact one. Then there is a difference between them:

$$E_j^n = f(x_j, t_n) - f_j^n. \tag{5.153}$$

where:

E_j^n – solution error at the node (j, n),
$f(x_j, t_n)$ – exact solution at the node (j, n),
f_j^n – approximate solution at the node (j, n).

Such error is introduced at every grid point (j, n) for $j = 1,2,3, \ldots, M$ and for $n = 0,1,2, \ldots$.

Let us consider the box scheme applied previously to solve the advection equation (5.105). Since $f_j^n = f(x_j, t_n) - E_j^n$ then substitution of this expression in Eq. (5.108) yields:

$$f(x_j, t_{n+1}) - E_j^{n+1} = C_a(f(x_{j-1}, t_n) - E_{j-1}^n) + (1 - C_a)(f(x_j, t_n) - E_j^n) \tag{5.154}$$

Since the exact solution $f(x, t)$ must satisfy the algebraic equation of the scheme (5.108):

$$f(x_j, t_{n+1}) = C_a \cdot f(x_{j-1}, t_n) + (1 - C_a) f(x_j, t_n) \tag{5.155}$$

then Eq. (5.154) is reduced to the following form:

$$E_j^{n+1} = C_a \cdot E_{j-1}^n + (1 - C_a) E_j^n \tag{5.156}$$

The same equation can be written for $j = 2,3, \ldots, M$ and for $n = 0,1,2, \ldots$. Note, that the system (5.156) has identical structure as the one provided by the up-wind scheme, i.e. as Eq. (5.108). This system is completed by the appropriate initial and boundary conditions. It is assumed that $E_j^0 = 0$ for $j = 2, 3, \ldots, M$ and $E_1^n = 0$ for $n = 0, 1, 2, \ldots$, since the exact initial and boundary conditions do not generate any error.

In the Neumann method the errors at the nodes at considered time level n are expanded in a finite complex Fourier series. The conclusions on stability or instability are deduced by examining how a single component of series behaves while passing from time level n to time level $n + 1$.

Using the Fourier series, the error E_j^n at the node x_j can be expressed as follows (Abbott and Basco 1989):

$$E_j^n = \sum_{k=1}^{K} A_k^n \cdot e^{i \cdot k \cdot m \cdot j \cdot \Delta x}, \tag{5.157}$$

where:

$j = 2,3, \ldots, M,$ – index of node,
$i = (-1)^{1/2}$ – imaginary unit,
n – index of time level,
k – index of Fourier component,
m – wave number,
Δx – space interval,
A_k^n – Fourier coefficient (amplitude of component kth at the time level n),
K – index of given finite value.

The wave number is given as:

$$m = \frac{2\pi}{\lambda} \tag{5.158}$$

where λ is the wave length. Let us multiply both sides of Eq. (5.158) by the space interval Δx:

$$m \cdot \Delta x = \frac{2\pi}{\lambda} \Delta x$$

and let us introduce a new variable N:

$$N = \frac{\lambda}{\Delta x} \tag{5.159}$$

which represents the number of grid intervals over one wavelength. Then Eq. (5.158) can be rearranged to the form:

$$\varphi = m \cdot \Delta x = \frac{2\pi}{N}, \tag{5.160}$$

where $\varphi = m \cdot \Delta x$ is dimensionless wave number. As the shortest waves represented at the considered grid points have wavelength $2\Delta x$, whereas the longest ones tend to infinity, then $2 \le N \le \infty$ implies that $0 \le \varphi \le \pi$.

For the linear problems as considered here, it is sufficient to examine the behavior of one Fourier component. Therefore, instead of Eq. (5.157), one can consider only one component of the sum corresponding, for example, to $k = 1$. For simplicity this index will be omitted. Taking into account Eq. (5.158) the chosen component can be written as:

$$E_j^n = A^n \cdot e^{i \cdot \varphi \cdot j}. \tag{5.161}$$

Note that the time is included in the amplitude A^n.

Equation (5.161) can be substituted into Eq. (5.156) yielding:

$$A^{n+1} \cdot e^{i \cdot \varphi \cdot j} = C_a \cdot A^n \cdot e^{i \cdot \varphi (j-1)} + (1 - C_a) A^n \cdot e^{i \cdot \varphi \cdot j}. \tag{5.162}$$

Dividing both sides by the factor $e^{i \cdot \varphi \cdot j}$ one obtains:

$$\frac{A^{n+1}}{A^n} = G, \tag{5.163}$$

where

$$G = 1 - C_a + C_a \cdot e^{-i \cdot \varphi} \tag{5.164}$$

G is interpreted as the amplification factor of the amplitude of kth component of Fourier representation of the error function, while passing from the time level n to $n + 1$. The error amplitude will not increase if the modulus of G is not greater than unity:

$$|G| \leq 1 \tag{5.165}$$

Using the Euler relation $e^{-i \cdot \varphi} = \cos\varphi - i \cdot \sin\varphi$, Eq. (5.164) is reformed to the following form:

$$G = 1 - C_a + C_a \cdot \cos\varphi - i \cdot C_a \cdot \sin\varphi \tag{5.166}$$

Since the amplification factor G is a complex number, therefore the condition (5.165) yields:

$$|G| = \left[(1 - C_a(1 - \cos\varphi))^2 + (C_a \cdot \sin\varphi)^2 \right]^{1/2} \leq 1. \tag{5.167}$$

This relation can be reduced to the following one:

$$4 \sin^2 \frac{\varphi}{2} \left(C_a^2 (1 + \cos^2 \frac{\varphi}{2}) - C_a \right) \leq 0. \tag{5.168}$$

Since φ ranges from 0 to π, then inequality (5.168) can be verified for its extreme values only. Note that for $\varphi = 0$ this relation is satisfied for any value of C_a, whereas setting $\varphi = \pi$ one obtains:

$$C_a (C_a - 1) \leq 0 \tag{5.169}$$

As the Courant number is always positive, then relation (5.169) will be true for:

$$C_a \leq 1 \tag{5.170}$$

The final conclusion is then following: to avoid amplification of the Fourier wave amplitude while solving the advection equation using the up-wind scheme, the Courant number cannot exceed unity. This condition implies the following restriction with regard to the applied time step:

$$\Delta t \leq \frac{\Delta x}{U} \tag{5.171}$$

As long as the value of time step Δt does not exceed this limit, we can be sure that the solution of the advection equation provided by the upwind scheme is still numerically stable. Since the stability requires satisfying the condition (5.171), then the up-wind scheme is conditionally stable. This property explains the oscillating results of computation provided by $C_a = 1.035$ and shown in Fig. 5.16.

Equation (5.170) is known as Courant-Friedrichs-Levy (CFL) condition (Potter 1973). It is valid for all hyperbolic partial differential equations solved by explicit schemes. The Courant number can be interpreted as relation between the propagation celerity in analytical solution (in this case it is the advection velocity U) and the celerity of propagation in the numerical solution $\Delta x/\Delta t$ (Abbott and Basco 1983).

Summarizing the above presented considerations, we can state that upwind scheme applied for the advection equation is only conditionally stable – the mesh dimensions must satisfy appropriate relations. Since from the preceding section this scheme is known to be consistent as well, then it is convergent if only the stability conditions is respected while computations. Then, using the Lax Equivalence Theorem we confirmed the convergence of the up-wind scheme. Note that the same was deduced from the numerical experiments.

As far as the numerical solution of the diffusion equation by the Crank-Nicolson finite element method (5.141) is considered, for uniform elements we have the following formula for any internal node:

$$\begin{aligned}\frac{1}{6}\frac{f_{j-1}^{n+1}-f_{j-1}^{n}}{\Delta t}+\frac{4}{6}\frac{f_{j}^{n+1}-f_{j}^{n}}{\Delta t}+\frac{1}{6}\frac{f_{j+1}^{n+1}-f_{j+1}^{n}}{\Delta t}+\\ -\frac{D}{2}\frac{f_{j-1}^{n+1}-2f_{j}^{n+1}+f_{j+1}^{n+1}}{\Delta x^2}-\frac{D}{2}\frac{f_{j-1}^{n}-2f_{j}^{n}+f_{j+1}^{n}}{\Delta x^2}=0\end{aligned} \tag{5.172}$$

This equation is rewritten in the form:

$$\begin{aligned}(1-3C_d)f_{j-1}^{n+1}+(4+6C_d)f_{j}^{n+1}+(1-3C_d)f_{j+1}^{n+1}=\\ (1+3C_d)f_{j-1}^{n}+(4-6C_d)f_{j}^{n}+(1+3C_d)f_{j+1}^{n}\end{aligned} \tag{5.173}$$

where C_d, sometimes called the diffusive Courant number, is given by:

$$C_d = \frac{\Delta t \cdot D}{\Delta x^2} \tag{5.174}$$

Using Eq. (5.161) one obtains the following formula for Fourier component representing the error function:

$$A^{n+1}\cdot e^{i\cdot\varphi\cdot j}\left((1-3C_d)e^{-i\cdot\varphi\cdot j}+(4+6C_d)+(1-3C_d)e^{i\cdot\varphi\cdot j}\right)=$$
$$=A^{n}\cdot e^{i\cdot\varphi\cdot j}\left((1+3C_d)e^{-i\cdot\varphi\cdot j}+(4-6C_d)+(1+3C_d)e^{i\cdot\varphi\cdot j}\right)\quad . \tag{5.175}$$

Dividing both sides of Eq. (5.175) by $A^n\cdot e^{i\cdot\varphi\cdot j}$ yields:

$$\frac{A^{n+1}}{A^n}=G=\frac{(1+3C_d)(e^{-i\cdot\varphi}+e^{i\cdot\varphi})+(4-6C_d)}{(1-3C_d)(e^{-i\cdot\varphi}+e^{i\cdot\varphi})+(4+6C_d)}\quad . \tag{5.176}$$

Since $\left(e^{i\cdot\varphi}+e^{-i\cdot\varphi}\right)/2=\cos\varphi$ then

$$G=\frac{(1+3C_d)\cos\varphi+(2-3C_d)}{(1-3C_d)\cos\varphi+(2+3C_d)}\quad . \tag{5.177}$$

The stability condition (5.165) requires:

$$-1\leq\frac{(1+3C_d)\cos\varphi+(2-3C_d)}{(1-3C_d)\cos\varphi+(2+3C_d)}\leq 1. \tag{5.178}$$

This relation must be satisfied for the extreme values of the dimensionless wave number φ, i.e. for $\varphi=0$ and $\varphi=\pi$ i.e. for $\cos\varphi=1$ and for $\cos\varphi=-1$. Substituting consecutively these values one can find out that relation (5.178) is satisfied for any value of C_d. Therefore the Crank–Nicolson finite element method applied for solving the diffusion equation is absolutely stable. This explains the results of solution obtained in Example 5.2.

References

Abbott MB, Basco DR (1989) Computational fluid dynamics. Longman Scientific and Technical, New York

Billingham J, King AC (2000) Wave motion. Cambridge University Press, Cambridge

Fletcher CAJ (1991) Computational techniques for fluid dynamics, vol. I. Springer-Verlag, Berlin

Gresho PM, Sani RL (1998) Incompressible flow and the finite-element method, Advection-diffusion, vol. 1. Wiley, Chichester, England

Hervouet JM (2007) Free surface flow – modeling with the finite element method. Wiley, England

Oden JT, Reddy JN (1976) An introduction to the mathematical theory of finite elements. Wiley, New York

Potter D (1973) Computational physics. Wiley, London

Rychtmyer RD, Morton KW (1967) Difference methods for initial-value problems. Interscience, New York

Szymkiewicz R, Mitosek M (2007) Numerical aspects of improvement of the unsteady pipe flow equations. Int. J. Numer. Methods Fluids 55:1039–1058

Weiyan T (1992) Shallow water hydrodynamics. Elsevier, Amsterdam

Zienkiewicz OC (1972) The finite element method in engineering science. McGraw-Hill, London

Chapter 6
Numerical Solution of the Advection Equation

6.1 Solution by the Finite Difference Method

6.1.1 Approximation with the Finite Difference Box Scheme

In order to show the specific problems related to the numerical solution of partial differential hyperbolic equations, let us solve the pure advection equation (5.3):

$$\frac{\partial f}{\partial t} + U\frac{\partial f}{\partial x} = 0 \tag{6.1}$$

where:

f – scalar function,
U – flow velocity assumed to be constant.

To solve this equations one can use a number of finite difference schemes. We begin with applying the implicit four point scheme commonly called the box scheme. This scheme has interesting history (see Abbott and Basco 1989), is well known and frequently used in open channel flow modeling. It is very flexible and particularly suitable for illustrating all the most important aspects of the numerical solution of hyperbolic equations.

The scheme uses grid points shown in Fig. 6.1. The derivatives are approximated inside the mesh, at point P. To this end the following formulas are used:

$$\left.\frac{\partial f}{\partial t}\right|_P = \psi\frac{f_{j-1}^{n+1} - f_{j-1}^{n}}{\Delta t} + (1-\psi)\frac{f_j^{n+1} - f_j^{n}}{\Delta t} \tag{6.2}$$

$$\left.\frac{\partial f}{\partial x}\right|_P = (1-\theta)\frac{f_j^{n} - f_{j-1}^{n}}{\Delta x} + \theta\frac{f_j^{n+1} - f_{j-1}^{n+1}}{\Delta x} \tag{6.3}$$

where:

j – index of cross-section,
n – index of time level,

R. Szymkiewicz, *Numerical Modeling in Open Channel Hydraulics*, Water Science and Technology Library 83, DOI 10.1007/978-90-481-3674-2_6,

Δx – spatial mesh dimension,
Δt – time step,
ψ – weighting parameter which ranges $\langle 0, 1\rangle$,
θ – weighting parameter which ranges $\langle 0, 1\rangle$,

Originally, in the box scheme the meaning of parameter ψ is opposite, since it denotes the distance from the cross-section $j-1$ to point P (Fig. 6.1) (Abbott and Basco 1989, Cunge et al. 1980). We change this convention to be compatible with the notation used in the lumped flood routing models presented in Chapter 9, in which the box scheme occurs in implicit way.

Substitution of Eqs. (6.2) and (6.3) in advection equation (6.1) yields:

$$\psi\frac{f_{j-1}^{n+1}-f_{j-1}^{n}}{\Delta t}+(1-\psi)\frac{f_{j}^{n+1}-f_{j}^{n}}{\Delta t}+ \\ +U\left((1-\theta)\frac{f_{j}^{n}-f_{j-1}^{n}}{\Delta x}+\theta\frac{f_{j}^{n+1}-f_{j-1}^{n+1}}{\Delta x}\right)=0 \quad \text{for} \quad j=2,3,\cdots,M \tag{6.4}$$

where M is the total number of cross-sections in the considered channel reach. Note, that with the assumption $U>0$ the value f_1^{n+1} is known from the boundary condition proscribed at the upstream end. Therefore in each equation of the system (6.4) only one unknown f_j^{n+1} exists. It can be easily isolated, giving:

$$f_j^{n+1}=\alpha\cdot f_{j-1}^{n}+\beta\cdot f_j^{n}+\gamma\cdot f_{j-1}^{n+1} \tag{6.5}$$

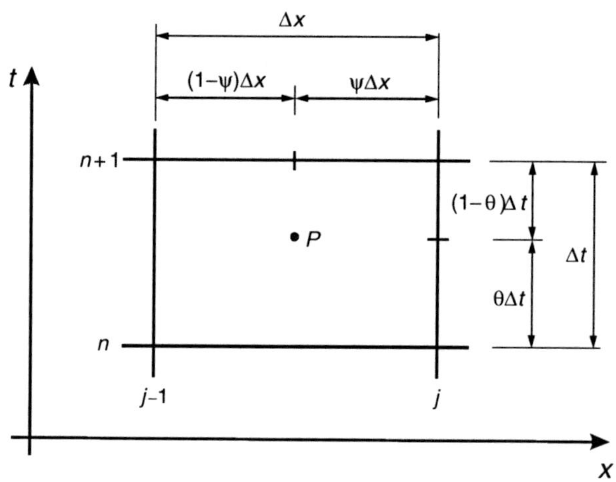

Fig. 6.1 Grid points for the box scheme

where:

$$\alpha = \frac{\psi + (1-\theta)C_a}{1 - \psi + \theta \cdot C_a}, \tag{6.6a}$$

$$\beta = \frac{1 - \psi - (1-\theta)C_a}{1 - \psi + \theta \cdot C_a}, \tag{6.6b}$$

$$\gamma = \frac{-\psi + \theta \cdot C_a}{1 - \psi + \theta \cdot C_a} \tag{6.6c}$$

The advective Courant number C_a was defined in preceding chapter (Eq. 5.109) as follows:

$$C_a = \frac{U \cdot \Delta t}{\Delta x} \tag{6.7}$$

Setting in Eq. (6.5) $j = 2, 3, \ldots, M$ one can compute the nodal values of the function f in all nodes at the time level t_{n+1}. Note that the computations must run in a strictly defined order – from the upstream end towards the downstream end and it is impossible to reverse the order of calculations. Such scheme is said to be implicit.

The box scheme involves two weighting parameters ψ and θ, which can take any value form the interval $\langle 0, 1 \rangle$. Moreover the mesh dimensions Δx and Δt are mutually related via the Courant number (6.7). One can suppose that the values of all mentioned parameters will determine the accuracy of the numerical solution produced by the box scheme. To examine this influence let us solve advection equation (6.1) for an arbitrary set of data.

Example 6.1 In a prismatic channel of length L the water flows with constant velocity U. Assume that:

- initial condition is following: $f(x,t = 0) = 0$ for $0 \le x \le L$,
- boundary condition is following:

$$f(x=0,t) = \begin{cases} F_m \cdot \frac{t}{T_m} & \text{for } 0 \le t \le T_m \\ F_m \cdot \left(2 - \frac{t}{T_m}\right) & \text{for } T_m \le t \le 2T_m \\ 0 & \text{for } t > 2T_m \end{cases}$$

where F_m is peak value and T_m is time to peak. Therefore we will examine advective transport of a scalar quantity represented by the function f, which has initial distribution in the form of a triangle with the height equal to F_m and the base equal to $2T_m$.

Knowing the analytical solution of the advection equation (see Section 5.1.2) we expect that the triangular distribution imposed at the upstream end will travel along the channel axis without any shape deformation. Then in each cross-section it will be shifted in time only.

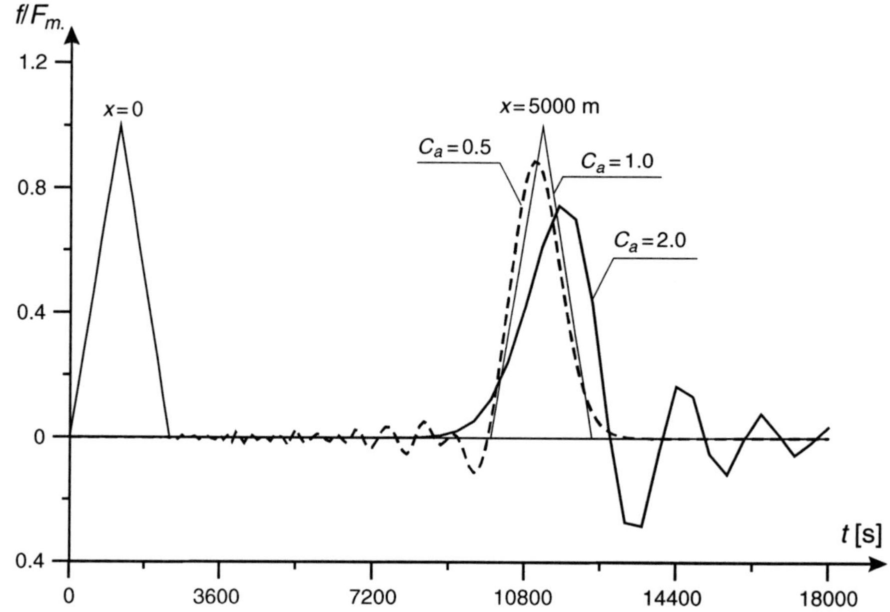

Fig. 6.2 Solution of advection equation using the box scheme for various values of the Courant number with $\psi = 0.5$ and $\theta = 0.5$

The following data are assumed: $U = 0.5$ m/s = const., $\Delta x = 100$ m = const., $F_m = 100$ and $T_m = 1{,}200$ s. As a control we take the cross-section distanced 5,000 m from the upstream end. Let us begin with assuming $\psi = 0$ and $\theta = 0$. For these values of the weighting parameters one obtains $\alpha = C_a$, $\beta = 1 - C_a$ and $\gamma = 0$, and consequently Eq. (6.5) turns into Eq. (5.108). This version of the box scheme coincides with the upwind scheme previously discussed in Section 5.2.3. The results of computations are shown in Figs. 5.15 and 5.16.

The results of numerical experiments carried out for other sets of the parameters ψ, θ and C_a are shown in Figs. 6.2, 6.3 and 6.4. One can see that the exact solution is given only for specific combinations of the parameters, as for instance $\psi = 0.5$, $\theta = 0.5$ and $C_a = 1$. For other sets one obtains either oscillating or damped solutions.

Since the only physical process represented in the solved equation is advection then observed effects in its solution must have the numerical roots. In order to explain such behavior of the box scheme let us analyze its numerical properties.

6.1.2 Stability Analysis of the Box Scheme

It is known from Section 5.4 that the local solution error satisfies the equation of the applied numerical scheme. Therefore, according to Eqs. (6.5) and (5.153) one can write:

$$E_j^{n+1} = \alpha \cdot E_{j-1}^n + \beta \cdot E_j^n + \gamma \cdot E_{j-1}^{n+1} \tag{6.8}$$

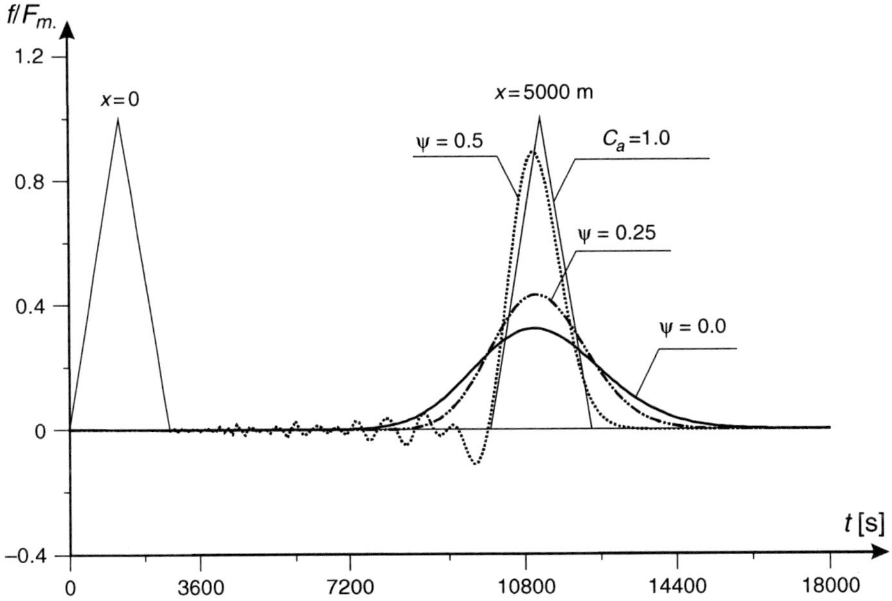

Fig. 6.3 Solution of advection equation using the box scheme for various values of ψ with $\theta = 0$ and $C_a = 0.5$

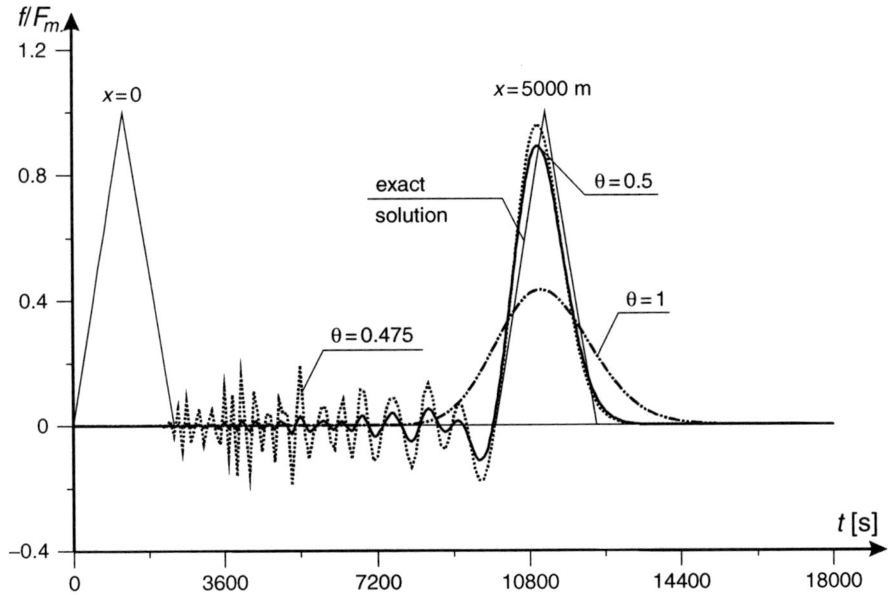

Fig. 6.4 Solution of advection equation using the box scheme for various values of θ with $\psi = 0.5$ and $C_a = 0.5$

With Fourier representation of the solution error (5.161), Eq. (6.8) takes the following form:

$$A^{n+1} \cdot e^{i \cdot \varphi \cdot j} = \alpha \cdot A^{n} \cdot e^{i \cdot \varphi (j-1)} + \beta \cdot A^{n} \cdot e^{i \cdot \varphi \cdot j} + \gamma \cdot A^{n+1} \cdot e^{i \cdot \varphi (j-1)}. \quad (6.9)$$

Dividing both sides by $e^{i \cdot \varphi \cdot j}$ one obtains:

$$\frac{A^{n+1}}{A^{n}} = G = \frac{\beta + \alpha \cdot e^{-i \cdot \varphi}}{1 - \gamma \cdot e^{-i \cdot \varphi}}, \quad (6.10)$$

where G is the amplification factor. The error amplitude should not increase, so it is required that $|G| \leq 1$.

Using the Euler relation $e^{-i \cdot \varphi} = \cos \varphi - i \cdot \sin \varphi$ (McQuarrie 2003), Eq. (6.10) is reformed to the following form:

$$G = \frac{\beta + \alpha \cdot \cos \varphi - i \cdot \alpha \cdot \sin \varphi}{1 - \gamma \cdot \cos \varphi + i \cdot \gamma \cdot \sin \varphi} \quad (6.11)$$

Arithmetic operations on complex numbers allow us to rewrite this equation as:

$$G = \frac{\beta + (\alpha - \beta \cdot \gamma) \cos \varphi - \alpha \cdot \gamma}{1 - 2\gamma \cdot \cos \varphi + \gamma^2} - \frac{(\alpha + \beta \cdot \gamma) \sin \varphi}{1 - 2\gamma \cdot \cos \varphi + \gamma^2} i \quad (6.12)$$

The amplification factor G is a complex number, therefore the stability condition $|G| \leq 1$ yields the following relation:

$$|G| = \left\{ \left[\frac{\beta + (\alpha - \beta \cdot \gamma) \cos \varphi - \alpha \cdot \gamma}{1 - 2\gamma \cdot \cos \varphi + \gamma^2} \right]^2 + \left[\frac{(\alpha + \beta \cdot \gamma) \sin \varphi}{1 - 2\gamma \cdot \cos \varphi + \gamma^2} \right]^2 \right\}^{1/2} \leq 1 \quad (6.13)$$

Since φ ranges from 0 to π, then inequality (6.13) can be verify for its extreme values only. Setting $\varphi = \pi$ one obtains:

$$\left(\frac{\beta - \alpha}{1 + \gamma} \right)^2 \leq 1 \quad (6.14)$$

This inequality is equivalent to the following relation:

$$-1 \leq \frac{\beta - \alpha}{1 + \gamma} \leq 1 \quad (6.15)$$

Substitution of the parameters α, β and γ defined by Eqs. (6.6a), (6.6b) and (6.6c) in Eq. (6.15) yields the following relation:

$$-1 \leq \frac{1 - 2\psi - 2(1 - \theta) C_a}{1 - 2\psi + 2\theta \cdot C_a} \leq 1 \quad (6.16)$$

which is equivalent to the following two inequalities:

$$-1 \leq 1 - \frac{2C_a}{1 - 2\psi + 2\theta \cdot C_a} \text{ and} \tag{6.17a}$$

$$1 - \frac{2C_a}{1 - 2\psi + 2\theta \cdot C_a} \leq 1. \tag{6.17b}$$

These relations will be satisfied for:

$$\theta \geq \frac{1}{2} \text{ and} \tag{6.18a}$$

$$\psi \leq \frac{1}{2} \tag{6.18b}$$

Both conditions ensure that the denominators in Eqs. (6.17a) and (6.17b) are always different from zero. Since the inequalities (6.17a) and (6.17b) are satisfied for any value of the Courant number, then the conditions (6.18a) and (6.18b) ensure unconditional stability of the box scheme.

Note that for $\psi = 0$ and $\theta = 0$ Eq. (6.5) becomes the upwind scheme, with very different numerical properties. Its stability was analyzed in Section 5.3.

6.2 Amplitude and Phase Errors

The numerical results presented in preceding section constitute a good starting point for the discussion of the numerical aspects of solution of the hyperbolic equations. The open channel flow equations, being of hyperbolic type like the advection equation, do not contain any mechanism of dissipation. The solution of such equation has the form of wave, with a specific amplitude and celerity. It is important that the applied numerical scheme changes neither the wave amplitude nor its phase celerity. This means that the numerical methods should not introduce into solution any artificial dissipation or dispersion. The numerical method which generates artificial dissipation is called a dissipative method, whereas the method which produces artificial dispersion is called a dispersive one. In the first case the wave amplitude is attenuated giving rise to unphysical smoothing, whereas in the second case the wave celerity is affected giving rise to the unphysical oscillations of the results. Due to numerical dispersion and dissipation the solution of hyperbolic equations is much more difficult and challenging than the solution of the elliptic equations. The same is true for parabolic equations containing significant advective term, like the advection-diffusion transport equation widely applied to model the motion of scalar quantity in open channels, where the turbulent flow of large Reynolds numbers usually takes place. Both phenomena, having purely numerical roots, can be examined by an appropriate numerical analysis. As an example, let us perform such analysis for the considered numerical solution of the pure advection equation using the difference box scheme.

The dissipative and dispersive properties of any numerical scheme are expressed by the amplitude and phase errors. Both errors can by estimated using the amplitude factor G and its modulus. To this end two coefficients of convergence R_1 and R_2 are defined. The coefficient R_1 is expressed by the ratio of the amplification factors for numerical solution and for exact one. For first time these coefficients were introduces by Leendertse (see Abbott and Basco 1989, Cunge et al. 1980, Liggett and Cunge 1975). Since for the exact solution this factor is equal to unity, then the coefficient R_1 is represented by the modulus of the amplification factor $|G|$. As we will see further, both coefficients of convergence are functions of the Courant number C_a and the variable N, being the number of space interval Δx per wavelength of kth Fourier component of the solution.

Let us begin with an analysis of the dissipative properties of the box scheme. Substitution of Eq. (5.160) in Eq. (6.13) yields the following formula

$$|G| = \left\{ \left[\frac{\beta + (\alpha - \beta \cdot \gamma)\cos\left(\frac{2\pi}{N}\right) - \alpha \cdot \gamma}{1 - 2\gamma \cdot \cos\left(\frac{2\pi}{N}\right) + \gamma^2} \right]^2 + \left[\frac{(\alpha + \beta \cdot \gamma)\sin\left(\frac{2\pi}{N}\right)}{1 - 2\gamma \cdot \cos\left(\frac{2\pi}{N}\right) + \gamma^2} \right]^2 \right\}^{1/2} \tag{6.19}$$

Equation (6.19) expresses the modulus of the amplification factor as a function of the parameter N, representing the relation between the wave length λ and the space interval Δx (see Eqs. (5.158), (5.159) and (5.160)).

Plotting this function for selected values of the Courant number and the weighting parameters, one can show the relation between the wave length and intensity of the wave amplitude attenuation. The plots in Figs. 6.5 and 6.6 show that for $C_a = 1$, $\psi = 0.5$ and $\theta = 0.5$ the box scheme is dissipation free since $|G| = 1$ for any

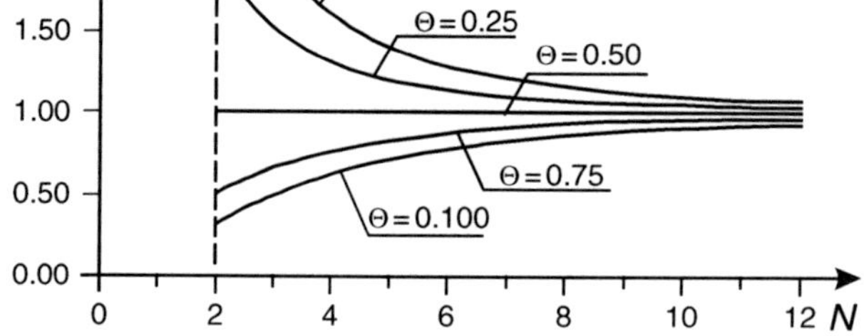

Fig. 6.5 Modulus of the amplification factor (6.19) versus the parameter N for advection equation solved by the box with $\psi = 0.5$ and $C_a = 2.0$

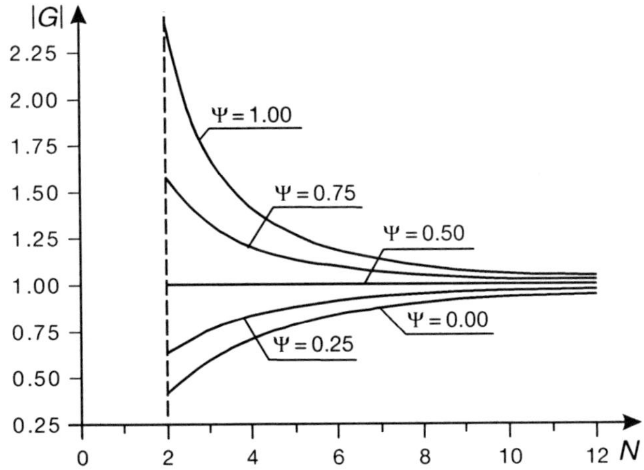

Fig. 6.6 Modulus of the amplification factor (6.19) versus the parameter N for the advection equation solved by the box scheme with $\theta = 0.5$ and $C_a = 2.0$

length of the wave. Consequently the wave amplitude is neither damped nor amplified. On the other hand $\theta \geq 0.5$ and $\psi \leq 0.5$ give $|G| > 1$, which means that the scheme increases the wave amplitude in time. Finally this process leads to numerical instability. This fact could be expected on the basis of the stability analysis carried out previously – compare Eqs. (6.18a) and (6.18b). For other values of the parameters the box scheme becomes dissipative. One can notice that the most intense damping is for short waves (small values of N) and large Courant numbers. Damping is reduced for increasing wavelength.

In a similar way can be analyzed the dissipative properties of the up-wind scheme. With $\psi = 0$ and $\theta = 0$ from Eqs. (6.6a), (6.6b) and (6.6c) results that $\alpha = C_a$, $\beta = 1 - C_a$ and $\gamma = 0$. Substituting these values in Eq. (6.19) and taking into account Eq. (5.160) one obtains the following formula

$$|G| = \left(\left[1 - C_a\left(1 - \cos\left(\frac{2\pi}{N}\right)\right)\right]^2 + \left[C_a \cdot \sin\left(\frac{2\pi}{N}\right)\right]^2\right)^{1/2}. \tag{6.20}$$

The dissipative behavior of the up-wind scheme is summarized graphically in Fig. 6.7.

This figure confirms the numerical stability conditions shown previously. As long as the Courant number does not exceed unity, the up-wind scheme damps the wave amplitude ensuring stable solution, since the modulus of its amplification factor does not exceed unity as well. At the same time we can observe (as previously) that short waves are more intensively damped than long ones. For $C_a = 1$ one obtains $|G| = 1$, which means that the wave amplitude is neither damped nor amplified. However, for greater values of C_a the scheme becomes unstable, as in this case the value of modulus of the amplification factor exceeds unity. Moreover, Fig. 6.7

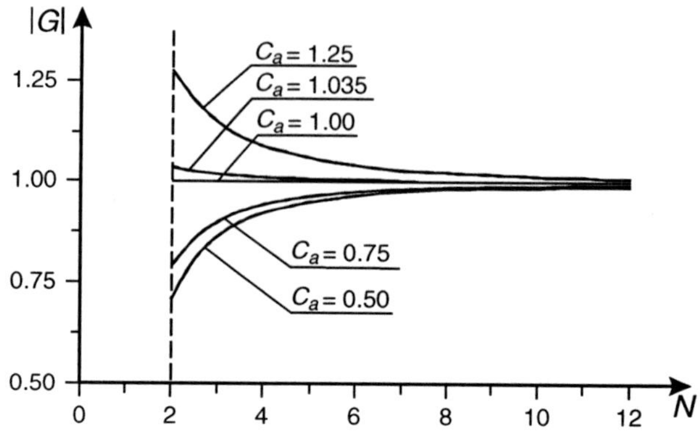

Fig. 6.7 Modulus of the amplification factor versus the parameter N for the advection equation solved by the up-wind scheme

shows that indeed, as it was mentioned previously, the waves represented on the grid cannot be shorter than $2\Delta x$. Relation (6.20) is valid for $N \geq 2$ only.

The second type of error is the phase error. This one can be represented using the amplitude factor G as well. Since it is a complex number:

$$G = \operatorname{Re} G + i \cdot \operatorname{Im} G \tag{6.21}$$

then the phase celerity of kth components of the Fourier series is defined by the angle ϕ as:

$$\tan(\phi) = \frac{\operatorname{Im} G}{\operatorname{Re} G} \tag{6.22}$$

For the box scheme (Eq. 6.12) one obtains:

$$\tan(\phi) = -\frac{(\alpha + \beta \cdot \gamma)\sin(\varphi)}{\beta + (\alpha - \beta \cdot \gamma)\cos(\varphi) - \alpha \cdot \gamma} \tag{6.23}$$

A similar expression can be derived for the exact solution of the pure advection equation. To this end let us write this solution in the following form:

$$f(x,t) = F(t)e^{i \cdot m \cdot x}. \tag{6.24}$$

Substituting Eq. (6.24) in Eq. (6.1) yields:

$$\frac{dF}{dt}e^{i \cdot m \cdot x} + i \cdot m \cdot U \cdot F \cdot e^{i \cdot m \cdot x} = 0. \tag{6.25}$$

This expression can be simplified to:

$$\frac{dF}{dt} = -i \cdot m \cdot U \cdot F. \tag{6.26}$$

Equation (6.26) has the following solution:

$$F(t) = e^{-i \cdot m \cdot U \cdot t}. \tag{6.27}$$

Now this result can be substituted in the solution (6.24):

$$f(x,t) = e^{i \cdot m(x - U \cdot t)}. \tag{6.28}$$

For the neighboring time levels t_n and $t_{n+1} = t_n + \Delta t$ we obtain respectively:

$$f^n = e^{i \cdot m(x - U \cdot t_n)}, \tag{6.29}$$

$$f^{n+1} = e^{i \cdot m(x - U(t_n + \Delta t))} = e^{i \cdot m \cdot x} e^{-i \cdot m \cdot U \cdot t} e^{-i \cdot m \cdot U \cdot \Delta t} \tag{6.30}$$

Dividing the respective sides of these equations yields:

$$\frac{f^{n+1}}{f^n} = e^{-i \cdot m \cdot U \cdot \Delta t} = e^{-i \cdot C_a \cdot m \cdot \Delta x} = e^{-i \cdot C_a \cdot \varphi}. \tag{6.31}$$

Let us compare this equation with Eq. (5.163). Equation (6.31) can be considered as the definition of the amplification factor for exact solution of the pure advection equation. Since $e^{-\varphi \cdot i} = \cos(\varphi) - i \cdot \sin(\varphi)$, then the factor takes the following form:

$$G^{\text{exact}} = \cos(C_a \cdot \varphi) - i \cdot \sin(C_a \cdot \varphi). \tag{6.32}$$

The corresponding angle ϕ^{exact} defining the exact phase celerity of wave is given by the following relation:

$$\tan(\phi^{\text{exact}}) = \frac{\text{Im}G^{\text{exact}}}{\text{Re}G^{\text{exact}}} = -\frac{\sin(C_a \cdot \varphi)}{\cos(C_a \cdot \varphi)} = \tan(-C_a \cdot \varphi). \tag{6.33}$$

Then one can write:

$$\phi^{\text{exact}} = -C_a \cdot \varphi = -C_a \cdot m \cdot \Delta x. \tag{6.34}$$

Now let us consider again Eq. (6.23) representing the phase celerity of the wave described by the advection equation solved by the box scheme. The numerical scheme is dispersion free, if for any Fourier component k the angle ϕ is constant. Conversely, if the component's phase celerity varies, the scheme is dispersive. Note that for the box scheme constant phase celerity is ensured by $\psi = 0.5$, $\theta = 0.5$

and $C_a = 1$. In such a case one obtains: $\alpha = 1$, $\beta = 0$ and $\gamma = 0$. Consequently Eq. (6.23) becomes:

$$\tan(\phi) = -\frac{\sin(\varphi)}{\cos(\varphi)} = \tan(-\varphi) = \tan(-m \cdot \Delta x) \tag{6.35}$$

which means that $\phi = -\varphi = -m{\cdot}\Delta x$. Therefore, for the mentioned values of the weighting parameters and for the Courant number equal to unity the box scheme is not dispersive. At the same time Eq. (6.35) coincides with Eq. (6.33) for the exact solution of the advection equation.

Obviously, in real-life cases the computations are performed with $C_a \neq 1$. This means that the dispersion error occurs always. It is interesting to know how the applied scheme changes the phase celerity comparing with the exact solution. Dispersive properties of any scheme are expressed by the coefficient R_2 defined as the ratio of the numerical phase celerity to the exact celerity:

$$R_2 = \frac{\phi}{\phi^{\mathrm{exact}}}. \tag{6.36}$$

Substitution of Eqs. (6.23) and (6.34) in Eq. (6.36) yields:

$$R_2 = \frac{\arctan\left(\dfrac{(\alpha - \beta\cdot\gamma)\sin(2\pi/N)}{\beta + (\alpha - \beta\cdot\gamma)\cos(2\pi/N) - \alpha\cdot\gamma}\right)}{2C_a \cdot \pi/N}. \tag{6.37}$$

where α, β and γ are given by Eqs. (6.6a), (6.6b) and (6.6c).

The coefficient R_2 given by Eq. (6.37) and calculated for $\psi = 0.5$ and $\theta = 0.5$ is shown in Fig. 6.8 as a function of the wavelength (expressed by N i.e. the number of

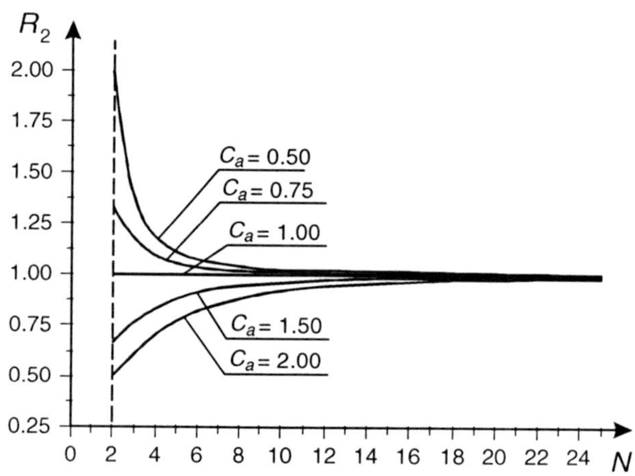

Fig. 6.8 Phase portraits for the box scheme with $\psi = 0.5$ and $\theta = 0.5$

space intervals Δx per wavelength) for various values of the Courant number. Note that the assumed values of the weighting parameters ensure that the box scheme is dissipation free. Figures 6.5 and 6.6 show that in this case the modulus of the amplification factor is constant and equal to unity.

In Fig. 6.8 we observe that R_2 decreases with increasing wavelength. The value of this coefficient, being significant for short waves, tends to unity for long waves. It means that the phase celerity of the long waves is less affected by the numerical approximation than the celerity of the short waves. The only case when the wave celerity is not affected corresponds to $C_a = 1$. Note that the Courant numbers greater than 1 cause that the numerical phase celerity is less than the celerity in the exact solution ($R_2 < 1$). For $C_a < 1$ the situation is opposite. The numerical phase celerity dominates over the one corresponding to the exact solution ($R_2 > 1$). This fact can be related to the solution of the advection equation using the box scheme presented in Fig. 6.2. For C_a>1 which corresponds to R_2<1, the oscillations occur behind the steep front of f, whereas for C_a<1, or R_2>1, the oscillations occur before the front of wave. Let us remember that the shortest waves possible to represent at the numerical grid are equal to $2\Delta x$. For this reason the coefficient R_2 cannot be resolved for $N < 2$.

The information on the dissipative and dispersive properties of the numerical schemes provided by the amplification factor G and coefficient R_2 is helpful in the interpretation of the results of computations. However, comparing the graphs as those presented in Figs. 6.7 and 6.8 provides only general (qualitative) information on the properties of the numerical schemes, but it does not allow a quantitative evaluation.

6.3 Accuracy Analysis Using the Modified Equation Approach

Since numerical methods are the main tool of solution of partial differential equations, the solution of any open channel flow problem having practical meaning will have approximate character. Therefore the question of great importance is to estimate the accuracy of the obtained solution. Usually, as it was shown in the preceding sections, the only available information on the accuracy concerns the order of approximation of the derivatives. One can say that our knowledge on applied numerical scheme as the order of approximation, its consistency and the stability condition is necessary but insufficient to explain and to predict its behavior. The examples of solution of the pure advection equation using the box scheme, presented in Section 5.2.3, confirm this conclusion.

The same is true for the coefficients of convergence, which characterize the amplitude and phase error. They explain rather general tendencies of the applied numerical method. As these coefficients are related to the wave lengths, information resulting from such an analysis cannot be used directly to assess the numerical errors generated while solving the differential equations (Fletcher 1991). However, there is a tool which can provide more information on the numerical properties of the applied method of solution and which can help us to better understand the obtained results. It is called the modified equation approach. For the first time the term "modified equation approach" was used in Section 5.4.2 in context of the consistency of numerical method. The method was proposed by Warming and Hyett (1974). Afterwards it was applied by many authors, e.g. Abbott and Basco (1989), Fletcher (1991). Such analysis provides large amount of information and additionally can suggests the way of increasing the solution accuracy. This method is very useful since it gives possibility to separate the numerical and physical effects in the obtained solution. It is particularly well suited for examining numerical dissipation and dispersion. Consequently this approach allows us to interpret correctly the results of computation. Moreover, in the case of hyperbolic equations the modified equation method allows us to draw practically full information on the applied numerical method, including sometimes the stability conditions as well.

The numerical dissipation and dispersion can be clearly showed while considering the numerical solution of the pure advection equation using the box difference scheme presented in Section 6.1. To this order let us come back to the Section 5.2 and let us take up again the accuracy of approximation of the derivatives.

The transformation of the partial differential equation into the system of algebraic equations is based on the Taylor series expansion (Korn and Korn 1968):

$$f(x+\Delta x) = f(x) + \sum_{m=1}^{\infty} \frac{(\Delta x)^m}{m!} \frac{\partial^m f(x)}{\partial x^m} \tag{6.38}$$

While approximating the derivative, the appropriate formulas are obtained owing to truncation of the Taylor series. This process introduces truncation error. For instance, the forward difference (Eq. 5.99):

$$\left.\frac{\partial f}{\partial x}\right|_x \approx \frac{f(x+\Delta x) - f(x)}{\Delta x} \tag{6.39}$$

is obtained from Eq. (6.38) in which the terms up to the one containing the first derivative only were taken into account:

$$f(x+\Delta x) = f(x) + \Delta x \frac{\partial f}{\partial x} + R. \tag{6.40}$$

In this equation R denotes the truncation terror:

$$R = \frac{\Delta x^2}{2!}\frac{\partial^2 f}{\partial x^2} + \frac{\Delta x^3}{3!}\frac{\partial^3 f}{\partial x^3} + \frac{\Delta x^4}{4!}\frac{\partial^4 f}{\partial x^4} + \cdots \tag{6.41}$$

The exact value of the derivative resulting from (6.40) is:

$$\left.\frac{\partial f}{\partial x}\right|_x = \frac{f(x+\Delta x) - f(x)}{\Delta x} + R(\Delta x) \tag{6.42}$$

As we know from Section 5.4.2, the term $R\,(\Delta x)$ informs us how the error caused by truncation varies with respect to the value of Δx. In this case we have:

$$R(\Delta x) = \frac{\Delta x}{2!}\frac{\partial^2 f}{\partial x^2} + \frac{\Delta x^2}{3!}\frac{\partial^3 f}{\partial x^3} + \frac{\Delta x^3}{4!}\frac{\partial^4 f}{\partial x^4} + \cdots \tag{6.43}$$

which means that the error caused by approximation of the 1st derivative using formula (6.39) vary linearly with the variation of Δx.

Usually the truncation error is related to the order of approximating formula. The latter is determined by the highest order of derivative accounted in the approximating formula. For instance, formula (6.39) approximates the 1st order derivative of $f(x)$ with accuracy of 1st order, since in the Taylor series (6.40) only the terms up to first order were taken. Unfortunately, as it was stated previously, the order of the method is a general information only and it says nothing about the nature of the error and its consequences for the solution. However, it appears that the truncation error determines the numerical properties of numerical scheme. To explain this fact, let us remember that for smooth functions the error provided by truncation of the Taylor series is determined by the value of the first term of the truncated part of series.

As an example consider the following function:

$$f(x) = e^x \tag{6.44}$$

The 1st derivative of $f(x)$ at point $x = 1$ is equal to 2.7183. If we calculate its approximated value using the formula (6.39) assuming $\Delta x = 0.1$, then we will obtain:

$$\left.\frac{\partial f}{\partial x}\right|_{x=1} \approx \frac{f(1.1) - f(1)}{0.1} = 2.8588$$

Therefore the approximation error is: $\delta = 2.8588 - 2.7183 = 0.1405$. This difference represents the value of the total truncated part of the Taylor series, i.e. it is simply $R(\Delta x)$. Now we can calculate the value of the first term of the truncated part for previously assumed $x = 1$ and $\Delta x = 0.1$. We obtain:

$$\left.\frac{\Delta x}{2!}\frac{\partial^2 f}{\partial x^2}\right|_{x=1} = 0.1359$$

One can see that indeed for smooth functions the value of the truncated part of the Taylor series is dominated by the value of its first term. In the considered case the first term is equal to 0.1359, while a sum of other terms is equal to 0.0046 only.

Fletcher (1991) shows convincingly that the kind of error dominating in the numerical solution depends on the order of derivative in the first term of the truncated part of the Taylor series expansion. He considered the propagation of a plane wave, which simultaneously is subjected to dissipation and dispersion. This wave is described by Eq. (5.50):

$$f_m(x,t) = A_m \cdot e^{-p(m)t} \cdot e^{i \cdot m \cdot (x - q(m) \cdot t)} \tag{6.45}$$

where:

A_m – amplitude of the mth component,
m – wave-number related to the components' wave length λ by formula $m = 2\pi/\lambda$,
$p(m)$ – dissipation parameter, which determines how rapidly the amplitude of the wave is attenuated,
$q(m)$ – wave propagation speed,
i – imaginary unit.

Assume that this wave is governed by the linear advection-diffusion equation:

$$\frac{\partial f}{\partial t} + U\frac{\partial f}{\partial x} - D\frac{\partial^2 f}{\partial x^2} = 0 \tag{6.46}$$

We showed previously, in Section 5.1.6, that for this equation Eq. (6.45) has the following parameters:

$$p(m) = D \cdot m^2 \text{ and} \tag{6.47a}$$

$$q(m) = U \tag{6.47b}$$

then it becomes Eq. (5.77):

$$f_m(x,t) = A_m \cdot e^{-D\cdot m^2\cdot t} \cdot e^{-i\cdot m(x-U\cdot t)} \tag{6.48}$$

We also found out earlier that Eq. (6.48) describes the wave, which propagates with constant celerity with simultaneously attenuated amplitude.

Now, assume that the wave (6.45) is governed by the linear Kortweg-de Vries equation (Billingham and King 2000, Whitham 1979):

$$\frac{\partial f}{\partial t} + U\frac{\partial f}{\partial x} + E\frac{\partial^3 f}{\partial x^3} = 0 \tag{6.49}$$

where E is the coefficient of dispersion. For his equation we obtain the following:

$$p(m) = 0 \text{ and} \tag{6.50a}$$

$$q(m) = U - E \cdot m^2 \tag{6.50b}$$

and consequently we have:

$$f_m(x,t) = A_m \cdot e^{-i\cdot m(x-(U-E\cdot m^2)t)} \tag{6.51}$$

In this case the attenuation does not exist. However the wave propagates with celerity depending on the coefficient of dispersion and on the wavelength. If there are waves of different length, they move at different speeds, which simply means that dispersion occurs. The advective velocity U is affected more intensive for short wave (large value of m) than for long ones depending on the sign of E.

Fletcher (1991) proposes to assume the following rule: the even order derivatives are associated with dissipation process, whereas the odd order derivatives are associated with dispersion process. This assumption has important meaning for the modified equation approach, since it can be related with the truncation error.

Now we can go back to the solution of advection equation using the difference box scheme given by Eq. (6.4). Following the consistency analysis described in Section 5.4.2, one can see that at each grid point (j, n) the box scheme modifies the advection equation to the following form:

$$\frac{\partial f}{\partial t} + U\frac{\partial f}{\partial x} = R(f) \tag{6.52}$$

where:

$$\begin{aligned} R(f) = {} & \frac{\Delta t^2}{2}\frac{\partial^2 f}{\partial t^2} + \frac{U\Delta x}{2}\frac{\partial^2 f}{\partial x^2} + (\psi\Delta x + (1-\theta)U\Delta t)\frac{\partial^2 f}{\partial x \partial t} + \\ & -\frac{\Delta t^2}{6}\frac{\partial^3 f}{\partial t^3} - \left(\psi\frac{\Delta x^2}{2} + (1-\theta)\frac{U\Delta x\Delta t}{2}\right)\frac{\partial^3 f}{\partial x^2 \partial t} + \\ & + \left(-\psi\frac{\Delta x\Delta t}{2} + (1-\theta)\frac{U\Delta t^2}{2}\right)\frac{\partial^3 f}{\partial x\partial t^2} - \frac{U\Delta x^2}{6}\frac{\partial^3 f}{\partial x^3} + \cdots \end{aligned} \tag{6.53}$$

To facilitate further analysis, all the time derivatives, except that of the 1st order, are eliminated from Eq. (6.53). To this order the advection equation is differentiated:

$$\frac{\partial^2 f}{\partial x \partial t} = -U\frac{\partial^2 f}{\partial x^2}, \tag{6.54a}$$

$$\frac{\partial^2 f}{\partial t^2} = U^2\frac{\partial^2 f}{\partial x^2}, \tag{6.54b}$$

$$\frac{\partial^3 f}{\partial x^2 \partial t} = -U\frac{\partial^3 f}{\partial x^3}, \tag{6.54c}$$

$$\frac{\partial^3 f}{\partial x \partial t^2} = U^2\frac{\partial^3 f}{\partial x^3}, \tag{6.54d}$$

$$\frac{\partial^3 f}{\partial t^3} = -U^3\frac{\partial^3 f}{\partial x^3}. \tag{6.54e}$$

Substitution of Eq. (6.54) in Eqs. (6.53) and (6.52) yields:

$$\frac{\partial f}{\partial t} + U\frac{\partial f}{\partial x} = D_n\frac{\partial^2 f}{\partial x^2} + E_n\frac{\partial^3 f}{\partial x^3} + \cdots \tag{6.55}$$

where:

$$D_n = \frac{U \cdot \Delta x}{2}((2\theta - 1)C_a + (1 - 2\psi)) \tag{6.56}$$

$$E_n = \frac{U \cdot \Delta x^2}{6}\left((2 - 3\theta)C_a^2 + 3(\psi + \theta - 1)C_a + (1 - 3\psi)\right) \tag{6.57}$$

Equation (6.55) is called modified equation (Fletcher 1991), whereas the coefficients (6.56) and (6.57) are called coefficients of numerical diffusion and numerical dispersion, respectively. For the considered PDE the form of the modified equation depends on the numerical method applied to solve it. Generally one can say that $R(f)$ contains all terms of the Taylor series neglected while approximation the derivatives in the solved equation.

The modified equation allows us to conclude on the properties of the applied scheme. For instance, in Section 5.4.2 such equation was used to prove consistency of the upwind scheme. This equation allows us to deduce other properties as well.

As far as the order of approximation by the box scheme is considered, from Eqs. (6.56) and (6.57) result that it depends on the values of weighting parameters. For $\psi = 0.5$ and $\theta = 0.5$ it represents accuracy of 2nd order with regard to both variables x and t. In this case the coefficient of numerical diffusion is cancelled as the space and time derivatives are approximated with the centered difference (see Fig. 6.1). For other values of θ and ψ the scheme is an approximation of 1st order.

The box scheme will be stable, if the coefficient of numerical diffusion is non-negative ($D_n \geqslant 0$), since only in such a case the problem of solution of Eq. (6.55) will be well-posed. For $D_n < 0$, this problem is ill–posed and the solution of Eq. (6.55) does not exist. The condition $D_n \geq 0$ will be always satisfied for $\theta \geq 0.5$ and $\psi \leq 0.5$, irrespectively of the value of the Courant number. This means that the box scheme is absolutely stable. For other values of the weighting parameters the scheme is conditionally stable or even unstable.

The modified equation (6.55) allows us to examine the dissipative and dispersive properties of the box scheme as well. In the modified equation, similarly to Eqs. (6.46) and (6.49), the terms with odd derivatives are associated with the numerical dispersion and the terms with even derivatives are associated with the numerical dissipation. Since for smooth functions the truncation error is determined by the first term of truncated part of the Taylor series, then the first term at the right hand side of Eq. (6.55) will determine the error dominating in the solution. If this term contains an odd derivative, the scheme will be dispersive, whereas if it contains an even derivative – the scheme will be dissipative. Numerical diffusion is associated with the derivative of the lowest even order. Consequently, according to the dominating term in the truncation error, we will obtain either smooth or oscillating solution. These effects have numerical nature and they are unrelated to the physical processes represented in the solved equation. They are particularly distinct when strong gradients of function f occurs in the solution.

If the applied method is dissipation free, oscillations will occur in the solution. They cannot be explained by the numerical instabilities as they can appear for numerically stable schemes as well. To eliminate them some numerical diffusion should be introduced. If it is sufficiently large, the oscillations are suppressed. At the same time the gradients of function f are reduced – the steep front is smoothed. However in the case of discontinuity of f, the truncation error is not dominated by the first term of the truncated part of the Taylor series, but rather by the next term. Consequently, the oscillations can occur even if the applied scheme is slightly dissipative. The problem of generation of the numerical diffusion is independent of the spatial dimensionality. For 2D and 3D problems tensors of numerical diffusion and dispersion appear at the place of scalar coefficients.

Using the modified equation (6.55) provided by the box scheme, one can carry out a full analysis and interpretation of the computational results displayed in Figs. 5.15, 5.16, 6.2, 6.3 and 6.4.

Let us begin with the solution obtained for $\theta = 0$ and $\psi = 0$, i.e. using the well known upwind scheme. For these values of the weighting parameters one obtains:

$$D_n = \frac{U \cdot \Delta x}{2}(1 - C_a) \tag{6.58a}$$

$$E_n = \frac{U \cdot \Delta x^2}{6}(2C_a^2 - 3C_a + 1) \tag{6.58b}$$

The upwind scheme ensures accuracy of order $O(\Delta t, \Delta x)$. Since in this case the term containing the derivative of 2nd order dominates in the truncation error, then in the numerical solution a dissipation error must occur. Consequently the modified equation (6.55) can be rewritten as:

$$\frac{\partial f}{\partial t} + U\frac{\partial f}{\partial x} = D_n\frac{\partial^2 f}{\partial x^2} \tag{6.59}$$

It means that solving the pure advection equation with upwind scheme, we are actually solving the advection-diffusion equation. The exact solution is provided only for $C_a = 1$. In this case, as results from Eq. (6.58a), one obtains $D_n = 0$. This means that the modified equation (6.59) becomes the original Eq. (6.1). Assuming $C_a < 1$, we have $D_n > 0$, which triggers numerical diffusion, giving rise to attenuation of the initially imposed distribution of f along the channel axis. This attenuation increases as the Courant number decreases, which can be observed in Fig. 5.15. In such a situation the numerical dispersion of the scheme is not significant, since the right hand side of Eq. (6.55) is dominated by its first term. It is worth to add that for $C_a > 1$ Eq. (6.59) cannot be solved, as for $D_n < 0$ the problem of its solution is ill-posed. Confirmation of this conclusion is given in Fig. 5.16, where for $C_a =$ 1.035 numerical instability occurred.

Now let us consider the case illustrated in Fig. 6.2. These results were provided by $\theta = 0{,}5$ and $\psi = 0{,}5$. For these values, the coefficients of numerical diffusion and dispersion in the modified equation (6.55) are given by:

$$D_n = 0 \tag{6.60a}$$

$$E_n = \frac{U \cdot \Delta x^2}{12}(C_a^2 - 1). \tag{6.60b}$$

Then this equation takes the following form:

$$\frac{\partial f}{\partial t} + U\frac{\partial f}{\partial x} = E_n \frac{\partial^3 f}{\partial x^3} \tag{6.61}$$

In such a way the Kortweg-de Vries equation (6.49) is obtained. Equation (6.61) will provide the exact solution only for $C_a = 1$, for which $E_n = 0$. For other values of the Courant number the error of numerical dispersion is generated and the scheme provides oscillating solution. The position of oscillations (before or behind the peak) varies and it depends on the sign of E_n, i.e. on the assumed value of the Courant number (Fig. 6.2). They will be present always.

Influence of the weighting parameter ψ is shown in Fig. 6.3. In this case $\theta = 0.5$ and $C_a = 0.5$ are assumed. For these values the modified equation contains:

$$D_n = \frac{U \cdot \Delta x}{2}(1 - 2\psi), \tag{6.62a}$$

$$E_n = \frac{U \cdot \Delta x^2}{12}\left(\frac{3}{4} - 3\psi\right) \tag{6.62b}$$

Setting $\psi = 0.5$ the numerical diffusion is eliminated. In such a case the right hand side of the modified equation is dominated by the dispersive term, since $E_n \neq 0$. Consequently, an oscillating solution is produced (Fig. 6.3). For $\psi < 0.5$ the box scheme provides the numerical diffusion. Therefore solution obtained with $\psi = 0.25$ and $\psi = 0.0$ is damped and similar to the one provided by the advection-diffusion equation. On the other hand, for $\psi > 0.5$ an unstable solution is generated, since $D_n < 0$.

The last example of calculation deals with $\psi = 0.5$ and $C_a = 0.5$. For these values Eqs. (6.56) and (6.57) give:

$$D_n = \frac{U \cdot \Delta x}{4}(2\theta - 1), \tag{6.63a}$$

$$E_n = \frac{U \cdot \Delta x^2}{8}(\theta - 1) \tag{6.63b}$$

In this version of the box scheme the numerical diffusion is cancelled by setting $\theta = 0.5$. However we obtain $E_n < 0$ which means that the scheme is dispersive. Consequently one can observe oscillations behind the peak (Fig. 6.4). Assuming $\theta > 0.5$ we introduce numerical diffusion, which is the most intense for $\theta = 1$. The obtained solution differs appreciably from the exact one. Assuming $\theta < 0.5$ we have $D_n < 0$. Then the solution should be unstable. Indeed, in Fig. 6.4 one can see the effects of numerical instability for $\theta = 0.475$.

Note that all the presented conclusions were drawn from the modified equation corresponding to the solved pure advection equation. Thus, the modified equation approach seems to be a robust tool for consistency, stability and accuracy analysis for the hyperbolic partial differential equations. In the next sections we will use it many times.

6.4 Solution of the Advection Equation with the Finite Element Method

6.4.1 Standard Finite Element Approach

Assume that the pure advection equation (6.1) with constant flow velocity is solved for $0 \le x \le L$ and $t \ge 0$ with appropriate auxiliary conditions. To solve this equation the Galerkin finite element method, presented in Section 5.3, is applied. The considered channel reach is divided into $M - 1$ elements of length Δx_j ($j = 1, 2, \ldots, M - 1$). The discrete solution domain is shown in Fig. 5.19. According to the Galerkin procedure the numerical solution must satisfy the condition (5.118), which for Eq. (6.1) is written as follows:

$$\int_{o}^{L} \Phi(f_a) \cdot \mathbf{N}(x)\; dx = \sum_{j=1}^{M-1} \int_{x_j}^{x_{j+1}} \left(\frac{\partial f_a}{\partial t} + U \frac{\partial f_a}{\partial x} \right) \mathbf{N}(x)\; dx = 0, \tag{6.64}$$

where:

N – vector of shape functions given by Eq. (5.125),
f_a – aproximation of function f defined by Eq. (5.116).

Let us calculate a single component of the above sum, corresponding to the element j:

$$I = \int_{x_j}^{x_{j+1}} \left(\frac{\partial f_a}{\partial t} + U \frac{\partial f_a}{\partial x} \right) \mathbf{N}(x)\; dx \tag{6.65}$$

A similar procedure was applied previously in Section 5.3.2 while solving the diffusion equation. Since in this element only two components of the vector $\mathbf{N}(x)$, i.e. $N_j(x)$ and $N_{j+1}(x)$, are non-zero then in Eq. (6.65) only two non-zero products, corresponding to these functions will occur. Therefore in element j we have to calculate the following two integrals:

$$I^{(j)} = \int_{x_j}^{x_{j+1}} \left(\frac{\partial f_a}{\partial t} + U \frac{\partial f_a}{\partial x} \right) N_j(x) \cdot dx, \tag{6.66a}$$

$$I^{(j+1)} = \int_{x_j}^{x_{j+1}} \left(\frac{\partial f_a}{\partial t} + U \frac{\partial f_a}{\partial x} \right) N_{j+1}(x) \cdot dx, \tag{6.66b}$$

Integration of the time derivative in Eq. (6.66a) was carried out in Section 5.3.2. It yields:

$$\int_{x_j}^{x_{j+1}} \frac{\partial f_a}{\partial t} N_j(x) \cdot dx = \frac{\Delta x_j}{3} \frac{df_j}{dt} + \frac{\Delta x_j}{6} \frac{df_{j+1}}{dt}, \tag{6.67}$$

whereas integration of the advective term gives:

$$\int_{x_j}^{x_{j+1}} U \frac{\partial f_a}{\partial x} N_j(x) \cdot dx = \frac{U}{\Delta x}(- f_j + f_{j+1}) \int_{x_j}^{x_{j+1}} N_j(x) \cdot dx = \frac{U}{2}(- f_j + f_{j+1}). \tag{6.68}$$

In a similar way is calculated integral (6.66b). Substituting the results of integration in Eqs. (6.66a) and (6.66b) one obtains the following:

$$I^{(j)} = \frac{\Delta x_j}{3} \frac{df_j}{dt} + \frac{\Delta x_j}{6} \frac{df_{j+1}}{dt} + \frac{U}{2}(- f_j + f_{j+1}) \tag{6.69a}$$

$$I^{(j+1)} = \frac{\Delta x_j}{6} f_j + \frac{\Delta x_j}{3} f_{j+1} + \frac{U}{2}(- f_j + f_{j+1}) \tag{6.69b}$$

The same equations are obtained for all elements ($j = 1, 2, \ldots, M - 1$). According to Eq. (6.64) they are assembled in a global system of equations:

– $j = 1$:

$$\frac{\Delta x_1}{3} \frac{df_1}{dt} + \frac{\Delta x_1}{6} \frac{df_2}{dt} + \frac{U}{2}(- f_1 + f_2) = 0, \tag{6.70a}$$

– $j = 2,\ldots, M-1$:

$$\frac{\Delta x_{j-1}}{6}\frac{df_{j-1}}{dt} + \left(\frac{\Delta x_{j-1}}{3} + \frac{\Delta x_j}{3}\right)\frac{df_j}{dt} + \frac{\Delta x_j}{6}\frac{df_{j+1}}{dt} + \\ + \frac{U}{2}(-f_{j-1} + f_j) + \frac{U}{2}(-f_j + f_{j+1}) = 0. \tag{6.70b}$$

– $j = M$:

$$\frac{\Delta x_{M-1}}{6}\frac{df_{M-1}}{dt} + \frac{\Delta x_{M-1}}{3}\frac{df_M}{dt} + \frac{U}{2}(-f_{M-1} + f_M) = 0. \tag{6.70c}$$

In matrix notation this system of ordinary differential equations takes the following form:

$$\mathbf{A}\frac{d\mathbf{f}}{dt} + \mathbf{C}\cdot\mathbf{f} = \mathbf{0}, \tag{6.71}$$

where:

$\frac{d\mathbf{f}}{dt} = \left(\frac{df_1}{dt}, \frac{df_2}{dt}, \ldots, \frac{df_M}{dt}\right)^T,$
$\mathbf{f} = (f_1, f_2, \ldots, f_M)^T,$
$\mathbf{A}, \mathbf{C}$ – tridiagonal matrices.

The integration of such a system of ODEs is described in Chapter 3. The application of generalized two-level scheme (3.71) yields the following system of equations:

$$(\mathbf{A} + \Delta t\cdot\theta\cdot\mathbf{C})\,\mathbf{f}_{n+1} = (\mathbf{A} - \Delta t(1-\theta)\mathbf{C})\,\mathbf{f}_n, \tag{6.72}$$

where:

n – index of time level,
θ – weighting parameter ranging from 0 to 1,
Δt – time step.

Let us consider an equation of system (6.72) corresponding to node j. For equally spaced nodes when $\Delta x = \text{const.}$, it takes the following form:

$$\left(\tfrac{1}{6} - \theta\tfrac{C_a}{2}\right) f_{j-1}^{n+1} + \frac{2}{3}f_j^{n+1} + \left(\tfrac{1}{6} + \theta\tfrac{C_a}{2}\right) f_{j+1}^{n+1} = \\ \left(\tfrac{1}{6} + (1-\theta)\tfrac{C_a}{2}\right) f_{j-1}^{n} + \frac{2}{3}f_j^{n} + \left(\tfrac{1}{6} - (1-\theta)\tfrac{C_a}{2}\right) f_{j+1}^{n} = 0. \tag{6.73}$$

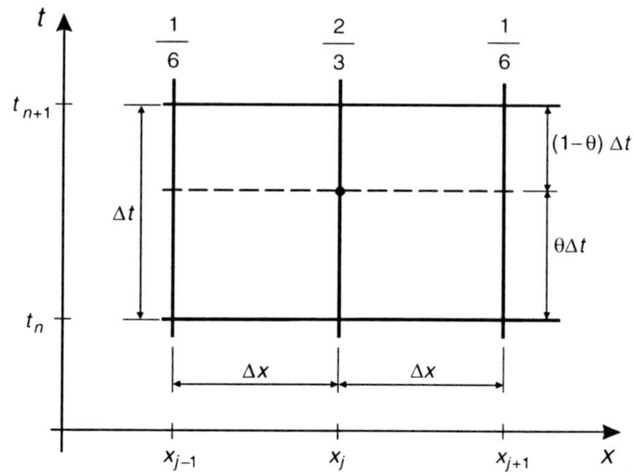

Fig. 6.9 Grid points for the method (6.73)

where C_a is the advective Courant number defined by Eq. (6.7). One can notice that Eq. (6.73) involves the grid points presented in Fig. 6.9. Appropriate equations for the extreme nodes $j = 1$ and $j = M$ are given by Eqs. (6.70a) and (6.70c).

The modified equation corresponding to Eq. (6.73) is as follows:

$$\frac{\partial f}{\partial t} + U\frac{\partial f}{\partial x} = D_n\frac{\partial^2 f}{\partial x^2} + E_n\frac{\partial^3 f}{\partial x^3} + \ldots \tag{6.74}$$

The coefficients of numerical diffusion D_n and dispersion E_n are given by:

$$D_n = \left(\theta - \frac{1}{2}\right) C_a \cdot \Delta x \cdot U, \tag{6.75a}$$

$$E_n = \frac{U \cdot \Delta x^2}{2}\left(\frac{2}{3} - \theta\right) C_a^2. \tag{6.75b}$$

The values of these coefficients depend on the chosen value of θ:

– for the implicit trapezoidal rule ($\theta = 1/2$):

$$D_n = 0, \tag{6.76a}$$

$$E_n = \frac{1}{12} U \cdot \Delta x^2 \cdot C_a^2, \tag{6.76b}$$

– for the Galerkin method ($\theta = 2/3$):

$$D_n = \frac{1}{6} U \cdot \Delta x \cdot C_a, \tag{6.77a}$$

$$E_n = 0 \tag{6.77b}$$

– for the implicit Euler method ($\theta = 1$):

$$D_n = \frac{1}{2} U \cdot \Delta x \cdot C_a, \tag{6.78a}$$

$$E_n = -\frac{1}{6} U \cdot \Delta x^2 \cdot C_a^2. \tag{6.78b}$$

From the above relations results that the finite element method applied to solve advection equation is dissipation free when the time integration is carried out using the implicit trapezoidal rule, whereas it generates numerical diffusion when the Galerkin or implicit Euler methods are used. Consequently one can expect an oscillatory solution in the first case and an artificially damped solution for the other two methods. The method is absolutely stable for $\theta \geq 1/2$, since this condition ensures that $D_n \geq 0$.

To illustrate the influence of the weighting parameter θ on the solution accuracy a propagation of the steep front of concentration governed by the pure advection transport equation is considered.

Example 6.2 In a straight open channel of length L the water flows with constant velocity U. Assume:

– initial condition: $f(x, t) = 0$ for $0 \leq x \leq L$
– boundary condition:

$$f(x = 0, t) = \begin{cases} 0 & \text{for} \quad t \leq 0 \\ 1 & \text{for} \quad t > 0 \end{cases}$$

The steep front of the function f entering the upstream end travels along channel axis with constant velocity. Assume $U = 0.5$ m/s = const. and $\Delta x = 100$ m = const.

In Fig. 6.10 this traveling front is shown after $t = 7{,}200$ s. The exact solution has the form of sharp front imposed as the boundary condition at $x = 0$ shifted along x

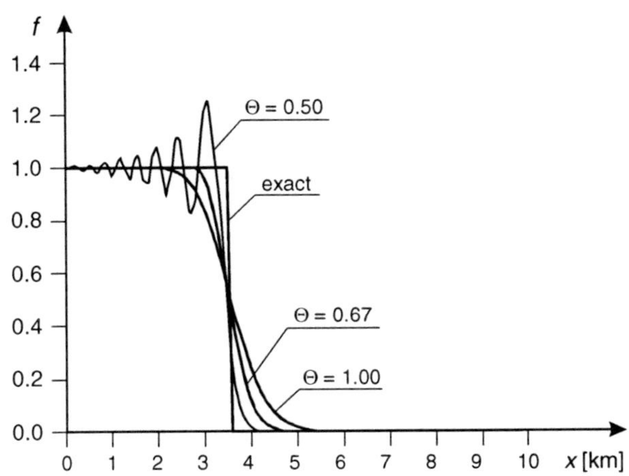

Fig. 6.10 Solution of the advection equation by the finite element method with various values of the weighting parameter θ

axis to the position $x = 0.5 \times 7{,}200 = 3{,}600$ m. This solution cannot be reproduced by the finite element method for any value of θ and for any value of the advective Courant number. As results from Eqs. (6.76a), (6.76b), (6.77a), (6.77b), (6.78a) and (6.78b), regardless of the values of these parameters the finite element method is always either dissipative or dispersive one. Therefore, either unphysical damping or oscillations will be always present in the numerical solution.

To improve the solution acuracy of the advection equation provided by the finite element method some modifications to its standard version were proposed. Below we present two ways of increasing of the accuracy approximation of the advection equation up to 3rd order with regard to both x and t variables.

6.4.2 Donea Approach

In the approach proposed by Donea (1984) the approximation accuracy of the time derivative in the advection equation (6.1) is increased. At first this equation is approximated with regard to time:

$$\left.\frac{\partial f}{\partial t}\right|^{n} + U\left.\frac{\partial f}{\partial x}\right|^{n} = 0 \tag{6.79}$$

For this purpose usually the time derivative is replaced by one of standard difference formula. However, Donea (1984) proposed a more accurate approximation resulting from the Taylor series expansion of the function $f(x, t)$ around the time level n. The Taylor formula:

$$f^{n+1} = f^{n} + \Delta t\left.\frac{\partial f}{\partial t}\right|^{n} + \left.\frac{\Delta t^{2}}{2}\frac{\partial^{2} f}{\partial t^{2}}\right|^{n} + \left.\frac{\Delta t^{3}}{6}\frac{\partial^{3} f}{\partial t^{3}}\right|^{n} + \cdots \tag{6.80}$$

retaining the terms up to 3th order allows us to express the first order derivative as follows:

$$\left.\frac{\partial f}{\partial t}\right|^{n} = \frac{f^{n+1} - f^{n}}{\Delta t} - \left.\frac{\Delta t}{2}\frac{\partial^{2} f}{\partial t^{2}}\right|^{n} - \left.\frac{\Delta t^{2}}{6}\frac{\partial^{2} f}{\partial t^{3}}\right|^{n} + O(\Delta t^{3}) \tag{6.81}$$

Equation (6.81) is approximation of 3rd order with regard to time. The time derivatives of the order higher than one are eliminated using the following expressions obtained from differentiation of the considered advection equation (6.1):

$$\frac{\partial^{2} f}{\partial t^{2}} = U^{2}\frac{\partial^{2} f}{\partial x^{2}}, \tag{6.82a}$$

$$\frac{\partial^{3} f}{\partial t^{3}} = U^{2}\frac{\partial^{2}}{\partial x^{2}}\left(\frac{\partial f}{\partial t}\right). \tag{6.82b}$$

Substitution of Eq. (6.81) with Eqs. (6.82a) and (6.82b) in Eq. (6.79) yields:

$$\frac{f^{n+1} - f^n}{\Delta t} - \frac{\Delta t}{2} U^2 \frac{\partial^2 f^n}{\partial x^2} - \frac{\Delta t^2}{6} U^2 \frac{\partial^2}{\partial x^2} \left(\frac{\partial f^n}{\partial t} \right) + U \frac{\partial f^n}{\partial x} = 0. \tag{6.83}$$

Approximation of the time derivative gives:

$$\frac{f^{n+1} - f^n}{\Delta t} - \frac{\Delta t}{2} U^2 \frac{\partial^2 f^n}{\partial x^2} - \frac{\Delta t^2}{6} U^2 \frac{\partial^2}{\partial x^2} \left(\frac{f^{n+1} - f^n}{\Delta t} \right) + U \frac{\partial f^n}{\partial x} = 0. \tag{6.84}$$

After regrouping one obtains:

$$f^{n+1} - \frac{\Delta t^2}{6} U^2 \frac{\partial^2 f^{n+1}}{\partial x^2} = f^n + \frac{\Delta t^2}{2} U^2 \frac{\partial^2 f^n}{\partial x^2} - \frac{\Delta t^2}{6} U^2 \frac{\partial^2 f^n}{\partial x^2} + \Delta t \cdot U \frac{\partial f^n}{\partial x} \tag{6.85}$$

If we introduce the advective Courant number C_a , Eq. (6.85) will take the final form:

$$f^{n+1} - \frac{C_a^2}{6} \Delta x^2 \frac{\partial^2 f^{n+1}}{\partial x^2} = f^n + \frac{C_a^2}{3} \Delta x^2 \frac{\partial^2 f^n}{\partial x^2} + C_a \cdot \Delta x \frac{\partial f^n}{\partial x} \tag{6.86}$$

Note that f^{n+1} and f^n must be still discretized with regard to x. To this end the Galerkin finite element method with linear trial functions is used. For uniform grid points the following algebraic equation is obtained:

$$\left(\frac{1}{6} - \frac{C_a^2}{6}\right) f_{j-1}^{n+1} + \left(\frac{2}{3} + \frac{C_a^2}{3}\right) f_j^{n+1} + \left(\frac{1}{6} - \frac{C_a^2}{6}\right) f_{j+1}^{n+1} =$$

$$= \left(\frac{1}{6} + \frac{C_a^2}{6} + \frac{C_a}{2}\right) f_{j-1}^n + \left(\frac{2}{3} - \frac{C_a^2}{3}\right) f_j^n + \left(\frac{1}{6} + \frac{C_a^2}{6} - \frac{C_a}{2}\right) f_{j+1}^n \tag{6.87}$$

Such equation is written for each internal node $j = 2,3,\ldots, M - 1$. For $j = 1$ the appropriate equation is given by the imposed boundary condition, whereas for $j = M$ Eq. (6.87) is modified according to the rules of the finite element method. Integration of the diffusive term in Eq. (6.86) for linear shape functions was explained in Section 5.3. Consequently one obtains a system of linear algebraic equations with tri-diagonal matrix. Its solution provides the nodal values of the function f at the time level n+1.

The Donea method used for solving the steep front propagation for the data from Example 6.2 provided much better results compared to the standard finite element method. In Fig. 6.11 one can see that the method produces results which coincide with the exact solution ($C_a = 1$) or which are only sightly different from the exact solution for other values of C_a.

The accuracy analsis using the modified equation approach shows that the method modifies the advection equation to the form of Eq. (6.74) in which $D_n = 0$ and $E_n = 0$ regardless of the value of Courant number. Therefore the method

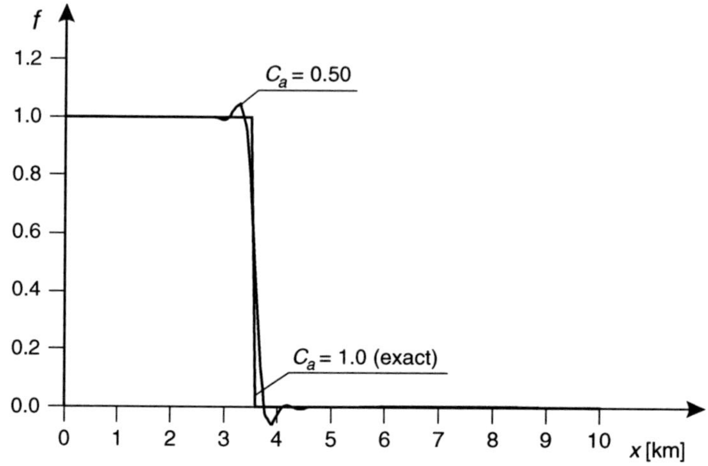

Fig. 6.11 Solution of the advection equation by the finite element method with Donea formulation

approximates the advection equation with accuracy of 3rd order with regard to x and t. For $C_a = 1$ all the terms of higher order in the modified equation dissapear as well, so the method provides the exact solution. The method is only conditionally stable, as for stability the condition $C_a \leq 1$ is required (Donea 1984).

6.4.3 *Modified Finite Element Approach*

6.4.3.1 The Concept of the Modified Finite Element Method

Another approach to improve the standard finite element method was proposed by Szymkiewicz (1995). It deals with modification of the process of integration and it leads to a more general form of algebraic equations approximating the governing equation. The accuracy analysis carried out using the modified equation approach provides information which suggests the way of increasing the solution accuracy up to 3rd order. The resulting formula comprises the standard finite element and finite difference schemes as particular cases.

According to the Galerkin procedure, described in details in Section 6.4.1, the solution of the advection equation must satisfy condition (5.118), which in this case takes the form of Eq. (6.64). In the standard approach the function $f(x, t)$ is approximated as follows:

$$f_a(x,t) = \sum_{j=1}^{M} f_j(t) N_j(x) \tag{6.88}$$

where:

$N_j(x)$ – vector of shape functions given by Eq. (5.125),
f_a – aproximation of function f defined by Eq. (5.116).
j – index of node,
M – number of grid points.

where $f_j(t)$ represents nodal value of function $f(x, t)$. When linear basis functions are applied in each element only two following integrals in Eq. (6.64) exist:

$$I_1 = \int_{x_j}^{x_{j+1}} f_a(x,t)N_j(x)dx = \left(\frac{2}{3}f_j(t) + \frac{1}{3}f_{j+1}(t)\right)\frac{\Delta x_j}{2} \tag{6.89a}$$

$$I_2 = \int_{x_j}^{x_{j+1}} f_a(x,t)N_{j+1}(x)dx = \left(\frac{1}{3}f_j(t) + \frac{2}{3}f_{j+1}(t)\right)\frac{\Delta x_j}{2} \tag{6.89b}$$

where Δx_j is the distance between nodes. This standard approach can be modified. Namely, the integral of the product of approximation of function and basis function in an element can be expressed as a product of certain average value of the function in the element and the integral of the basis function in this element, i.e.:

$$I_1 = \int_{x_j}^{x_{j+1}} f_a(x,t)N_j(x)dx = f_c(t)\int_{x_j}^{x_{j+1}} N_j(x)dx = f_c(t)\frac{\Delta x_j}{2} \tag{6.90a}$$

$$I_2 = \int_{x_j}^{x_{j+1}} f_a(x,t)N_{j+1}(x)dx = f_c(t)\int_{x_j}^{x_{j+1}} N_{j+1}(x)dx = f_c(t)\frac{\Delta x_j}{2} \tag{6.90b}$$

The weighted average $f_c(t)$ in the element can be expressed using the following formulas:

– for Eq. (6.90a):

$$f_c(t) = \omega \cdot f_j(t) + (1-\omega)f_{j+1}(t) \tag{6.91a}$$

– for Eq. (6.90b):

$$f_c(t) = (1-\omega)f_j(t) + \omega \cdot f_{j+1}(t) \tag{6.91b}$$

where ω is a weighting parameter ranging from 0 to 1. Equations (6.90a) and (6.90b) can be rewritten as follows:

$$I_1 = (\omega \cdot f_j(t) + (1-\omega)f_{j+1}(t))\frac{\Delta x_j}{2} \tag{6.92a}$$

$$I_2 = ((1-\omega)f_j(t) + \omega \cdot f_{j+1}(t))\frac{\Delta x_j}{2} \tag{6.92b}$$

6.4.3.2 Solution of the Advection Equation Using the Modified Finite Element Method

Calculation of a single integral in expression (6.64) over an element is carried out in the following way:

$$\int_{x_j}^{x_{j+1}} \left(\frac{\partial f_c}{\partial t} + U \frac{\partial f_a}{\partial x} \right) N_j(x) dx = \left(\omega \frac{df_j}{dt} + (1 - \omega) \frac{df_{j+1}}{dt} \right) \frac{\Delta x_j}{2} + U \left(-f_j + f_{j+1} \right) \tag{6.93a}$$

$$\int_{x_j}^{x_{j+1}} \left(\frac{\partial f_c}{\partial t} + U \frac{\partial f_a}{\partial x} \right) N_{j+1}(x) dx = \left((1 - \omega) \frac{df_j}{dt} + \omega \frac{df_{j+1}}{dt} \right) \frac{\Delta x_j}{2} + U \left(-f_j + f_{j+1} \right) \tag{6.93b}$$

In these expressions the subscript a denotes approximation according to the formula (6.88) while the subscript c denotes approximation by Eqs. (6.91a) and (6.91b). It means that the alternative way of approximation of equations is applied only to the time derivative, while the derivative of 1st order with regard to x is approximated using the standard approach.

When all integrals in each element are assembled according to Eq. (6.64), the global system of ordinary differential equations over time of dimension $M \times M$ is obtained. For equally spaced nodes this system has the following form:

– for $j = 1$

$$\omega \frac{df_j}{dt} + (1 - \omega) \frac{df_{j+1}}{dt} + \frac{U}{\Delta x} (- f_j + f_{j+1}) = 0, \tag{6.94a}$$

– for $j = 2, 3, \ldots, M - 1$

$$\frac{1 - \omega}{2} \frac{df_{j-1}}{dt} + \omega \frac{df_j}{dt} + \frac{1 - \omega}{2} \frac{df_{j+1}}{dt} + \frac{U}{2\Delta x} (- f_{j-1} + f_{j+1}) = 0, \tag{6.94b}$$

– for $j = M$

$$(1 - \omega) \frac{df_{j-1}}{dt} + \omega \frac{df_j}{dt} + \frac{U}{\Delta x} (- f_{j-1} + f_j) = 0, \tag{6.94c}$$

Equation (6.94) can be written in a more compact matrix form:

$$\mathbf{A} \frac{d\mathbf{f}}{dt} + \mathbf{C} \cdot \mathbf{f} = 0 \tag{6.95}$$

where:

$\mathbf{A}$ – constant tri-diagonal matrix,
$\mathbf{C}$ – constant tri-diagonal matrix,
$\mathbf{f} = (f_1, f_2, \ldots, f_M)^T$ – vector of unknowns comprising nodal values of f,
$\frac{d\mathbf{f}}{dt} = \left(\frac{df_1}{dt}, \frac{df_2}{dt}, \ldots, \frac{df_M}{dt}\right)^T$ – vector of time derivatives,
T – symbol of transposition.

The system (6.95), similarly to the one obtained for the standard finite element method, is solved using formula (3.71) with the weighting parameter θ. Its application for (6.95) yields:

$$(\mathbf{A} + \Delta t \cdot \theta \cdot \mathbf{C})\,\mathbf{f}^{n+1} = (\mathbf{A} - \Delta t\,(1-\theta)\mathbf{C})\,\mathbf{f}^{n}, \tag{6.96}$$

where:

n – index of time level,
θ – weighting parameter ranging from 0 to 1,
Δt – time step.

The equation for node j takes the following form:

$$\begin{gathered}\left(\frac{1-\omega}{2} - \theta\frac{C_a}{2}\right) f_{j-1}^{n+1} + \omega \cdot f_j^{n+1} + \left(\frac{1-\omega}{2} + \theta\frac{C_a}{2}\right) f_{j+1}^{n+1} = \\ = \left(\frac{1-\omega}{2} + (1-\theta)\frac{C_a}{2}\right) f_{j-1}^{n} + \omega \cdot f_j^{n} + \left(\frac{1-\omega}{2} - (1-\theta)\frac{C_a}{2}\right) f_{j+1}^{n}\end{gathered}\ . \tag{6.97}$$

where C_a is the advective Courant number ($C_a = U{\cdot}\Delta t/\Delta x$). Equation (6.97) involves grid nodes presented in Fig. 6.12.

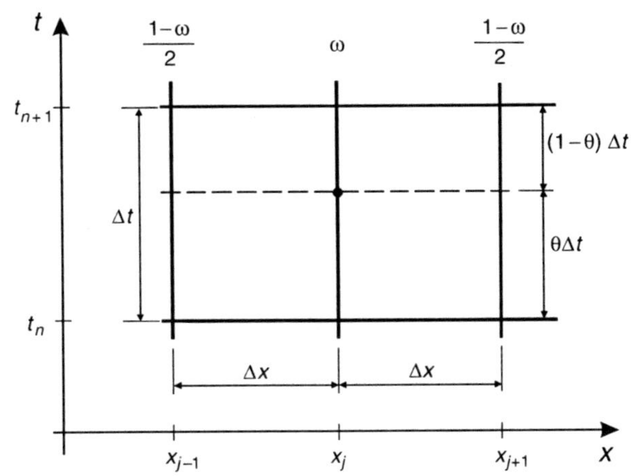

Fig. 6.12 Numerical grid for the modified finite element method

Table 6.1 Particular cases of the method depending on the values ω and θ

ω	θ	Method
1	1/2	FD Crank-Nicolson
1	1	FD implicit Euler
1	0	FD explicit Euler
2/3	1/2	FE Crank-Nicolson
2/3	1	FE implicit Euler

It can be seen that the modified FE approach yields a six-point implicit scheme with two weighting parameters. For some particular values of ω and θ one obtains well known methods of numerical solution of the advection equation. Some of them are listed in Table 6.1.

6.4.3.3 Stability Analysis of the Modified Finite Element Method

Stability analysis for numerical solution of the linear advection equation solved by the modified finite element method is carried out using the Neumann method. In this approach the solution error $E(x, t)$ described by the same formula as the considered scheme, is expanded in the Fourier series (see Section 5.4.3). Since we consider the linear equation, the properties of the method can be deduced on the basis of the behaviour of one component of series.

Substitution of Eq. (5.161) in Eq. (6.97) yields:

$$\begin{aligned}&(1-\omega)A^{n+1}\frac{1}{e^{i\cdot\varphi}}+2\omega\cdot A^{n+1}+(1-\omega)A^{n+1}\cdot e^{i\cdot\varphi}+\theta\cdot C_a\left(A^{n+1}\cdot e^{i\cdot\varphi}-A^{n+1}\frac{1}{e^{i\cdot\varphi}}\right)\\&=(1-\omega)A^{n}\frac{1}{e^{i\cdot\varphi}}+2\omega\cdot A^{n}+(1-\omega)A^{n}\cdot e^{i\cdot\varphi}-(1-\theta)C_a\left(A^{n}\cdot e^{i\cdot\varphi}-A^{n}\frac{1}{e^{i\cdot\varphi}}\right)\end{aligned} \tag{6.98}$$

where

$i = (-1)^{1/2}$ – imaginary unit,
j – index of node,
A^n – Fourier coefficient accounting for time variability,
n – index of time level,
$\phi = m\cdot\Delta x$ – dimensionless wave number given by Eq. (5.160),

Using the following Euler relations (Korn and Korn 1968):

$$\sin(\varphi) = \frac{e^{i\cdot\varphi}-e^{-i\cdot\varphi}}{2i}, \tag{6.99a}$$

$$\cos(\varphi) = \frac{e^{i\cdot\varphi}+e^{-i\cdot\varphi}}{2}, \tag{6.99b}$$

Equation (6.98) is reformed to the formula representing the amplification factor:

$$\frac{A^{n+1}}{A^n} = G = \frac{[\omega + (1-\omega)\cos(\varphi)]^2 + \theta^2 C_a^2 \sin^2(\varphi) - \theta\, C_a^2 \sin^2(\varphi)}{[\omega + (1-\omega)\cos(\varphi)]^2 + \theta^2 C_a^2 \sin^2(\varphi)} + \\ - \frac{[\omega + (1-\omega)\cos(\varphi)]\, C_a \sin(\varphi)}{[\omega + (1-\omega)\cos(\varphi)]^2 + \theta^2 C_a^2 \sin^2(\varphi)}\, i \tag{6.100}$$

where G is the amplification factor in the form of complex number. The condition of numerical stability (Potter 1973) $|G| \leq 1$ implies the following condition:

$$\left(1 + \frac{(1-2\theta)\, C_a^2 \cdot \sin^2(\varphi)}{[\omega + (1-\omega)\cos(\varphi)]^2 + \theta^2 \cdot C_a^2 \cdot \sin^2(\varphi)}\right)^{1/2} \leq 1. \tag{6.101}$$

Since

$$\omega + (1-\omega)\cos(\varphi) = \cos^2\left(\frac{\varphi}{2}\right) + (2\omega - 1)\sin^2\left(\frac{\varphi}{2}\right)$$

the condition (6.101) can be rewritten as:

$$\left(1 + \frac{(1-2\theta)\, C_a^2 \cdot \sin^2(\varphi)}{\left[\cos^2\left(\frac{\varphi}{2}\right) + (2\omega-1)\sin^2\left(\frac{\varphi}{2}\right)\right]^2 + \theta^2 \cdot C_a^2 \cdot \sin^2(\varphi)}\right)^{1/2} \leq 1. \tag{6.102}$$

The dimensionless wave number ϕ varies in the range $0 \leq \varphi \leq \pi$, so $\cos^2(\phi/2)$, $\sin^2(\phi/2)$ and $\sin^2(\phi)$ vary as well. The conditions:

$$\theta \geq 1/2 \text{ and} \tag{6.103a}$$

$$\omega \geq 1/2 \tag{6.103b}$$

ensure that relation (6.102) is satisfied for any Courant number. Therefore the conditions (6.103a) and (6.103b) guarantee unconditional stability.

6.4.3.4 Accuracy Analysis Using the Modified Equation Approach

The numerical properties of the methods listed in Table 6.1 are well known and we also know that they are rather usuitable to solve the advection equation. However, it appears that using the accuracy analysis carried out by the modified equation approach it is possible to reach a higher order of approximation compared to the methods given in Table 6.1. Let us carry out the accuracy analysis for system (6.97). If all nodal values of function f are expanded in Taylor series around the node $(j, n+1)$ and substituted in Eq. (6.97), then after regrouping the following modified equation is obtained:

$$\frac{\partial f}{\partial t} + U\frac{\partial f}{\partial x} = D_n \frac{\partial^2 f}{\partial x^2} + E_n \frac{\partial^3 f}{\partial x^3} + \ldots \tag{6.104}$$

The coefficients of numerical diffusion D_n and dispersion E_n are given by:

$$D_n = \left(\theta - \frac{1}{2}\right) C_a \cdot \Delta x \cdot U, \tag{6.105}$$

$$E_n = \frac{U \cdot \Delta x^2}{2} \left(\frac{2}{3} - \omega + \left(\theta - \frac{2}{3}\right) C_a^2\right). \tag{6.106}$$

From the modified equation results that the finite element method is of 2nd order of accuracy with regard to both x and t for $\theta = 1/2$ only. For other values of θ it is of 1st order of accuracy with regard to t. However, from formula (6.106) results that the possible order of accuracy can be increased. Namely, for $\theta = 1/2$ and for

$$\omega = \frac{2}{3} - \frac{C_a^2}{6} \tag{6.107}$$

the terms of 2nd and 3rd order in the modified equation (6.104) are cancelled. It means that in this case the proposed version of the finite element method ensures an accuracy of 3rd order with regard to x as well as to t. This holds for $C_a \leq 1$ only because of the stability condition (6.103b). Consequently one can expect that appropriate value of the weighting parameter ω will provide more accurate numerical solution than the one given by the standard versions of finite element and finite difference method.

To illustrate the effect of improvment of the finite element method the advective transport equation is solved.

Example 6.3 In a straight open channel of length L the water flows with constant velocity U. Assume that:

– initial condition is given by the Gauss distribution

$$f(x, t = 0) = \frac{1}{\sqrt{2\pi}} \exp\left(-\frac{(x - \mu)^2}{2\sigma^2}\right) \quad \text{for} \quad 0 \leq x \leq L,$$

where μ is mean value and and σ is standard deviation;
– boundary condition is following: $f(x = 0, t) = 0 \quad \text{for} \quad t \geq 0.$

The initially imposed distribution of the function f along x axis travels towards downstream channel end with constant velocity. The following set of data is accepted: $U = 0.5$ m/s, $\Delta x = 100$ m, $\Delta t = 100$ s, $\mu = 1{,}400$ m and $\sigma = 264$ m;

In Fig. 6.13 this distribution is shown for $t = 14{,}400$ s. The exact solution perfectly agrees with the numerical solution provided by the modified finite element method. This result confirms that the accuracy of the standard finite element method can be appreciably improved.

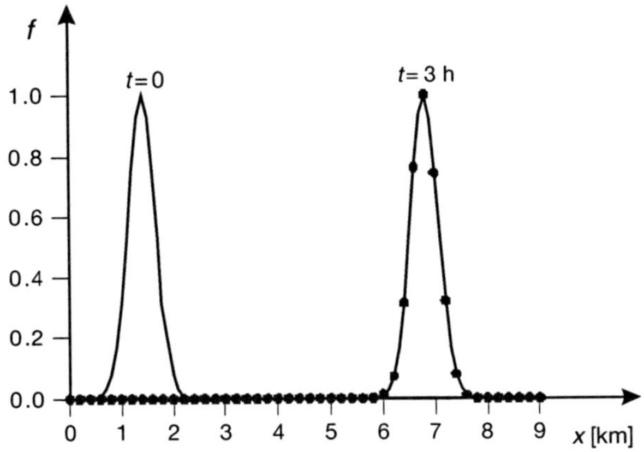

Fig. 6.13 Solution of the advection equation by the modified finite element method for $\theta = 0.5$, $C_a = 0.5$ and $\omega = 0.625$ (exact – *solid line*, numerical – *dotted line*)

6.5 Numerical Solution of the Advection Equation with the Method of Characteristics

6.5.1 Problem Presentation

The difficulties arising in numerical solution of the pure advection equation using the methods based on the approximation of the derivatives were the motivation to seek alternative approaches. One of them is the method of characteristics, which for a long time was the main tool for solving hyperbolic equations. In this section we present a couple of algorithms based on this approach.

Again, we consider the advection equation (6.1) with $U =$ const. > 0. It is solved in the following domain: $0 \leq x \leq L$ and $t \geq 0$. The initial condition is: $f(x, t = 0) = f_i\,(x)$ for $0 \leq x \leq L$. With assumed $U > 0$ the boundary condition is imposed at the upstream end ($x = 0$) only: $f(x = 0, t) = f_o(t)$ for $t \geq 0$. For the constant velocity the characteristics of Eq. (6.1) are straight lines in the (x, t) plane defined by Eq. (5.9):

$$dx = U \cdot dt. \tag{6.108}$$

Let us cover the domain of solution with a mesh having the dimensions $\Delta x \times \Delta t$, as presented in Fig. 6.14.

If we assume such values of the mesh dimensions Δx and Δt that they satisfy the characteristic equation (6.108):

$$\Delta x = U \cdot \Delta t \tag{6.109}$$

then the diagonals of the mesh will coincide with the characteristics (Fig. 6.14). From the definition of characteristics results that the value of f is constant along a characteristic. Thus, the values at the new time level can be calculated as:

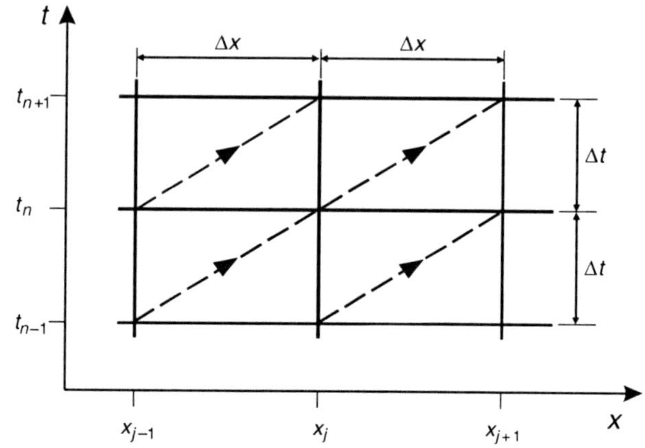

Fig. 6.14 Grid points applied in the method of characteristics

$$f(x_{j+1}, t_{n+1}) = f(x_j, t_n) \quad \text{or} \tag{6.110}$$

$$f_{j+1}^{n+1} = f_j^n \tag{6.111}$$

Equation (6.110) implies that the particle which at the time t_n is in the node (x_j, t_n) after the time step Δt will travel to the node (x_{j+1}, t_{n+1}). Then, if the flow velocity U is constant, numerical solution of the advection equation by the method of characteristics is equivalent to the exact solution.

6.5.2 Linear Interpolation

Unfortunately, in natural channels the flow velocity varies is space ($U = U(x)$) or even in time as well ($U = U(x, t)$). In such a situation it is impossible to construct such a grid which satisfies Eq. (6.108). Consequently the characteristics will no longer coincide with the diagonals of mesh and they will intersect the time level n between the nodes as it is presented in Fig. 6.15.

As the function f does not vary along the characteristic, its value at the node $(j, n + 1)$ is equal to the one at the point of intersection x^*:

$$f_j^{n+1} = f(x^*, t_n) \tag{6.112}$$

where

$$x^* = x_j - U\,\Delta t. \tag{6.113}$$

The value of f at x^* can be calculated, for instance, using the linear interpolation between the nodes j and $j - 1$. This yields:

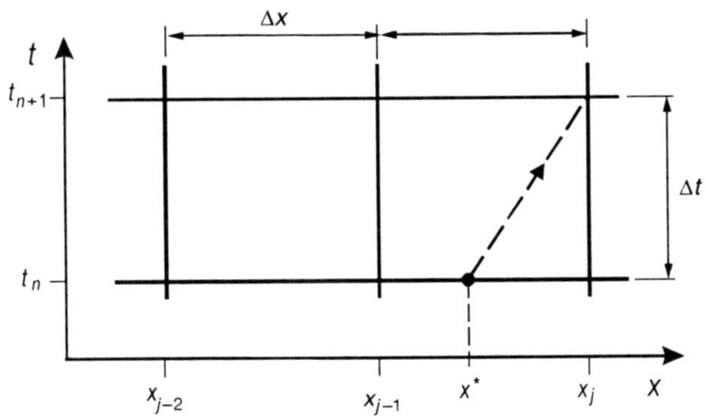

Fig. 6.15 Characteristic of the advection equation going through grid point $(j, n + 1)$ and passing between the nodes

$$f^* = f_j^n - \frac{U \cdot \Delta t}{\Delta x}(f_j^n - f_{j-1}^n). \tag{6.114}$$

and consequently:

$$f_j^{n+1} = f_j^n - \frac{U \cdot \Delta t}{\Delta x}(f_j^n - f_{j-1}^n). \tag{6.115}$$

According to Eq. (5.109), which defines the Courant number we have:

$$\frac{U \cdot \Delta t}{\Delta x} = C_a \tag{6.116}$$

Substitution of Eq. (6.116) in Eq. (6.115) yields:

$$f_j^{n+1} = C_a f_{j-1}^n + (1 - C_a) f_j^n \tag{6.117}$$

Note that the method of characteristics with the interpolation coincides with the well known upwind scheme, being on the other hand, a particular case of the finite difference box scheme. This scheme was discussed in details in the preceding chapter and it has been shown to generate strong numerical diffusion. The example of its application is given in Figs. 5.15 and 5.16, whereas its numerical properties were described in Section 5.4.

6.5.3 Quadratic Interpolation

Since linear interpolation appeared unsuccessful, a natural way of increasing of the solution accuracy seems to apply an interpolating polynomial of higher degree say of 2nd degree. For this purpose assume a local system of co-ordinates (X, Y) at node $j-2$ (Fig. 6.15). In this system we define the following polynomial:

$$Y(X) = A \cdot X^2 + B \cdot X + C \quad \text{for} \quad 0 \le X \le 2\Delta x. \tag{6.118}$$

Using nodes $j-2, j-1$ and j one can express its coefficients as:

$$A = \frac{1}{2\Delta x^2}(f_{j-2}^n - 2f_{j-1}^n + f_j^n), B = \frac{1}{\Delta x}(-1.5f_{j-2}^n + 2f_{j-1}^n - 0.5f_j^n), C = f_{j-2}^n$$

Then Eq. (6.118) takes the following form:

$$Y(X) = \frac{1}{2\Delta x^2}(f_{j-2}^n - 2f_{j-1}^n + f_j^n)X^2 + \frac{1}{\Delta x}(-1.5f_{j-2}^n + 2f_{j-1}^n - 0.5f_j^n)X + f_{j-2}^n \quad (6.119)$$

One can see in Fig. 6.15 that we are looking for the value of $Y(X)$ at the point X^*, which is equal to:

$$X^* = 2\Delta x - U\cdot\Delta t = \Delta x\left(2 - \frac{U\cdot\Delta t}{\Delta x}\right) = \Delta x\,(2 - C_a) \quad (6.120)$$

where C_a is the advective Courant number defined by Eq. (6.116). Since the function f is constant along the characteristics, then its nodal value f_j^{n+1} is the same as at the point X^*, i.e.:

$$\begin{aligned} Y(X^*) = f_j^{n+1} &= \left(-0.5C_a + 0.5C_a^2\right)f_{j-2}^n + \left(2C_a - C_a^2\right)f_{j-1}^n \\ &+ \left(1 - 1.5C_a + 0.5C_a^2\right)f_j^n \end{aligned} \quad (6.121)$$

As the boundary condition is given at node $j = 1$, then this equation is not self starting and an apparent node $j = 0$ must be added, where $f_0^n = f_1^n$ is assumed. As it will be shown in the next example, this method also produces significant numerical errors.

6.5.4 Holly–Preissmann Method of Interpolation

A very interesting approach to solve the problem of interpolation was proposed by Holly and Preissmann (1977). The interpolation is performed between two nodes $j-1$ and j using a polynomial of 3rd degree. To determine this polynomial the values of function f as well as its spatial derivative are used.

Knowing the function f and its derivative at the time level t_n at the nodes $j-1$ and j, (which makes together 4 values), one can determine the polynomial of 3rd degree. This polynomial is given by the formula:

$$Y(\alpha) = A\cdot C_a^3 + B\cdot C_a^2 + C\cdot C_a + D, \quad (6.122)$$

in which C_a is the well known Courant number (Eq. 6.116), considered here as an independent variable ranging from 0 to 1. The coefficients A, B, C and D can be determined using the following conditions, satisfied by the polynomial (6.122):

$$Y(1) = f_{j-1}^n, \quad (6.123a)$$

$$Y(0) = f_j^n, \tag{6.123b}$$

$$Y'(1) = \left.\frac{df}{dx}\right|_{j-1}^{n}, \tag{6.123c}$$

$$Y'(0) = \left.\frac{df}{dx}\right|_{j}^{n} \tag{6.123d}$$

These conditions allow us to express the interpolating polynomial as:

$$Y(\alpha) = f_j^{n+1} = a_1 \cdot f_{j-1}^n + a_2 \cdot f_j^n + a_3 \left.\frac{\partial f}{\partial x}\right|_{j-1}^{n} + a_4 \left.\frac{\partial f}{\partial x}\right|_{j}^{n}, \tag{6.124}$$

where:

$$a_1 = C_a^2\,(3 - 2C_a)\,, \quad a_2 = 1 - C_a^2\,(3 - 2C_a)\,, \quad a_3 = C_a^2\,(1 - C_a)\;\Delta x,$$

$$a_4 = -C_a\,(1 - C_a)^2\;\Delta x$$

In order to calculate the derivatives of the function f Holly and Preissmann (1977) assumed that the derivative $\partial f/\partial x$ is transported similarly to the function f with the advection velocity U. Therefore it satisfies the following advection equation:

$$\frac{\partial}{\partial t}\left(\frac{\partial f}{\partial x}\right) + U\frac{\partial}{\partial x}\left(\frac{\partial f}{\partial x}\right) = 0. \tag{6.125}$$

Substitution of the previously determined coefficients A, B, C and D in equation for $Y'(\alpha)$ obtained by differentiation of Eq. (6.122), yields:

$$Y'(\mathrm{C_a}) = \left.\frac{\partial f}{\partial x}\right|_{j}^{n+1} = b_1\,\cdot f_{j-1}^n + b_2 \cdot f_j^n + b_3 \left.\frac{\partial f}{\partial x}\right|_{j-1}^{n} + b_4 \left.\frac{\partial f}{\partial x}\right|_{j}^{n}, \tag{6.126}$$

where

$$b_1 = \frac{6C_a(\mathrm{C_a} - 1)}{\Delta x}, \quad b_2 = -\frac{6C_a(\mathrm{C_a} - 1)}{\Delta x}, \quad b_3 = C_a(3C_a - 2),$$

$$b_4 = (\mathrm{C_a} - 1)(3C_a - 1)$$

This method provides relatively low numerical diffusion. It is only conditionally stable and can be applied for $C_a \leq 1$. Moreover, the method requires as the initial condition not only the values of the unknown function, but also its derivative, which can be problematic in some cases.

6.5.5 Interpolation with Spline Function of 3rd Degree

Interpolation of the function f can be carried out using a spline function of 3rd degree (Szymkiewicz 1993). An important feature of this approach is that no limitation of the time step Δt exists.

The spline function of 3rd degree is composed of the segments which is a polynomial of 3rd degree assuming the values of interpolated function at nodes. This function is the most smooth curve among all polynomials of 3rd degree. It is important that the spline function tends to the interpolated one when the distances between the nodes are reduced. The interpolating function is expressed as follows:

$$Y(x) = f_j + \alpha_j(x - x_j) + \beta_j(x - x_j)^2 + \gamma_j(x - x_j)^3 \text{ for } x_j \le x \le x_{j+1} \tag{6.127}$$

Where:

f_j – nodal value of the function f,
α_j, β_j, γ_j – coefficients of the polynomial to be determined,
$j = 1, 2, \ldots, M - 1$ – index of node,
M – total number of nodes.

If the values of function f are given at all nodes at the time level t_n (Fig. 6.16), then the polynomial (6.127) can be determined.

It means that the coefficients α_j, β_j, γ_j for $j = 1,2, \ldots, M - 1$ can be calculated. Detailed description of the spline function and the method of its determination is given by Stoer and Bulirsch (1980). The algorithm is very simple and it reduces itself to the solution of the system of linear algebraic equations with tri-diagonal matrix. Knowing the interpolating polynomial, one can use it to compute the values of function f at the points of intersection x_j^* of the backward characteristics going through the nodes at the time level t_{n+1} (Fig. 6.16). In such a way the problem is solved, since we have:

$$Y(x_j^*) = f_{j+1}^{n+1} = f_j^n + \alpha_j(x_j^* - x_j) + \beta_j(x_j^* - x_j)^2 + \gamma_j(x_j^* - x_j)^3 \text{ for } j = 1,2,\ldots,M-1\,. \tag{6.128}$$

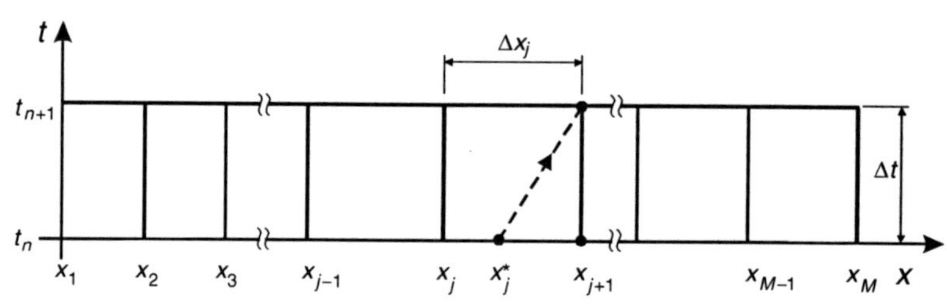

Fig. 6.16 Grid points with the backward characteristics

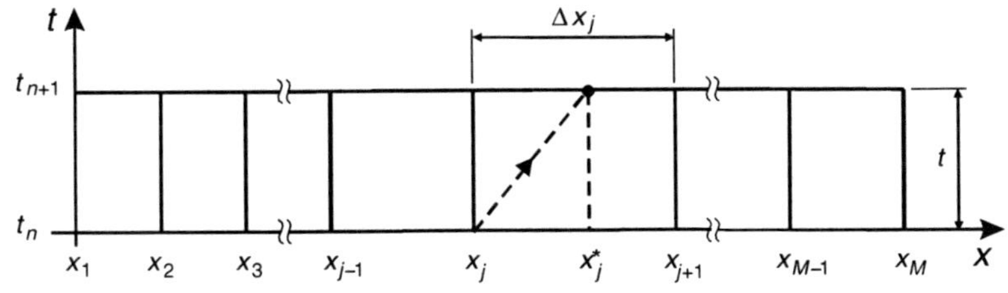

Fig. 6.17 Grid points with the forward characteristics

The value of f_1^{n+1} is given by the imposed boundary condition.

The presented approach makes use of the backward characteristics, which start from the nodes at the time level t_{n+1} and intersect the time level t_n . An alternative method would be to apply the forward characteristics starting from the nodes at the time level t_n ,which intersect the time level t_{n+1} (Fig. 6.17).

The appropriate formula can be derived in a similar way as presented previously. The only difference is that this time the characteristics are going through the nodes at the time level n and they intersect the time level $n + 1$, so we must interpolate at this level.

To illustrate the presented techniques of interpolation, they are applied to solve the same case of the advection transport.

Example 6.4 In straight open channel of length L the water flows with constant velocity U. Assume the following auxiliary conditions:

– initial condition is determined by the Gauss distribution

$$f(x, t = 0) = \frac{1}{\sqrt{2\pi}} \exp\left(-\frac{(x-\mu)^2}{2\sigma^2}\right) \quad \text{for} \quad 0 \leq x \leq L,$$

with the mean value μ and the standard deviation σ;
– boundary condition: $f(x = 0,t) = 0 \quad \text{for} \quad t > 0$.

Therefore, the initial distribution of the function f in form of the Gauss distribution moves along the channel axis with constant velocity. Computations are carried out for $U = 0.5$ m/s, $\Delta x = 100$ m, $\mu = 1{,}400$ m, $\sigma = 264$ m and for various values of time step Δt.

In Fig. 6.18 this traveling distribution is shown after $t = 9{,}600$ s. The exact solution corresponds to the initially imposed distribution, with the center of gravity shifted to the position $x = \mu + U \cdot T = 1{,}400 + 0.5 \cdot 9{,}600 = 6{,}200$ m. This solution is provided by all method of interpolation for $C_a = 1$. For other values of the Courant number the solution accuracy depends on the applied method of interpolation.

As it could be expected, the linear interpolation provides strongly smoothed solution for the Courant number less than unity (Fig. 6.18). It is because of the numerical

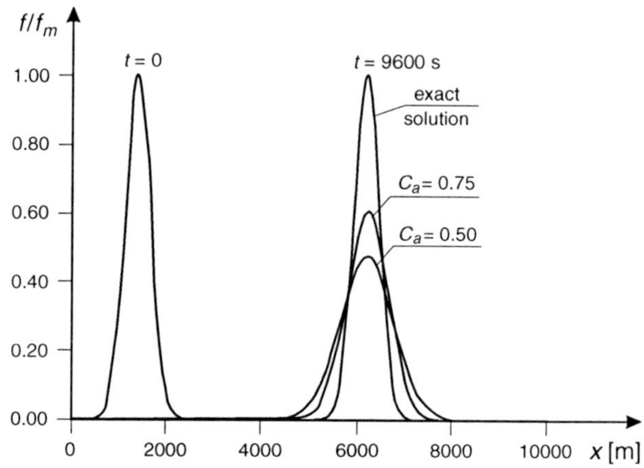

Fig. 6.18 Solution of the advection equation by the method of characteristics with linear interpolation

diffusion generated by the method, which disappears for $C_a = 1$ only. It should be added that for $C_a > 1$ the linear interpolation produces unstable results.

Increasing of the degree of interpolating polynomial does not improve the accuracy solution. As one can see in Fig. 6.19, that the polynomial of 2nd degree introduces numerical dispersion and oscillations appear. The same is true for polynomials of higher degrees (Cunge et al. 1980).

Thus, one can suppose that the standard manner of interpolation cannot be successful and some special interpolating techniques should rather be applied. This suggestion is confirmed by the results displayed in Figs. 6.20 .

One can see that the method of interpolation proposed by Holly and Preissmann (1977) ensures an excellent agreement with the exact solution. The difference in peak obtained for $C_a = 0.5$ is 1.6% only. Similarly accurate solution is provided by the interpolation with the spline function.

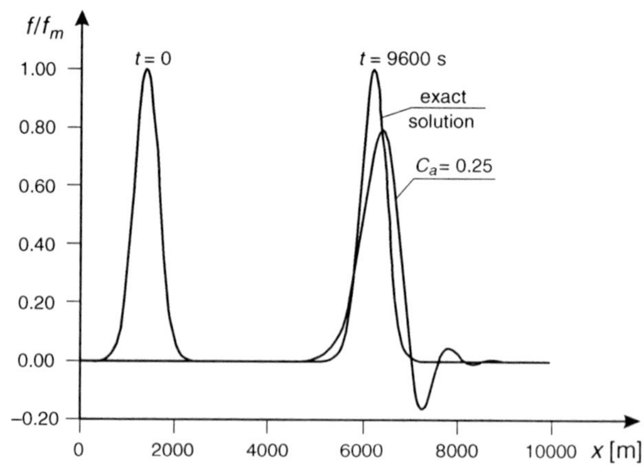

Fig. 6.19 Solution of the advection equation by the method of characteristics with quadratic interpolation

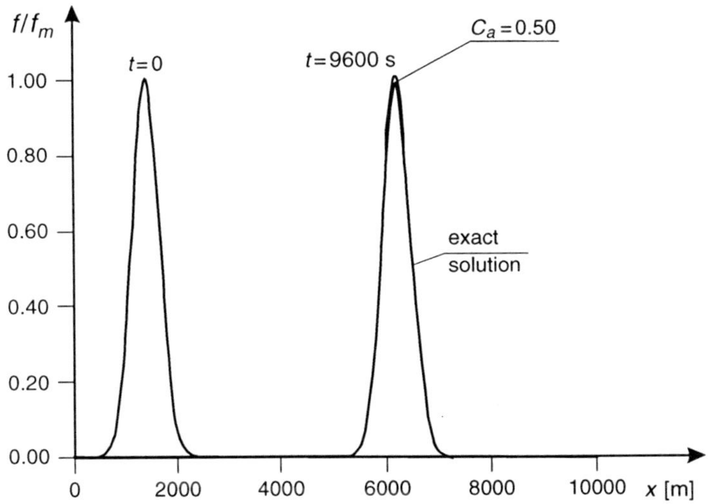

Fig. 6.20 Solution of the advection equation by the method of characteristics with interpolation by the Holly–Preissmann approach

References

Abbott MB, Basco DR (1989) Computational fluid dynamics. Longman Scientific and Technical, New York

Billingham J, King AC (2000) Wave motion. Cambridge University Press, Cambridge

Cunge J, Holly FM, Verwey A (1980) Practical aspects of computational river hydraulics. Pitman Publishing, London

Donea J (1984) A Taylor–Galerkin method for convective transport problems. Int. J. Num. Meth. Engng. 20:101–119

Fletcher CAJ (1991) Computational techniques for fluid dynamics, vol. I. Springer-Verlag, New York

Holly FM jr, Preissmann A (1977) Accurate calculation of transport in two dimensions. J. Hydr. Engng. ASCE 103 (11):1259–1277

Korn GA, Korn TM (1968) Mathematical handbook for scientists and engineers, 2nd edn. McGraw-Hill, New York

Liggett JA, Cunge JA (1975) Numerical methods of solution of the unsteady flow equations. In: Mahmood K, Yevjewich V (eds) Unsteady flow in open channels. Water Resources Publications, Fort Collins, Colorado, USA

McQuarrie DA (2003) Mathematical methods for scientists and engineers. University Science Books, Sausalito

Potter D (1973) Computational physics. Wiley, London

Stoer J, Bulirsch R (1980) Introduction to numerical analysis. Springer-Verlag, New York

Szymkiewicz R (1993) Solution of the advection-diffusion equation using the spline function and finite elements. Commun. Numer. Methods Engng. 9 (4):197–206

Szymkiewicz R (1995) Method to solve 1D unsteady transport and flow equations. J. Hydr. Engng. ASCE 121 (5):396–403

Warming RF, Hyett BJ (1974) The modified equation approach to the stability and accuracy analysis of finite-difference methods. J. Comput. Phys. 14:159–179

Whitham GB (1979) Lectures on wave propagation. Tata Institute of Fundamental Research, Bombay

Chapter 7
Numerical Solution of the Advection-Diffusion Equation

7.1 Introduction to the Problem

Comparing the advection-diffusion equation (1.157) with the pure advection equation one can notice, that only the diffusive term differs one another equation. However this term introduces an essential difference between both equations. First of all the transport equation becomes a partial differential equation of 2nd order, whereas the advection equation is of 1st order. In addition, the type of equation changes as well. Instead of a hyperbolic equation we face a parabolic one. Consequently the auxiliary conditions should be reformulated as the parabolic equation needs two boundary conditions imposed at both upstream and downstream channel ends. More information on the properties of advection-diffusion equation were given in Section 5.1.6.

Numerical experiments show that the diffusive term has very positive influence on the numerical solution process. This term has dissipative character so it posses smoothing properties giving rise to positive numerical effects. If the transport is diffusion dominated, no numerical difficulties are expected. In such a case to solve the advection-diffusion equation one can choose any scheme of the finite difference or finite element method. If the applied scheme is dissipation-free, its dispersivity will be neutralized by the diffusion term. Conversely, a dissipative scheme will reinforce the smoothing effect caused by the diffusion term. On the other hand, for the advection dominated transport all numerical problems presented in Chapter 6 related to the pure advection equation, occur. For this reason the essential question is to distinguish, which of the process is dominating.

A criterion, which allows us to define the participation of both forms of transport is called the Peclet number (Patankar 1980). For the grid points spaced with Δx interval this number is given by the following relation:

$$P = \frac{U \cdot \Delta x}{D} \tag{7.1}$$

R. Szymkiewicz, *Numerical Modeling in Open Channel Hydraulics*, Water Science and Technology Library 83, DOI 10.1007/978-90-481-3674-2_7,

where:

P – Peclet number,
U – advection velocity,
D – coefficient of diffusion,
Δx – mesh dimension.

The Peclet number can be considered as the relation between the advective and diffusive Courant numbers:

$$P = \frac{C_a}{C_d} \tag{7.2}$$

where:

C_a – the advection Courant number given by Eq. (5.109),
C_d – the diffusive Courant number given by Eq. (5.174).

Since this number is similar to the Reynolds number, very often it is called the cell Reynolds number (Fletcher 1991).

From the definition of the Peclet number (7.1) results that:

- $P = \infty$ for pure advection equation, when $D = 0$,
- $P = 0$ for pure diffusion equation, when $U = 0$.

Numerical difficulties arise for large value of the Peclet number when the transport is dominated by advection. They have the same roots as those connected with the hyperbolic equation. If in the advection-diffusion transport equation the natural physical dissipative process represented by the diffusive term is relatively weak, there is no possibility to suppress the oscillations. On the other hand, if we introduce some numerical dissipation, it can dominate the physical one. Consequently, in the numerical solution too strong attenuation will be observed. The trouble with the numerical diffusion is that it produces in the solution the advection–diffusion equation the same effects as the physical diffusion.

During the last 30 years many numerical techniques and algorithms were proposed. In these approaches both eulerian and lagrangian representation of the transport equation are used. One can say that the methods basing on the direct approximation of the partial differential equations like the finite difference method or the finite element method applied in their standard forms, do not lead to effective algorithms. Some of the more accurate approaches are presented in the following sections.

7.2 Solution by the Finite Difference Method

7.2.1 Solution Using General Two Level Scheme with Up-Winding Effect

Many finite difference schemes have been proposed to solve the advection-diffusion equation – for a review see e.g. Fletcher (1991). They have different properties and not all of them are equally useful. Generally it can be said that nearly all difference schemes work well for diffusion-dominated transport. Conversely, if the transport is advection-dominated, i.e. for a large Peclet number, practically all of them suffer from poor accuracy. Since the difference methods are based on the Taylor series expansion, a truncation error, greater or smaller, occurs always. Consequently, numerical schemes will always generate a dissipation and dispersion errors. The problems related to the solution of advection-diffusion equation with the finite difference method will be illustrated using a general two-level approximating formula. Note that the box scheme used previously for solving the advection equation is not appropriate for the advection-diffusion equation, since it is not capable to approximate the diffusive term.

Consider the advection-diffusion equation (1.157) without source term:

$$\frac{\partial f}{\partial t} + U\frac{\partial f}{\partial x} - D\frac{\partial^2 f}{\partial x^2} = 0, \tag{7.3}$$

with constant flow velocity U and coefficient of diffusion D are assumed. Equation (7.3) is solved in the following domain: $0 \leq x \leq L$, $t \geq 0$, where L is length of the channel reach. The following auxiliary conditions are assumed:

- initial condition: $f(x, t = 0) = f_i(x)$ for $0 \leq x \leq L$
- boundary conditions: $f(x = 0, t) = f_o(t)$ for $t \geq 0$

$$f(x = L, t) = f_L(t) \text{ for } t \geq 0 \text{ or } \partial f/\partial x|_{x=L} = \varphi_L(t) \text{ for } t \geq 0$$

where: $f_i(x), f_o(t), f_L(t), \varphi_L(t)$ are known.

Equation (7.3) is approximated on the numerical grid shown in Fig. 7.1 using previously presented approximating formulas for the derivatives involved in this equation. They are following:

- for 1st order time derivative (Eq. 5.99)

$$\left.\frac{df}{dt}\right|_n \approx \frac{f^{n+1} - f^n}{\Delta t}, \tag{7.4}$$

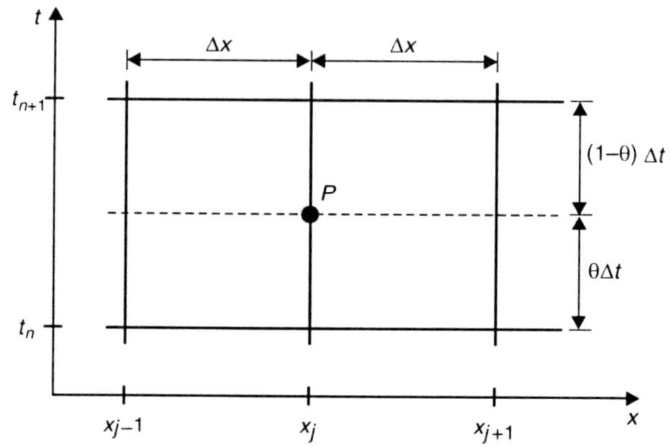

Fig. 7.1 Grid points for general two levels difference scheme

– for 1st order spatial derivative (Eq. 5.100)

$$\left.\frac{\partial f}{\partial x}\right|_j \approx \frac{-\eta \cdot f_{j-1} + (2\eta - 1)f_j + (1-\eta)f_{j+1}}{\Delta x} \tag{7.5}$$

– for 2nd order spatial derivative (Eq. 5.86)

$$\left.\frac{\partial^2 f}{\partial x^2}\right|_j \approx \frac{f_{j-1} - 2f_j + f_{j+1}}{\Delta x^2}\ . \tag{7.6}$$

where Δt is time step and Δx is spatial step, whereas n and j are the indices of time level and node, respectively.

Let us apply these formulas for Eq. (7.3) and assume that it is approximated at point P (Fig. 7.1). The position of this point with respect to the time axis is determined by the parameter θ ranging form 0 to 1.

Substitution of Eqs. (7.4), (7.5) and (7.6) in Eq. (7.1) yields:

$$\begin{aligned}\frac{f_j^{n+1} - f_j^n}{\Delta t} &+ U\left[(1-\theta)\frac{-\eta \cdot f_{j-1}^n + (2\eta-1)f_j^n + (1-\eta)f_{j+1}^n}{\Delta x} + \right.\\ &\left. + \theta\,\frac{-\eta \cdot f_{j-1}^{n+1} + (2\eta-1)f_j^{n+1} + (1-\eta)f_{j+1}^{n+1}}{\Delta x}\right] + \\ &- D\left[(1-\theta)\frac{f_{j-1}^n - 2f_j^n + f_{j+1}^n}{\Delta x^2} + \theta\,\frac{f_{j-1}^{n+1} - 2f_j^{n+1} + f_{j+1}^{n+1}}{\Delta x^2}\right] = \\ &0 \text{ for } j = 2,3,\ldots, M-1\end{aligned} \tag{7.7}$$

where:

η – the weighting parameter ranging from 0 to 1,
θ – the weighting parameter ranging from 0 to 1,
M – number of grid points.

Setting $j = 2,3, \ldots, M-1$ one obtains a system of algebraic equation, which must be completed by the prescribed boundary conditions. In matrix notation this system is rewritten as:

$$\mathbf{A} \cdot \mathbf{f} = \mathbf{R} \tag{7.8}$$

where:

$$\mathbf{A} = \begin{bmatrix} b_1 & c_1 & & & \\ a_2 & b_2 & c_2 & & \\ & a_3 & b_3 & c_3 & \\ & & & \ddots & \\ & & & a_M & b_M \end{bmatrix}, \quad \mathbf{f} = \begin{Bmatrix} f_1^{n+1} \\ f_2^{n+1} \\ f_3^{n+1} \\ \vdots \\ f_{M-1}^{n+1} \end{Bmatrix}, \quad \mathbf{R} = \begin{Bmatrix} R_1 \\ R_2 \\ R_3 \\ \vdots \\ R_{M-1} \end{Bmatrix}.$$

The matrix coefficients are as follows:

$$b_1 = 1, c_1 = 0$$

$$a = \theta\,(-\eta \cdot C_a - C_d),\ b = 1 + \theta\,(C_a\,(2\eta - 1) + 2C_d),\ c = \theta\,(C_a\,(1 - \eta) - C_d)$$

$$b_M = 1, c_M = 0$$

whereas the elements of the vector of the right hand side are:

$$R_1 = f_0\,(t_{n+1})$$

$$R_j = f_j^n - C_a\,(1-\theta)\left(-\eta \cdot f_{j-1}^n + (2\eta - 1) f_j^n + (1-\eta) f_{j+1}^n\right)$$
$$- C_d\,(1-\theta)\left(f_{j-1}^n - 2f_j^n + f_{j+1}^n\right) \text{ for } j = 2, 4, \ldots, M-1$$

$$R_M = f_L\,(t_{n+1})$$

In the above equations

$$C_a = \frac{U \cdot \Delta t}{\Delta x} \quad \text{and} \tag{7.9}$$

$$C_d = \frac{D \cdot \Delta t}{\Delta x^2} \tag{7.10}$$

represent the previously introduced advective and diffusive Courant numbers, respectively.

Note, that for specific values of the weighting parameters determine some well known schemes are obtained. For instance $\theta = 1/2$ and $\eta = 1/2$ correspond to the Crank-Nicolson finite difference scheme; $\theta = 1$ and $\eta = 1/2$ – to the fully implicit centered difference scheme; $\theta = 0$ and $\eta = 1$ – to the upwind scheme. Thus, Eq. (7.7) can be considered as general two-level difference scheme.

The accuracy analysis performed using the modified approach, shows that Eq. (7.7) modifies the advection-diffusion equation to the following form:

$$\frac{\partial f}{\partial t} + U\frac{\partial f}{\partial x} - D\frac{\partial^2 f}{\partial x^2} = D_n\frac{\partial^2 f}{\partial x^2} + E_n\frac{\partial^3 f}{\partial x^3} + \cdots \tag{7.11}$$

The coefficients of numerical diffusion D_n and dispersion E_n are defined as follows:

$$D_n = \frac{U \cdot \Delta x}{2}\left((2\theta - 1)\ C_a + (2\eta - 1)\right) \tag{7.12}$$

$$E_n = \frac{U \cdot \Delta x^2}{2}\left(-\frac{1}{3} + C_a\,(2\eta - 1)\ (1 - \theta) + C_a^2\left(\theta - \frac{2}{3}\right)\right) \tag{7.13}$$

The amplification factor G is expressed as:

$$G = \frac{1 - (1-\theta)\ C_a\,[(1-2\eta)\ (\cos\ (\varphi) - 1) + i\ \sin\ (\varphi)]}{1 + \theta\ C_a\,[(1-2\eta)\ (\cos\ (\varphi) - 1) + i\ \sin\ (\varphi)]} \tag{7.14}$$

where:

$i = (-1)^{1/2}$ imaginary unit,
$\varphi = 2\pi/N$ dimensionless wave number,
N – number of grid intervals per wavelength.

The condition of numerical stability $|G| \le 1$ requires that:

$$\theta \ge 1/2 \text{ and} \tag{7.15a}$$

$$\eta \ge 1/2 \tag{7.15b}$$

For the values of weighting parameters satisfying these relations the scheme is absolutely stable.

Let us remember that formula (7.5) describes all the standard approximations of 1st order derivative, because $\eta = 0$ gives the forward difference, $\eta = 0.5$ – the centred difference, whereas $\eta = 1$ – the backward difference. Since η can take values from 0 to 1, then Eq. (7.5) is capable to increase the up-wind effect in a gradual manner.

As it can be seen, for $\theta = 0.5$ and $\eta = 0.5$ the coefficient of numerical diffusion (Eq. 7.12) disappears, whereas the coefficient of numerical dispersion (Eq. 7.13) is still present. This means that if the physical diffusion in Eq. (7.3) is not strong enough, oscillations will occur. Therefore in case of weak physical diffusion introduction of some artificial diffusion may be necessary.

Example 7.1 In prismatic open channel of length L the water flows with constant velocity U. The initial condition is determined by the Gauss distribution:

$$f(x,t=0) = \frac{1}{\sqrt{2\pi}} \exp\left(-\frac{(x-\mu)^2}{2\sigma^2}\right) \quad \text{for} \quad 0 \le x \le L, \tag{7.16}$$

The following Dirichlet boundary conditions are imposed at both channel ends:

$$f(x=0,t) = 0 \quad \text{for} \quad t > 0$$

$$f_L(t) = 0 \quad \text{for } L \to \infty \quad \text{and} \quad t > 0.$$

Assume that: $U = 0.5$ m/s = const., $D = 0.25$ m^2/s = const., $\Delta x = 100$ m = const. and $\Delta t = 25$ s ($C_a = 0{,}125$), whereas the mean value is $\mu = 1{,}400$ m and the standard deviation is $\sigma = 264$ m. In Fig. 7.2 the distributions of the function f along channel axis for $t = 7{,}200$ s is shown. One can see that the solution obtained for $\theta = 0.5$ and $\eta = 0.5$ i.e. for the fully centered scheme is oscillating. Such behavior could be expected, because in this case the value of Peclet number is $P = 200$, so the transport is advection-dominated.

To suppress the oscillations, at first $\theta = 1$ is assumed. This increases the dissipativity of the scheme by introducing numerical diffusion, however it appears insufficient to ensure smooth solution. To suppress entirely the oscillations it is necessary to increase the value of the second weighting parameter η as well, $\eta = 0.60$ should be assumed.

Note that to provide smooth solution it was necessary to introduce only small additional up-wind effect. This is an interesting feature of the applied scheme.

7.2.2 The Difference Crank-Nicolson Scheme

Let us consider the Crank-Nicolson scheme, which is the most commonly applied method for the solution of 1D diffusion equation. We will investigate its behavior in the case of the advection-diffusion equation. Setting in Eq. (7.7) $\theta = 0.5$ and $\eta = 0.5$ one obtains:

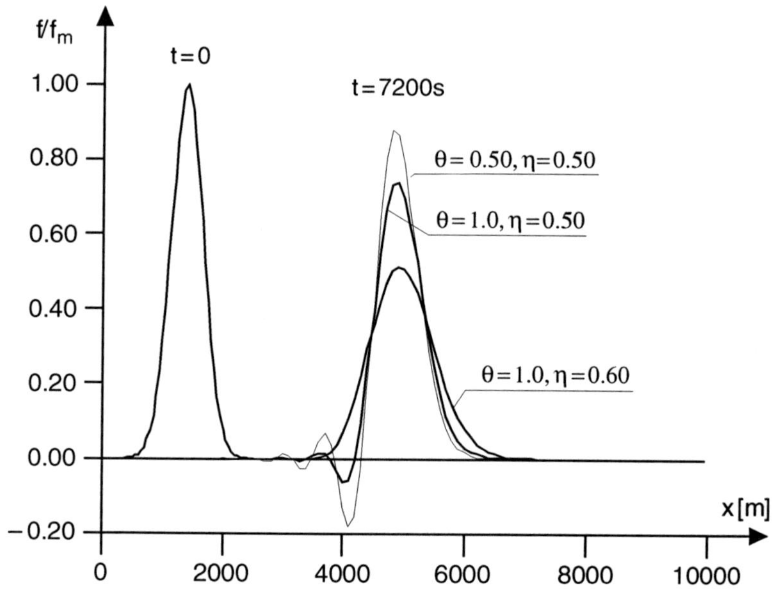

Fig. 7.2 Solution of the advection-diffusion equation by the difference scheme with $D = 0.25$ m^2/s and $C_a = 0.125$ for various weighting parameters θ and η.

$$\frac{f_j^{n+1} - f_j^n}{\Delta t} + \frac{U}{4}\left(\frac{f_{j+1}^n - f_{j-1}^n}{\Delta x} + \frac{f_{j+1}^{n-1} - f_{j-1}^{n+1}}{\Delta x}\right) + \\ - \frac{D}{2}\left(\frac{f_{j-1}^n - 2f_j^n + f_{j+1}^n}{\Delta x^2} + \frac{f_{j-1}^{n+1} - 2f_j^{n+1} + f_{j+1}^{n+1}}{\Delta x^2}\right) = 0 \text{ for } j=2,3,\ldots,M-1 \tag{7.17}$$

Collecting the similar terms and substituting of the Courant numbers C_a and C_d one obtains:

$$-\left(\frac{C_a}{4} + \frac{C_d}{2}\right) f_{j-1}^{n+1} + (1 + C_d) f_j^{n+1} \left(\frac{C_a}{4} - \frac{C_d}{2}\right) f_{j+1}^{n+1} = \\ = \left(\frac{C_a}{4} + \frac{C_d}{2}\right) f_{j-1}^n + (1 - C_d) f_j^n - \left(\frac{C_a}{4} - \frac{C_d}{2}\right) f_{j+1}^n \quad \text{for } j = 2,3, \ldots, M - 1. \tag{7.18}$$

This system of equation is completed using two equations given by the assumed boundary conditions. To illustrate the properties of the Crank-Nicolson scheme, let us solve the advection-diffusion equation (7.3).

Example 7.2 We use the same data as in Example 7.1 except the coefficient of diffusion. Numerical tests carried out for various values of the diffusion coefficient D allow us to notice that some of them generate oscillating solution. However, for increasing value of D the oscillations are systematically reduced and finally they

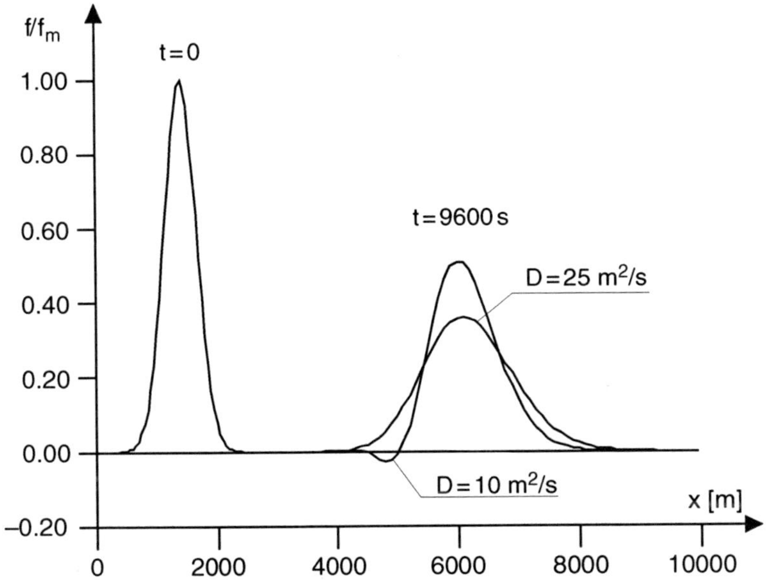

Fig. 7.3 Solution of the advection-diffusion equation by the difference Crank-Nicolson scheme with $C_a = 1.5$ for various coefficients of diffusion

disappear. This tendency is illustrated in Fig. 7.3. Increasing of the diffusion coefficient gives rise to increased smoothing of the results.

One can suppose, that the Crank-Nicolson scheme becomes effective tool for solving the advection-diffusion equation for sufficiently large participation of a diffusion in the transport process.

Explanation of these features of the Crank-Nicolson scheme can be obtained by the analysis of the modified equation (7.11). For the assumed values of the weighting parameters $\theta = 0.5$ and $\eta = 0.5$ one obtains:

$$D_n = 0 \text{ and} \tag{7.19}$$

$$E_n = \frac{U \cdot \Delta x^2}{6}\left(1 + \frac{C_a^2}{2}\right). \tag{7.20}$$

It means that the Crank-Nicolson scheme is dissipation free, since both derivatives of 1st order were approximated with the centred difference. The scheme does not generate numerical diffusion ($D_n = 0$). However, it is always dispersive, because E_n cannot be cancelled. Consequently, since no smoothing mechanism exists, the solution of the advection equation by the Crank-Nicolson scheme will be always oscillating.

On the other hand, while solving the advection-diffusion equation, a smoothing of solution is caused by the physical dissipation represented in the equation by the diffusion term. The intensity of smoothing increases with increasing of the coefficient of diffusion D. For sufficiently large value of D the wiggles are suppressed completely. An explanation of this property, resulting from analysis of the eigenvalues of the matrix of system (7.8) approximating the advection-diffusion equation, is given by Fletcher (1991). The eigenvalues of the tridiagonal matrix of the system (7.8) are related to a, b and c as:

$$\lambda_j = b + 2\sqrt{a \cdot c}\ \cos\left(\frac{j \cdot \pi}{M-1}\right) \qquad \text{for} \quad j = 1, 2, \ldots, M-1 \tag{7.21}$$

For solution to be spatially non-oscillatory, the real eigenvalues are required. Therefore the condition:

$$a \cdot c \geq 0 \tag{7.22}$$

must be respected to ensure smooth solution. For the system (7.18), obtained from (7.8) with $\theta = 0.5$ and $\eta = 0.5$ corresponding to the Crank-Nicolson scheme, relation (7.22) takes the following form:

$$\left(\frac{C_a}{4} + \frac{C_d}{2}\right)\left(-\frac{C_a}{4} + \frac{C_d}{2}\right) \geq 0. \tag{7.23}$$

This inequality will be satisfied for:

$$\frac{C_a}{C_d} \leq 2. \tag{7. 24}$$

Accordingly to Eq. (7.2), the ratio C_a/C_d defines the Peclet number. Therefore, the Crank-Nicolson scheme will provide highly accurate solution of the advection-diffusion equation on condition that the Peclet number is not greater than 2 ($P \leq 2$).

7.2.3 Numerical Diffusion Versus Physical Diffusion

A problem of great importance while solving the advection-diffusion transport equation is that the numerical error in form of the numerical diffusion gives rise to the same symptoms in solution as the physical diffusion. Therefore it is very important to distinguish in solution the effects of both processes and not to confuse them. To this order the modified equation approach should be applied as very helpful tool. This problem deals first of all with the advection dominated transport. For diffusion dominated transport the significance of numerical diffusion is relatively low. This fact is illustrated by the example presented below.

Example 7.3 The advection-diffusion equation is solved for a straight prismatic channel in which the water flows at constant velocity U. The channel is divided into intervals of constant length Δx. The following initial-boundary conditions are specified:

– at $t = 0$: $f(x, t) = 0$ for $x \geq 0$,
– at $x = 0$ the function f jumps immediately from 0 to 1 and remains constant i.e.:

$$f(x = 0,t) = \begin{cases} 0 \text{ for } & t = 0 \\ 1 \text{ for } & t > 0 \end{cases}$$

– at the downstream end located at $x = L \to \infty$ the function is still equal to zero: $f(x) = 0$.

These conditions describe the propagation of a steep front, which simultaneously is subjected to the diffusion. Analytical solution of the advection-diffusion equation is following (Chanson 2004, Elliot and James 1984):

$$f(x,t) = \frac{1}{2}\operatorname{erfc}\left(\frac{x - U \cdot t}{\sqrt{4D \cdot t}}\right) + \frac{1}{2}\exp\left(\frac{U \cdot x}{D}\right)\operatorname{erfc}\left(\frac{x + U \cdot t}{\sqrt{4D \cdot t}}\right) \tag{7.25}$$

where erfc(·) is the complementary function to the error function erf(·): erfc(·) =1 – erf(·). For more information about the error function see for example McQuarrie (2003).

The same problem is solved using the difference scheme (7.7) with $\theta = 0$ and $\eta = 1.0$. It means that the following explicit scheme is applied:

$$\frac{f_j^{n+1} - f_j^n}{\Delta t} + U\frac{f_j^n - f_{j-1}^n}{\Delta x} - D\frac{f_{j+1}^n - 2f_j^n + f_{j-1}^n}{\Delta x^2} = 0 \quad \text{for } j = 2,3, \cdots M - 1 \tag{7.26}$$

The advection-diffusion equation is solved analytically and numerically for $U = 0.01$ m/s and $\Delta x = 1$ m. In Figs. 7.4 and 7.5 both solutions are compared for $t = 3{,}000$ s.

Calculations were carried out with $\Delta t = 10$ s for $D = 0.002$ m²/s and $D = 0.02$ m²/s. This corresponds to the Pecet numbers equal to 5 and 0.5, respectively. Although in both cases one can notice the presence of the numerical diffusion, it is more significant for $P = 5$, i.e. for advection-dominated transport. For greater

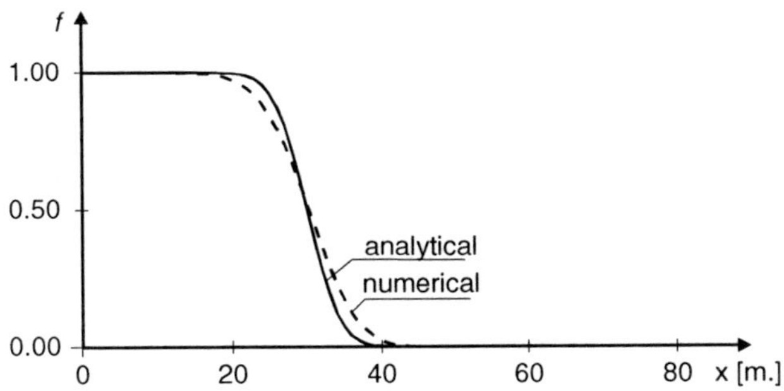

Fig. 7.4 Analytical and numerical solution of the advection-diffusion equation at $t = 3{,}000$ s and for $D = 0.002$ m²/s, i.e. for $P = 5$

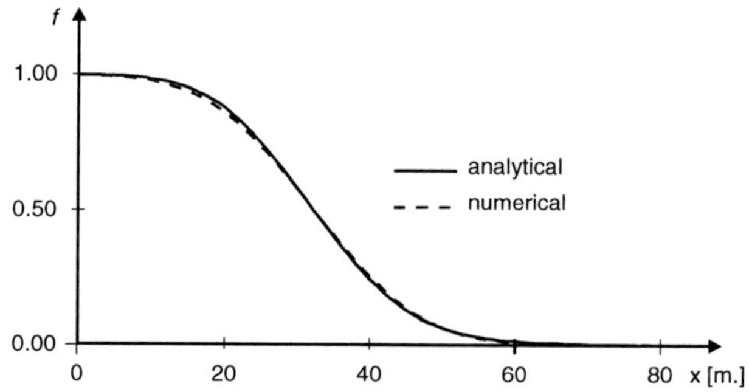

Fig. 7.5 Analytical and numerical solution of the advection-diffusion equation at $t = 3{,}000$ s and for $D = 0.02$ m^2/s, i.e. for $P = 0.5$

coefficient of diffusion D the influence of the numerical diffusion is relatively small. Note that for $\theta = 0$ and $\eta = 1.0$ the coefficient of numerical diffusion (7.12) given as:

$$D_n = \frac{U \cdot \Delta x}{2}(1 - C_a) \tag{7.27}$$

for the assumed data is equal to 0.0045 m^2/s. Then its relative influence on the considered transport process varies with variation of the coefficient of physical diffusion D.

To solve the advection-diffusion equation dissipation-free numerical methods should be applied if only possible. As it was shown previously, these methods provide accurate solutions for the Peclet number not greater than 2. Therefore, if the computation can be carried out with low value of the Peclet number, then the best choice is the application of a non-dissipative scheme as the Crank-Nicolson one. In opposite case, when the dissipative numerical method is applied to solve the advection-diffusion equation, we have to remember that in the solution a superposition of the physical and numerical diffusion will take place. We must be able to separate the effects of both processes. This question is particularly important if the value of coefficient of physical diffusion is determined by optimization methods in which the observed and computed functions at the downstream end are compared. In such a situation one can obtain a false value of searched coefficient of physical diffusion. Let us discuss this question using the results of the presented below examples of solution.

Example 7.4 The advection-diffusion equation (7.3) is solved for a straight prismatic channel in which the water flows at constant velocity U. The following initial-boundary conditions are specified:

– at $t = 0$:

$$f(x,t=0) = \exp\left(-\frac{(x-\mu)^2}{2\sigma^2}\right) \quad \text{for} \quad x \geq 0,$$

where μ is mean value, whereas σ is standard deviation.
– at $x = 0$ and at $x = L$ is assumed: $f(x = 0, t) = 0, \quad f(x = L, t) = 0$ for $t \geq 0$.

The equation is solved using the difference scheme (7.7) with $\eta = 0.5$, which generates the numerical diffusion characterized by the following coefficient (Eq. 7.12):

$$D_n = \frac{U \cdot \Delta x}{2} \cdot (2\theta - 1) \cdot C_a \tag{7.28}$$

Note, that for $\theta = 0.5$ Eq. (7.7) becomes the Crank-Nicolson scheme, whereas for $\theta = 1$ we obtain the fully implicit scheme.

Equally spaced grid is assumed. The computations are performed for the following set of data: $L = 15$ km, $U = 0.5$ m/s, $\Delta x = 50$ m, $\Delta t = 50$ s, $\mu = 1400$ m, $\sigma = 264$ m and $D = 12.5$ m^2/s. The obtained results are displayed in Fig. 7.6.

The Crank-Nicolson scheme ($\theta = 0.5$) does not generate numerical diffusion – from Eq. (7.28) one obtains $D_n = 0$. However, the physical diffusion is so strong, that it is capable to ensure smooth solution. Remember that the Crank-Nicolson scheme provides highly accurate solution for $P \leq 2$. In the considered case this restriction is respected.

If we increase the value of the weighting parameter, let's say assuming $\theta = 1$, then we obtain more damped solution. This is because of the additional numerical diffusion generated by the method. For the assumed data Eq. (7.28) gives

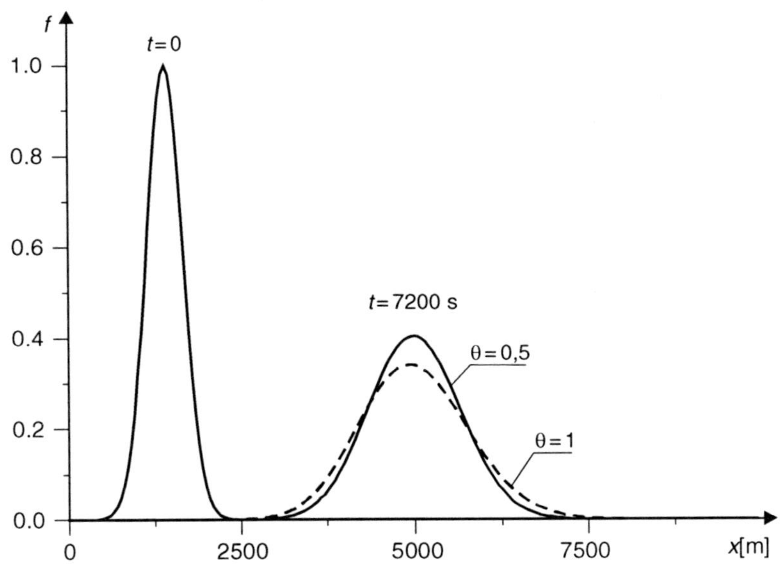

Fig. 7.6 Numerical solution of the advection-diffusion equation using the difference scheme (7.7) for various values of the weighting parameter θ

$D_n = 6.25$ m^2/s. In other words, while solving the advection-diffusion equation (7.3) using the dissipative scheme, as a matter of fact we obtain a solution, which coincides with the exact solution of the following equation:

$$\frac{\partial f}{\partial t} + U\frac{\partial f}{\partial x} - (D + D_n)\frac{\partial^2 f}{\partial x^2} = 0 \tag{7.29}$$

If we apply a dissipative method to solve the advection-diffusion equation, then as results form Eq. (7.29), we always reinforce the diffusive transport and consequently we increase smoothing.

While the numerical dispersion generated by the applied scheme can be identified relatively easily owing to the oscillating solution, identification of the numerical diffusion needs some additional analysis. Note that the lack of evaluation of the numerical diffusion can lead to wrong conclusions on the strength of physical diffusion, since in the solution we observe common effect of both natural and artificial dissipative processes. We must be aware that the numerical diffusion generated by the standard numerical methods for typically used mesh dimensions Δx and Δt, can be of the order similar to the physical one, i.e. $D_n \approx D$. This conclusion is illustrated in the example presented below.

Example 7.5 In straight channel of length L, having constant bed slope and constant cross-sections, the water flows with constant velocity U. Assume that initially at $t = 0$, the concentration along channel axis is equal to zero: $f(x, t = 0) = 0$ for $0 \leq x \leq L$. At the upstream end $x = 0$ the following boundary condition is imposed:

$$f(x = 0,t) = \begin{cases} f_m \frac{t}{t_m} & \text{for } 0 \leq t \leq t_m \\ f_m\left(2 - \frac{t}{t_m}\right) & \text{for } t_m \leq t \leq 2t_m \\ 0 & \text{for } t > 2t_m \end{cases}$$

This means that a triangular distribution of concentration with peak f_m occurring at t_m is assumed. At the downstream end $x = L$ the Neumann boundary condition $\partial f/\partial x|_{x=L} = 0$ is imposed.

The advection-diffusion equation is solved using the difference scheme (7.7) with $\eta = 0.5$ for the following data: $L = 5{,}000$ m, $\Delta x = 50$ m, $U = 0.5$ m/s, $f_m = 1$ and $t_m = 1{,}200$ s. Computations were performed for the two following sets of parameters:

(1) $\theta = 1$, $C_a = 1$, $D = 5$ m^2/s;
(2) $\theta = 1$, $C_a = 1.4$, $D = 0$ m^2/s.

The results of calculation obtained for these sets of data are shown in Fig. 7.7. They represent the graph of f in time at position $x = 2{,}500$ m.

As one can see, the results are practically identical for two different sets of data. This fact can be easily explained. Let us calculate the coefficient of numerical diffusion according to Eq. (7.28) for the first set of data:

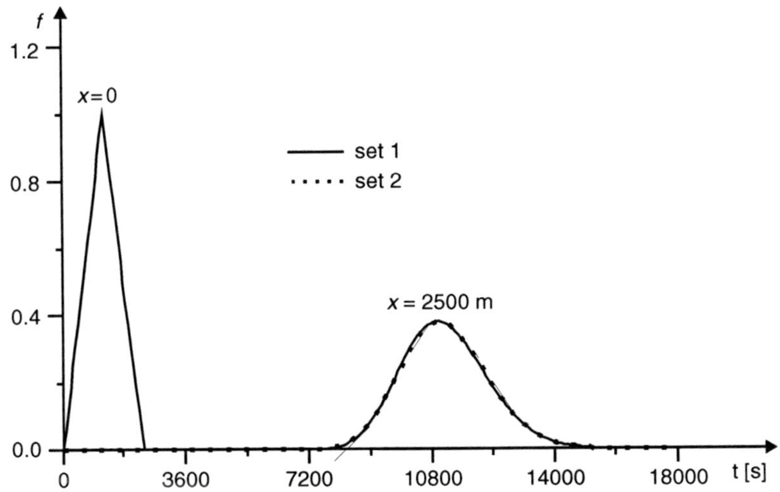

Fig. 7.7 Equivalent effect of the physical and numerical diffusion

$$D_n = \frac{0.5 \cdot 50}{2}(2 \cdot 1 - 1) \cdot 1 = 12.5\,\mathrm{m}^2/\mathrm{s}$$

This means that the diffusive transport was artificially increased and the obtained solution coincides with the exact solution of the advection-diffusion equation produced by a non-dissipative method for the following value of diffusion coefficient: $D + D_n = 5 + 12.5\ \mathrm{m^2/s} = 17.5\ \mathrm{m^2/s}$.

In the second set of data the physical diffusion is neglected ($D = 0$), whereas the values of θ and C_a are such, that the scheme generates numerical diffusion with $D_n = 17.5\ \mathrm{m^2/s}$. Obviously, in this case the results are the same as previously. Actually, each set of data which satisfies the condition: $D + D_n = 17.5\ \mathrm{m^2/s}$ will provide the same numerical solution.

In such a situation one can expect problems while trying to identify the real value of the coefficient D from the results of field experiments, applying the optimization approach. Since one can obtain the same results for practically infinite number of combinations of the values of parameters D, θ and C_a, then the determination of the extreme point of the objective function may be impossible. This problem may be solved on condition that we are capable to separate the effect of numerical diffusion from the effect of physical diffusion in the solution. As we showed previously, this can be done via the analysis of the modified equation only.

7.2.4 The QUICKEST Scheme

One of the most frequently applied finite difference approaches is the QUICKEST scheme (Quadratic Upstream Interpolation for Convective Kinematics with

Estimated Streaming Terms). This method approximates the advection-diffusion equation with accuracy of 3rd order.

The QUICKEST scheme was at first proposed by Leonard (1979). To approximate the advective term apart from the grid points the intermediate points $j-1/2$ and $j+1/2$ are used. At these point the values of function f are determined via quadratic interpolation based on the nodes $j-2$, $j-1$ and j in the first case, and $j-1$, j and $j+1$ in second case, respectively.

An equivalent scheme was derived by Basco (Abbott and Basco 1989), who introduced appropriate difference operators approximating the terms of the advection-diffusion equation. This variant, being more often used, will be presented below.

Let us approximate the equation of advection-diffusion with explicit centred difference scheme, including the truncation error. One obtains:

$$\begin{aligned}
&\frac{f_j^{n+1}-f_j^n}{\Delta t}+U\frac{f_{j+1}^n-f_{j-1}^n}{2\Delta x}=D\frac{f_{j+1}^n-2f_j^n+f_{j-1}^n}{\Delta x^2}+\\
&\quad+\left[\frac{\Delta t}{2}\frac{\partial^2 f}{\partial t^2}+\frac{\Delta t^2}{3!}\frac{\partial^3 f}{\partial t^3}+\frac{\Delta t^3}{4!}\frac{\partial^4 f}{\partial t^4}+\dots\right]_j^n+\\
&\quad+U\left[\frac{\Delta x^2}{3!}\frac{\partial^3 f}{\partial x^3}+\frac{\Delta x^4}{5!}\frac{\partial^5 f}{\partial x^5}+\dots\right]_j^n+\\
&\quad-D\left[\frac{2\Delta x^2}{4!}\frac{\partial^4 f}{\partial x^4}+\frac{2\Delta x^4}{6!}\frac{\partial^6 f}{\partial x^6}+\dots\right]_j^n.
\end{aligned} \tag{7.30}$$

Equation (7. 30) is simplified by:

- neglecting all terms with spatial derivatives of order higher than 3,
- replacing all derivatives with regard to time by the spatial derivatives using the expression resulting from advection-diffusion equation (7.3).

Taking into account these assumptions Eq. (7.30) is rearranged to the following form:

$$\begin{aligned}
f_j^{n+1}&=f_j^n-\frac{C_d}{2}\left(f_{j+1}^n-f_{j-1}^n\right)+C_d\left(f_{j+1}^n-2f_j^n+f_{j-1}^n\right)+\\
&+C_a\left[\frac{C_a}{2}\Delta x^2\frac{\partial^2 f}{\partial x^2}+\frac{\Delta x^3}{6}\left(1-C_a^2-6C_d\right)\frac{\partial^3 f}{\partial x^3}\right]_j^n.
\end{aligned} \tag{7.31}$$

The differential operators are approximated by the formulas:

$$\left.\frac{\partial^2 f}{\partial x^2}\right|_j^n\approx\frac{f_{j+1}^n-2f_j^n+f_{j-1}^n}{\Delta x^2}, \tag{7.32}$$

$$\left.\frac{\partial^3 f}{\partial x^3}\right|_j^n \approx \left.\frac{\partial^3 f}{\partial x^3}\right|_{j-\frac{1}{2}}^n = \frac{f_{j+1}^n - 3_j^n + 3f_{j-1}^n - f_{j-2}^n}{\Delta x^3}. \tag{7.33}$$

Substitution of Eqs. (7.32) and (7.33) in Eq. (7. 31) yields the final form of the scheme:

$$\begin{aligned} f_j^{n+1} = & f_j^n + \left[C_d\left(1-C_a\right) - \tfrac{C_a}{6}\left(C_a^2 - 3C_a + 2\right)\right] f_{j+1}^n + \\ & - \left[C_d\left(2-3C_a\right) - \tfrac{C_a}{2}\left(C_a^2 - 2C_a - 1\right)\right] f_j^n \\ & + \left[C_d\left(1-3C_a\right) - \tfrac{C_a}{2}\left(C_a^2 - C_a - 2\right)\right] f_{j-1}^n \\ & + \left[C_d \cdot C_a + \tfrac{C_a}{6}\left(C_a^2 - 1\right)\right] f_{j-2}^n. \end{aligned} \tag{7.34}$$

As in the formula for node j the values of f coming from preceding nodes are involved, the method is not self-starting.

Note that Eq. (7.34) with $U = 0$ ($C_a = 0$) coincides with the explicit difference scheme for diffusion equation (Eq. (7.7) with $\theta = 0$), whereas for $D = 0$ and $C_a = 1$ one obtains the well known formula $f_j^{n+1} = f_{j-1}^n$, coinciding with the upwind scheme for the advection equation (Eq. 5.108) with $C_a = 1$). Leonard (1979) showed that the region of stability of method (7.34) varies according to the relation between C_a and C_d.

7.3 Solution Using the Modified Finite Element Method

Let us reconsider the advection-diffusion equation (Eq. 7.3) but this time with variable advection velocity $U = U(x)$ and constant coefficient of diffusion $D = \text{const}$. Of course all discussed methods can be developed for variable flow velocity. However, as we will see soon, the modified element method allows taking into account $U(x)$ in a natural way.

To solve Eq. (7.3) the modified Galerkin finite element method, presented in the Section 6.4.3, is applied. For some particular choices of the weighting parameters this approach reduces to the well known finite element or finite difference methods. Assume that the segment of a channel is divided with M nodes into $M-1$ elements of length Δx_j ($j = 1,2, \ldots, M-1$). The discretized solution domain is shown in Fig. 5.19.

Accordingly to the Galerkin procedure the numerical solution must satisfy the condition (5.118). For Eq. (7.3) with previously proposed modification this condition takes the following form (Szymkiewicz 1993):

$$\sum_{j=1}^{M-1} \int_{x_j}^{x_{j+1}} \left(\frac{\partial f_c}{\partial t} + U_c \frac{\partial f_a}{\partial x} - D \frac{\partial^2 f_a}{\partial x^2} \right) \mathbf{N}(x) \cdot dx = 0, \tag{7.35}$$

where $\mathbf{N}(x)$ is the vector of shape functions having components given by Eq. (5.125). In Eq. (7.35) the subscript c denotes approximation of function according to formula (6.91), whereas a denotes the approximation of the function according to formula (6.88).

Let us calculate a single component of the above sum, corresponding to the element j. Since in this element only two components of the vector $\mathbf{N}(x)$, i.e. $N_j(x)$ and $N_{j+1}(x)$, are non-zero, then in Eq. (7.35) only two following non-zero products will occur:

$$I^{(j)} = \int_{x_j}^{x_{j+1}} \left(\frac{\partial f_c}{\partial t} + U_c \frac{\partial f_a}{\partial x} - D \frac{\partial^2 f_a}{\partial x^2} \right) N_j(x) \cdot dx, \tag{7.36a}$$

$$I^{(j+1)} = \int_{x_j}^{x_{j+1}} \left(\frac{\partial f_c}{\partial t} + U_c \frac{\partial f_a}{\partial x} - D \frac{\partial^2 f_a}{\partial x^2} \right) N_{j+1}(x) \cdot dx, \tag{7.36b}$$

Calculation of the integrals (7.36) is carried out step by step as it was shown previously:

- for the time variation term in Section 6.4.3.2
- for the advective term in Sections 6.4.1 and 6.4.3.2,
- for the diffusive term in Section 5.3.2 (Eqs. 5.136 and 5.139).

Finally, the integral (7.36a) gives:

$$\begin{aligned} I^{(j)} = {} & \omega \frac{\Delta x_j}{2} \frac{df_j}{dt} + (1-\omega) \frac{\Delta x_j}{2} \frac{df_{j+1}}{dt} + \\ & + \frac{(\omega \cdot U_j + (1-\omega) U_{j+1})}{2} (-f_j + f_{j+1}) - \frac{D}{\Delta x_j} (-f_j + f_{j+1}) + D \left. \frac{df_j}{dx} \right|_j \end{aligned} \tag{7.37a}$$

whereas the integral (7.36b) gives:

$$\begin{aligned} I^{(j+1)} = {} & (1-\omega) \frac{\Delta x_j}{2} f_j + \omega \frac{\Delta x_j}{2} f_{j+1} + \\ & + \frac{((1-\omega) U_j + \omega \cdot U_{j+1})}{2} (-f_j + f_{j+1}) + \frac{D}{\Delta x_j} (-f_j + f_{j+1}) \ - D \left. \frac{df}{dx} \right|_{j+1} \end{aligned} \tag{7.37b}$$

where ω is the weighting parameter ranging from 0 to 1.

Equations similar to (7.37a) and (7.37b) are obtained for all elements ($j = 1, 2, \ldots, M-1$). According to Eq. (7.35) they should be assembled leading to the following global system of ordinary differential equations:

- for $j = 1$

$$\begin{aligned} & \omega \frac{\Delta x_j}{2} \frac{df_j}{dt} + (1-\omega) \frac{\Delta x_j}{2} \frac{df_{j+1}}{dt} + \frac{(\omega \cdot U_j + (1-\omega) U_{j+1})}{2} (-f_j + f_{j+1}) \\ & - \frac{D}{\Delta x_j} (-f_j + f_{j+1}) + D \left. \frac{df_j}{dx} \right|_j \end{aligned} \tag{7.38a}$$

– for $j = 2, 3, \ldots, M-1$

$$(1-\omega)\frac{\Delta x_{j-1}}{2}\frac{df_{j-1}}{dt} + \omega\left(\frac{\Delta x_{j-1}}{2} + \frac{\Delta x_j}{2}\right)\frac{df_j}{dt} + (1-\omega)\frac{\Delta x_j}{2}\frac{df_{j+1}}{dt} +$$
$$+\tfrac{(1-\omega)U_{j-1}+\omega\cdot U_j}{2}(-f_{j-1}+f_j) + \tfrac{\omega\cdot U_j+(1-\omega)U_{j+1}}{2}(-f_j+f_{j+1}) +$$
$$+\tfrac{D}{\Delta x_{j-1}}(-f_{j-1}+f_j) - \tfrac{D}{\Delta x_j}(-f_j+f_{j+1}) = 0 \tag{7.38b}$$

– for $j = M$

$$(1-\omega)\frac{\Delta x_{j-1}}{2}f_{j-1} + \omega\frac{\Delta x_{j-1}}{2}f_j +$$
$$+\tfrac{((1-\omega)U_{j-1}+\omega\cdot U_j)}{2}(-f_{j-1}+f_j) + \tfrac{D}{\Delta x_{j-1}}(-f_{j-1}+f_j) - D\left.\tfrac{df}{dx}\right|_j. \tag{7.38c}$$

As the Dirichlet boundary conditions at the upstream and downstream ends are imposed, the fluxes at these endpoints disappear. Consequently Eq. (7.38) can be rewritten in the following matrix notation:

$$\mathbf{A}\frac{d\mathbf{f}}{dt} + (\mathbf{B}+\mathbf{C})\,\mathbf{f} = 0 \tag{7.39}$$

where:

A– constant three-diagonals matrix given by the time variable term,
B– variable three-diagonals matrix given by the advective term,
C– constant three-diagonals matrix given by the diffusion term,
$\mathbf{f} = (f_1, f_2, \ldots, f_M)^T$ – vector of unknowns set up from nodal values of f,
$\frac{d\mathbf{f}}{dt} = \left(\frac{df_1}{dt}, \frac{df_2}{dt}, \ldots, \frac{df_M}{dt}\right)^T$ – vector of time derivatives,
T– symbol of transposition.

All matrices are of dimensions of $(2M) \times (2M)$.

The initial-value problem is solved using the two levels method (3.71) described in Section 3.2. This method used for Eq. (7.39) yields the following system of algebraic equations:

$$(\mathbf{A} + \Delta t\cdot\theta\cdot(\mathbf{B}_{n+1}+\mathbf{C}))\,\mathbf{f}_{n+1} = (\mathbf{A} - \Delta t(1-\theta)(\mathbf{B}_n+\mathbf{C}))\,\mathbf{f}_n, \tag{7.40}$$

where:

n – index of time level,
θ – weighting parameter,
Δt – time step.

After introducing the imposed boundary conditions this system is solved using an appropriate method giving approximated value of f at next time level.

Example 7.6 In a straight channel of length L having constant bed slope and constant cross-sections, the water flows with constant velocity. Assume that initially at $t = 0$, the concentration along channel axis is equal to zero: $f(x, t = 0) = 0$ for $0 \leq x \leq L$. At the upstream end $x = 0$ the following boundary condition is imposed:

$$f(x = 0,t) = \begin{cases} 0 & \text{for } t \leq 0 \\ 1 & \text{for } t > 0 \end{cases}$$

At the downstream end $x = L$ the Neumann boundary condition

$$\left.\frac{\partial f}{\partial x}\right|_{x=L} = 0$$

is assumed. Therefore this example deals with the propagation of the sharp front of concentration caused simultaneously by advection and diffusion (without the source terms). In this case the advective-diffusive transport equation has the analytical solution given by (7.25).

The following data are assumed: $L = 5000$ m, $U = 0.5$ m/s, $\Delta x = 100$ m, $\Delta t = 100$ s, which gives $C_a = 0.5$, whereas the coefficient of diffusion is $D = 10$ m^2/s. It means that in this case the Peclet number is relatively high ($P = 40$). This indicates that advection dominates in the transport process.

In the Fig. 7.8 a comparison between analytical and numerical solution is presented. The numerical solution is obtained with the following values of the weighting parameters: $\theta = 0.5$ and $\omega = 0.625$ (calculated using Eq. (6.107)). Very

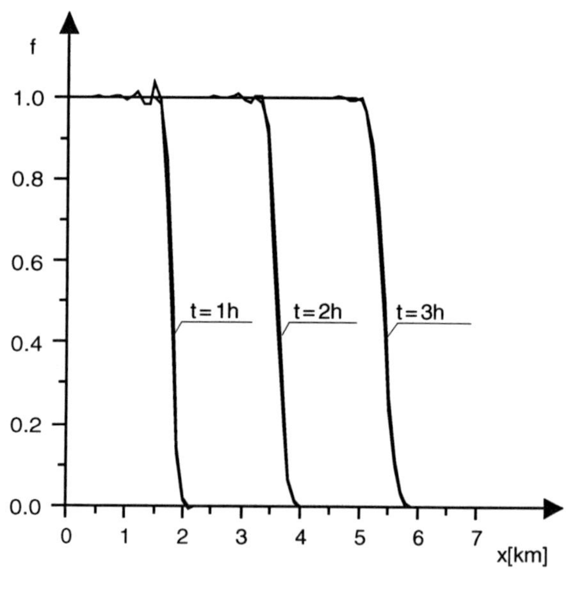

Fig. 7.8 Analytical and numerical solution of the advection-diffusion equation obtained for various time with $P = 40$

small differences between the solutions confirm good properties of the proposed method. In spite of large value of the Peclet number only insignificant wiggles, quickly disappearing with time, occurred.

7.4 Solution of the Advection-Diffusion Equation with the Splitting Technique

The splitting technique (also called factorized steps or decomposition method) is not a numerical method of solution of the partial differential equations. It should be rather considered as an approach that simplifies a complex problem by its decomposition. The concept of such approach is to split in each time step the governing equation or the system of equations into a sequence of simpler problems. Next, the obtained equations are solved using the appropriate numerical methods. The splitting process can be carried out in different ways. For example a 2D or 3D problem can be split with regard to the independent spatial variables in order to be solved as a sequence of 1D problems. The governing equation can be decomposed with regard to time as well. An example of such approach is well known Alternating Direction Implicit (ADI) method commonly used to solve 2D diffusion equation in which to reach solution at new time level we ought to perform the calculations in two stages with a half time step at each one (Abbott and Basco 1989). At every stage 1D diffusion equations are solved. Such approach leads to simpler and more efficient algorithms, which save the computational time and memory.

The decomposition can be also performed with regard to the physical processes present in the governing equation. For instance in the case of advection-diffusion equation we deal with a superposition of two physically different processes: advection and diffusion. Therefore this equation can be split in two equations representing the advective part and the diffusive part respectively.

Theoretical background of the factorized steps technique was given by Yanenko (1971). A comprehensive and up-to-date information on the splitting technique for hyperbolic problems is given by LeVeque (2002). The method and its implementation is presented by Cunge et al. (1980). Many examples of its implementation for solving both hydrodynamics and transport equations were published in numerous papers as well. In this section we will limit our consideration to the problem of 1D advection-diffusion transport equation.

Let us consider the advection-diffusion equation with the source term (1.157):

$$\frac{\partial f}{\partial t} + U\frac{\partial f}{\partial x} - D\frac{\partial^{2} f}{\partial x^2} + \delta = 0. \tag{7.41}$$

This equation can be rewritten as follows:

$$\frac{\partial f}{\partial t} = F \tag{7.42}$$

where the term F contains all terms of Eq. (7.41) except the time derivative:

$$F = -U\frac{\partial f}{\partial x} + D\frac{\partial^2 f}{\partial x^2} - \delta. \tag{7.43}$$

Equation (7.42) integrated over a single time step $\langle t, t + \Delta t \rangle$ gives:

$$f_{t+\Delta t} = f_t + \Delta t \int_t^{t+\Delta t} F \cdot dt, \tag{7.44}$$

where f_t and $f_{t+\Delta t}$ denote the values of function f at the time levels t and $t+ \Delta t$ respectively. Numerical methods differ in the way of approximation of the differential operators in F and in the quadrature method used to calculate the integral in Eq. (7.44).

From Eq. (7.43) results that the new variable F can be presented as the sum of three components representing the advective and diffusive transport as well as the source term:

$$F = F^{(1)} + F^{(2)} + F^{(3)}, \tag{7.45}$$

where:

$$F^{(1)} = -U\frac{\partial f}{\partial x}, \qquad F^{(2)} = D\frac{\partial^2 f}{\partial x^2}, \quad F^{(3)} = -\delta.$$

Substitution of Eq. (7.45) in Eq. (7.44) yields:

$$\begin{aligned} f_{t+\Delta t} &= f_t + \Delta t \int_t^{t+\Delta t} (F^{(1)} + F^{(2)} + F^{(3)})dt = \\ &= f_t + \Delta t \int_t^{t+\Delta t} F^{(1)}dt + \Delta t \int_t^{t+\Delta t} F^{(2)}dt + \Delta t \int_t^{t+\Delta t} F^{(3)}dt \end{aligned} \tag{7.46}$$

Let us exclude the first two terms of the right side hand and denote them as:

$$f^{(1)}_{t+\Delta t} = f_t + \Delta t \int_t^{t+\Delta t} F^{(1)}dt. \tag{7.47}$$

Then Eq. (7.46) will take a more compact form:

$$f_{t+\Delta t} = f^{(1)}_{t+\Delta t} + \Delta t \int_t^{t+\Delta t} F^{(2)}dt + \Delta t \int_t^{t+\Delta t} F^{(3)}dt. \tag{7.48}$$

The same can be applied to Eq. (7.48). Introducing a new variable:

$$f_{t+\Delta t}^{(2)} = f_{t+\Delta t}^{(1)} + \Delta t \int_{t}^{t+\Delta t} F^{(2)} dt \tag{7.49}$$

allows us to rewrite Eq. (7.48) as follows:

$$f_{t+\Delta t}^{(3)} = f_{t+\Delta t}^{(2)} + \Delta t \int_{t}^{t+\Delta t} F^{(3)} dt. \tag{7.50}$$

As it was afore-mentioned, in each time step Δt the solution of the advection-diffusion equation (7.41) is performed in three stages. In the first stage the pure advection equation (7.47) is solved. Next, the diffusion equation (7.49) is solved using the results obtained in the first stage. As the last stage of the calculation, Eq. (7.50) representing the source term is solved. This algorithm can be written in a more general form. Solution of Eq. (7.41) in the time interval $\langle t, t+\Delta t\rangle$ is obtained as a results of the following sequential solutions:

$$\frac{\partial f^{(1)}}{\partial t} = F^{(1)} \text{ with the initial condition } f_t^{(1)} = f_t, \tag{7.51a}$$

$$\frac{\partial f^{(2)}}{\partial t} = F^{(2)} \text{ with the initial condition } f_t^{(2)} = f_{t+\Delta t}^{(1)}. \tag{7.51b}$$

$$\frac{\partial f^{(3)}}{\partial t} = F^{(3)} \text{ with the initial condition } f_t^{(3)} = f_{t+\Delta t}^{(2)}. \tag{7.51c}$$

The searched value of the function f at the time level $t + \Delta t$ is given as follows:

$$f_{t+\Delta t} = f_{t+\Delta t}^{(3)}. \tag{7.52}$$

Note that in each time step the result obtained from the preceding stage are used as initial condition for the next one.

The splitting process can introduce an error into solution. As shows LeVeque (2002) this error does not exist for linear problems, whereas it occurs for nonlinear equations. Consequently, for linear problems the order of computational stages preformed in a single time step does not matter.

The advantage of the splitting technique is the possibility of using different numerical methods for solving each part of the transport equation. To solve the pure advection equation the method of characteristics with appropriate technique of interpolation is very often used, whereas the diffusion equation can be solved using a number of well known numerical methods. For instance Cunge et al. (1980) suggest the Holly-Preissmann approach for solution of the advective part (see Section 6.5.4) and an explicit difference scheme for the diffusive part of the governing equation. Good results can be also achieved with other methods (Szymkiewicz 1995).

For illustration of this approach the equation of advection-diffusion transport with a source term describing the decay of the transported matter is solved. If we assume the steady state and the source term in the simplest form $\delta = \kappa \cdot f$ (κ is constant of decay), then the governing equation (7.41) is reduced to the following ordinary differential equation:

$$U\frac{df}{dx} - D\frac{d^2f}{dx^2} + \kappa \cdot f = 0. \tag{7.53}$$

For constant value f_0 imposed at $x = 0$ Eq. (7.53) has the exact solution (Elliot and James 1984):

$$f(x) = f_0 \cdot \exp\left(\frac{U \cdot x}{2D}\left(1 - \left(1 + 4\frac{K \cdot D}{U^2}\right)^{1/2}\right)\right) \text{ for } x > 0 \tag{7.54}$$

where:

f_0 – concentration imposed at the upstream end,
κ – constant of decay.

In the example presented below we solve Eq. (7.41) using the splitting technique for such initial and boundary conditions which ensures the solution equivalent to Eq. (7.54).

Example 7.7 In a straight channel of length L in which the water flows with constant velocity U let us solve the advection-diffusion transport equation (7.41):

$$\frac{\partial f}{\partial t} + U\frac{\partial f}{\partial x} - D\frac{\partial^2 f}{\partial x^2} + \delta = 0. \tag{7.55}$$

Assume that initially at $t = 0$, the concentration along channel axis is equal to zero: $f(x, t = 0) = 0$ for $0 \leq x \leq L$. At the upstream end $x = 0$ the following boundary condition is imposed:

$$f(x = 0,t) = \begin{cases} 0 & \text{for} \quad t \leq 0 \\ 1 & \text{for} \quad t > 0. \end{cases}$$

Therefore at $x = 0$ a jump of function f is imposed. At the downstream end $x = L$ the Neumann boundary condition $(\partial f/\partial x)_{x=L} = 0$ is assumed.

The following numerical methods are applied to solve Eq. (7.55):

- the method of characteristics with interpolation via spline functions of 3rd degree for the advection term,
- the Crank-Nicolson difference scheme for the diffusion term,
- the implicit trapezoidal rule for the source term.

Equation (7.55) is solved for the following data: $L = 12$ km, $U = 0{,}5$ m/s, $x = 100$ m, $t = 100$ s, $D = 10$ m^2/s and $\kappa = 0.0004$ s^{-1}. Since the function imposed at the upstream end does not vary in time, the solution tends to the steady state. For

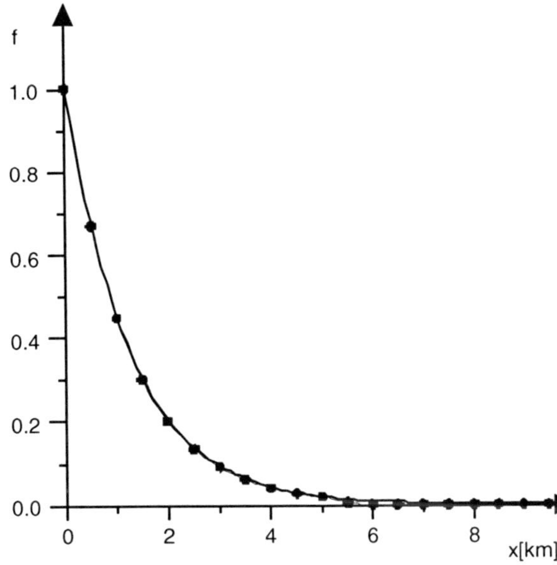

Fig. 7.9 Analytical (*solid line*) and numerical (*dotted line*) solutions of the advection-diffusion equation with source term after $t = 3$ h

the assumed data it is reached after $t = 3$ h. The function $f(x)$ corresponding to this state is shown in Fig. 7.9. For comparison the exact solution (7.54) of Eq. (7.53) is plotted as well. It can be seen that very good agreement between both solutions is obtained.

Example 7.8 Assuming that dissolved mater is fully mixed in the river cross section, the evaluation of the biochemical oxygen demand (*BOD*) and the dissolved oxygen (*DO*) in flowing stream can be described by the system of two 1D equations (Elliot and James 1984):

$$\frac{\partial B}{\partial t} + U \frac{\partial B}{\partial x} - \frac{1}{A} \frac{\partial}{\partial x} \left(D \cdot A \frac{\partial B}{\partial x} \right) + (K_1 + K_3)\, B = 0 \tag{7.56}$$

$$\frac{\partial C}{\partial t} + U \frac{\partial C}{\partial x} - \frac{1}{A} \frac{\partial}{\partial x} \left(D \cdot A \frac{\partial C}{\partial x} \right) - K_2 (C_s - C) + K_1 \cdot B + \phi = 0 \tag{7.57}$$

where:

t – time [s],
x – position [m],
B – *BOD* concentration of [mgl^{-1}],
C – concentration of *DO* of [mgl^{-1}]],
U – average velocity of water flow [$\mathrm{m\ s}^{-1}$],
E – coefficient of longitudinal dispersion [$\mathrm{m}^2\ \mathrm{s}^{-1}$],
A – wetted cross-sectional area [m^2]
K_1 – the *BOD* reaction rate [s^{-1}],
K_2 – the reaeration rate coefficient [s^{-1}],
K_3 – the rate coefficient for removal of *BOD* by sedimentation and adsorption [s^{-1}],

C_s – the saturated dissolved oxygen concentration [mgl^{-1}],
ϕ– the net removal rate of dissolved oxygen for all processes other then biochemical oxidation.

One can add that finding appropriate formulas for all biochemical processes represented in the source terms in these equations is a complex task and their mathematical description is still subject of research. We neglect this problem here since the source terms rather do not complicate the numerical solution of advective-diffusive transport equation.

The required auxiliary conditions are formulated as follows:

- the initial conditions: for $t = 0$ $B(x, t) = B_i(x)$ and $C(x, t) = C_i(x)$ for $0 \leq x \leq L$ are given,
- the boundary conditions:
  at the upstream end $B(x = 0, t) = B_0(t)$ and $C(x = 0, t) = C_0(t)$ for $t \geq 0$ are imposed
  at the downstream end $(\partial B/\partial x)_{x=L} = \phi_B(t)$ and $(\partial C/\partial x)_{x=L} = \phi_C(t)$ for $t \geq 0$ are imposed.

Solving numerically the initial-boundary problem for Eqs. (7.56) and (7.57), the distribution of the dissolved oxygen and the biochemical oxygen demand along considered channel reach is obtained. In the considered example it is assumed that initially $BOD(x, t = 0) = 0$ and $DO(x, t = 0) = 5\,\text{mg dm}^{-1}$. The time constant load causes that at point $x = 0.8\,\text{km}$ the BOD increases immediately to $20\,\text{mg dm}^{-1}$. The computations were performed for $K_1 = 1.25 \cdot 10^{-4}\,\text{s}^{-1}$, $K_2 = 5 \cdot 10^{-6}\,\text{s}^{-1}$, $K_3 = 5 \cdot 10^{-4}\,\text{s}^{-1}$, $C_s = 5\,\text{mg l}^{-1}$, $D = 10\,\text{m}^2/\text{s}$, $\Delta t = 100\,\text{s}$, $L = 15\,\text{km}$, $\Delta x = 100$ m and $U = 0.25$ m/s. The results obtained for $t = 10\,\text{h}$ are presented in Fig. 7.10.

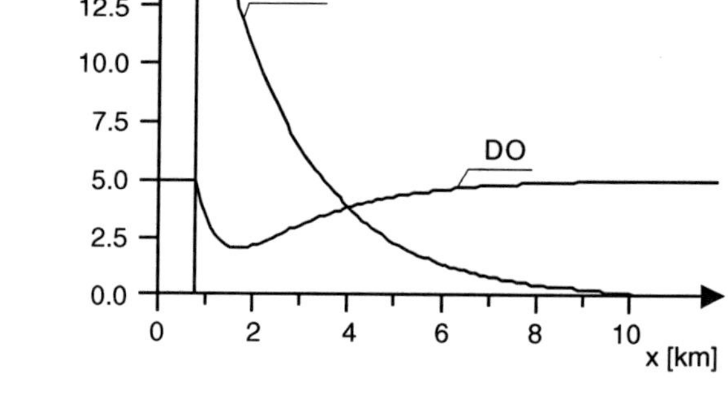

Fig. 7.10 Steady distribution of BOD and DO along channel downstream of the point of release the pollutant obtained via solution of Eqs. (7.56) and (7.57)

One can notice that the computed distributions $BOD(x)$ and $DO(x)$ are typical for the flow downstream of the point of discharging when the waste water is released with a load constant in time.

7.5 Solution of the Advection-Diffusion Equation Using the Splitting Technique and the Convolution Integral

7.5.1 Governing Equation and Splitting Technique

In certain cases, when the flow in open channel is steady and uniform, 1D linear advective-diffusive equation describing the transport of pollutants can be solved without the approximation of derivatives. In this approach one can use the exact solution of the advective-diffusive equation with constant coefficients obtained for the upstream boundary condition in form of the Dirac delta function and the downstream end tending to infinity. Therefore instead of a system of algebraic equations given by the finite difference or element method, an integral of convolution must be calculated numerically. For this purpose any quadrature method can be used. The error dependent on the accuracy of applied method is much smaller than the one caused by the truncation of Taylor series. Consequently the proposed approach is capable of producing highly accurate solution, because it does not generate any numerical dissipation or dispersion (Szymkiewicz and Weinerowska 2005).

In Chapter 1 the following equation describing transport of a passive substance dissolved in water in open channel was derived (Eq. 1.153):

$$\frac{\partial(A\cdot f)}{\partial t}+\frac{\partial(Q\cdot f)}{\partial x}-\frac{\partial}{\partial x}\left(D\cdot A\frac{\partial f}{\partial x}\right)-\delta=0 \tag{7.58}$$

where:

t – time,
x – spatial coordinate,
f – concentration,
D – coefficient of longitudinal dispersion,
A – cross-sectional area,
Q – discharge,
δ – source term.

Equation (7.58) is solved in domain: $0\le x\le L$ and $t\ge 0$ (L – length of channel reach). After developing the derivatives in the two first terms and taking into account the continuity equation (1.80) one obtains:

$$\frac{\partial f}{\partial t}+\frac{Q}{A}\frac{\partial f}{\partial x}-\frac{1}{A}\frac{\partial}{\partial x}\left(D\cdot A\frac{\partial f}{\partial x}\right)+\frac{q}{A}f-\delta=0 \tag{7.59}$$

where q is lateral inflow. This equation can be rewritten as follows:

$$\frac{\partial f}{\partial t} + U\frac{\partial f}{\partial x} - \frac{1}{A}\frac{\partial}{\partial x}\left(D \cdot A\frac{\partial f}{\partial x}\right) - \varphi = 0 \tag{7.60}$$

where:

$U = Q/A$ – cross sectional average velocity,
$\varphi = \delta - q \cdot f/A$ – modified source term.

To solve the above equation an approach based on the splitting technique described in preceding section is applied. Equation (7.60) can be rewritten in the form

$$\frac{\partial f}{\partial t} + F^{(1)} + F^{(2)} = 0 \tag{7.61}$$

where

$$F^{(1)} = U\frac{\partial f}{\partial x} - \frac{1}{A}\frac{\partial}{\partial x}\left(D \cdot A\frac{\partial f}{\partial x}\right) \text{ and } F^{(2)} = -\varphi$$

At every time step Δt Eq. (7.60) is solved in two stages. In the first stage the advective-diffusive transport is considered:

$$\frac{\partial f^{(1)}}{\partial t} + U\frac{\partial f^{(1)}}{\partial x} - \frac{1}{A}\frac{\partial}{\partial x}\left(D \cdot A\frac{\partial f^{(1)}}{\partial x}\right) = 0 \tag{7.62}$$

with the initial condition $f^{(1)}(t) = f(t)$. In the second stage the source term is taken into account:

$$\frac{\partial f^{(2)}}{\partial t} = \varphi \tag{7.63}$$

with the initial condition $f^{(2)}(t) = f^{(1)}(t + \Delta t)$. Finally one obtains: $f(t + \Delta t) = f^{(2)}(t + \Delta t)$. The above equations are solved using the convolution approach and the finite difference method, respectively.

7.5.2 Solution of the Advective-Diffusive Equation by Convolution Approach

Let us consider an advective-diffusive transport equation in the form:

$$\frac{\partial f}{\partial t} + U\frac{\partial f}{\partial x} - D\frac{\partial^2 f}{\partial x^2} = 0 \tag{7.64}$$

with constant coefficients (U = const., D = const.). For the initial condition $f(x, t = 0) = 0$ and the boundary conditions $f(x = 0, t) = \delta(t), f(x \to \infty, t) = 0$ for $t \geq 0$ the following exact solution is obtained (Eagleson 1970):

$$f(x,t) = \frac{1}{(4\pi \cdot D)^{1/2}} \frac{x}{t^{3/2}} \exp\left(-\frac{(U \cdot t - x)^2}{4D \cdot t}\right) \tag{7.65}$$

It holds for $t > 0$ and $x > 0$. The function $\delta(t)$ imposed at the upstream end represents the Dirac delta function (McQuarrie 2003).

A river reach bounded by cross-sections $x = 0$ and $x = x_1$ is considered as dynamic linear and time invariant system. Therefore it can be described by a convolution integral:

$$f_1(t) = \int_0^P w(\tau) f_0(t - \tau) d\tau \tag{7.66}$$

where P is memory of the system. Equation (7.66) represents the way of transformation of function $f_0(t)$ into $f_1(t)$ (Fig. 7.11). More information on the Dirac delta function and the convolution integral is given in Section 9.5.

Equation (7.66) allows us to calculate the function $f_1(t)$ at any point $x_1 > 0$ for any function $f_0(t)$ imposed at upstream boundary $x = 0$, on condition that the function $w(t)$ is given.

Equation (7.66) says, that an output at time t is determined by an input taken from time interval $\langle t - P, t\rangle$ because for $t \geq P$ the function $w(t)$ insignificantly differs to zero. In this case for $x = x_1$ Eq. (7.65) becomes an impulse response function for the considered river reach:

$$w(t) = \frac{1}{(4\pi \cdot D)^{1/2}} \frac{x_1}{t^{3/2}} \exp\left(-\frac{(U \cdot t - x_1)^2}{4D \cdot t}\right) \tag{7.67}$$

Application of Eq. (7.66) will be successful on condition that the mass balance is satisfied. This requirement implies the following condition:

$$\int_0^P w(t) dt = 1 \tag{7.68}$$

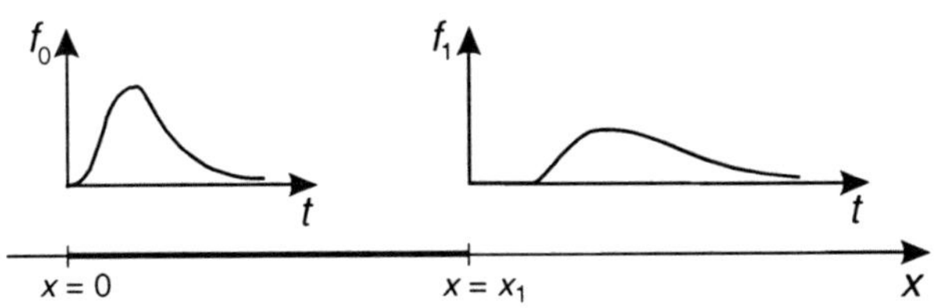

Fig. 7.11 A channel reach of length x_1 as a system transforming $f_0(t)$ into $f_1(t)$

As the convolution integral can be calculated numerically using any quadrature method (Press et al. 1992), this relation can be used to determine suitable value of system memory and time step $\Delta\tau$ accepted to compute the integral in Eq. (7.66). We will come back to the Dirac delta function and the convolution technique in Chapter 9, while discussing the linear lumped flood routing models.

Let us apply the described approach to solve the simplest case of advective-diffusive transport for a steady uniform flow in straight and long channel.

Example 7.9 In open channel a steady uniform flow takes place. The flow velocity U is constant. The initial concentration is equal to zero, whereas at $x = 0$ the following function $f_0(t)$ is assumed:

$$f_0(t) = \begin{cases} 0 \text{ for} & t < 600 \text{ s} \\ f_m \text{ for } 600\,\text{s} & \leq t \leq 1800 \text{ s} \\ 0 \text{ for} & t > 1800 \text{ s} \end{cases}$$

It means that a rectangular distribution of concentration is imposed. It is a very challenging test for the method of solution.

Assuming $U = 0.5$ m s^{-1} compute the functions $f_i(t)$ ($j = 1, 2$) at the points $x_1 = 1{,}500$ m and $x_2 = 3{,}000$ m for two values of diffusion coefficient: $D = 0.0005$ m^2 s^{-1} and 0.5 m^2 s^{-1}.

The obtained solutions presented in Fig. 7.12 are close to the exact one. One can suppose that the error resulting from trapezoidal rule used to integrate a convolution is not significant. The calculations were carried out for $\Delta\tau = 2$ s. The Peclet number (Eq. 7.1) for the data used here is equal to $(U|D)\Delta x$=1,000Δx. It means a total domination of advection in the transport process, even for the smallest value of Δx applied in practice.

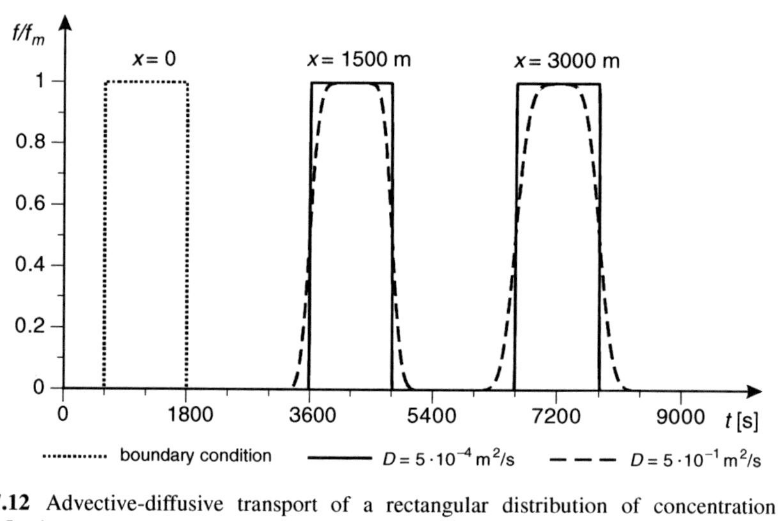

Fig. 7.12 Advective-diffusive transport of a rectangular distribution of concentration with $U = 0.5$ m/s

7.5.3 Solution of the Advective-Diffusive Equation with Variable Parameters and Without Source Term

The example of solution presented above concerned a steady uniform flow. As a matter of fact, steady flow in a natural stream is spatially varied. The velocity and cross-sectional areas vary along x axis. Such case of transport is described by Eq. (7.60). Neglecting the source term it can be rewritten as:

$$\frac{\partial f}{\partial t} + U\frac{\partial f}{\partial x} - \frac{1}{A}\frac{\partial}{\partial x}\left(D \cdot A\frac{\partial f}{\partial x}\right) = 0 \tag{7.69}$$

where:

$U = U(x)$ – cross-sectional average velocity,
$A = A(x)$ – area of wetted cross-section.

Both functions $U(x)$ and $A(x)$ can be obtained from the solution of the system of equations for steady gradually varied flow described in Chapter 4.

The convolution approach, which holds for steady uniform flow, can be extended to solve the transport equation with variable coefficients, i.e. for steady gradually varied flow. To this end the diffusive term in Eq. (7.69) is differentiated:

$$\frac{\partial f}{\partial t} + \left(U - \frac{\partial D}{\partial x} - \frac{D}{A}\frac{\partial A}{\partial x}\right)\frac{\partial f}{\partial x} - D\frac{\partial^2 f}{\partial x^2} = 0 \tag{7.70}$$

In this equation the advective velocity and the coefficient of diffusion vary in space. To solve Eq. (7.70) using the presented method, one can "freeze" locally the velocity and the diffusivity. For this purpose a channel of length L is divided into M intervals of length Δx_i as in the finite difference or finite element method (Fig. 7.13a).

Each channel reach is considered as a dynamic system in which both the advective velocity and the coefficient of diffusion are assumed constant. Let us introduce new variables defined as follows:

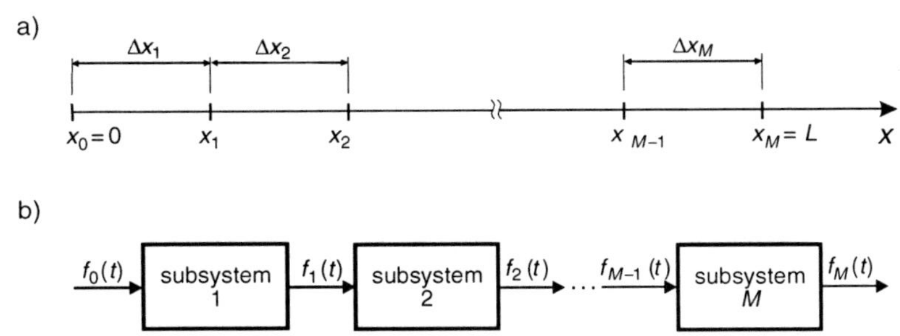

Fig. 7.13 A channel divided into intervals of length Δx_i (**a**) represented by M subsystems (**b**)

$$u_j = \frac{1}{2}\left[\left(U - \frac{\partial D}{\partial x} - \frac{D}{A}\frac{\partial A}{\partial x}\right)_{j-1} + \left(U - \frac{\partial D}{\partial x} - \frac{D}{A}\frac{\partial A}{\partial x}\right)_j\right] \approx$$
$$\approx \frac{1}{2}\left[\left(U_{j-1} - \frac{D_j - D_{j-1}}{\Delta x_j} - \frac{D_{j-1}}{A_{j-1}}\frac{A_j - A_{j-1}}{\Delta x_j}\right) + \left(U_j - \frac{D_j - D_{j-1}}{\Delta x_j} - \frac{D_j}{A_j}\frac{A_j - A_{j-1}}{\Delta x_j}\right)\right] \tag{7.71}$$

$$d_j = \frac{1}{2}(D_{j-1} + D_j) \tag{7.72}$$

where:

u_j – modified average advective velocity between nodes $j-1$ and j,
d_j – average coefficient of diffusion between nodes $j-1$ and j.

For each segment $j = 1, 2, \ldots, M$ one obtains the following impulse response:

$$w_j(t) = \frac{1}{(4\pi \cdot d_j)^{1/2}} \frac{\Delta x_j}{t^{3/2}} \exp\left(-\frac{(u_j \cdot t - \Delta x_j)^2}{4d_j \cdot t}\right) \tag{7.73}$$

Consequently the function $f_j(t)$ for consecutive nodes ($j = 1, 2, 3, \ldots$) can be calculated as follows:

$$f_j(t) = \int_o^{P_j} w_j(\tau) f_{j-1}(t - \tau) d\tau \tag{7.74}$$

where P_j is memory of system j. Therefore a channel of length L has been divided into M subsystems of length Δx_j in series, where the output from preceding reach is the input for next one (Fig. 7.13b). This procedure is illustrated by solution of Eq. (7.69) for steady gradually varied flow.

Example 7.10 In a trapezoidal channel of width $B = 10$ m, bank slope $1 : 1.5$, bed slope $s = 0.0005$, the Manning coefficient $n_M = 0.025$ the discharge rate is $Q = 18.487$ m^3 s^{-1}. The normal depth corresponding to this value of Q is $H_n = 1.5$ m. At $x = 0$ the water level was raised by a dam to $H_o = 5$ m (Fig. 7.14). For these data the flow profile $h(x)$ and the flow velocity $U(x)$ are computed with $\Delta x = -300$ m $=$ const. for 25 nodes using approach presented in Chapter 4. The calculated depths vary from 5 m at $x = 0$ to 1.502 m at $x = -7{,}200$ m. The corresponding cross-sectional areas vary from 87.5 m^2 to 18.43 m^2 and velocities from 0.091 to 0.43 m s^{-1} respectively.

Next at the upstream end ($x = -7{,}200$ m) the rectangular distribution of concentration as in Example 7.9, is specified. The function $f_{25}(t)$ calculated at the section of a dam for constant coefficient of diffusion $D = 0.0075$ m^2 s^{-1} is presented in Fig. 7.15.

The initial rectangular distribution of function f travels along x axis with insignificant deformation. It results from varied velocity $U(x)$ and small value of D. Due to the lack of numerical dissipation and dispersion the obtained solution is free of oscillations and it keeps very strong gradients. The error of solution can be caused

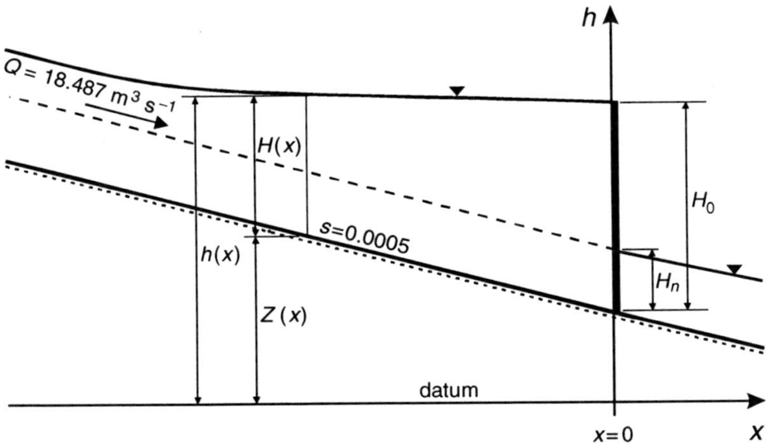

Fig. 7.14 A steady varied flow in a channel

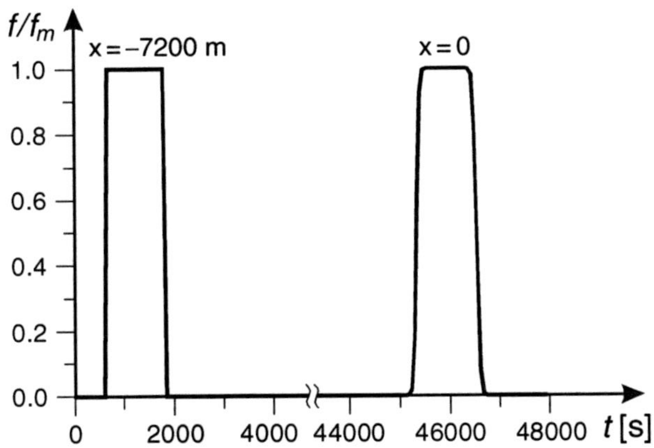

Fig. 7.15 Advective-diffusive transport with varied flow velocity $U = U(x)$

by numerical calculation of the convolution and by averaging of the flow velocities between the nodes on each intervals Δx_i ($i = 1, 2, \ldots, 25$) only. Because the initial distribution is changed insignificantly and the mass balance is satisfied perfectly, one can suppose that this error is relatively small. Note that for assumed set of data the Peclet number is very high ($P = 17{,}200$) and no other method can ensure the solution of Eq. (7.69) with similar accuracy.

7.5.4 Solution of the Advective-Diffusive Equation with Source Term

The simplest form of transport equation with source term is as follows:

$$\frac{\partial f}{\partial t} + U\frac{\partial f}{\partial x} - D\frac{\partial^2 f}{\partial x^2} - \delta = 0 \tag{7.75}$$

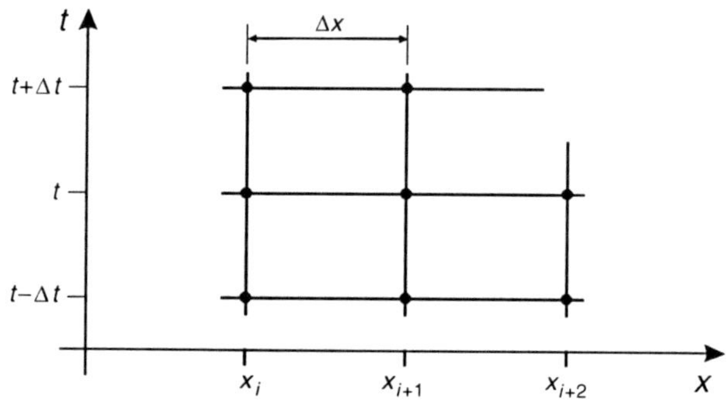

Fig. 7.16 The grid points applied to solve Eq. (7.75)

The domain of solution ($0 \leq x \leq L$ and $t \geq 0$) is covered by grid points as in the finite difference method. The mesh having dimensions $\Delta x \times \Delta t$ is presented in Fig. 7.16.

A constant flow velocity U between cross-section x_j and x_{j+1} is assumed. Let us assume that the concentration f is known at all time levels preceding t. It is known at the points $x_1, x_2, \ldots, x_j$ at time level $t+ \Delta t$ as well. The aim is to calculate the concentration f at point $(x_{j+1}, t+ \Delta t)$. According to Eqs. (7.62) and (7.63) in each time step the following equations should be solved:

$$\frac{\partial f^{((1)}}{\partial t} + U\frac{\partial f^{(1)}}{\partial x} - D\frac{\partial^2 f^{(1)}}{\partial x^2} = 0 \tag{7.76}$$

with $f^{(1)}(t) = f(t)$ and

$$\frac{\partial f^{(2)}}{\partial t} = \delta \tag{7.77}$$

with $f^{(2)}(t) = f^{(1)}(t + \Delta t)$.

In the first stage the value of concentration at node $(x_{i+1}, t + \Delta t)$ resulting from advection and diffusion processes is calculated. For this purpose the following equation is used:

$$f_{j+1}^{(1)}(t + \Delta t) = \int_0^{p_j} w_j(\tau) f_j(t - \tau) d\tau \tag{7.78}$$

This integration can be carried out as previously by the trapezoidal rule with the integration step equal to $\Delta\tau$. For the time between the time levels, $f_j(t - \tau)$ is calculated using linear interpolation.

To cover the distance between the cross-section x_j and x_{j+1} a particle of dissolved matter needs time equal to

$$\Delta T = \frac{\Delta x}{U} \tag{7.79}$$

During this time the pollutant decays and the effect of this process must be taken into account in the second step of calculation. Namely Eq. (7.77) should be solved. To this end the trapezoidal rule is applied:

$$f_{j+1}^{(2)}(t+\Delta t) = f_{j+1}^{(1)}(t+\Delta t) + \frac{\Delta T}{2}(\varphi_{j+1}^{(1)}(t+\Delta t) + \varphi_{j+1}^{(2)}(t+\Delta t)) \tag{7.80}$$

where:

j– index of cross-section,
ΔT– time of particle travelling from cross-section j to $j+1$,
Δt–mesh dimension in t direction.

As the source term is dependent on the concentration f, Eq. (7.80) usually is non-linear. Consequently an iterative method must be applied to solve it.

7.5.5 Solution of the Advective-Diffusive Equation in an Open Channel Network

The convolution approach can be also applied to solve the transport equation in a channel network, with variable coefficients U and D and lateral inflow. Similarly to the steady gradually varied flow in a network, considered in Chapter 4, the transport equation must be discretized for each branch of the network and additional conditions, resulting from conservation principles, must be imposed at junctions. For a junction of two channels showed in Fig. 7.17a they are:

$$f_k = \frac{f_i\,Q_i + f_j\,Q_j}{Q_i + Q_j}, \tag{7.81}$$

whereas for a bifurcation of a channel as in Fig. 7.17b, the following relations are valid:

$$f_k = f_i \text{ and} \tag{7.82a}$$

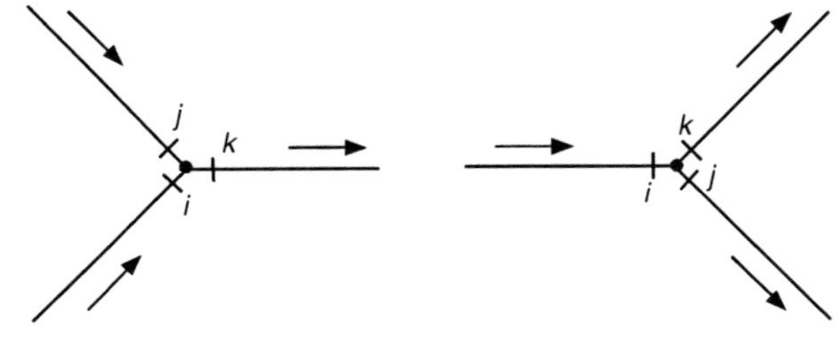

Fig. 7.17 Junction (**a**) and bifurcation (**b**) of the channels (*arrows* indicate positive flow direction)

$$f_k = f_j \tag{7.82b}$$

Equations (7.81), (7.82a) and (7.82b) allow us to solve the advective-diffusive transport equation for the entire network.

Example 7.11 The SGVF is considered in the looped channel network showed in Fig. 7.18. It consists of 10 branches having trapezoidal cross-sections. The characteristics of the considered network are following:

- bed width 5.0 m;
- side slope 1:1.5;
- bed slope 0.0005;
- Manning coefficient $n_M = 0.035$;
- length: channels no 1,2,3,8,9,10 – $L = 500$ m, channels no 4,5,6,7 – $L = 1{,}000$ m.

Each channel is divided into intervals of constant length $\Delta x = 50$ m. The total number of nodes is equal to 159. The bed elevation at the upstream end (point *a*) is 7,000 m, whereas at the downstream end (point *d*) it is equal to 8,500 m. The boundary conditions are specified in terms of water levels at the upstream end (point *a*) and at the downstream end (point *d*) of network. They are as follows: $H_a = 7{,}750$ m, $H_d = 7{,}500$ m.

The results of calculations for the network shown in Fig. 7.18 are presented in Table 7.1.

Now let us apply the convolution approach to solve the advective-diffusive transport equation for a steady uniform flow in considered channel network. The flow velocity $U(x)$ and cross-sectional areas $A(x)$ are known since they were calculated previously using the SGVF equation. The initial concentration is assumed equal to zero along all branches of the network. At the beginning of the channel number 1 i.e. at point *a* (Fig. 7.18), the following function $f_a(t)$ is imposed:

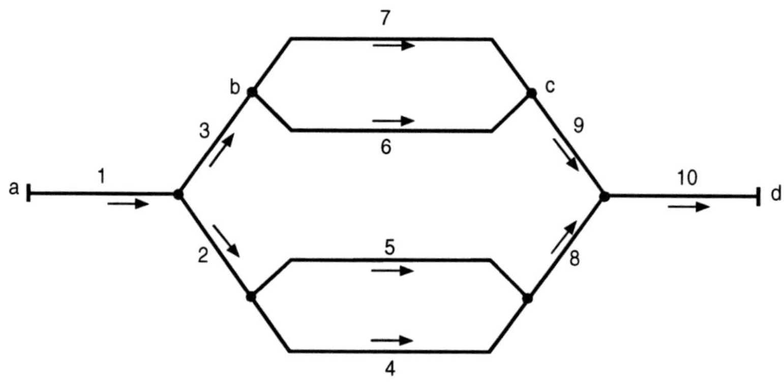

Fig. 7.18 Looped channel network

Table 7. 1 Results of solution for network in Fig. 7.18

Channel	Discharge [m³/s]	Upstream water level [m]	Downstream water level [m]
1	9,706	7,750	7,575
2	4,853	7,575	7,544
3	4,853	7,575	7,544
4	2,427	7,544	7,536
5	2,427	7,544	7,536
6	2,427	7,544	7,536
7	2,427	7,544	7,536
8	4,853	7,536	7,526
9	4,853	7,536	7,526
10	9,706	7,526	7,500

$$f_1(t) = \begin{cases} 0 \text{ for} & t < 0.5\ h \\ F_m \text{ for } 0.5\ h & \leq t \leq\ 1.5\ h \\ 0 \text{ for} & t > 1.5\ h \end{cases}$$

It means that a rectangular distribution of concentration is assumed (Fig. 7.19, point a).

The functions $f(t)$ at the points b, c and d were calculated for the value of diffusion coefficient equal to $D = 0.00005\ \text{m}^2\ \text{s}^{-1}$. The source term in the following simplest formula is used:

$$\delta = \kappa \cdot f \tag{7.83}$$

where $\kappa = 0.00005\ \text{s}^{-1}$ is the constant of decay.

In Fig. 7.19 the time distribution of concentration at the selected points of the network displayed in Fig. 7.18 is presented. The travelling rectangular distribution of concentration is simultaneously subjected to smoothing caused by diffusion and reduction of its height caused by the source term. Since the value of diffusion coefficient is very low, one can expect that imposed at the upstream end the rectangular

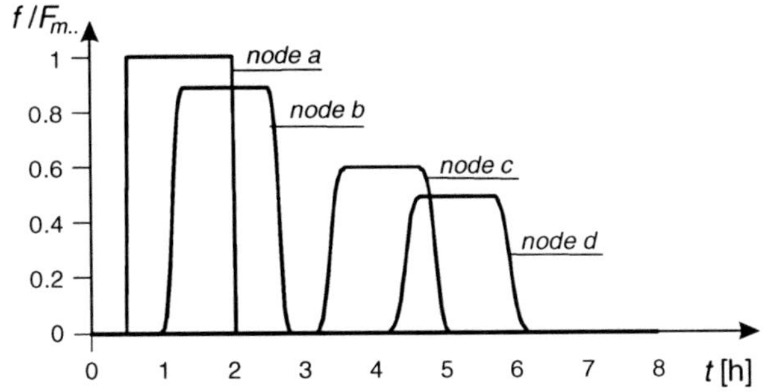

Fig. 7.19 Advective-diffusive transport of rectangular distribution of concentration in the looped network at its selected points computed with $D = 0.00005\ \text{m}^2/\text{s}$ and $\kappa = 0.00005\ \text{s}^{-1}$

distribution of concentration will be smoothed insignificantly. Indeed, in the concentration distributions calculated at points *b*, *c* and *d* one can notice, that only the corners are rounded off although the imposed distribution of concentration travels more than 6 h. It can be also seen that the error generated by the implicit trapezoidal method used to integrate a convolution is very low. The calculations were carried out for $\Delta\tau = 0.1$ s and $\Delta t = 100$ s. In the considered example the Peclet number (Eq. 7.2) is greater than 750,000. It indicates that advection dominates in the transport process. Note that in the calculated distributions no oscillations, typical for the finite difference or finite element method, are observed.

The obtained solution presented in Fig. 7.19 seems to be close to the exact one, since in this case the source term has the linear form. Consequently the splitting technique, applied for solution of the transport equation, does not generate any additional error.

References

Abbott MB, Basco DR (1989) Computational fluid dynamics. Longman Scientific and Technical, New York

Chanson H (2004) The hydraulics of open channel flow: An introduction, 2nd edn. Elsevier, Oxford

Cunge J, Holly FM, Verwey A (1980) Practical aspects of computational river hydraulics. Pitman Publishing, London

Eagleson PS (1970) Dynamic hydrology. McGraw-Hill, New York

Elliot DI, James A (1984) Models of water quality in rivers. In: James A (ed.) An introduction to water quality modeling. Wiley, New York

Fletcher CAJ (1991) Computational techniques for fluid dynamics, vol. I. Springer-Verlag, Berlin

Leonard BP (1979) A stable and accurate convective modelling procedure based on quadratic upstream interpolation. Comp. meth. in Appl. Mech. and Engng. 19:59–98

LeVeque RJ (2002) Finite volume methods for hyperbolic problems. Cambridge University Press, Cambridge

McQuarrie DA (2003) Mathematical methods for scientists and engineers. University Sci-ence Books, Sausalito

Patankar SV (1980) Numerical head transfer and fluid flow. Hemisphere Publishing Corpo-ration, McGraw-Hill, Washington

Press WH, Teukolsky SA, Veterling WT, Flannery BP (1992). Numerical recipes in C. Cambridge University Press, Cambridge

Szymkiewicz R (1993) Solution of the advection-diffusion equation using the spline function and finite elements. Commun. Numer. Methods Engng. 9 (4):197–206

Szymkiewicz R (1995) Method to solve 1D unsteady transport and flow equations. J. Hydr. Engng. ASCE 121 (5):396–403

Szymkiewicz R, Weinerowska K (2005) Analytical – numerical approach to solve the transport equation for steady gradually varied flow in open channel. Far East J. Appl. Math. 19 (24): 213–228

Yanenko NN (1971) The method of fractional steps. Springer New York, Berlin, Heidelberg.

Chapter 8
Numerical Integration of the System of Saint Venant Equations

8.1 Introduction

While the equations describing unsteady flow in open channels were derived by Barré de Saint Venant as early as in 1871 (Chanson 2004), for a long time they could not be used successfully in engineering practice. The reasons for such situation were both mathematical complexity of the equations and specific properties of open channels. To overcome these difficulties the hydrologists were looking for simpler equations describing unsteady flow. The simplifications of Saint Venant equations led to the well known models of kinematic wave and diffusive wave. These equations require less data and they can be solved using simpler methods, sometimes even analytical ones. It is important to note that both simplified equations are still widely applied in hydraulic engineering.

The equations of unsteady flow, being a system of quasi-linear partial differential hyperbolic equations, need initial and boundary conditions, which in practice can be given in numerical form only. On the other hand the data characterizing the natural open channels are given numerically as well. These circumstances cause that the equations of unsteady flow can be solved only using the numerical methods. Thus it is obvious that wide and effective application of the system of Saint Venant equations in hydraulic engineering was closely connected with the progress in informatics and computer technique. Simply, the numerical approach indispensable in open channel flow modeling was possible because of wide availability of the computers.

At the very beginning the method of characteristics was used for solving the system of Saint Venant equations. Its detailed description for the flows with free surface is given by Abbott (1979). However, because of some difficulties caused by nonlinearity of equations as well as for required non-equally spaced nodes typical for rivers this method was practically gave up. Currently in open channel modeling the finite difference method dominates. During the last 50 years many difference schemes were proposed. For instance Cunge et al. (1980) present the following list of the numerical schemes possible for solving the system of Saint Venant equations: Lax scheme, leap-frog scheme, Abbot-Ionescu scheme, Delf Hydraulic Laboratory scheme, Vasiliev scheme, Gunaratman-Perkins scheme, Preissmann scheme. The

R. Szymkiewicz, *Numerical Modeling in Open Channel Hydraulics*, Water Science and Technology Library 83, DOI 10.1007/978-90-481-3674-2_8,

first two schemes are explicit, the others are implicit. Except the Abbot-Ionecu scheme working on so called staggered grid, all of them use the non-staggered grid. This means that in each node the values of both functions Q and h are calculated.

It is commonly accepted that the most robust scheme is the four point implicit difference scheme or more precisely – the Preissmann scheme (Abbott and Basco 1989, Cunge et al. 1980). Its alternative name is the box scheme. After Cunge et al. (1980) the reasons of its widespread use are follows:

- it works on non-staggered grid, which allows us to calculate both unknowns in the same nodes. This is important in natural rivers.
- it relates the variables coming from neighboring nodes only, what allows us using variable space interval Δx without affecting the accuracy of approximation.
- it ensures approximation of 1st order of accuracy and for particular case of 2nd order.
- it gives exact solution of the linear wave equations for properly chosen values of Δx and Δt, making possible the comparison of exact and numerical solutions.
- it is implicit and absolutely stable so it does not require limiting of the value of time step, whereas the imposed boundary conditions are introduced readily.

Fascinating history of the box scheme is given by Abbott and Basco (1989). Note, that the box scheme was used in Section 6.1 to solve the pure advection equation.

As it was mentioned previously, the dominating method for solving of the unsteady flow equations is the finite difference method. The finite element method, very effective for 2D and 3D flow problems, is rather seldom applied in open channel hydraulics. Some attempts (Cooley and Moin 1976) seemed to suggest that this method has no distinct advantages compared to the finite difference method, which did encourage its application. However, modifying the standard finite element method one can obtain a numerical solver for the Saint Venant equations as effective as the Preissmann scheme (Szymkiewicz 1991, 1995). It has the form of an implicit two level six point scheme, which for properly chosen weighting parameters represents even 3rd order of accuracy (Szymkiewicz 1995).

In the next sections we will present the solution of unsteady flow equations using both mentioned approaches, i.e. the implicit four point difference scheme and the modified finite element method.

8.2 Solution of the Saint Venant Equations Using the Box Scheme

8.2.1 Approximation of Equations

Consider the system of Saint Venant equations in the form of Eqs. (1.87) and (1.88):

$$\frac{\partial Q}{\partial t} + \frac{\partial}{\partial x}\left(\frac{\beta \cdot Q^2}{A}\right) + g \cdot A \frac{\partial h}{\partial x} + \frac{g \cdot n_M^2}{R^{4/3}} \frac{|Q|\, Q}{A} = 0, \tag{8.1}$$

$$\frac{\partial h}{\partial t} + \frac{1}{B}\frac{\partial Q}{\partial x} = \frac{q}{B}. \tag{8.2}$$

This system is solved for $0 \le x \le L$ and $t \ge 0$. Let us assume subcritical flow in the considered channel reach. In such a case the following initial and boundary conditions must be prescribed: $h(x,t=0) = h_i(t)$ and $Q(x,t=0) = Q_i(t)$ *for* $0 \le x \le L$, $h(x=0,t) = h_0(t)$ *or* $Q(x=0,t) = Q_0(t)$ and $h(x=L,t) = h_L(t)$ *or* $Q(x = L,t) = Q_L(t)$ for $t \ge 0$.

The solution domain is covered with a grid of dimensions $\Delta x_i \times \Delta t$ ($i = 1, 2, \ldots, M$) as presented in Fig. 4.11. Let us consider a single mesh as in Fig. 8.1, containing 4 nodes $(j-1, n)$, (j, n), $(j-1, n+1)$ and $(j, n+1)$.

Approximation of the derivative is carried out at point P, which is located in the middle of the interval Δx_i. Note that the Preissmann scheme corresponds to the box scheme with $\psi = 0.5$, so point P can move along t axis only, in a way controlled by the weighting parameter θ.

The value of an arbitrary function $f_P\,(x, t)$ at point P is approximated as follows (Cunge et al. 1980):

$$f_P \approx \frac{1}{2}\left(\theta \cdot f_j^{n+1} + (1-\theta)f_j^n\right) + \frac{1}{2}\left(\theta \cdot f_{j+1}^{n+1} + (1-\theta)f_{j+1}^n\right), \tag{8.3}$$

where:

θ – weighting parameter ranging from 0 to 1,
j – index of cross-section,
n – index of time level.

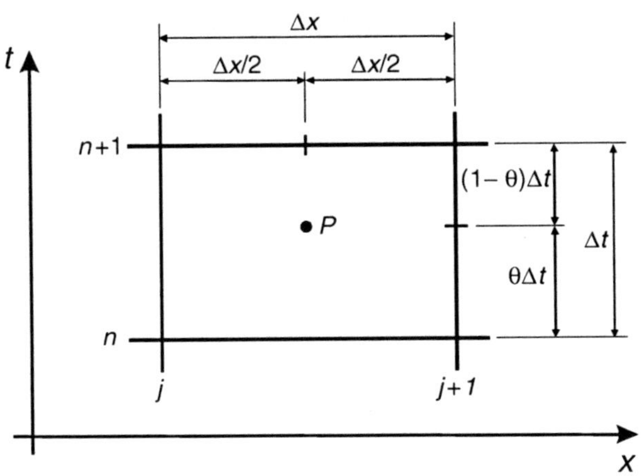

Fig. 8.1 Grid point for the Preissmann scheme

Appropriate approximating formulas for the derivatives coincide with those given by Eqs. (6.2) and (6.3) for $\psi = 0.5$:

$$\left.\frac{\partial f}{\partial t}\right|_p \approx \frac{1}{2}\left(\frac{f_j^{n+1} - f_j^n}{\Delta t} + \frac{f_{j+1}^{n+1} - f_{j+1}^n}{\Delta t}\right), \tag{8.4}$$

$$\left.\frac{\partial f}{\partial x}\right|_p \approx (1-\theta)\frac{f_j^n - f_{j+1}^n}{\Delta x_j} + \theta\frac{f_j^{n+1} - f_{j+1}^{n+1}}{\Delta x_j}, \tag{8.5}$$

Let us apply these formulas to Eqs. (8.1) and (8.2). The dynamic equation becomes:

$$\begin{aligned}
&\frac{1}{2}\frac{Q_j^{n+1} - Q_j^n}{\Delta t} + \frac{1}{2}\frac{Q_{j+1}^{n+1} - Q_{j+1}^n}{\Delta t} + \frac{(1-\theta)}{\Delta x_j}\left(\left(\frac{\beta \cdot Q^2}{A}\right)_{j+1}^n - \left(\frac{\beta \cdot Q^2}{A}\right)_j^n\right) \\
&+ \frac{\theta}{\Delta x_j}\left(\left(\frac{Q^2}{A}\right)_{j+1}^{n+1} - \left(\frac{Q^2}{A}\right)_j^{n+1}\right) + \\
&+ g \cdot A_p\left((1-\theta)\frac{h_{j+1}^n - h_j^n}{\Delta x_j} + \theta\frac{h_{j+1}^{n+1} - h_j^{n+1}}{\Delta x_j}\right) + \left(\frac{g \cdot n_M^2}{R^{4/3}}\frac{|Q|\,Q}{A}\right)_P = 0
\end{aligned} \tag{8.6a}$$

The continuity equation (8.2) is approximated in a similar way:

$$\frac{1}{2}\frac{h_j^{n+1} - h_j^n}{\Delta t} + \frac{1}{2}\frac{h_{j+1}^{n+1} - h_{j+1}^n}{\Delta t} + \frac{1}{B_P}\left((1-\theta)\frac{Q_{j+1}^n - Q_j^n}{\Delta x_j} + \theta\frac{Q_{j+1}^{n+1} - Q_j^{n+1}}{\Delta x_j}\right) = \frac{q_P}{B_P}. \tag{8.6b}$$

In both equations index P means that a function or algebraic expression is approximated accordingly to the formula (8.3). Therefore we have:

$$A_p = \frac{1}{2}\left(\theta \cdot A_j^{n+1} + (1-\theta)A_j^n\right) + \frac{1}{2}\left(\theta \cdot A_{j+1}^{n+1} + (1-\theta)A_{j+1}^n\right), \tag{8.7a}$$

$$q_p = \frac{1}{2}\left(\theta \cdot q_j^{n+1} + (1-\theta)q_j^n\right) + \frac{1}{2}\left(\theta \cdot q_{j+1}^{n+1} + (1-\theta)q_{j+1}^n\right), \tag{8.7b}$$

$$B_p = \frac{1}{2}\left(\theta \cdot B_j^{n+1} + (1-\theta)B_j^n\right) + \frac{1}{2}\left(\theta \cdot B_{j+1}^{n+1} + (1-\theta)B_{j+1}^n\right), \tag{8.7c}$$

$$\begin{aligned}
\left(\frac{g \cdot n_M^2}{R^{4/3}}\frac{|Q|Q}{A}\right)_P = &\frac{1}{2}\left[\theta\left(\frac{g \cdot n_M^2}{R^{4/3}}\frac{|Q|\,Q}{A}\right)_j^{n+1} + (1-\theta)\left(\frac{g \cdot n_M^2}{R^{4/3}}\frac{|Q|Q}{A}\right)_j^n\right] + \\
&+\frac{1}{2}\left[\theta\left(\frac{g \cdot n_M^2}{R^{4/3}}\frac{|Q|\,Q}{A}\right)_{j+1}^{n+1} + (1-\theta)\left(\frac{g \cdot n_M^2}{R^{4/3}}\frac{|Q|\,Q}{A}\right)_{j+1}^n\right] = 0
\end{aligned} \tag{8.7d}$$

In Eqs. (8.6a) and (8.6b) the nodal values of the water levels h_j^{n+1} and h_{j+1}^{n+1} as well as the flow rates Q_j^{n+1} and Q_{j+1}^{n+1} are unknowns. Such a pair of equations can be written for each space interval i.e. for j=1, 2, 3, . . ., $M-1$. Then we obtain a set of 2($M-1$) algebraic equations with 2M unknowns, representing the nodal values of function h and Q. Two boundary conditions allow us to complete the system:

– for node $j = 1$:

$$\delta_0 \cdot h_1^{n+1} + (1-\delta_0)Q_1^{n+1} = \delta_0 \cdot h_0(t_{n+1}) + (1-\delta_0)Q_0(t_{n+1}), \tag{8.8}$$

– for node $j = M$:

$$\delta_L \cdot h_M^{n+1} + (1-\delta_L)Q_M^{n+1} = \delta_L \cdot h_L(t_{n+1}) + (1-\delta_L)Q_L(t_{n+1}). \tag{8.9}$$

In the above equations δ_0 and δ_L are integer numbers that can take the value of 0 or 1. The value of 1 designates the boundary condition imposed in the form of water stage (function $h_0(t)$ or $h_L(t)$), whereas the value of 0 corresponds to the flow discharge (function $Q_0(t)$ or $Q_L(t)$).

Finally, the system of non-linear algebraic equations of dimension ($2M \times 2M$) is obtained. This system can be presented in more compact matrix form:

$$\mathbf{F(X)} = 0, \tag{8.10}$$

where:

$\mathbf{X}=(h_1, Q_1, \ldots, h_j, Q_j, \ldots, h_M, Q_M)^T$ – vector of unknowns,
$\mathbf{F}=(F_1, F_2, \ldots, F_j, F_{j+1}, \ldots, F_{2M-1}, F_{2M})^T$– vector of equations.

The equations of his system are as follows:

$$F_1(h_1^{n+1},Q_1^{n+1}) = \delta \cdot h_1^{n+1} + (1-\delta)Q_1^{n+1} - \delta \cdot h_0(t_{n+1}) - (1-\delta)Q_0(t_{n+1}) = 0,$$
$$\vdots \tag{8.11a}$$

$$\left.\begin{aligned} F_{2j-1}\left(h_j^{n+1}, Q_j^{n+1}, h_{j+1}^{n+1}, Q_{j+1}^{n+1}\right) &= 0 \\ F_{2j}\left(h_j^{n+1}, Q_j^{n+1}, h_{j+1}^{n+1}, Q_{j+1}^{n+1}\right) &= 0 \end{aligned}\right\} \quad j = 1, \ldots, M-1, \tag{8.11b, c}$$
$$\vdots$$

$$F_{2M}\left(h_M^{n+1},Q_M^{n+1}\right) = \delta h_M^{n+1} + (1-\delta)\, Q_M^{n+1} - \delta h_L\,(t_{n+1}) - (1-\delta)\, Q_L\,(t_{n+1}) = 0. \tag{8.11d}$$

The first and the last equation result from the boundary conditions, whereas the other ones represent Eqs. (8.6a) and (8.6b) discretized using the Preissmann scheme.

The system (8.10) must be solved using an iterative method. For this purpose one can use the Newton method presented in Section 2.3.2. The iteration process has the following form:

$$\mathbf{J}^{(k)} \cdot \Delta\,\mathbf{X}^{(k+1)} = -\mathbf{F}^{(k)}, \tag{8.12}$$

where

$\Delta \mathbf{X}^{(k+1)} = \mathbf{X}^{(k+1)} - \mathbf{X}^{(k)}$ – correction vector,
$\mathbf{J}^{(k)}$ – Jacobian matrix,
k – index of iteration.

The Jacobian matrix is given as:

$$\mathbf{J} = \begin{bmatrix} \frac{\partial F_1}{\partial x_1} & \frac{\partial F_1}{\partial x_2} & \cdots & \frac{\partial F_1}{\partial x_i} & \cdots & \frac{\partial F_1}{\partial x_N} \\ \frac{\partial F_2}{\partial x_1} & \frac{\partial F_2}{\partial x_2} & \cdots & \frac{\partial F_2}{\partial x_i} & \cdots & \frac{\partial F_2}{\partial x_N} \\ \vdots & \vdots & & \vdots & & \vdots \\ \frac{\partial F_i}{\partial x_1} & \frac{\partial F_i}{\partial x_2} & \cdots & \frac{\partial F_i}{\partial x_i} & \cdots & \frac{\partial F_i}{\partial x_N} \\ \vdots & \vdots & & \vdots & & \vdots \\ \frac{\partial F_N}{\partial x_1} & \frac{\partial F_N}{\partial x_2} & \cdots & \frac{\partial F_N}{\partial x_i} & \cdots & \frac{\partial F_N}{\partial x_N} \end{bmatrix}, \tag{8.13}$$

where $N = 2M$ is the dimension of considered system of equations.

One can notice that Eqs. (8.6a) and (8.6b) include the unknown values from neighboring nodes only. Therefore in each algebraic equation only 4 unknowns exist. This means that the coefficients related to other unknowns are equal to zero and consequently the Jacobian matrix is banded. Its structure depends on the imposed boundary conditions as shown in Fig. 8.2.

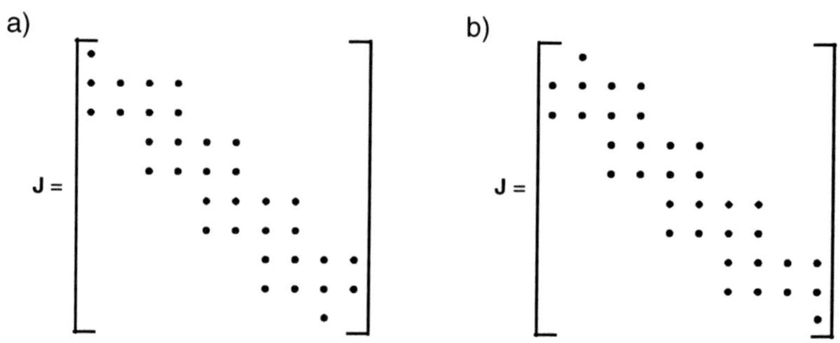

Fig. 8.2 Structure of matrix **J** for the system (8.10) corresponding to $M = 5$ for various boundary conditions: (**a**) $\delta_0 = 1$ and $\delta_L = 0$; (**b**) $\delta_0 = 0$ and $\delta_L = 1$

One can see that the bandwidth of the Jacobian matrix is equal to 5.

The following two issues are related to the numercal solution of the nonlinear system (8.10):

- initial evaluation of solution,
- criterion for convergence.

As the first approximation of searched solution $\mathbf{X}^{(k=0)}$at the time level $n + 1$ usually the solution obtained for previous level n is assumed, i.e.:

$$(Q_j^{n+1})^{(k=0)} = Q_j^n \quad \text{and} \quad (h_j^{n+1})^{(k=0)} = h_j^n \ \text{ for } j = 1, 2, \ldots\ldots, M. \tag{8.14}$$

As a criterion for convergence one can use the following ones:

$$\left|Q_j^{(k+1)} - Q_j^{(k)}\right| \leqslant \varepsilon_Q \quad \text{and} \quad \left|h_j^{(k+1)} - h_j^{(k)}\right| \leqslant \varepsilon_h \quad \text{for} \quad j = 1, 2, \ldots, M, \tag{8.15}$$

where ε_Q is the assumed tolerance for the flow discharge Q and ε_h is the assumed tolerance for the water stage h. Numerical experiments show that usually it is enough to perform $2 \div 3$ iterations to obtain the accuracy satisfying for practical purposes.

The Newton method requires that the system of linear algebraic equations is solved in each iteration. To this purpose a modified Gauss elimination or **LU** decomposition algorithm should be used, which takes advantage of the banded matrix.

The Preissmann scheme involves the weighting parameter θ, which generally ranges from 0 to 1. To recognize the influence of this parameter on the scheme properties a stability analysis is carried out. Since the system of Saint Venant equations is non-linear, then such analysis using Neumann approach is performed for the following linearized system of wave equations:

$$\frac{\partial H}{\partial t} + \overline{H}\frac{\partial U}{\partial x} = 0, \tag{8.16}$$

$$\frac{\partial U}{\partial t} + g\frac{\partial H}{\partial x} = 0, \tag{8.17}$$

where $\overline{H}$ is a constant (average) depth. Such an analysis was carried out previously in Section 6.1.2 for the pure advection equation solved using the box scheme. In this case the stability analysis, although very similar, is slightly more difficult, because it deals with the system of equations. On the other hand, this analysis has been presented by others authors, e.g. Liggett and Cunge (1975), Cunge et al. (1980), Abbott and Basco (1989). For this reason only the final conclusions are presented here. The Preissmann scheme applied for solving the Saint Venant equations is stable for any Courant number defined as:

$$C_r = \frac{(g \cdot \overline{H})^{1/2}\Delta t}{\Delta x}, \tag{8.18}$$

on condition that

$$\theta \geq \frac{1}{2}. \tag{8.19}$$

Since this scheme is stable for any Courant number, relation (8.19) constitutes the condition of absolute stability.

In the afore-mentioned publications one can find analysis of the amplitude and phase errors as well. In next section we will take up an accuracy analysis using the modified equation approach. As we will see this approach can be consider as an alternative way of investigation of the numerical schemes.

8.2.2 Accuracy Analysis Using the Modified Equation Approach

Accuracy analysis is performed for the linear equations (8.16) and (8.17). Approximation of these equations using the Preissmann scheme with uniformly spaced nodes leads to the following algebraic equations:

$$\frac{1}{2}\frac{H_j^{n+1}-H_j^n}{\Delta t}+\frac{1}{2}\frac{H_{j+1}^{n+1}-H_{j+1}^n}{\Delta t}+\overline{H}(1-\theta)\frac{U_{j+1}^n-U_j^n}{\Delta x}+\overline{H}\cdot\theta\frac{U_{j+1}^{n+1}-U_j^{n+1}}{\Delta x}=0, \tag{8.20}$$

$$\frac{1}{2}\frac{U_j^{n+1}-U_j^n}{\Delta t}+\frac{1}{2}\frac{U_{j+1}^{n+1}-U_{j+1}^n}{\Delta t}+g(1-\theta)\frac{H_{j+1}^n-H_j^n}{\Delta x}+g\cdot\theta\frac{H_{j+1}^{n+1}-H_j^{n+1}}{\Delta x}=0. \tag{8.21}$$

Using the Taylor series expansion the nodal values of function H and U are expressed in terms of their values at node $(j+1,\ n+1)$ (Fig. 8.1). Taking into account the terms up to 3rd order one obtains:

$$H_j^{n+1}\approx H_{j+1}^{n+1}-\Delta x\left.\frac{\partial H}{\partial x}\right|_{j+1}^{n+1}+\frac{\Delta x^2}{2}\left.\frac{\partial^2 H}{\partial x^2}\right|_{j+1}^{n+1}-\frac{\Delta x^3}{6}\left.\frac{\partial H}{\partial x^3}\right|_{j+1}^{n+1}, \tag{8.22a}$$

$$\begin{aligned}H_j^n\approx{}&H_{j+1}^{n+1}-\Delta x\left.\frac{\partial H}{\partial x}\right|_{j+1}^{n+1}+\Delta t\left.\frac{\partial H}{\partial t}\right|_{j+1}^{n+1}+\frac{\Delta x^2}{2}\left.\frac{\partial^2 H}{\partial x^2}\right|_{j+1}^{n+1}+\\&+\Delta x\Delta t\left.\frac{\partial^2 H}{\partial x\partial t}\right|_{j+1}^{n+1}+\frac{\Delta t^2}{2}\left.\frac{\partial^2 H}{\partial t^2}\right|_{j+1}^{n+1}-\frac{\Delta x^3}{6}\left.\frac{\partial^3 H}{\partial x^3}\right|_{j+1}^{n+1}-\frac{\Delta t^3}{6}\left.\frac{\partial^3 H}{\partial t^3}\right|_{j+1}^{n+1}+\\&-\frac{\Delta x^2\Delta t}{2}\left.\frac{\partial^3 H}{\partial x^2\partial t}\right|_{j+1}^{n+1}-\frac{\Delta x\Delta t^2}{2}\left.\frac{\partial^3 H}{\partial x\partial t^2}\right|_{j+1}^{n+1}\end{aligned} \tag{8.22b}$$

$$H_{j+1}^{n} \approx H_{j+1}^{n+1} - \Delta t \left.\frac{\partial H}{\partial t}\right|_{j+1}^{n+1} + \frac{\Delta t^2}{2}\left.\frac{\partial^2 H}{\partial t^2}\right|_{j+1}^{n+1} - \frac{\Delta t^3}{6}\left.\frac{\partial^3 H}{\partial t^3}\right|_{j+1}^{n+1}. \tag{8.22c}$$

Similar formulas can be written for the nodal values of function U. Substitution in Eqs. (8.20) and (8.21) and regrouping yields the following modified equations:

$$\begin{aligned}\frac{\partial H}{\partial t} + \overline{H}\frac{\partial U}{\partial x} = {} & \frac{\Delta t}{2}\frac{\partial^2 H}{\partial t^2} - \frac{\Delta t^2}{6}\frac{\partial^3 H}{\partial t^3} + \frac{\Delta x}{2}\frac{\partial^2 H}{\partial x \partial t} + \\ & - \frac{\Delta x^2}{4}\frac{\partial^3 H}{\partial x^2 \partial t} - \frac{\Delta x \Delta t}{4}\frac{\partial^3 H}{\partial x \partial t^2} + \overline{H}\frac{\Delta x}{2}\frac{\partial^2 U}{\partial x^2} - \overline{H}\frac{\Delta x^2}{6}\frac{\partial^3 U}{\partial x^3} + \\ & - \overline{H}(1-\theta)\Delta t\frac{\partial^2 U}{\partial x \partial t} - \overline{H}(1-\theta)\frac{\Delta x \Delta t}{2}\frac{\partial^3 U}{\partial x^2 \partial t} - \overline{H}(1-\theta)\frac{\Delta t^2}{2}\frac{\partial^3 U}{\partial x \partial t^2}\end{aligned} \tag{8.23}$$

$$\begin{aligned}\frac{\partial U}{\partial t} + g\frac{\partial H}{\partial x} = {} & \frac{\Delta t}{2}\frac{\partial^2 U}{\partial t^2} - \frac{\Delta t^2}{6}\frac{\partial^3 U}{\partial t^3} + \frac{\Delta x}{2}\frac{\partial^2 U}{\partial x \partial t} - \frac{\Delta x^2}{4}\frac{\partial^3 U}{\partial x^2 \partial t} + \\ & - \frac{\Delta x \cdot \Delta t}{4}\frac{\partial^3 U}{\partial x \partial t^2} + g\frac{\Delta x}{2}\frac{\partial^2 H}{\partial x^2} - g\frac{\Delta x^2}{6}\frac{\partial^3 H}{\partial x^3} + g(1-\theta)\Delta t\frac{\partial^2 H}{\partial x \partial t} + \\ & - g(1-\theta)\frac{\Delta x \cdot \Delta t}{2}\frac{\partial^3 H}{\partial x^2 \partial t} - g(1-\theta)\frac{\Delta t^2}{2}\frac{\partial^3 H}{\partial x \partial t^2}\end{aligned} \tag{8.24}$$

In these equations written for node $(j+1, n+1)$, the node indices were omitted for simplicity. From wave equations (8.16) and (8.17) the following relations are derived:

$$\frac{\partial^2 H}{\partial x \partial t} = -\overline{H}\frac{\partial^2 U}{\partial x^2}, \tag{8.25}$$

$$\frac{\partial^2 H}{\partial t^2} = g \cdot \overline{H}\frac{\partial^2 H}{\partial x^2}, \tag{8.26}$$

$$\frac{\partial^2 U}{\partial x\, \partial t} = -g\,\frac{\partial^2 H}{\partial x^2}, \tag{8.27}$$

$$\frac{\partial^2 U}{\partial t^2} = g \cdot \overline{H}\frac{\partial^2 U}{\partial x^2}, \tag{8.28}$$

$$\frac{\partial^3 H}{\partial x^2 \partial t} = -\overline{H}\frac{\partial^3 U}{\partial x^3}, \tag{8.29}$$

$$\frac{\partial^3 H}{\partial t^3} = -g \cdot \overline{H}^2\frac{\partial^3 U}{\partial x^3}, \tag{8.30}$$

$$\frac{\partial^3 H}{\partial x \partial t^2} = g \cdot \overline{H} \frac{\partial^3 H}{\partial x^3}, \tag{8.31}$$

$$\frac{\partial^3 U}{\partial x^2 \partial t} = -g \frac{\partial^3 H}{\partial x^3}, \tag{8.32}$$

$$\frac{\partial^3 U}{\partial t^3} = -g^2 \cdot \overline{H} \frac{\partial^3 H}{\partial x^3}, \tag{8.33}$$

$$\frac{\partial^3 U}{\partial x \partial t^2} = g \cdot \overline{H} \frac{\partial^3 U}{\partial x^3}, \tag{8.34}$$

These relations allow us to eliminate the time-derivatives of the order higher than one from Eqs. (8.23) and (8.24) and the modified equations take the final form:

$$\frac{\partial H}{\partial t} + \overline{H} \frac{\partial U}{\partial x} = D_n \frac{\partial^2 H}{\partial x^2} + E_{n1} \frac{\partial^3 H}{\partial x^3} + E_{n2} \frac{\partial^3 U}{\partial x^3} \ldots, \tag{8.35}$$

$$\frac{\partial U}{\partial t} + g \frac{\partial H}{\partial x} = D_n \frac{\partial^2 U}{\partial x^2} + E_{n1} \frac{\partial^3 U}{\partial x^3} + E_{n3} \frac{\partial^3 H}{\partial x^3} + \ldots, \tag{8.36}$$

where the coefficient of numerical diffusion D_n and the coefficients of numerical dispersion E_{n1}, E_{n2} and E_{n3} are given as follows:

$$D_n = \left(\theta - \frac{1}{2}\right) g \cdot \overline{H} \cdot \Delta t, \tag{8.37}$$

$$E_{n1} = \frac{g \cdot \overline{H} \cdot \Delta x \cdot \Delta t}{2} \left(\frac{1}{2} - \theta\right), \tag{8.38}$$

$$E_{n2} = \frac{\overline{H} \cdot \Delta x^2}{6} \left((3\theta - 2)C_r^2 + \frac{1}{2}\right). \tag{8.39}$$

$$E_{n3} = \frac{g \cdot \Delta x^2}{6} \left((3\theta - 2)C_r^2 + \frac{1}{2}\right). \tag{8.40}$$

The Courant number C_r is defined by Eq. (8.18).

The modified equations (8.35) and (8.36) allow us to draw some conclusions on the numerical properties of the Preissmann scheme:

1. This scheme applied for solving the unsteady flow equations leads to the system of algebraic equations which satisfies the consistency condition. We can see that for Δx, $\Delta t \rightarrow 0$ all terms at the right handside of Eqs. (8.35) and (8.36) tend to zero. Consequently in the limit the modified equations at each node of the numerical grid become the original wave equation.
2. The initial-boundary value problem for the modified system of equations limited to the terms of second order (parabolic ones):

$$\frac{\partial H}{\partial t} + \overline{H}\frac{\partial U}{\partial x} = D_n\frac{\partial^2 H}{\partial x^2}, \tag{8.41}$$

$$\frac{\partial U}{\partial t} + g\frac{\partial H}{\partial x} = D_n\frac{\partial^2 U}{\partial x^2}, \tag{8.42}$$

is ill posed for D_n<0 i.e. for $\theta < 1/2$. The solution does not exist, which means that the scheme is unstable. The solution problem is well posed if $D_n \geq 0$. This happens for $\theta \geq 0.5$ only. Therefore in such a case the Preissman scheme is absolutely stable. Since at the same time it is consistent, then for $\theta \geq 0.5$ the scheme is convergent. Note, that this conclusion on stability condition coincides with the conclusion presented previously using the stability analysis.

3. The diffusive term disappears for $\theta = 0.5$. In such a case the scheme becomes an approximation of 2nd order and consequently it is dissipation free. Therefore in the modified equations at its right side hand the terms of 3rd order, connected with dispersion, dominate. Indeed, for $\theta = 0.5$ we obtain $D_n = 0$ and simultaneously

$$E_{n1} = 0, \tag{8.43a}$$

$$E_{n2} = \frac{g \cdot \Delta x^2}{12}(1-C_r^2). \tag{8.43b}$$

Consequently we can expect unphysical oscillations in the numerical solution. Note that for C_r= 1 both dispersive terms disappear and the modified equations becomes the original equation. In such a case the Preissmann scheme is non-dissipative and non-dispersive, producing an exact solution of the wave equations (8.16) and (8.17).

4. The numerical diffusion will occur for $\theta > 0.5$ giving $D_n > 0$. It means that the initial-boundary value problem for Eqs. (8.41) and (8.42) is well posed and its solution exists always. Therefore the scheme is absolutely stable regardless of the assumed mesh dimensions and average depth $\overline{H}$. However this scheme, being an approximation of 1st order, generates the numerical diffusion which will affect the numerical solution. Its magnitude depends on the values of mesh dimensions and the weighting parameter θ. Therefore the scheme is dissipative. It eliminates the oscillations caused by dispersivity, but at the same time it gives rise to unphysical smoothing, especially significant when strong gradients occur in the solution.

These conclusions are confirmed by the following numerical example.

Example 8.1 Assume a straight horizontal channel in which the water stays initially at rest with constant depth $H(x, t) = 0.85$ m. At the boundary $x = 0$ the following function is imposed:

$$H(x = 0, t) = \begin{cases} 0.85 \text{ m} & \text{for} \quad t \leq 0 \\ 0.95 \text{ m} & \text{for} \quad t > 0 \end{cases}$$

This boundary condition generates a jump of depth propagating along the channel. The second boundary condition imposed far away from upstream end (L = 200 m) is: $H(x = L, t) = 0.85$ m. Additionally, it is assumed that $\overline{H} = 0.90$ m and $g = 10$ m/s^2. For these data the wave celerity is $(g\cdot\overline{H})^{1/2} = 3$ m/s. A uniform grid with $\Delta x = 1.50$ m is used. The results of computations carried out for various values of the Courant number and for various values of the weighting parameter θ are displayed in Figs. 8.3, 8.4 and 8.5.

They agree with the conclusions resulting from the accuracy analysis. The Preissmann scheme provides exact solution for C_r=1 and $\theta = 0.5$. For other Courant numbers oscillations appear in both $H(x, t)$ and $U(x, t)$ (Figs. 8.3 and 8.4). Such results could be expected, since for $\theta = 0.5$ and $C_r \neq 1$ the scheme, being dissipation free, becomes dispersive.

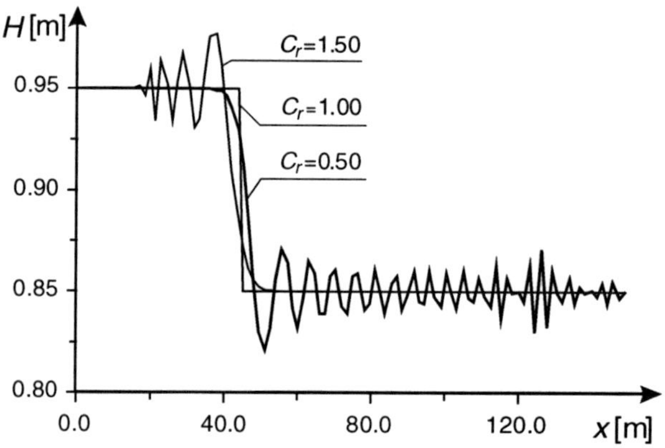

Fig. 8.3 Depth $H(x)$ at $t = 15$ s calculated using the Preissmann scheme for $\theta = 0.50$

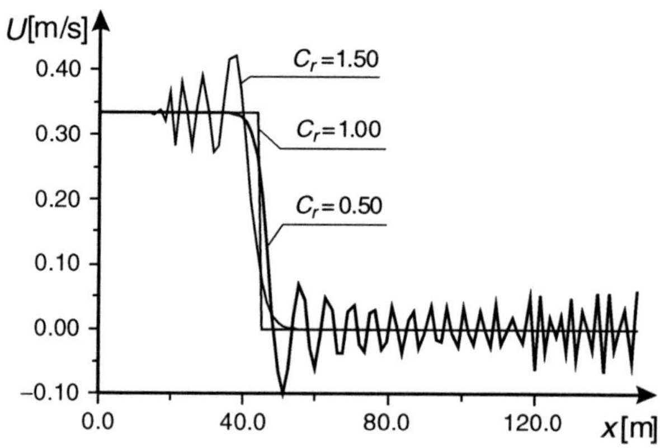

Fig. 8.4 Flow velocity $U(x)$ at $t = 15$ s calculated using the Preissmann scheme for $\theta = 0.50$

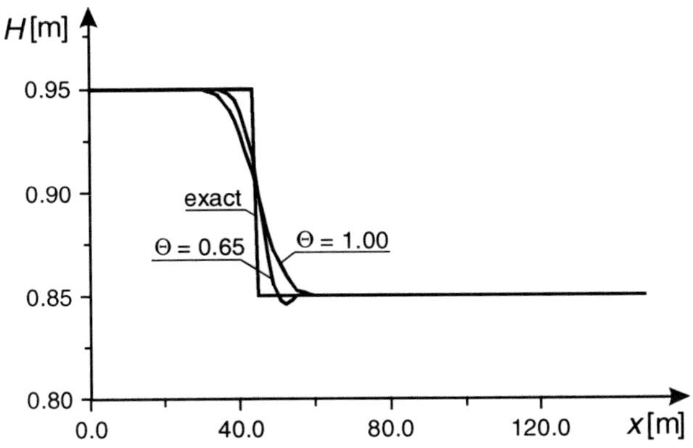

Fig. 8.5 Depth $H(x)$ at $t = 15$ s calculated using the Preissmann scheme for $C_r = 0.5$

To eliminate the effect of dispersivity some numerical diffusion can be introduced into the solution. Indeed, as it is shown in Fig. 8.5, taking $\theta = 0.65$ we significantly reduce the oscillations, whereas $\theta = 1.0$ eliminates them completely. Obviously, due to this diffusion the sharp discontinuity in H and U gradually disappears.

The results of calculations should be rather smooth, however without to strong smoothing. The experiences of many authors (for instance Cunge et al. 1980) suggest that the best results are given by the weighting parameter θ close to 0.67, which ensures smooth solution in the case of gentle flood waves.

8.3 Solution of the Saint Venant Equations Using the Modified Finite Element Method

8.3.1 Spatial and Temporal Discretization of the Saint Venant Equations

Although the finite difference method is dominating in modelling of 1D open channel flow, the finite element method can be applied as well. Moreover, it appears that this method can yield a scheme representing higher accuracy comparing with well known difference schemes. This approach previously used to solve the 1D transport equations in open channel (Szymkiewicz, 1995) can be applied to solve the unsteady flow equations as well. The modification deals with the process of integration and leads to a more general form of algebraic equations approximating the governing equations.

In Section 6.4 the finite element method was used to solve the advection equation. However to obtain satisfying results the standard version of the finite element method had to be modified. This approach yields a six-point implicit scheme with two weighting parameters. Its particular cases are the standard finite element method and the well-known finite difference schemes. An accuracy analysis carried out using the modified equation approach shows that by proper choice of the values of two weighting parameters an accuracy of 3rd order can be obtained. For better understanding the method let us remember some basic information presented previously in Section 6.4.

Assume a channel reach of length L, which using M nodes is divided into $M - 1$ elements. According to the Galerkin procedure, described in details by Zienkiewicz (1972) and previously presented in Section 6.4, solution of considered partial differential equation has to satisfy the following condition:

$$\int_0^L \Phi\left(f_a, \ldots\right) \mathbf{N}(x)\, dx = \sum_{j=1}^{M-1} \int_{x_j}^{x_{j+1}} \Phi\left(f_a, \ldots\right) \mathbf{N}(x)\, dx = 0 \tag{8.44}$$

where:

Φ – symbolic representation of solved equation,
f_a – approximation of any function $f(x, t)$ occurring in equation,
$\mathbf{N}(x)$ – vector of linear basis functions,
L – length of channel reach,
j – index of node,
M – number of grid points.

For the system of Saint Venant equations the condition (8.44) reads:

$$\sum_{j=1}^{M-1} \int_{x_j}^{x_{j+1}} \left(\frac{\partial Q}{\partial t} + \frac{\partial}{\partial x}\left(\frac{\beta \cdot Q^2}{A}\right) + g \cdot A \frac{\partial h}{\partial x} + g \cdot (n_M)^2 \frac{|Q|\, Q}{R^{4/3} \cdot A} \right) \mathbf{N}(x)\, dx = 0, \tag{8.45a}$$

$$\sum_{j=1}^{M-1} \int_{x_j}^{x_{j+1}} \left(\frac{\partial h}{\partial t} + \frac{1}{B}\frac{\partial Q}{\partial x} - \frac{q}{B} \right) \mathbf{N}(x)\, dx = 0. \tag{8.45b}$$

In the standard approach the function $f(x, t)$ is approximated as follows:

$$f_a(x,t) = \sum_{j=1}^{M} N_j(x) f_j(t), \tag{8.46}$$

where $f_j(t)$ represents nodal value of function $f(x, t)$. When linear basis functions are applied in each element only two following integrals in Eq. (8.44) exist:

$$\int_{x_j}^{x_{j+1}} f_a(x,t) N_j(x) dx = \left(\frac{2}{3} f_j(t) + \frac{1}{3} f_{j+1}(t)\right) \frac{\Delta x_j}{2}, \tag{8.47a}$$

$$\int_{x_j}^{x_{j+1}} f_a(x,t) N_{j+1}(x) dx = \left(\frac{1}{3} f_j(t) + \frac{2}{3} f_{j+1}(t)\right) \frac{\Delta x_j}{2}, \tag{8.47b}$$

where Δx_j is the distance between nodes. Integration of Eqs. (8.45a) and (8.45b) term by term gives a system of ordinary differential equations over time. This procedure has been described in details for the advection equation in Section 6.4. As we remember, such approach appears ineffective. The same is for the Saint Venant equations as well. The standard version of the finite element method does not ensure satisfactory results. Spatial oscillations of the type "$2\Delta x$" are usually observed in the water level and discharge. In Fig. 8.6 an example of oscillating results is displayed. This is a longitudinal profile of the water stages calculated using the standard form of the finite element method. The same can be observed for the discharges as well. Such oscillations are typical for approximation of the advective term using the central difference when the diffusion, physical or numerical, is too weak to counteract local convective instabilities. They should not be regarded as instability in the Neumann sense (Abbott and Basco 1989). The reason of these oscillations is the applied approximation, which for the non-linear terms generates a large number of products of the nodal values of dependent variable (Fletcher 1991). To eliminate the oscillations usually some numerical dissipativity is introduced (see Katopodes 1984).

As it was shown in Section 6.4, the standard approach can be modified when the linear trial functions are assumed. Namely the integral of the product of approximation of function and basis function in an element can be expressed as a product

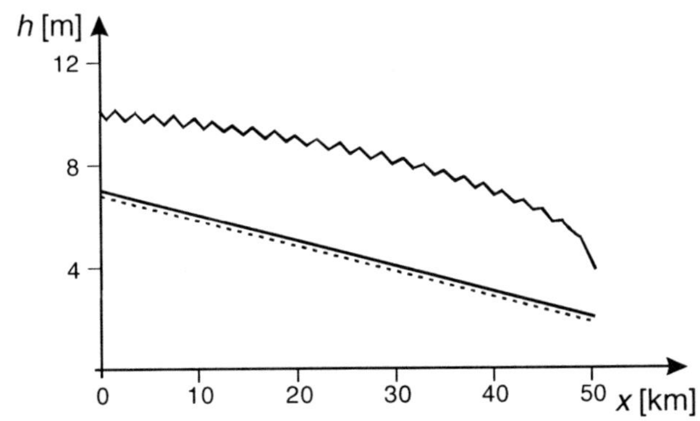

Fig. 8.6 Oscillating water level along channel computed using the standard finite element method

of certain average value of the function in the element and the integral of the basis function in this element. Therefore we have:

$$\int_{x_j}^{x_{j+1}} f_a(x,t)N_j(x)dx = f_c(t)\int_{x_j}^{x_{j+1}} N_j(x), \tag{8.48a}$$

$$\int_{x_j}^{x_{j+1}} f_a(x,t)N_{j+1}(x)dx = f_c(t)\int_{x_j}^{x_{j+1}} N_{j+1}(x). \tag{8.48b}$$

The weighted average values of function $f(t)$ in element j are defined as follows:

$$f_c(t) = \omega \cdot f_j(t) + (1-\omega) f_{j+1}(t) \quad \text{for Eq. (8.48a)} \tag{8.49a}$$

$$f_c(t) = (1-\omega) f_j(t) + \omega \cdot f_{j+1}(t) \quad \text{for Eq. (8.48b)} \tag{8.49b}$$

where ω is weighting parameter ranging from 0 to 1. Note that assuming $\omega = 2/3$ one obtains the standard finite element method (Eqs. 8.47a and 8.47b), whereas $\omega = 1/2$ corresponds to approximation using the algebraic average taken from the nodal values in an element. This concept applied for open channel unsteady flow and transport equations (Szymkiewicz 1995) appeared effective, since it leads to the scheme having the numerical properties comparable with the finite difference Preissmann scheme.

Calculation of one integral in expression (8.45a) over an element of length $\Delta x_j = x_{j+1} - x_j$ is carried out in the following way:

$$\int_{X_j}^{X_{j+1}} \left(\frac{\partial Q_c}{\partial t} + \frac{\partial}{\partial x}\left(\frac{\beta \cdot Q^2}{A}\right)_a + g \cdot A_c \frac{\partial h_a}{\partial x} + \left(g \cdot n_M^2 \frac{Q\,|Q|}{R^{4/3} \cdot A}\right)_c \right) N_j(x)dx =$$
$$= \left(\omega \frac{dQ_j}{dt} + (1-\omega)\frac{dQ_{j+1}}{dt}\right)\frac{\Delta x_j}{2} + \left(-\left(\frac{\beta \cdot Q^2}{A}\right)_j + \left(\frac{\beta \cdot Q^2}{A}\right)_{j+1}\right) +$$
$$+ g\left(\omega \cdot A_j + (1-\omega) A_{j+1}\right)\left(-h_j + h_{j+1}\right) +$$
$$+ \left(\omega \cdot \left(g \cdot n_M^2 \frac{Q\,|Q|}{R^{4/3} \cdot A}\right)_j + (1-\omega)\left(g \cdot n_M^2 \frac{Q\,|Q|}{R^{4/3} \cdot A}\right)_{j+1}\right)\frac{\Delta x_j}{2} \tag{8.50a}$$

$$\int_{x_j}^{x_{j+1}} \left(\frac{\partial h_c}{\partial t} + \frac{1}{B_c}\frac{\partial Q_a}{\partial x} - \frac{q_c}{B_c} \right) N_j(x)dx =$$
$$= \left(\omega \frac{dh_j}{dt} + (1-\omega)\frac{dh_{j+1}}{dt} \right) \frac{\Delta x_j}{2} + \frac{-Q_j + Q_{j+1}}{\omega \cdot B_j + (1-\omega) B_{j+1}} - \frac{\omega \cdot q_j + (1-\omega) q_{j+1}}{\omega \cdot B_j + (1-\omega) B_{j+1}} \frac{\Delta x_j}{2} \tag{8.50b}$$

$$\int_{x_j}^{x_{j+1}} \left(\frac{\partial Q_c}{\partial t} + \frac{\partial}{\partial x}\left(\frac{Q^2}{A}\right)_a + g \cdot A_c \frac{\partial h_a}{\partial x} + \left(g \cdot n_M^2 \frac{Q\,|Q|}{R^{4/3} \cdot A} \right)_c \right) N_{j+1}(x)dx =$$
$$= \left((1-\omega)\frac{dQ_j}{dt} + \omega \frac{dQ_{j+1}}{dt} \right) \frac{\Delta x_j}{2} + \left(-\left(\frac{\beta \cdot Q^2}{A}\right)_j + \left(\frac{\beta \cdot Q^2}{A}\right)_{j+1} \right) +$$
$$+ g\left((1-\omega) A_j + \omega \cdot A_{j+1}\right)\left(-h_j + h_{j+1}\right) +$$
$$+ \left((1-\omega)\left(g \cdot n_M^2 \frac{Q\,|Q|}{R^{4/3} \cdot A} \right)_j + \omega \cdot \left(g \cdot n_M^2 \frac{Q\,|Q|}{R^{4/3} \cdot A} \right)_{j+1} \right) \frac{\Delta x_j}{2} \tag{8.51a}$$

$$\int_{x_j}^{x_{j+1}} \left(\frac{\partial h_c}{\partial t} + \frac{1}{B_c}\frac{\partial Q_a}{\partial x} - \frac{q_c}{B_c} \right) N_{j+1}(x)dx =$$
$$= \left((1-\omega)\frac{dh_j}{dt} + \omega \frac{dh_{j+1}}{dt} \right) \frac{\Delta x_j}{2} + \frac{-Q_j + Q_{j+1}}{(1-\omega) B_j + \omega \cdot B_{j+1}} - \frac{(1-\omega) q_j + \omega \cdot q_{j+1}}{(1-\omega) B_j + \omega \cdot B_{j+1}} \frac{\Delta x_j}{2} \tag{8.51b}$$

In these expressions the subscript a denotes approximation according to formula (8.46) while the subscript c denotes approximation by Eqs. (8.49a) and (8.49b). It can be seen that the proposed modification does not concern the derivatives with regard to x, which are approximated using the standard approach.

When all integrals in each element are summed up according to Eq. (8.44), the global system of ordinary differential equations over time is obtained:

– for $j = 1$

$$\omega \frac{\Delta x_j}{2}\frac{dQ_j}{dt} + (1-\omega)\frac{\Delta x_j}{2}\frac{dQ_{j+1}}{dt} - \left(\frac{\beta \cdot Q^2}{A}\right)_j + \left(\frac{\beta \cdot Q^2}{A}\right)_{j+1} +$$
$$+ g\left(\omega \cdot A_j + (1-\omega) A_{j+1}\right)\left(-h_j + h_{j+1}\right) + \tag{8.52a}$$
$$+ \omega \cdot \tfrac{\Delta x_j}{2} \left(g \cdot n_M^2 \frac{Q\,|Q|}{R^{4/3} \cdot A} \right)_j + (1-\omega)\tfrac{\Delta x_j}{2} \left(g \cdot n_M^2 \frac{Q\,|Q|}{R^{4/3} \cdot A} \right)_{j+1} = 0$$

$$\omega \frac{\Delta x_j}{2} \frac{dh_j}{dt} + (1-\omega)\frac{\Delta x_j}{2}\frac{dh_{j+1}}{dt} + \frac{-Q_j + Q_{j+1}}{\omega \cdot B_j + (1-\omega) B_{j+1}} - \frac{\omega \cdot q_j + (1-\omega) q_{j+1}}{\omega \cdot B_j + (1-\omega) B_{j+1}} \frac{\Delta x_j}{2} = 0 \tag{8.52b}$$

– for $j = 2, \ldots, M-1$

$$\begin{aligned}
&(1-\omega)\frac{\Delta x_{j-1}}{2}\frac{dQ_{j-1}}{dt} + \omega\left(\frac{\Delta x_{j-1}}{2} + \frac{\Delta x_j}{2}\right)\frac{dQ_j}{dt} + (1-\omega)\frac{\Delta x_j}{2}\frac{dQ_{j+1}}{dt} + \\
&-\left(\frac{\beta \cdot Q^2}{A}\right)_{j-1} + \left(\frac{\beta \cdot Q^2}{A}\right)_{j+1} + g\left(\omega \cdot A_{j-1} + (1-\omega) A_j\right)\left(-h_{j-1} + h_j\right) + \\
&+g\left(\omega \cdot A_j + (1-\omega) A_{j+1}\right)\left(-h_j + h_{j+1}\right) + (1-\omega)\frac{\Delta x_{j-1}}{2}\left(g \cdot n_M^2 \frac{Q\,|Q|}{R^{4/3} \cdot A}\right)_{j-1} + \\
&+\omega\left(\frac{\Delta x_{j-1}}{2} + \frac{\Delta x_j}{2}\right)\left(g \cdot n_M^2 \frac{Q\,|Q|}{R^{4/3} \cdot A}\right)_j + (1-\omega)\frac{\Delta x_j}{2}\left(g \cdot n_M^2 \frac{Q\,|Q|}{R^{4/3} \cdot A}\right)_{j+1} = 0
\end{aligned} \tag{8.52c}$$

$$\begin{aligned}
&(1-\omega)\frac{\Delta x_{j-1}}{2}\frac{dh_{j-1}}{dt} + \omega\left(\frac{\Delta x_{j-1}}{2} + \frac{\Delta x_j}{2}\right)\frac{dh_j}{dt} + (1-\omega)\frac{\Delta x_j}{2}\frac{dh_{j+1}}{dt} + \\
&+\frac{-Q_{j-1} + Q_j}{\omega \cdot B_{j-1} + (1-\omega) B_j} + \frac{-Q_j + Q_{j+1}}{\omega \cdot B_j + (1-\omega) B_{j+1}} + \\
&-\frac{\omega \cdot q_{j-1} + (1-\omega) q_j}{\omega \cdot B_{j-1} + (1-\omega) B_j}\frac{\Delta x_{j-1}}{2} - \frac{\omega \cdot q_j + (1-\omega) q_{j+1}}{\omega \cdot B_j + (1-\omega) B_{j+1}}\frac{\Delta x_j}{2} = 0
\end{aligned} \tag{8.52d}$$

– for $j = M$

$$\begin{aligned}
&(1-\omega)\frac{\Delta x_{j-1}}{2}\frac{dQ_{j-1}}{dt} + \omega\frac{\Delta x_{j-1}}{2}\frac{dQ_j}{dt} + \\
&-\left(\frac{\beta \cdot Q^2}{A}\right)_{j-1} + \left(\frac{\beta \cdot Q^2}{A}\right)_j + g\left((1-\omega) A_{j-1} + \omega \cdot A_j\right)\left(-h_{j-1} + h_j\right) + \\
&+(1-\omega)\frac{\Delta x_{j-1}}{2}\left(g \cdot n_M^2 \frac{Q\,|Q|}{R^{4/3} \cdot A}\right)_{j-1} + \omega\frac{\Delta x_{j-1}}{2}\left(g \cdot n_M^2 \frac{Q\,|Q|}{R^{4/3} \cdot A}\right)_j = 0
\end{aligned} \tag{8.52e}$$

$$(1-\omega)\frac{\Delta x_{j-1}}{2}\frac{dh_{j-1}}{dt}+\omega\frac{\Delta x_{j-1}}{2}\frac{dh_j}{dt}+$$
$$+\frac{-Q_{j-1}+Q_j}{(1-\omega)\,B_{j-1}+\omega\cdot B_j}-\frac{(1-\omega)\,q_{j-1}+\omega\cdot q_j}{(1-\omega)\,B_{j-1}+\omega\cdot B_j}\frac{\Delta x_{j-1}}{2}=0 \tag{8.52f}$$

It can be rewritten in the matrix form, typical for the finite element method:

$$\mathbf{S}\frac{d\mathbf{X}}{dt}+\mathbf{C}\cdot\mathbf{X}=\mathbf{0} \tag{8.53}$$

where:

$\mathbf{X}=(Q_1(t),\, h_1(t),\, \ldots,\, Q_M(t),\, h_M(t))^T$ – vector of unknowns set up from nodal values of Q and h,
$\frac{d\mathbf{X}}{dt}=\left(\frac{dQ_1}{dt},\frac{dh_1}{dt},\ldots,\frac{dQ_M}{dt},\frac{dh_M}{dt}\right)^T$ – vector of time derivatives,
T– symbol of transposition
S – constant matrix, symmetrical and banded,
C – variable matrix, asymmetrical and banded.

The matrices **S** and **C** have dimensions of (2M) × (2M) with bandwidth equal to 8.

The system (8.53) is integrated over time using the method (3.71):

$$\mathbf{X}_{n+1}=\mathbf{X}_n+\Delta t\left(\theta\frac{d\mathbf{X}_{n+1}}{dt}+(1-\theta)\frac{d\mathbf{X}_n}{dt}\right) \tag{8.54}$$

where:

θ – weighting parameter ranging from 0 to 1,
Δt – time step,
n – index of time level.

This leads to the system of non-linear algebraic equations:

$$(\mathbf{S}+\Delta t\cdot\theta\cdot\mathbf{C}_{n+1})\mathbf{X}_{n+1}=(\mathbf{S}-\Delta t(1-\theta)\mathbf{C}_n)\mathbf{X}_n \tag{8.55}$$

The system (8.55) has to be completed by the boundary conditions. They are as follows:

- at the upstream end of a channel ($j=1$) the water level $h_1(t)$ or the flow discharge $Q_1(t)$ for $t>0$ are imposed,
- at the downstream end of a channel ($j=M$) the water level $h_M(t)$ or the flow discharge $Q_M(t)$ for $t>0$ are imposed.

The system (8.55) has dimensions of $(2M) \times (2M)$, whereas its matrix of coefficients is banded with bandwidth equal to 8. To solve this non-linear system an iterative method like the Newton method has to be used.

8.3.2 Stability Analysis of the Modified Finite Element Method

Stability analysis is carried out for the simplified linear form of the governing equations, i.e. for the wave equations (8.16) and (8.17). Approximation of this system for the equally spaced nodes ($\Delta x = \text{const.}$) yields the following system of algebraic equations:

$$\frac{1-\omega}{2}\frac{H_{j-1}^{n+1}-H_{j-1}^{n}}{\Delta t}+\omega\frac{H_{j}^{n+1}-H_{j}^{n}}{\Delta t}+\frac{1-\omega}{2}\frac{H_{j+1}^{n+1}-H_{j+1}^{n}}{\Delta t}+ \\ +\theta\frac{\overline{H}}{2\Delta x}\left(-U_{j-1}^{n+1}+U_{j+1}^{n+1}\right)+(1-\theta)\frac{\overline{H}}{2\Delta x}\left(-U_{j-1}^{n}+U_{j+1}^{n}\right)=0 \tag{8.56a}$$

$$\frac{1-\omega}{2}\frac{U_{j-1}^{n+1}-U_{j-1}^{n}}{\Delta t}+\omega\frac{U_{j}^{n+1}-U_{j}^{n}}{\Delta t}+\frac{1-\omega}{2}\frac{U_{j+1}^{n+1}-U_{j+1}^{n}}{\Delta t}+ \\ +\theta\frac{g}{2\Delta x}\left(-H_{j-1}^{n+1}+H_{j+1}^{n+1}\right)+(1-\theta)\frac{g}{2\Delta x}\left(-H_{j-1}^{n}+H_{j+1}^{n}\right)=0 \tag{8.56b}$$

for $j = 2, 3, \ldots, M-1$

Note that the considered method corresponds to the mesh presented in Fig. 8.7.

The presented method involves two weighting parameters ω and θ. Taking $\omega = 2/3$ the finite element method in its standard form is obtained, whereas for $\omega = 1$ this method coincides with the finite difference one. When $\theta = 1/2$ the method becomes

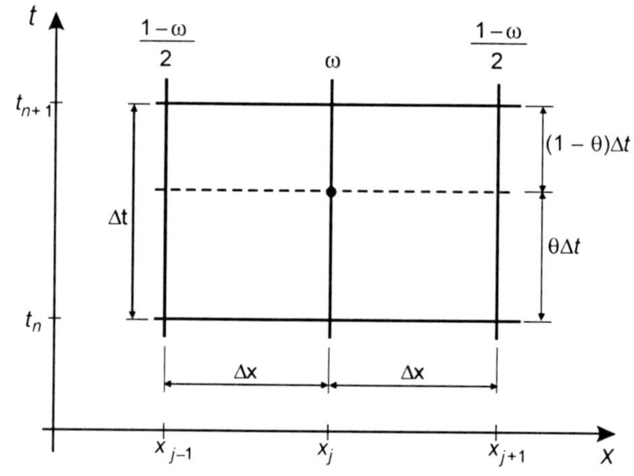

Fig. 8.7 Numerical grid for the modified finite element method

the Crank-Nicolson scheme (equivalent to the implicit trapezoidal rule), whereas with $\theta = 1$, it is the implicit Euler scheme.

The stability analysis is carried out by the Neumann method (Potter 1973). As we know from Section 5.4.3, this approach requires expanding the unknown functions $H(x, t)$ and $U(x, t)$ in Fourier series and examining the behaviour of its components. Since the system of equations (8.52a) and (8.52b) is linear, then one can examine a single kth component only. According to Eqs. (5.157) and (5.161), this component for both functions U and H are given as:

$$U_j^n = u_k^n \cdot e^{i \cdot k \cdot m \cdot j \cdot \Delta x} = u^n \cdot e^{i \cdot \varphi \cdot j}, \tag{8.57}$$

$$H_j^n = h_k^n \cdot e^{i \cdot k \cdot m \cdot j \cdot \Delta x} = h^n \cdot e^{i \cdot \varphi \cdot j}, \tag{8.58}$$

where:

$i = (-1)^{1/2}$,
m – wave number,
$\varphi = m \cdot \Delta x$ – dimensionless wave number,
j – index of node,
Δx – spatial mesh dimension.

The variables u_k^n and h_k^n are the coefficients of kth components at time level n. Since we assumed that only a single component is examined, then index k can be omitted as it was done in Eqs. (8.57) and (8.58).

Let us multiply both sides of Eqs. (8.56a) and (8.56b) by $2\Delta t$ and rewrite them gathering the values from time level $n + 1$ at left side and from time level n at right side:

$$\begin{aligned} &(1-\omega)\, H_{j-1}^{n+1} + 2\omega \cdot H_j^{n+1} + (1-\omega)\, H_{j+1}^{n+1} + \theta \frac{\Delta t \cdot \overline{H}}{\Delta x} \left(-U_{j-1}^{n+1} + U_{j+1}^{n+1}\right) = \\ &= (1-\omega)\, H_{j-1}^{n} + 2\omega \cdot H_j^{n} + (1-\omega)\, H_{j+1}^{n} - (1-\theta) \frac{\Delta t \cdot \overline{H}}{\Delta x} \left(-U_{j-1}^{n} + U_{j+1}^{n}\right) \end{aligned} \tag{8.59a}$$

$$\begin{aligned} &(1-\omega)\, U_{j-1}^{n+1} + 2\omega \cdot U_j^{n+1} + (1-\omega)\, U_{j+1}^{n+1} + \theta \frac{\Delta t \cdot g}{\Delta x} \left(-H_{j-1}^{n+1} + H_{j+1}^{n+1}\right) = \\ &= (1-\omega)\, U_{j-1}^{n} + 2\omega \cdot U_j^{n} + (1-\omega)\, U_{j+1}^{n} - (1-\theta) \frac{\Delta t \cdot g}{\Delta x} \left(-H_{j-1}^{n} + H_{j+1}^{n}\right). \end{aligned} \tag{8.59b}$$

Substitution of Eqs. (8.57) and (8.58) in Eqs. (8.59a) and (8.59b) yields:

$$
\begin{aligned}
&(1-\omega)\,h^{n+1}\cdot e^{i\cdot\varphi(j-1)}+2\omega\cdot h^{n+1}\cdot e^{i\cdot\varphi\cdot j}+(1-\omega)\,h^{n+1}\cdot e^{i\cdot\varphi(j+1)}+\\
&\qquad+\theta\frac{\Delta t\cdot\overline{H}}{\Delta x}\left(-u^{n+1}\cdot e^{i\cdot\varphi(j-1)}+u^{n+1}\cdot e^{i\cdot\varphi(j+1)}\right)=\\
&=(1-\omega)\,h^{n}\cdot e^{i\cdot\varphi(j-1)}+2\omega\cdot h^{n}\cdot e^{i\cdot\varphi\cdot j}+(1-\omega)\,h^{n}\cdot e^{i\cdot\varphi(j+1)}+\\
&\qquad-(1-\theta)\frac{\Delta t\cdot\overline{H}}{\Delta x}\left(-u^{n}\cdot e^{i\cdot\varphi(j-1)}+u^{n}\cdot e^{i\cdot\varphi(j+1)}\right)
\end{aligned}
\tag{8.60a}
$$

$$
\begin{aligned}
&(1-\omega)\,u^{n+1}\cdot e^{i\cdot\varphi(j-1)}+2\omega\cdot u^{n+1}\cdot e^{i\cdot\varphi\cdot j}+(1-\omega)\,u^{n+1}\cdot e^{i\cdot\varphi(j+1)}+\\
&\qquad+\theta\frac{\Delta t\cdot g}{\Delta x}\left(-h^{n+1}\cdot e^{i\cdot\varphi(j-1)}+h^{n+1}\cdot e^{i\cdot\varphi(j+1)}\right)=\\
&=(1-\omega)\,u^{n}\cdot e^{i\cdot\varphi(j-1)}+2\omega\cdot u^{n}\cdot e^{i\cdot\varphi\cdot j}+(1-\omega)\,u^{n}\cdot e^{i\cdot\varphi(j+1)}+\\
&\qquad-(1-\theta)\frac{\Delta t\cdot g}{\Delta x}\left(-h^{n}\cdot e^{i\cdot\varphi(j-1)}+h^{n}\cdot e^{i\cdot\varphi(j+1)}\right)
\end{aligned}
\tag{8.60b}
$$

After dividing both equations by the factor $e^{i\cdot\varphi}$ and appropriate transformations one obtains:

$$
\begin{aligned}
&\left((1-\omega)\,e^{-i\cdot\varphi}+2\omega+(1-\omega)\,e^{i\cdot\varphi}\right)h^{n+1}+\theta\tfrac{\Delta t\cdot\overline{H}}{\Delta x}\left(-e^{-i\cdot\varphi}+e^{i\cdot\varphi}\right)u^{n+1}=\\
&=\left((1-\omega)\,e^{-i\cdot\varphi}+2\omega+(1-\omega)\,e^{i\cdot\varphi}\right)h^{n}-(1-\theta)\tfrac{\Delta t\cdot\overline{H}}{\Delta x}\left(-e^{-i\cdot\varphi}+e^{i\cdot\varphi}\right)u^{n}
\end{aligned}
\tag{8.61a}
$$

$$
\begin{aligned}
&\left((1-\omega)\,e^{-i\cdot\varphi}+2\omega+(1-\omega)\,e^{i\cdot\varphi}\right)u^{n+1}+\theta\frac{\Delta t\cdot g}{\Delta x}\left(-e^{-i\cdot\varphi}+e^{i\cdot\varphi}\right)h^{n+1}=\\
&=\left((1-\omega)\,e^{-i\cdot\varphi}+2\omega+(1-\omega)\,e^{i\cdot\varphi}\right)u^{n}-(1-\theta)\frac{\Delta t\cdot g}{\Delta x}\left(-e^{-i\cdot\varphi}+e^{i\cdot\varphi}\right)h^{n}
\end{aligned}
\tag{8.61b}
$$

Using the following Euler relations (Korn and Korn 1968):

$$
\sin(\varphi)=\frac{e^{i\cdot\varphi}-e^{-i\cdot\varphi}}{2i}\quad\text{and}
\tag{8.62}
$$

$$
\cos(\varphi)=\frac{e^{i\cdot\varphi}+e^{-i\cdot\varphi}}{2}
\tag{8.63}
$$

Equations (8.61a) and (8.61b) are rewritten as:

$$
\begin{aligned}
&(\omega+(1-\omega)\cos(\varphi))\,h^{n+1}+i\cdot\theta\frac{\Delta t\cdot\overline{H}}{\Delta x}\sin(\varphi)\,u^{n+1}=\\
&\qquad=(\omega+(1-\omega)\cos(\varphi))\,h^{n}-i\cdot(1-\theta)\frac{\Delta t\cdot\overline{H}}{\Delta x}\sin(\varphi)\,u^{n}
\end{aligned}
\tag{8.64a}
$$

$$
\begin{aligned}
&(\omega+(1-\omega)\cos(\varphi))\,u^{n+1}+i\cdot\theta\frac{\Delta t\cdot g}{\Delta x}\sin(\varphi)\,h^{n+1}=\\
&\qquad=(\omega+(1-\omega)\cos(\varphi))\,u^{n}-i\cdot(1-\theta)\frac{\Delta t\cdot g}{\Delta x}\sin(\varphi)\,h^{n}
\end{aligned}
\tag{8.64b}
$$

In matrix notation this system of equations can be written as follows:

$$\mathbf{A} \cdot \mathbf{F}^{n+1} = \mathbf{B} \cdot \mathbf{F}^n \tag{8.65}$$

where:

$$\mathbf{F}^{n+1} = \begin{pmatrix} h^{n+1} \\ u^{n+1} \end{pmatrix}, \tag{8.66a}$$

$$\mathbf{F}^{n} = \begin{pmatrix} h^{n} \\ u^{n} \end{pmatrix}, \tag{8.66b}$$

$$\mathbf{A} = \begin{bmatrix} (\omega + (1-\omega)\cos(\varphi)) & i \cdot \theta \frac{\Delta t \cdot \overline{H}}{\Delta x} \sin(\varphi) \\ i \cdot \theta \frac{\Delta t \cdot g}{\Delta x} \sin(\varphi) & (\omega + (1-\omega)\cos(\varphi)) \end{bmatrix}, \tag{8.66c}$$

$$\mathbf{B} = \begin{bmatrix} (\omega + (1-\omega)\cos(\varphi)) & -i \cdot (1-\theta) \frac{\Delta t \cdot \overline{H}}{\Delta x} \sin(\varphi) \\ -i \cdot (1-\theta) \frac{\Delta t \cdot g}{\Delta x} \sin(\varphi) & (\omega + (1-\omega)\cos(\varphi)) \end{bmatrix} \tag{8.66d}$$

or in alternative form as:

$$\mathbf{F}^{n+1} = \mathbf{G} \cdot \mathbf{F}^n. \tag{8.67}$$

The amplification matrix $\mathbf{G}$ is given by:

$$\mathbf{G} = \mathbf{A}^{-1} \cdot \mathbf{B} = \begin{bmatrix} g_{11} & g_{12} \\ g_{21} & g_{22} \end{bmatrix}. \tag{8.68}$$

To find this matrix at first let us calculate the inverse matrix $\mathbf{A}^{-1}$, which is defined as follows:

$$\mathbf{A}^{-1} = \frac{1}{D} \begin{bmatrix} a_{22} & -a_{12} \\ -a_{21} & a_{11} \end{bmatrix} \tag{8.69}$$

where a_{11}, a_{12}, a_{21} and a_{22} are the elements of matrix $\mathbf{A}$, whereas $D = a_{11} \cdot a_{22} - a_{12} \cdot a_{21}$ is its determinant. Using the Courant number defined by Eq. (8.18) this determinant is written as follows:

$$\begin{aligned} D = (\omega + (1-\omega)\cos(\varphi))^2 + \theta^2 \frac{\Delta t^2 \cdot g \cdot \overline{H}}{\Delta x^2} \sin^2(\varphi) = \\ = (\omega + (1-\omega)\cos(\varphi))^2 + \theta^2 \cdot C_r^2 \cdot \sin^2(\varphi) \end{aligned} \tag{8.70}$$

The matrix $\mathbf{A}^{-1}$ is given

$$\mathbf{A}^{-1} = \frac{1}{D} \begin{bmatrix} (\omega + (1-\omega)\cos(\varphi)) & -i \cdot \theta \dfrac{\Delta t \cdot \overline{H}}{\Delta x} \sin(\varphi) \\ -i \cdot \theta \dfrac{\Delta t \cdot g}{\Delta x} \sin(\varphi) & (\omega + (1-\omega)\cos(\varphi)) \end{bmatrix} \tag{8.71}$$

Multiplication $\mathbf{A}^{-1} \cdot \mathbf{B}$ yields the following elements of matrix $\mathbf{G}$:

$$g_{11} = g_{22} = \frac{(\omega + (1-\omega)\cos(\varphi))^2 - \theta(1-\theta) C_r^2 \cdot \sin^2(\varphi)}{D} \tag{8.72a}$$

$$g_{12} = -\frac{(\omega + (1-\omega)\cos(\varphi))\,\theta \dfrac{\Delta t \cdot \overline{H}}{\Delta x} \sin(\varphi)}{D} i \tag{8.72b}$$

$$g_{11} = -\frac{(\omega + (1-\omega)\cos(\varphi))\,\theta \dfrac{\Delta t \cdot g}{\Delta x} \sin(\varphi)}{D} i \tag{8.72c}$$

The numerical scheme written in general form (8.67) will be stable if the modulus of the greatest eigenvalue of the amplification matrix **G** is not greater than unity. The eigenvalue of matrix **G** can be found directly from its definition (Potter 1973):

$$\det(\mathbf{G} - \lambda \cdot \mathbf{I}) = 0 \tag{8.73}$$

where:

λ – eigenvalue
I – identity matrix.

Condition (8.73) rewritten as:

$$\begin{vmatrix} g_{11} - \lambda & g_{12} \\ g_{21} & g_{11} - \lambda \end{vmatrix} = 0 \tag{8.74}$$

leads to the following equation:

$$(g_{11} - \lambda)^2 - g_{12} \cdot g_{21} = 0. \tag{8.75}$$

This equation has two roots:

$$\lambda_{1,2} = g_{11} \pm (g_{12} \cdot g_{21})^{1/2}. \tag{8.76}$$

Substitution of the appropriate expressions into Eq. (8.76) shows that both eigenvalues being complex numbers are identical:

$$\lambda = \frac{(\omega + (1-\omega)\cos(\varphi))^2 - \theta(1-\theta) C_r^2 \cdot \sin^2(\varphi)}{(\omega + (1-\omega)\cos(\varphi))^2 + \theta^2 \cdot C_r^2 \cdot \sin^2(\varphi)} + \\ + \frac{(\omega + (1-\omega)\cos(\varphi))\, C_r \cdot \sin(\varphi)}{(\omega + (1-\omega)\cos(\varphi))^2 + \theta^2 \cdot C_r^2 \cdot \sin^2(\varphi)} i \tag{8.77}$$

To rewrite Eq. (8.77) in a more compact form let us introduce a new variable r defined as follows:

$$r = \frac{C_r \cdot \sin(\varphi)}{\omega + (1-\omega)\cos(\varphi)} = \frac{C_r \cdot 2\sin\left(\frac{\varphi}{2}\right)\cos\left(\frac{\varphi}{2}\right)}{\omega + (1-\omega)\left(\cos^2\left(\frac{\varphi}{2}\right) - \sin^2\left(\frac{\varphi}{2}\right)\right)} =$$
$$= \frac{C_r \cdot 2\sin\left(\frac{\varphi}{2}\right)\cos\left(\frac{\varphi}{2}\right)}{\cos^2\left(\frac{\varphi}{2}\right) + (2\omega - 1)\sin^2\left(\frac{\varphi}{2}\right)} = \frac{2C_r \cdot \tan\left(\frac{\varphi}{2}\right)}{1 + (2\omega - 1)\tan^2\left(\frac{\varphi}{2}\right)} \tag{8.78}$$

Using this relation, after some appropriate manipulations the eigenvalue (8.77) can be reformed to the following form:

$$\lambda = 1 - \frac{r^2 \cdot \theta}{r^2 \cdot \theta^2 + 1} + \frac{r}{r^2 \cdot \theta^2 + 1} i \tag{8.79}$$

The numerical stability requires (Potter 1972):

$$|\lambda| \leq 1. \tag{8.80}$$

Since λ is a complex number, then the relation (8.80) takes the following form:

$$|\lambda| = \left(1 - \frac{2\theta - 1}{\theta^2 + 1/r^2}\right)^{1/2} \leq 1 \tag{8.81}$$

Note that condition (8.81) will be satisfied if only we assume $\theta \geq 0.5$. On the other hand one can see, that the factor r given by Eq. (8.78) will have the denominator always different from zero for any dimensionless wave number ($0 \leq \varphi \leq \pi$), if only $\omega \geq 0.5$. Summarizing, one can find out that the modified finite element method applied for solution of the linear wave equations is stable for:

$$\theta \geq 0.5 \quad \text{and} \tag{8.82}$$

$$\omega \geq 0.5 \tag{8.83}$$

As the relations (8.82) and (8.83) are satisfied for any Courant number, these conditions ensure unconditional stability. The amplitude and phase portraits for the method will be presented in the next subsection.

8.3.3 Numerical Errors Generated by the Modified Finite Element Method

To examine more exactly the considered numerical method the modified equation approach is used. To carry out an accuracy analysis using this approach all nodal values of functions U and H in the algebraic equations (8.57a) and (8.57b) are replaced by Taylor series expansions around the node $(n, j+1)$ (Fig. 8.7). Next, the obtained relations are rearranged so that they contain only spatial derivatives. Finally the following system of equations is obtained:

$$\frac{\partial U}{\partial t} + g\frac{\partial H}{\partial x} = D_n\frac{\partial^2 U}{\partial x^2} + E_{n1}\frac{\partial^3 H}{\partial x^3} + \cdots \tag{8.84}$$

$$\frac{\partial H}{\partial t} + \overline{H}\frac{\partial U}{\partial x} = D_n\frac{\partial^2 H}{\partial x^2} + E_{n2}\frac{\partial^3 U}{\partial x^3} + \cdots \tag{8.85}$$

In Eqs. (8.84) and (8.85) the coefficient of numerical diffusion D_n is given by the expression:

$$D_n = \left(\theta - \frac{1}{2}\right) g \cdot \overline{H} \cdot \Delta t \tag{8.86}$$

whereas the coefficients of numerical dispersion E_{n1} and E_{n2} are defined as follows:

$$E_{n1} = \frac{g \cdot \Delta x^2}{6}((2 - 3\omega) - (2 - 3\theta)C_r) \tag{8.87a}$$

$$E_{n2} = \frac{\overline{H} \cdot \Delta x^2}{6}((2 - 3\omega) - (2 - 3\theta)C_r) \tag{8.87b}$$

The modified equations can be used to deduce practically full information on the numerical properties of the considered method, including consistency and the order of accuracy.

From relations (8.84), (8.85), (8.86) and (8.87) results that the finite element method does not generate any numerical diffusion for $\theta = 0.5$, i.e. when it represents an approximation of the 2nd order with regard to t. It can be shown that for ω=0.5 and $C_r = 1$ all terms of 3rd and higher order disappear. In this case the method ensures an exact solution. For $C_r \neq 1$ unphysical oscillations can appear in the solution because of dispersivity ($D_n = 0$ and $E_{n1}, E_{n2} \neq 0$). The method is unstable for $\theta < 0.5$. In this case we have $D_n < 0$. It means that the initial-value problem for the system of Eqs. (8.84) and (8.85) limited to the terms of 2nd order is ill-posed and consequently its numerical solution does not exist. Modified finite element method ensures a stable solution for $\theta > 0.5$, because with $D_n > 0$ the initial-boundary problem for Eqs. (8.84) and (8.85) is well posed and its solution exists always. However, it contains a dissipation error because in this case the method produces a numerical diffusion. It increases with the increase in Δt giving rise to smoothing solution and reduction of the pressure gradients.

As it was shown before, the finite element method represents 2nd order of accuracy with regard to x as well as with regard to t for $\theta = 0.5$ only. For other values of θ this method is only of 1st order of accuracy with regard to t. However the order of accuracy can be further increased. For $\theta = 0.5$ the numerical diffusion is cancelled ($D_n = 0$), while the dispersion coefficients are:

$$E_{n1} = \frac{g \cdot \Delta x^2}{6}\left(2 - 3\omega - \frac{C_r}{2}\right), \tag{8.88a}$$

$$E_{n2} = \frac{\overline{H} \cdot \Delta x^2}{6}\left(2 - 3\omega - \frac{C_r}{2}\right) \tag{8.88b}$$

Note that assuming:

$$\omega = \frac{2}{3} - \frac{C_r^2}{6} \tag{8.89}$$

we obtain $E_{n1} = 0$ and $E_{n2} = 0$ as well. Therefore with $\theta = 0.5$ and ω given by Eq. (8.89) the terms of 2nd and 3rd order in the system of modified equations (8.84) and (8.85) are cancelled. In this case the proposed version of the finite element method ensures 3rd order accuracy with regard to both x and t. This holds for $C_r \leq 1$ only because of the condition of stability (8.82).

The conclusions presented above can be illustrated by plotting of the eigenvalue modulus of amplification matrix (8.81), which is also known as the amplitude portrait (Abbott and Basco 1989). Since $\varphi = m\Delta x = 2\pi/N$($N$ = number of computational intervals Δx per wavelength corresponding to the considered component of Fourier series – see Section 5.4.3), the modulus of eigenvalue is a function of N: $|\lambda| = f(N)$. In Fig. 8.8 the moduli of eigenvalues of amplification matrix calculated for $\omega = 1/2$, $C_r = 1$ and for various values of θ are shown. One can notice that for $\theta = 0.5$ we obtain $|\lambda| = 1$. It means that the considered scheme does not change the wave amplitude. It is neither damped nor amplified. For $\theta < 0.5$ one obtains $|\lambda| > 1$. In this case the wave amplitude is amplified and consequently the scheme becomes unstable. Note that this situation corresponds to negative values of the coefficient of numerical diffusion D_n. The consequences of this fact were discussed earlier. For $\theta > 0.5$ we obtain $|\lambda| < 1$, what means that the amplitude of the propagating wave is

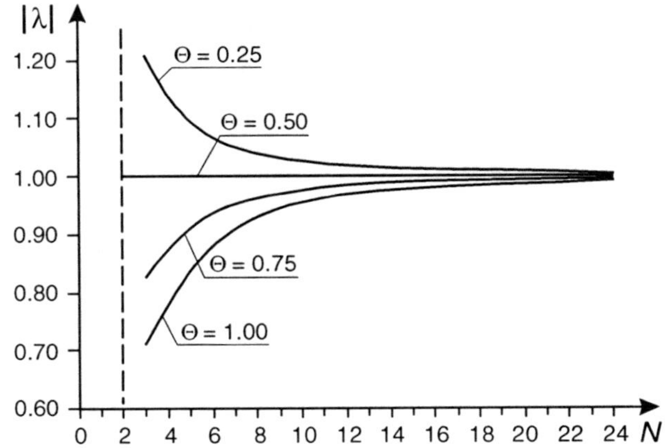

Fig. 8.8 Amplitude portrait for different θ

damped. Therefore the scheme ensures stable solution. However, at the same time it generates numerical diffusion, which affects the propagating wave. As one can see the shorter waves (small value of N) are more affected than the longer ones.

The advantages of the proposed approach are better illustrated by the second coefficient of convergence $R(N)$ called the phase portrait, which represents the dispersive properties of the scheme. For the considered modified finite element method the phase celerity of considered kth Fourier component is expressed by the following equation:

$$\tan(\phi) = \frac{\mathrm{Im}\lambda}{\mathrm{Re}\lambda} = \frac{r}{r^2 \cdot \theta(\theta - 1) + 1}. \tag{8.90}$$

Therefore we can find ϕ as:

$$\phi = \arctan\left(\frac{r}{r^2 \cdot \theta(\theta - 1) + 1}\right). \tag{8.91}$$

On the other hand, this component propagates with the following exact celerity:

$$\phi^{\text{exact}} = C_r \cdot \varphi = 2C_r \cdot \pi/N. \tag{8.92}$$

Then the coefficient of convergence R_2 defined as the ratio of the two celerities, is:

$$\frac{\phi}{\phi^{\text{exact}}} = \frac{\arctan\left(\frac{r}{r^2 \cdot \theta(\theta - 1) + 1}\right)}{2C_r \cdot \pi/N}. \tag{8.93}$$

The factor r given by formula (8.78) expressed in terms of the number of space intervals per wavelength is given as follows:

$$r = \frac{2C_r \cdot \tan\left(\frac{\pi}{N}\right)}{1 + (2\omega - 1)\tan^2\left(\frac{\pi}{N}\right)} \tag{8.94}$$

Let us calculate the value of tan (ϕ) given by Eq. (8.90) for $\theta = 0.5$, $\omega = 0.5$ and $C_r = 1$. One obtains:

$$\tan(\phi) = \frac{2\tan\left(\frac{\varphi}{2}\right)}{1 - \tan^2\left(\frac{\varphi}{2}\right)} = \tan(\varphi). \tag{8.95}$$

Since the dimensionless wave number is $\varphi = m.\Delta x$ then one can write that:

$$\phi = \varphi = m \cdot \Delta x. \tag{8.96}$$

This relation indicates that each Fourier component propagates with constant speed because the product $m{\cdot}\Delta x$ is constant for the considered component. Therefore for this case no variation of the wave speed occurs. In other words, the scheme is not dispersive. As for $\theta = 0.5$ it is dissipation free as well, then the assumed set of the parameters: θ =0.5, ω =0.5 and C_r= 1 ensures exact solution of the system of equations (8.16) and (8.17). For other values of ω and C_r the scheme becomes dispersive.

The plots presenting dispersive properties are shown in Fig. 8.9. First of all one can see that indeed for $\theta = 0.5$, $\omega = 0.5$ and C_r= 1 the coefficient of convergence R_2 is equal to unity for any wave length. The graphs indicate that modification of the finite element method significantly improves its numerical properties. One can notice, that for ω defined by Eq. (8.89) the wave speed of the shortest waves are affected only. There is a remarkable difference comparing with $\omega = 2/3$ (corresponding to the standard version of the finite element method) and with $\omega = 1$ (corresponding to the finite difference method). This is obvious, because the six-point implicit scheme with two weighting parameters ensures an approximation of 3rd order with regard to x and to t for $C_r \leq 1$. Therefore one can expect that this scheme is able to ensure higher accuracy of the solution, especially when large gradients occur.

The analysis of numerical properties was carried out for a system of linear equations, while the Saint-Venant equations are nonlinear. Nevertheless, such an analysis seems very useful, since it provides some indications with regard to the stability, dissipation and dispersion of the nonlinear problem. In principle, it is recommended to perform the analysis of the linearized versions of equations to demonstrate the stability of numerical scheme (Fletcher 1991). In practice, the final verification of stability and accuracy can be carried out only by means of numerical experiments.

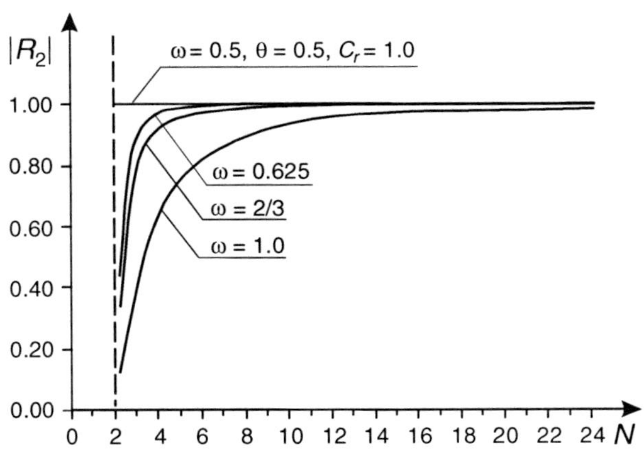

Fig. 8.9 Phase portraits for the modified finite element method with $\theta = 0.5$, $C_r = 0.5$ and for different values of ω

8.4 Some Aspects of Practical Application of the Saint Venant Equations

8.4.1 Formal Requirements and Actual Possibilities

When we undertake an attempt to apply the system of Saint Venant equations to model a real flow phenomenon in a river, it is immediately noticed that some requirements resulting from the formal analysis usually cannot be fulfilled directly. For instance, for subcritical flow at the downstream end of a channel the flow discharge or the water stage must be proscribed as boundary condition. However, if we consider a typical river reach this requirement can be fulfilled rather seldom. More often we must find some formulas relating both mentioned functions to use them as equivalent form of required boundary condition. On the other hand, except for artificial prismatic channels, we can have some difficulties with accurate and effective calculation of the cross-sectional parameters especially when they have complex forms and composite roughness. Another problem appears when we try to extrapolate the Saint Venant equations validity applying them beyond the acceptable limits. This happens for instance, when these equations are used to model the river flow, which is actually two dimensional or which is one-dimensional, but rapidly varied. Yet another challenge appear, when we consider unsteady flow in an open channel network with different hydraulic structures inside. One can say generally, that successful implementation of the Saint Venant equations for actual cases of river flow requires overcoming many various difficulties resulting from the difference between real physical worlds and idealized one, assumed during their derivation and solution.

Many valuable information and recommendations dealing with practical aspects are coming from experiences owing to long time of application of the Saint Veneant equations in hydraulic engineering. For instance, comprehensive explanations and discussions of many practical aspects are given by Cunge et al. (1980). Similar helpful practical recommendations can be found in many chapters of the book edited by Mahmood and Yevjewich (1975). These publications covering practically whole area of the problem, although they were edited years ago, are still valid and valuable for engineers. Some practical advises are available in other publications as well.

For these reasons below we only briefly discuss some practical questions connected with the implementation of the unsteady flow equations in form of the computer codes supporting hydraulic engineering practice.

8.4.2 Representation of the Channel Cross-Section

Let us consider the simplest case of unsteady flow, which takes place in a single channel reach as shown in Fig. 8.10a. A natural feature of rivers is their irregular longitudinal and transversal shape. The cross-sections geometry can be measured in the field only at selected, suitable locations. Consequently, the computational cross-sections are non-uniformly spaced along the channel axis (Fig. 8.10a). From this point of view both previously described numerical solution methods are suitable,

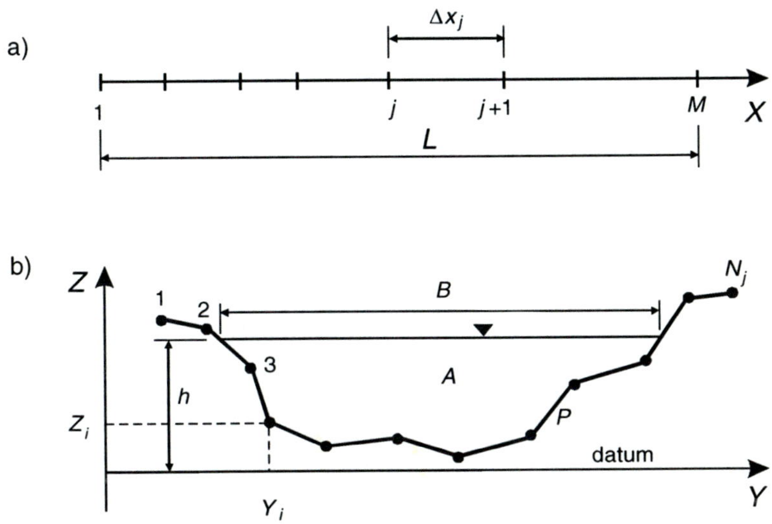

Fig. 8.10 Discretization of channel reach (**a**) and numerical representation of a cross-section (**b**)

since they work on non-uniform grids. The river cross sections are defined numerically by sets of coordinate pairs (Y_i, Z_i) related to a local coordinate system as presented in Fig. 8.10b.

Numerical solution of the Saint Venant equations requires that the wetted cross sectional area A, wetted perimeter P and top width B are specified as functions of the water stage h. A convenient approach is to tabulate the parameters in each cross-section for h ranging from Z_{min} to Z_{max} with appropriate an interval Δh. During the computations the actual values of parameters are calculated via linear interpolation:

$$f(h) = h_i + \frac{f_{i+1} - f_i}{h_{i+1} - h_i}(h - h_i) \quad \text{for} \quad h_i \le h < h_{i+1} \tag{8.97}$$

where f represents A, P, or B. Sometimes, when the cross-sections are relatively regular, one can approximate them using the standard shapes such as trapezoidal or triangular. Another possible approach is to fit an analytical formula to the tabulated data using the least-square method. In such a situation we have:

$$\begin{aligned} &A = F_1(h), P = F_2(h), B = F_3(h) \\ &\text{for } Z_{min} \le h \le Z_{max} \\ &\text{where } F_1, F_2, F_3 \text{ are given functions.} \end{aligned} \tag{8.98}$$

Very often in hydraulic practice compound channel cross-sections are encountered, for example if the river has flood plains as presented in Fig. 8.11. When the water stage exceeds the level of flood plain, the flow process becomes more complicated, since it takes place in parts of the channel having different hydraulic properties. If 1D Saint Venant equations are used to model the unsteady flow they

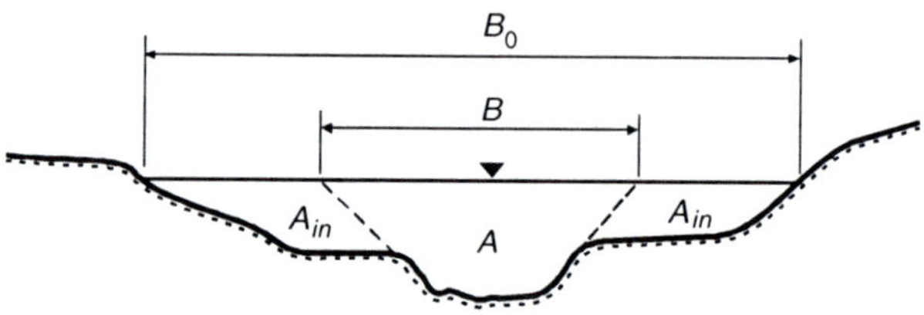

Fig. 8.11 Channel of compound section

should be modified and special treatment of the cross-section is needed. Such modification proposed by Abbott and Ionescu (1967) bases on distinction of two parts of channel cross-section playing different roles in flow process. The first one called active, is connected with the main channel and it is involved in the dynamic equation. The second part of section is called inactive, since it serves as a reservoir, which stocks the water only. This part which together with the active part constitutes the whole area of section, is involved in the continuity equation.

The original Saint Venant equations are modified as follows:

$$\frac{\partial Q}{\partial t} + \frac{\partial}{\partial x}\left(\frac{\beta \cdot Q^2}{A}\right) + g \cdot A\frac{\partial h}{\partial x} + \frac{g \cdot n_M^2}{R^{4/3}}\frac{|Q|\,Q}{A} = 0, \tag{8.99}$$

$$\frac{\partial h}{\partial t} + \frac{1}{B_0}\frac{\partial Q}{\partial x} = \frac{q}{B_0}. \tag{8.100}$$

It seems that nowadays such approaches for solving the unsteady flow in channel with flood plain became less interesting. Actually, such case of flow should be considered rather as a 2D process with the river bed being partially and temporarily dry. It can be modelled with one of the widely available computer codes for 2D unsteady flow.

Very often the Saint Venant equations are solved for a storm sewer network. Such system is constituted by closed conduits where the flow is usually free-surface (Fig. 8.12a). However, every now and then, after heavy rain, the conduits are filled

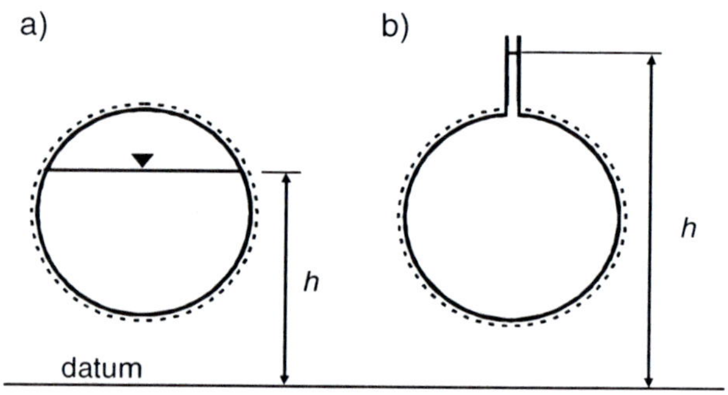

Fig. 8.12 Flow in closed conduit: (**a**) with free surface, (**b**) pressurized

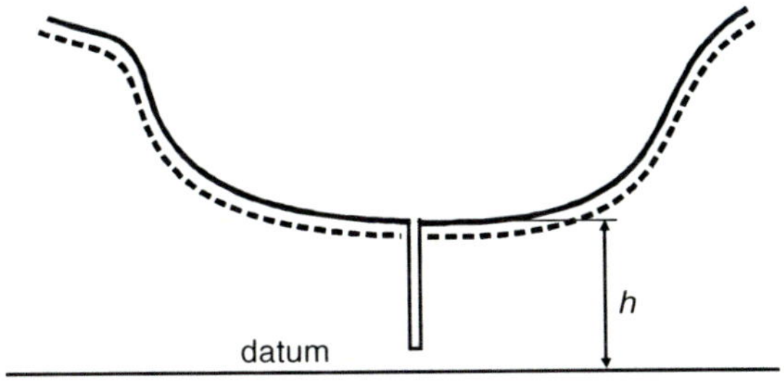

Fig. 8.13 The "Abbott slot" in a channel with dry bed

to the top and pressurized flow occurs. Since it happens only temporarily one can apply so called "Preissmann slot" (Cunge et al. 1980) shown in Fig. 8.12b. Owing to this concept it is possible to continue the computation using the same mathematical model, i.e. the system of Saint Venant equations. When the water level in conduit reaches its top, then further increasing of pressure does not cause increasing of the cross-section parameters. They are still constant, corresponding to the conduit entirely filled. Assumed width of slot should be relatively small (after Cunge et al. (1980) – of the order of 0.01 m) so that the mass balance remains not affected.

A similar idea can be also applied in open channels with temporary dry bed. This can arise in the channels of watering systems or in the vicinity of a pump station removing water from the polder. We have to remember that the equations of unsteady flow are valid for positive values of depth. In such a case one can apply a concept opposite to the "Preissmann slot", called "Abbott slot" (Abbott and Basco 1984). The slot goes down from the level of bottom as presented in Fig. 8.13. This technique allows us to continue the computations even if a channel bed becomes temporarily dry.

8.4.3 Initial and Boundary Conditions

As explained previously the uniqueness of solution of the system of Saint Venant equations requires us to impose appropriate initial conditions and boundary conditions.

The initial conditions give information on the state of channel flow at the beginning of the considered time interval, in which the equations will be solved. Their actual form can be different, depending on the considered situation:

1. In the case of nearly horizontal channel (e.g. in the lower part of a river), one can assume the hydrostatic state. The water is at rest, so the water stages over the entire channel reach are constant, whereas the flow discharge is equal to zero. The flow starts due to the imposed boundary conditions. Such an initial condition is rather unrealistic, but its influence disappears after some time, depending of the celerity of the propagating wave.
2. If the considered channel is prismatic, then the required water levels can be obtained via solution of the equation for steady uniform flow. For instance for

given flow discharge the Manning formulae is solved with regard to the normal depth. Alternatively, for given depth using the same formulae one can calculate the flow discharge.

3. For channels with variable cross-sections the water levels are found via solution of the equations of steady gradually varied flow. This is carried out for flow discharge given at one end of the considered reach.
4. The initial conditions can be given as the results of computation of unsteady flow carried out previously and stopped at the chosen moment to restart later.

As far as the boundary conditions are considered, for subcritical flow in a single channel at each end either $Q(t)$ or $h(t)$ must be specified. It should be remember that specification of the same type of function (Q or h) at both ends is not recommended, since in such a case the solution strongly depends of the initial conditions (Cunge et al. 1980).

At the end through which the water enters the channel reach one function $Q(t)$ or $h(t)$ must be imposed. They are obtained as the result of observation or as the effect of hydrological forecast for given time period. At the downstream end, through which the water leaves the channel reach, the situation is more complicated. There are only particular cases when the required functions $Q_L(t)$ or $h_L(t)$ are available.

If the considered river flows into the sea then at its downstream end one can specify the function $h_L(t)$ (Fig. 8.14). The water level in the sea is not determined by the river flow but rather by other factors acting on the sea surface, so $h_L(t)$ is independent information. In a particular case this condition can take the form $h_L(t) = \text{const}$.

If a dam is located at the downstream end (Fig. 8.15) then in this section neither flow rate $Q_L(t)$ nor the water stage $h_L(t)$ are known.

However we know discharge equation for the spillway, which relates both functions:

$$Q_L = K \cdot L_S \cdot \sqrt{2g} \cdot (h_L - h_S)^{3/2} \tag{8.101}$$

where:

Q_L – discharge,
K – flow coefficient of the spillway,

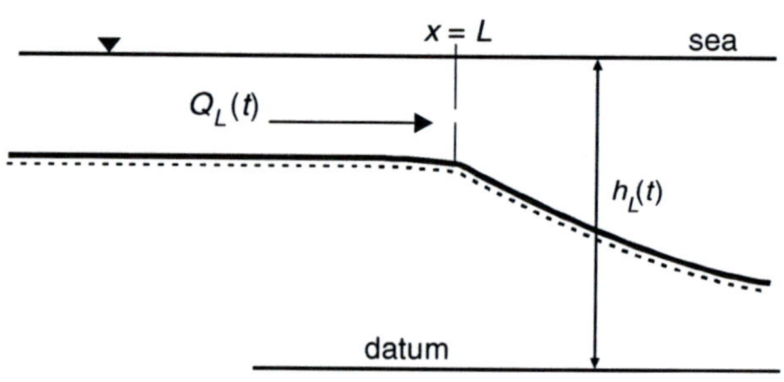

Fig. 8.14 Schematic of river flowing into the sea

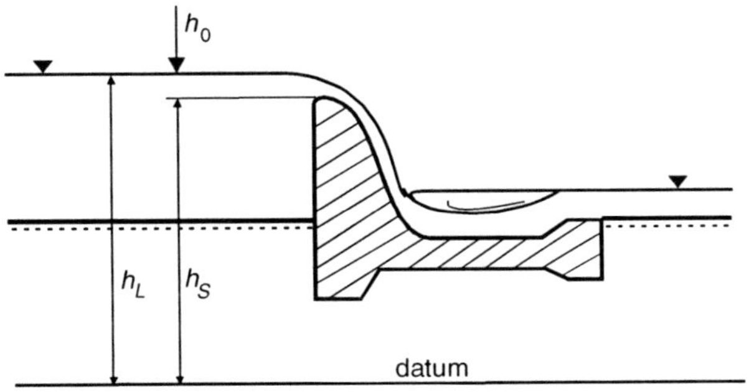

Fig. 8.15 Flow over spillway of a dam

L_S – length of the spillway,
h_L – water level with respect to assumed datum,
h_S – elevation of spillway crest above datum.

Such relation is an appropriate boundary condition, since it closes the final system of algebraic equations, which must be solved to obtain the searched values at the next time level.

Sometimes the river flows into a lake or reservoir having relatively small dimensions, so the inflow has a significant effect on the water stage in the recipient (Fig. 8.16). In such a case one can use the storage equation, which relates the unknown functions (see Section 1.7):

$$\frac{dh_L}{dt} = \frac{1}{F(h_L)} \left(Q_L(t) + q(t) + P(t) - E(t) + \cdots \right) \tag{8.102}$$

where:

h_L – water level in recipient above assumed datum,
F – area of recipient at the water level,
Q_L – inflow coming from river,
q – runoff from catchment,

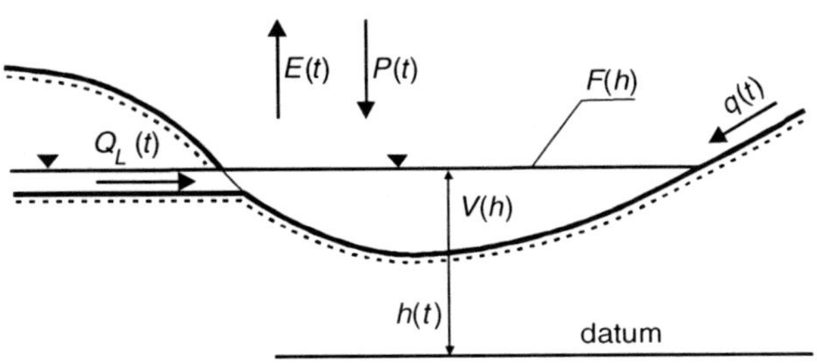

Fig. 8.16 River flowing into the lake

E – evaporation from water surface,
P – rainfall at surface of recipient.

Note, that in this case the additional relation has form of an ordinary differential equation. This equation, together with an appropriate initial condition of the type $h_L(t = 0) = h_{L0}$ (h_{L0} is given) must be solved simultaneously with the final system of algebraic equations given by the Preissmann difference scheme or by the finite element method. In the latter case this ordinary differential equation is simply incorporated into the system of ordinary differential equations obtained after spatial discretization of the governing equations.

In most cases, however, the downstream end of the considered channel reach is an ordinary cross-section in which is impossible to know the required function $Q_L(t)$ or $h_L(t)$. In such a case one can use relation between both mentioned functions in the form of rating curve

$$Q_L = f(h_L) \tag{8.103}$$

However, such an approach gives rise to some problems. Relation (8.103) is available as single valued function, whereas during the unsteady flow the hysteresis of this function is generated. Therefore introducing of Eq. (8.103) as boundary condition at the downstream end $x = L$ must be considered as a false independent information. To decrease or even to eliminate influence of inappropriate condition, the considered channel reach is artificially extended as in Fig. 8.17).

Therefore the boundary condition in the form of rating curve instead of $x = L$ is imposed at $x = L_1 > L$. Singh et al. (1997) examined influence of the new position of downstream end on results computed at $x = L$. They related the required distance $L_1 - L$ (Fig. 8.17) with the channel properties.

Alternative relation, which can be used, is the Manning formulae:

$$Q_L = \frac{1}{n_M} S^{1/2} \cdot R^{2/3} \cdot A \tag{8.104}$$

However, this implies the same problems as the previously used rating curve. Then Eq. (8.103) should be rather imposed at the end of extended channel reach. If we set this condition at $x = L$, variation of the energy line slope should be taken into account. This can be done accordingly to the recommendation given by Fread (1993). It is suggested to compute S using an approximate form of the dynamic equation (8.1). With the finite difference approximation this equation can

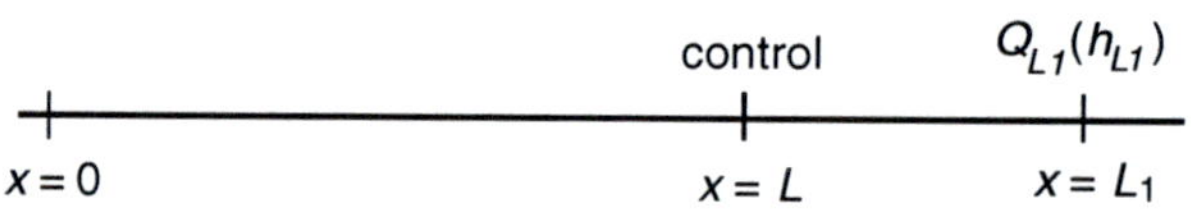

Fig. 8.17 Extending of river reach

be expressed as:

$$S \approx -\frac{Q_M^n - Q_M^{n-1}}{g \cdot A_M^n \cdot \Delta t} - \frac{\left(Q^2/A\right)_M^n - \left(Q^2/A\right)_M^{n-1}}{g \cdot A_M^n \cdot \Delta x_{M-1}} - \frac{h_M^n - h_M^{n-1}}{\Delta x_{M-1}} \tag{8.105}$$

where:

n – index of time level,
M – index of last node,
Δt – time step,
Δx – spatial interval.

This approach accounts for the hysteresis of the relation $Q(h)$ existing in the unsteady flow, since in Eq. (8.105) the terms responsible for this phenomenon are included.

8.4.4 Unsteady Flow in Open Channel Network

Very often in engineering practice we face more complex problem dealing with flow through a channel network. The problem is similar to the one presented in Section 4.4 for the gradually varied flow. As an elementary example let us consider the simplest case of network presented in Fig. 8.18. It consists of 3 branches (a, b and c) and one junction.

Assume that all branches have the same number of nodes which are numbered as presented. The total number of nodes is 15.

In such a case, using the finite difference or finite element method, we can set up appropriate systems of algebraic equations for each branch, as it was done previously for a single channel. Next we have to specify the boundary conditions at all ends of the pendant branches of considered network (node 1, 6 and 15). However to obtain the global system for the whole network, which will be solvable, we must introduce additional equations for the channel junction (Fig. 8.19):

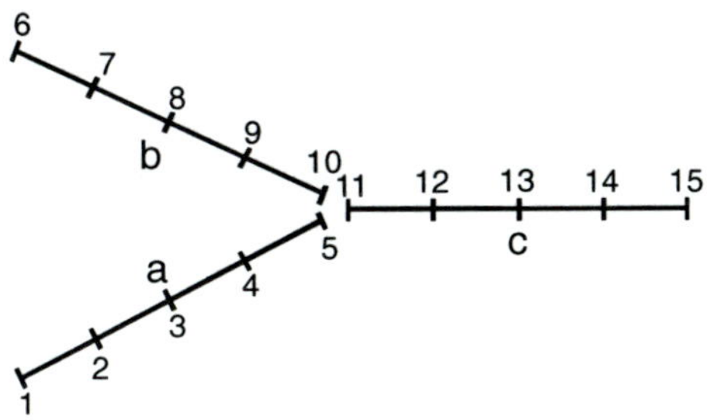

Fig. 8.18 Channel network containing three arms

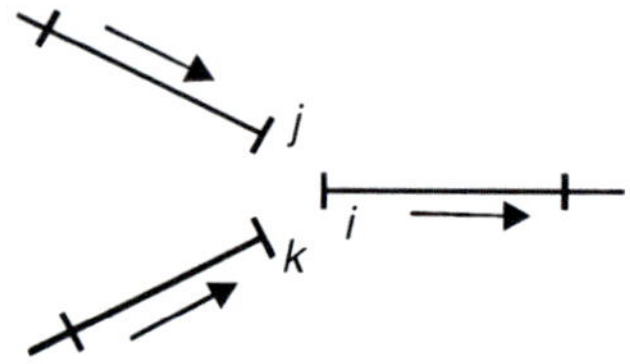

Fig. 8.19 Junction of three channels

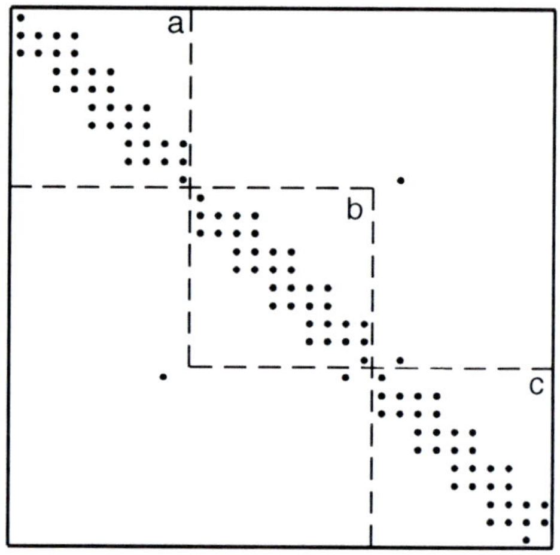

Fig. 8.20 Structure of matrix for system of equations for network presented in Fig. 8.18

$$Q_i = Q_j + Q_k, \tag{8.106a}$$

$$h_i = h_j, \tag{8.106b}$$

$$h_i = h_k \tag{8.106c}$$

The first relation is the continuity equation, whereas latter ones represent a simplified form of energy equation written for junction with the assumption that the head velocities and the local loses are neglected.

These additional equations close the global system of algebraic equations. For the Preissmann scheme the matrix of coefficients is shown in Fig. 8.20.

The matrix contains three blocks corresponding to consecutive braches a, b and c, which are linked by additional relations written for junction. Consequently it is very sparse and banded, however its bandwidth is much larger than the one obtained for a single channel (see Fig. 8.2). Note that the bandwidth depends on the numbering of nodes.

Example 8.2 This example deals with unsteady flow through the network of channels shown in Fig. 8.21 (Szymkiewicz 1991). This network is composed of:

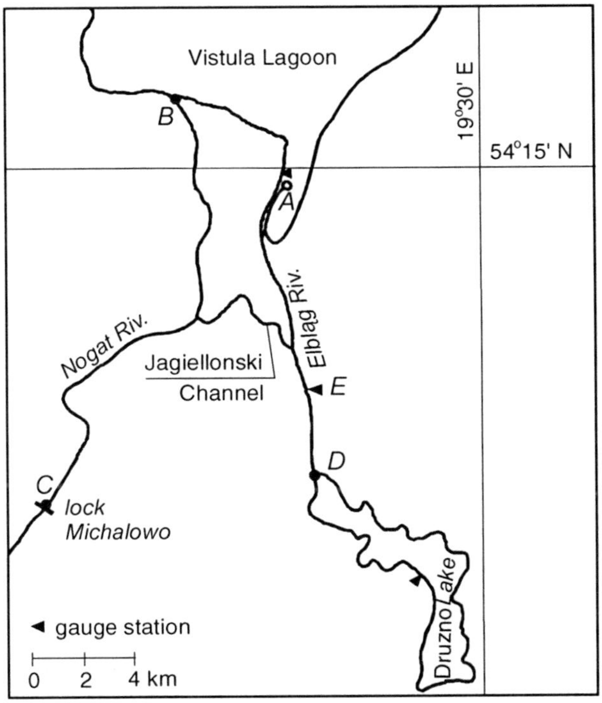

Fig. 8.21 Channel network considered in example (Szymkiewicz 1991)

1. The Elblag River of length 18.7 km, of width 35–64 m (at the water level) and of depth 2.0–4.25 m, connecting the Vistula Lagoon with Lake Druzno.
2. The Nogat River of length 22.7 km, of width ~125 m and depth ~2.5 m, being regulated cut-off arm of the Vistula River.
3. The Jagiellonski Channel of length 5.73 km, width 36 m and depth 2.70 m connecting both mentioned rivers.

Unsteady flow in the network is generated mainly by the changes of water level in Vistula Lagoon. These changes caused by wind acting on water surface and having amplitudes of ~3 m (±1.5 m from the mean sea level) give rise to the flow in variable directions: in or out the lake Druzno.

The initial condition corresponds to the hydrostatic state of the channel system. For $t = 0$, $h(x, t) = h_0 =$ const. and $Q(x, t) = 0$ are assumed. The boundary conditions are specified as follows:

– At points A and B the function $h_L(t)$ representing the water level elevation in the Vistula Lagoon is imposed: $h_A(t) = h_B(t) = h_L(t)$;
– At point C a lock practically cuts off the flow, so $Q_C(t) \sim 0$;

– At point D the Elblag River falls into Lake Druzno. At this point neither the function $h_D(t)$ nor $Q_D(t)$ is known. However, for the lake one can assume equation relating both mentioned functions in the form of the storage equation (8.102). Neglecting the influence of evaporation and rainfall on the lake's surface, this equation is rewritten as:

$$\frac{dh_D}{dt} = \frac{1}{F(h_D)}(Q_D(t) + q(t)) \tag{8.107}$$

where:

h_D – water level in recipient above assumed datum,
F – area of Lake Druzno at the water level h_D,
Q_D – inflow coming from Elblag River,
q – runoff from catchment's area,

For Eq. (8.107) the following initial condition is imposed: $h_D(t = 0) = h_0$.

Lake Druzno is very shallow (mean depth ~1.20 m) and surrounded by flood banks. Its water surface area varies with water level accordingly to the following function:

$$F(h) = \begin{cases} 13.0 & [\mathrm{km}^2] \text{ for } & h_D \le 0 \\ 13.0 + 200 \cdot h_D^2 & [\mathrm{km}^2] \text{ for } & 0 \le h_D \le 0.3 \text{ m} \\ 27.0 & [\mathrm{km}^2] \text{ for } & h_D > 0.3 \text{ m} \end{cases}$$

The largest area of the water surface occurs, when the water level in the lake reaches the foot of flood banks. This takes place for $h_D = 0.30$ m.

The results of calculations for the set of hydrological data coming from time period 11.10.1985 to 18.10.1985 are displayed in Fig. 8.22. The system of Saint Venent equations was solved using the modified finite element method for the following numerical data: $\Delta x = 0.700 \div 1.000$ km, $\Delta t = 1{,}800$s.

Example 8.3 This example deals with unsteady flow through the network of channels existing at a real polder in the delta of Vistula River. The considered network is composed of 5 channels as shown in Fig. 8.23. The channels are fed by many small drainage ditches. Next, the water gravitationally flows towards the end of channel number 5. At this point the pump station, equipped with 4 pump units, each of discharge 0.4 m^3/s, removes the water out of polder. The pump station is controlled automatically When the pumps are working it sometimes happens that the water can be completely removed from some reaches of channels so their beds become partially dry. In such a situation it is necessary to use the "Abbott slot". This afore mentioned technique allows us to continue the calculations even in the case when the water level temporarily is dropped below the channel bed.

The initial condition is given by the hydrostatic state of the channel system. It is assumed that at $t = 0$ the water is in rest with $h(x, t) = 4.75\ m = \text{const.}$ and $Q(x, t) = 0$. The boundary conditions are specified as follows:

– At the upstream ends of all pendant branches the condition $Q(t) = 0$ for $t \ge 0$ is imposed.

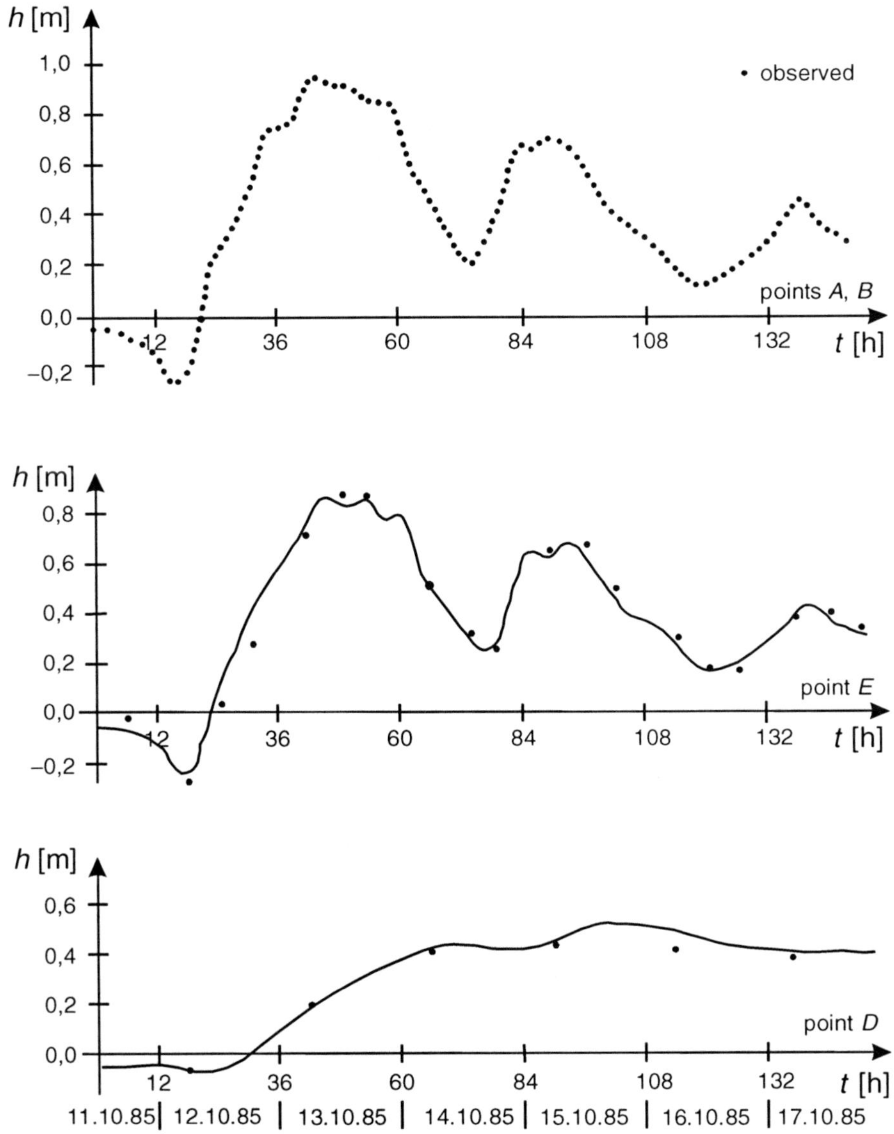

Fig. 8.22 Observed and calculated water surface level (Szymkiewicz 1991)

- At the downstream end of channel number 5 the flow discharge is forced by the pump station, then at this node $Q(t)=Q_{\text{pump}}(t)$ is specified. This function can change with a jump equal to 0.40 m^3/s ranging from 0 to 1.6 m^3/s.
- All channels are fed by lateral inflow caused by rainfall.

The system of Saint Venent equations was solved using the finite element method for the following numerical data: Δx=0.700 ÷ 1.000 km, Δt = 360 s.

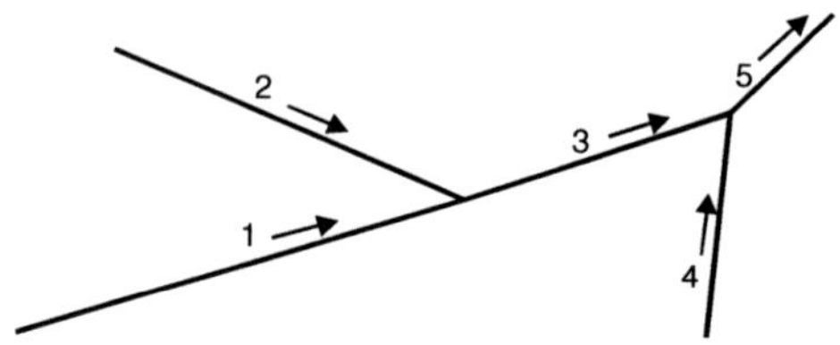

Fig. 8.23 Polder's open channel network

The result of calculations displayed in Figs. 8.24 and 8.25 show the flow profiles in channels 2 and 4 plotted for selected times. One can see that working pumps remove systematically the water from polder so the water level is decreased. After 12 h the lower part of channel 4 becomes dry. The length of this part is increasing. The same is observed in the upper part of channel 2. However the computations are continued successfully thanks to the applied "Abbott slot".

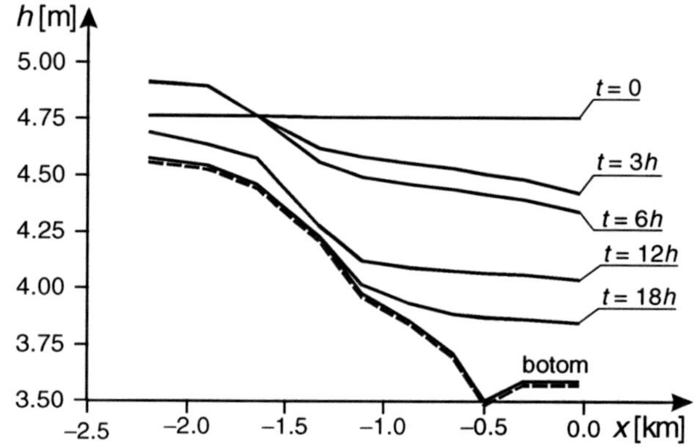

Fig. 8.24 The water profiles in channel 2

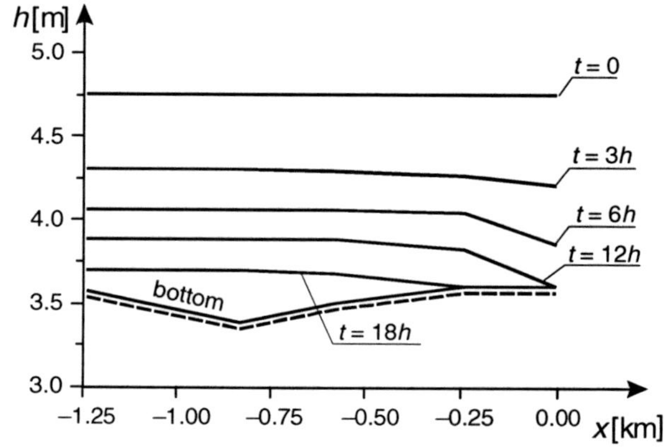

Fig. 8.25 The water profiles in channel 4

8.5 Solution of the Saint Venant Equations with Movable Channel Bed

8.5.1 Full System of Equations for the Sediment Transport

In preceding sections of this chapter we considered unsteady flow in open channel assuming that the channel bed is fixed. In fact there is mutual relation between the flowing water and the bed. Since in alluvial stream its bed is built of mineral grains then sufficiently large flow velocity forces their movement. They are washed away and, depending of their dimensions and the flow characteristics, they move as suspended in the water or they are transported along the bed in direction of flow (Fig. 8.26). The first kind of solid transport is called suspended-load, whereas the second one is called bed-load (Yalin and da Silva 2001). The motion of each kind of sediment is governed by different transport mechanisms.

When the bed material is washed it is said the erosion takes place. If the forces which cause movement of the grains are decreasing, then they settle. In such a case it is said that deposition takes place. Both processes change the flow conditions in a stream influencing the sediment transport as well. This suggests that taking into account the sediment transport we improve description of the flow process in open channel. However as a matter of fact the problem of sediment transport is one of the most complicated and challenging tasks for both scientists and engineers. This is caused by 3D spatially and random character of main factors determining solid transport. The readers who are interested in this problem can find many important publications presenting theoretical and practical aspects of problem (Yalin and da Silva 2001, Chanson 2004 among others). Some software packages, sometimes very sophisticated, supporting engineering practice, are available as well. In this section we will present some basic aspects of computation unsteady flow in channels with moving bed. These considerations will be limited to the 1D case only.

The modeling is based on the sediment transport continuity equation. In its general form this equation is derived from the mass conservation principle applied for control volume limited by control surface during two phase unsteady motion. The

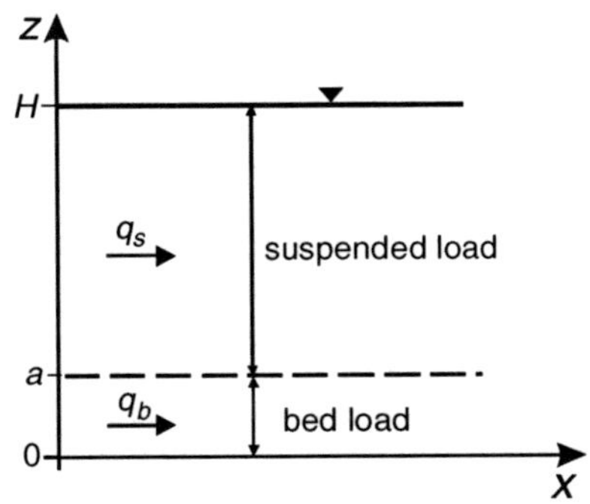

Fig. 8.26 Schematic representation of sediment transport zones: from 0 to a – bed load, from a to H – suspended load

one-dimensional form of this equation can be expressed as follows:

$$(1-p)\frac{\partial Z}{\partial t} + \frac{\partial q}{\partial x} = 0 \tag{8.108}$$

where:

x – space co-ordinate,
t – time,
Z – bed elevation above the assumed datum,
q – total (suspended and bed load) specific volumetric rate,
p – porosity.

For the first time this equation was derived by Exner (Cunge et al. 1980, Yalin and da Silva 2001) as a result of investigation of the bed forms. Equation (8.108) allows us to describe the evolution of the longitudinal bed profile in time, $Z(x, t)$. The variable q represents the sediment discharge rate and depends on the bed material and flow parameters. In the derivation of Eq. (8.108) the following assumptions were made:

- Lateral inflow of sediment does not exist,
- Distribution of the flow velocity and sediment concentration is uniform over the cross-section (Fig. 8.27),
- sediment travels in the direction of x axis ($q > 0$) and sediment transport per unit channel width is uniform

$$q = C \cdot U \cdot H \tag{8.109}$$

where U, C are averaged flow velocity and concentration over cross-section,

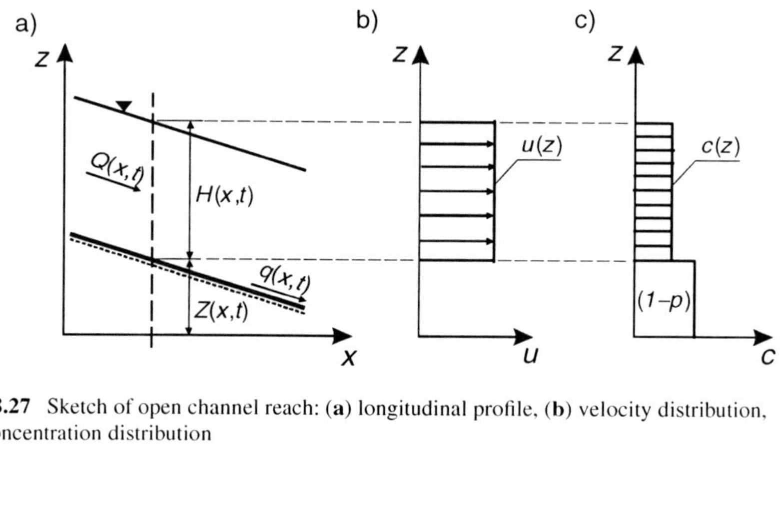

Fig. 8.27 Sketch of open channel reach: (**a**) longitudinal profile, (**b**) velocity distribution, (**c**) concentration distribution

- total specific volumetric rate represents the combined effect of the suspended load q_s and the bed load q_b

$$q = q_s + q_b \tag{8.110}$$

Generally this equation takes into account both suspended load and bed load although the time derivative of the suspended load was neglected (see for instance Yalin and da Silva 2001).

There are many formulas for determining the bed load rate q_b (see for example: Chanson 2004). One of the most frequently used is the empirical formula of Meyer-Peter and Muller, based on the results of field experiments:

$$q_b = 8\sqrt{g\frac{\gamma_s - \gamma_w}{\gamma_w}d_m^3}\left[\left(\frac{k_b}{k_s}\right)^{3/2}\frac{\gamma_w \cdot R \cdot S}{(\gamma_s - \gamma_w)\,d_m} - 0.047\right]^{3/2} \tag{8.111}$$

where:

q_b – specific volumetric bed-load per 1 m of channel width [m^3/s·m],
γ_s, γ_w – specific weight of sediment and water respectively [N/m^3],
R – hydraulic radius ,
S – energy grade line slope,
d_m – typical grain size (usually d_{50}),
k_d – Strickler coefficient being the inverse of Manning coefficient n_M :

$$k_d = \frac{1}{n_M} \tag{8.112}$$

k_s – granular roughness of bed surface given by:

$$k_s = \frac{21}{d_{50}^{1/6}} \tag{8.113}$$

Let us integrate the Exner equation (8.108) over a channel reach of length L:

$$\frac{d}{dt}\int_0^L (1-p)\,Z\,(x,t)dx = q\,(x=0,t) - q\,(x=L,t) \tag{8.114}$$

Note that the net flux of sediment material through the endpoints $x = 0$ and $x = L$ which bound channel reach causes changes of bed position (Fig. 8.28). Erosion will occur when the flux through the upstream end is less than one through the downstream end, i.e.

$$q(x=0,t) < q(x=L,t) \text{ thus } \partial q/\partial x > 0. \tag{8.115}$$

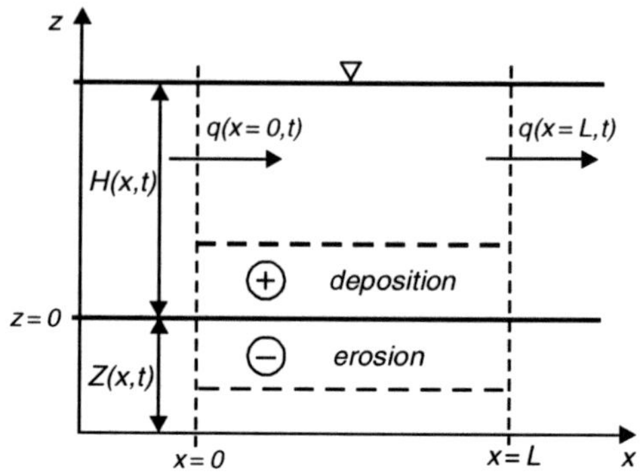

Fig. 8.28 Erosion and deposition caused by the net sediment material flux

Conversely, deposition will take place when:

$$q(x = 0, t) > q(x = L, t) \quad \text{thus} \quad \partial q / \partial x < 0. \tag{8.116}$$

Assume that the solution of the Exner equation is a continuous function. This allows us to express Eq. (8.108) as follows:

$$\frac{\partial Z}{\partial t} + \frac{1}{1 - p} \frac{\partial q}{\partial Z} \frac{\partial Z}{\partial x} = 0 \tag{8.117}$$

Introduction of a new variable:

$$C(Z) = \frac{1}{1 - p} \frac{\partial q}{\partial Z} \tag{8.118}$$

gives an alternate form of Eq. (8.117):

$$\frac{\partial Z}{\partial t} + C(Z) \frac{\partial Z}{\partial x} = 0 \tag{8.119}$$

One can see that the Exner equation is a nonlinear advection equation with advective velocity $C(Z)$. This velocity defines how quickly any disturbance of bed elevation propagates. Since the advection velocity is positive then each disturbance will propagate towards the downstream end. Of course for constant C any bed form will travel without shape deformation.

In the Exner equation the specific volumetric bed-load q or the advective velocity C depend on the hydraulic parameters of flowing stream, i.e. discharge rate and depth. These parameters are available from the solution of unsteady flow equations. For this purpose we must use the Saint Venant equations discussed previously. Neglecting the lateral inflow and introducing the depth $H(x, t)$ as dependent variable we obtain from Eqs. (1.87) and (1.88):

$$\frac{\partial Q}{\partial yt} + \frac{\partial}{\partial x}\left(\frac{\beta \cdot Q^2}{A}\right) + g \cdot A\frac{\partial H}{\partial x} + g \cdot A\frac{\partial Z}{\partial x} + g \cdot A \cdot S = 0, \tag{8.120}$$

$$\frac{\partial H}{\partial t} + \frac{1}{B}\frac{\partial Q}{\partial x} = 0. \tag{8.121}$$

Note that in this case the bed elevation above the assumed datum Z, being a function of x co-ordinate and time t, is an additional dependent variable. Therefore the system of Saint Venant equation must be completed with additional equation, i.e. with the previously presented Exner equation. In such a way one obtains a system of three partial differential equations. Simplified forms of the sediment transport model were also proposed – see e.g. Cunge et al. (1980).

8.5.2 Initial and Boundary Conditions for the Sediment Transport Equations

The equations describing the unsteady sediment transport should be solved for appropriately imposed auxiliary conditions on the limits of the following domain of integration: $0 \le x \le L$ and $t \ge 0$. The auxiliary conditions for the sediment transport model can be properly specified on condition that the structure of characteristics of considered system is well recognized. To this end let us consider the system of Saint Venant equations in the form of Eqs. (1.77) and (1.78) completed by the Exner equation written in the form of Eq. (8.119). Then we have:

$$\frac{\partial U}{\partial t} + U\frac{\partial U}{\partial x} + g\frac{\partial H}{\partial x} + g\frac{\partial Z}{\partial x} = -g \cdot S, \tag{8.122a}$$

$$\frac{\partial H}{\partial t} + \frac{\partial}{\partial x}(U \cdot H) = 0, \tag{8.122b}$$

$$\frac{\partial Z}{\partial t} + C\frac{\partial Z}{\partial x} = 0. \tag{8.122c}$$

A similar analysis was presented by Cunge et al. (1980). These authors used the Exner equation in the form of Eq. (8.108). We apply a slightly different approach, which will allow us to provide the equations of characteristic directions explicitly.

To find of the characteristics it is required at first to rewrite the system (8.122) in matrix notation. This system should be presented in the form of Eq. (5.14). Next the condition (5.15) is used. For the system (8.122) this condition takes the following form:

$$\det\begin{bmatrix} 1 & 0 & 0 & U & g & g \\ 0 & 1 & 0 & H & U & 0 \\ 0 & 0 & 1 & 0 & 0 & C \\ dt & 0 & 0 & dx & 0 & 0 \\ 0 & dt & 0 & 0 & dx & 0 \\ 0 & 0 & dt & 0 & 0 & dx \end{bmatrix} = 0. \tag{8.123}$$

Equation (8.123) leads to equation which determines the characteristics of system. Expanding the determinant (8.123) yields:

$$\left(\frac{dx}{dt}\right)^3 + (2U + C)\left(\frac{dx}{dt}\right)^2 + \left(2U \cdot C + U^2 - g \cdot H\right)\left(\frac{dx}{dt}\right) - \left(U^2 \cdot C - g \cdot H \cdot C\right) = 0 \tag{8.124}$$

This equation is of third degree with regard to *dx*/*dt*. Its roots determine the characteristics of system (8.122). To find all roots it is helpful to notice that the polynomial (8.124) has, among others, the following real root:

$$\frac{dx}{dt} = C. \tag{8.125}$$

Thus Eq. (8.124) can be expressed as a product of two polynomials of lower degrees. Dividing this polynomial by the following one:

$$\left(\frac{dx}{dt} - C\right) \tag{8.126}$$

yields its equivalent form:

$$\left(\frac{dx}{dt} - C\right)\left(\left(\frac{dx}{dt}\right)^2 - 2U\left(\frac{dx}{dt}\right) + \left(U^2 - g \cdot H\right)\right) = 0. \tag{8.127}$$

The obtained polynomial of 2nd degree is well known. It has been obtained previously (Eq. 5.24), while analyzing the characteristics of the Saint Venant equations. Its two real roots are given by Eqs. (5.26a) and (5.26b). Therefore Eq. (8.127) can be rewritten in the following form:

$$\left(\frac{dx}{dt} - C\right)\left(\frac{dx}{dt} - \left(U + \sqrt{g \cdot H}\right)\right)\left(\frac{dx}{dt} - \left(U - \sqrt{g \cdot H}\right)\right) = 0 \tag{8.128}$$

From this equation results that polynomial (8.124) has three following roots:

$$\left.\frac{dx}{dt}\right|_1 = C, \tag{8.129a}$$

$$\left.\frac{dx}{dt}\right|_2 = U + \sqrt{g \cdot H}, \tag{8.129b}$$

$$\left.\frac{dx}{dt}\right|_3 = U - \sqrt{g \cdot H} \tag{8.129c}$$

These roots determine three families of characteristics of the considered system (8.122). Since all of them are real, the system is of hyperbolic type.

The problem of solution of the system (8.122) will be well posed if appropriate initial and boundary conditions on the limits of solution domain are prescribed. In

Chapter 5 we found out, that for the hyperbolic equations the following rule is valid: at every boundary of the considered solution domain it is necessary to impose as many additional conditions as many characteristics enter the solution domain from this boundary. Taking into account this rule and Eqs. (8.129a), (8.129b) and (8.129c) for subcritical flow $\left(U < \sqrt{g \cdot H}\right)$ we must impose:

- The initial conditions in the form of three functions: $H(x,\ t = 0) = H_i(x)$, $Q(x, t = 0) = Q_i(x)$ and $Z(x, t = 0) = Z_i(x)$ for $0 \leq x \leq L$, since three characteristics are going through the time boundary $t = 0$;
- The boundary conditions in the form of two functions (one for flow and one for sediment transport) at the upstream channel end and one function (for flow only) at the downstream end.

For supercritical flow $\left(U < \sqrt{g \cdot H}\right)$ the same initial conditions must be prescribed, whereas the boundary conditions are defined in another way. At the upstream end all three functions must be imposed while at the downstream end the boundary condition is not specified at all.

If the sediment transport in open channel network is considered then additional equations have to be introduced for the junctions of channels. They are analogous to those presented in Section 4.4.2.

8.5.3 Numerical Solution of the Sediment Transport Equations

There are two main ways of numerical treatment of the system of sediment transport equations (Cunge et al. 1980). In the first approach the system constituted by the Saint Venant equations (Eqs. 8.120 and 8.121) and the Exner equation (Eq. 8.119) is solved as three coupled equations, with imposed initial and boundary conditions. All three equations are approximated using the finite difference or element technique leading to the global system of algebraic equations. Usually the difference box scheme is applied. This scheme is particularly suitable for this purpose as it works well for both the Saint Venant equations (Preissmann scheme – see Section 8.2) and the pure advection equation (see Chapter 6). Of course the final system of algebraic equation given by this approach is larger than the one obtained for the Saint Venant equations.

Another possible approach, allows us to decrease the computational effort by splitting the considered problem in each time step into two smaller problems. The governing system of equations is uncoupled and in each time step the equations for unsteady flow and one for the sediment transport are solved separately but simultaneously. Then we have to use directly the solution techniques presented in the preceding sections of this chapter for the unsteady flow as well as those for the pure advection equations described in Chapter 6. Usually for both problems the same grid points should be used. In such a way in each time step instead of one relatively large system of algebraic equations two much smaller systems are solved.

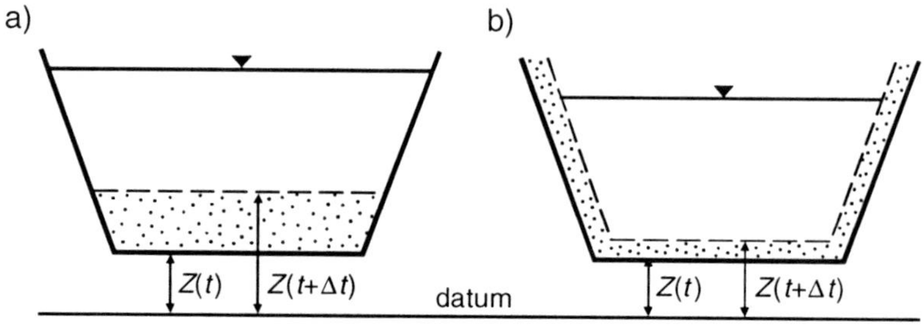

Fig. 8.29 Possible distribution of settled sediment over trapezoidal cross-section: deposition of sediment on the bottom only (**a**) and on the bottom and sides (**b**)

The problem of great importance is the relation between the volumetric sediment rate which gives rise to erosion or deposition and time deformation of the shape of channel's cross-sections. Various approaches are possible (Cunge et al. 1980). For the simplest case as in prismatic channel one can assume that the whole sediment is eroded from or deposed at the bottom only (Fig. 8.29a) or it is distributed over the bottom as well as over the channel sides (Fig. 8.29a). The problem of sediment distribution becomes more complex for an arbitrary shape of the channel cross-section.

Example 8.4 A channel reach of length $L = 50$ km and of bed slope $s = 0.4 \cdot 10^{-3}$is ended by a dam. Assuming that the channel is rectangular of width $B = 50$ m and of Manning roughness coefficient $n_M = 0.025$, compute the time evolution of the bed elevation $Z(x,t)$ in the reservoir caused by the transported sediment.

The sediment flow rate is calculated using the Meyer-Peter formulae (8.111) for sand characterized by typical grain size $d_m = 1$ mm, whereas the porosity is assumed to be $p = 0.4$. The system of Saint Venant equations (8.120) and (8.121) and the Exner equation (8.108), are solved uncoupled using the Preissmann scheme.

The initial conditions at $t = 0$ are determined assuming steady gradually varied flow with constant discharge $Q_i(x) = 330$ m^3/s for $0 \leq x \leq L$. The flow profile $h(x, t = 0)$ is found from the solution of the steady gradually varied flow equation (Eq. 4.2) with the initial condition $h(x = L) = 10$ m being the water level close to a dam. It enables us to define the initial flow depth $H(x, t = 0)$ for the assumed initial bed profile $Z_i(x, t = 0)$.

At $x = 0$ two following boundary conditions are assumed:

– flow rate given by the formulae

$$Q_0(t) = Q_i + Q_m \left(\frac{t}{t_m}\right)^2 \exp\left(1 - \left(\frac{t}{t_m}\right)^2\right) \tag{8.130}$$

with Q_m= 1,150 m^3/s andt_m= 25 days is imposed.
– bed elevation during the transition of the flood wave given by

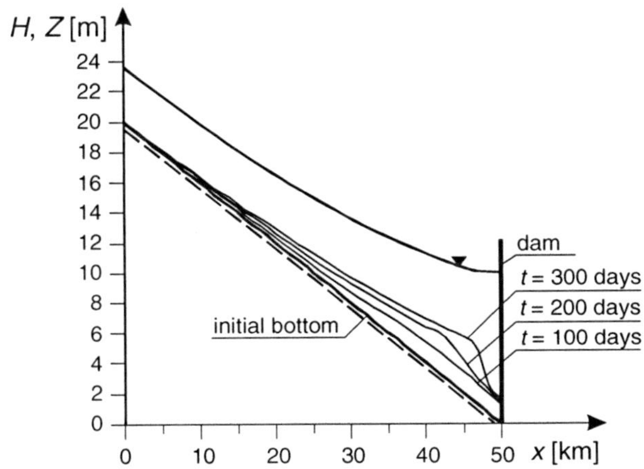

Fig. 8.30 Evolution of the bed elevation in reservoir due to transported sediment

$$Z_0(t) = Z_i + Z_m \left(\frac{t}{t_m}\right)^2 \exp\left(1 - \left(\frac{t}{t_m}\right)^2\right) \tag{8.131}$$

where the assumed peak value $Z_m = 0.5$ m corresponds to time t_m.

At the downstream end $x = L$, where a dam exists, a constant water level $h(x = L, t) = 10$ m is imposed.

The channel reach is divided into intervals of constant length $\Delta x = 1{,}000$ m. The computations are performed for the time step $\Delta t = 2{,}200$ s with the value of weighting parameter $\theta = 0.65$ (see Sections 6.2 and 8.2). The computed results are displayed in Fig. 8.30.

Note that the obtained results represent qualitative agreement with observed effects of grains settlement in the reservoir. Systematical decrease of the flow velocity in the reservoir towards a dam increases deposition of solid material transported by the river. Consequently the reservoir is loosing its usable storage with time, since it is continuously fulfilled by entering sediment.

8.6 Application of the Saint Venant Equations for Steep Waves

8.6.1 Problem Presentation

The system of Saint Venant equations was derived with the assumption that the flow is gradually varied. However in many cases we face another type of flow. Sometimes the flow profile in open channel varies locally so rapidly that in fact a discontinuity occurs. Such very steep wave in the form of surge can occur in many circumstances.

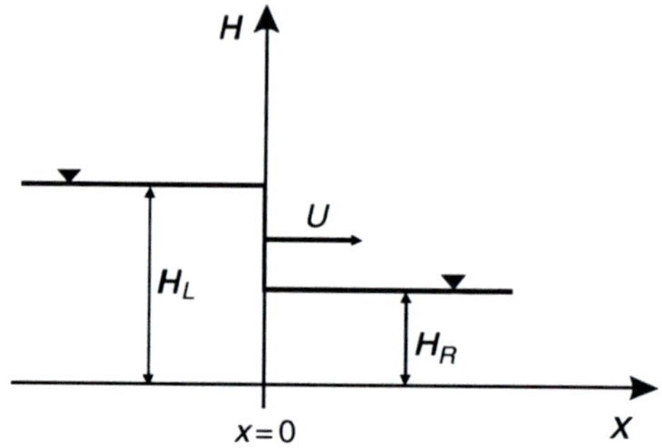

Fig. 8.31 Propagating wave with discontinuity

For instance, the wave of surge type can be generated by the operation of hydro–power plant, when the turbines are started or stopped suddenly. Another reason of surges can be sudden opening or closing of a gate. The water waves arising in such situations can be idealized as in Fig. 8.31.

Similarly, very steep wave in the form of surge can occur when a tide enters the river mouth. Usually the surge generated by tide is termed a bore (Chanson 2004). The difference in nomenclature does not mean the differences in principles of propagation of bores and surges. Another type of shock wave occurs when a dam separating different water levels is suddenly removed. Analytical analysis of idealized propagation of bore and dam-break wave is given by Billingham and King (2000).

For those types of steep waves the system of Saint Venant equations is formally not valid, since in its derivation the vertical acceleration has been omitted. Note that not every wave caused by dam-break has to be considered as strictly unsteady flow with discontinuities. Since the formation of the breach (the opening in a dam which forms as a dam fails) needs some time, the outflow from reservoir is more evenly distributed in time (French 1985). This fact makes reasonable using even such a simple model as the storage equation (1.120). Such approach was proposed and developed by Fread (1993).

On the other hand, in hydraulic engineering when the main subject of interest is flow considered in large scale, the local untypical phenomena can be less important and their internal structure may be neglected. For instance, although the internal structure of the hydraulic jump can be effectively reproduced by solving the Reynolds averaged Navier-Stokes equations and continuity equation completed with the appropriate model of turbulence (Cippada et al. 1994), such approach is not applied in open channel flow modeling. As the length of the hydraulic jump is small comparing with the spatial dimension of practically applied grid, it is spatially reduced to a point, in which the water surface is discontinuous. It appears that with such approach the Saint Venant equations can be used as an effective model of flow. However some additional conditions are required.

First of all the system of unsteady flow equations must be expressed in particular, so called conservative form. This means that the form of equations have to satisfy the integral form of conservation principles, from which they were derived. Note that the system of Saint Venant equations can be written in integral form as well as in differential form (Chanson 2004, Cunge et al. 1980). Both representations are equivalent as long as smooth solutions are considered. The difference becomes noticeable when discontinuities occur. In Chapter 1 the Saint Venant equations have been derived directly from the fundamental differential equations of the fluid mechanics and we used them in the differential form. However one can expect that the differential equation does not hold near discontinuity in the solution (LeVeque 2002).

The second condition concerns the applied numerical method. It is assumed that standard finite difference methods based on the Taylor series expansion, can break down near discontinuity arising in the solution (LeVeque 2002). An appropriate method should ensure required solution accuracy. Particularly it is expected that the considered conservation principles will be satisfied on the numerical grid. This requirement is satisfied particularly exactly by the finite volume method based on the equations written in integral form. However, to solve successfully some flow problems of shock wave type one can use the finite difference methods as well. The flow of shock wave type means that we consider flow with large gradients rather than containing pure discontinuity.

8.6.2 Conservative Form of the Saint Venant Equations

Let us remember the system of the Saint Venant equations (1.77) and (1.78) obtained for a wide rectangular channel and written per unit width:

$$\frac{\partial H}{\partial t} + U\frac{\partial H}{\partial x} + H\frac{\partial U}{\partial x} = 0, \tag{8.132}$$

$$\frac{\partial U}{\partial t} + U\frac{\partial U}{\partial x} + g\frac{\partial H}{\partial x} = g(s - S), \tag{8.133}$$

where:

x – space co-ordinate,
t – time,
U – cross-sectional average flow velocity,
H – flow depth,
s – bed slope,
S – energy grade line slope'
g – acceleration due to gravity.

Equations (8.132) and (8.133) represent the mass and momentum conservation laws respectively. These so called the 1D shallow water equations with source term are written in the non-conservative form. It means that these equations do not satisfy the integral form of conservation law from which they were derived.

By a simple transformation the equations (8.132) and (8.133) can be reformed to the following equivalent conservative form:

$$\frac{\partial H}{\partial t} + \frac{\partial (U \cdot H)}{\partial x} = 0 \tag{8.134}$$

$$\frac{\partial (U \cdot H)}{\partial t} + \frac{\partial \left(U^2 \cdot H\right)}{\partial x} + \frac{1}{2} g \frac{\partial H^2}{\partial x} = g \cdot H(s - S). \tag{8.135}$$

To show the difference between both forms of the Saint Venant equations let us integrate the continuity equation (8.132) with regard to x over considered channel reach of length L. Integration performed by parts leads to the following expression:

$$\begin{aligned} \frac{\partial}{\partial t} \int_0^L H \cdot dx = {} & [U \cdot H]_0 - [H \cdot U]_L + \\ & + \left([U \cdot H]_0 - [H \cdot U]_L + \int_0^L H \cdot \frac{\partial U}{\partial x} \cdot dx + \int_0^L U \cdot \frac{\partial H}{\partial x} \cdot dx \right) \end{aligned} \tag{8.136}$$

Similar integration of Eq. (8.134) over considered channel reach this time gives:

$$\frac{\partial}{\partial t} \int_0^L H \cdot dx = [U \cdot H]_0 - [H \cdot U]_L \tag{8.137}$$

Comparing Eq. (8.136) with Eq. (8.137) we notice that although in both cases the continuity equation was integrated the obtained results differ significantly. Equation (8.137) expresses strictly the law of mass conservation, since it shows that time variation of water capacity stored by a channel reach of length L is caused by the net flux through its endpoints only. We conclude that the continuity equation (8.134) is written in the conservative form. Conversely, Eq. (8.132) is not written in the conservative form since Eq. (8.136) obtained by its integration does not represent the mass conservation principle. Besides the first two terms at right hand side representing net flux through the channel endpoints, we have additional terms written in brackets. The same is for the dynamic equation. Consequently, if we integrate numerically the system (8.132) and (8.133) the errors in the mass and momentum balances can occur. These errors, being insignificant for smooth solution, can be appreciable when strong gradients occur, particularly when the discontinuities in solution arise.

A rough evaluation of the difference between conservative and non-conservative forms of the 1-D shallow water equations can be carried on the example of the numerical solution of the dam break problem. We shall consider an idealized case of dam break when the vertical wall separating two different water levels in a channel is suddenly removed. For simplicity we will solve the homogeneous version of the systems (8.132) and (8.133) as well as of (8.134) and (8.135). It means that the unsteady flow will take place in a horizontal and frictionless channel.

Example 8.5 Let us solve subsequently the following systems of equations:

$$\frac{\partial H}{\partial t} + U\frac{\partial H}{\partial x} + H\frac{\partial U}{\partial x} = 0 \tag{8.138}$$

$$\frac{\partial U}{\partial t} + U\frac{\partial U}{\partial x} + g\frac{\partial H}{\partial x} = 0 \tag{8.139}$$

and

$$\frac{\partial H}{\partial t} + \frac{\partial q}{\partial x} = 0 \tag{8.140}$$

$$\frac{\partial q}{\partial t} + \frac{\partial}{\partial x}\left(\frac{q^2}{H}\right) + \frac{1}{2}g\frac{\partial H^2}{\partial x} = 0 \tag{8.141}$$

where $q = U/H$ is the flow rate per unit channel width. A wide rectangular channel reach of length L is divided into two parts by the dam located at its mid-length $x = L/2$ (Fig. 8.32).

Both systems will be solved for the same initial and boundary conditions corresponding to the dam-break problem. They are specified as follows:

- at $t = 0$ the water is at rest filling the channel up to H_u= 3.0 m above the bottom at the left side of the dam ($0 \le x \le L/2$) and up to H_d = 0.5 m above the bottom at its right side ($L/2 < x \le L$);
- at the upstream end $x = 0$ the flow depth is constant and corresponds to the initially imposed value: $H(x = 0, t) = 3.0$ m for $t \ge 0$;

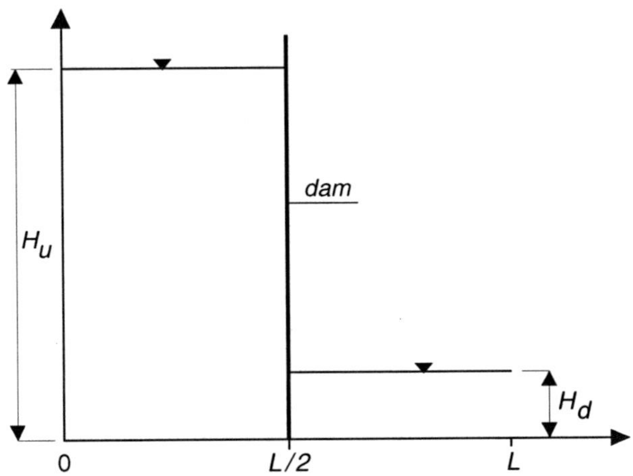

Fig. 8.32 Initial water levels at both sides of the dam

– at the downstream end $x = L$ a constant depth corresponding to the initial one H_d is imposed: $H(x = L, t) = 0.5$ m for $t \geq 0$ (L is assumed large enough that the downstream end will not be reached by propagating wave).

Both systems of equations are solved using the difference Preissmann scheme, which in Section 8.2 was applied for solving the system of Saint Venant equations for non-prismatic channel. The computations carried out for $L = 1{,}250$ m, $\Delta x = 0.5$ m, $\Delta t = 0.10$ s and $\theta = 0.67$ confirmed that appreciable difference between the conservative and non-conservative forms of solved equations exists. Evaluation of this difference is performed in terms of the balance of the water volume stored in considered channel, which must be constant during the flow process. After very short time $t = 120$ s the system written in the non-conservative form generates the balance error of magnitude $\sim$0.18%, whereas for the system in conservative form this error practically does not exist.

The mass and momentum balance errors can be avoided on condition that the non-linear equations of unsteady flow will be written in proper conservative form. This is true for all non-linear transport equations. Comprehensive discussion on the solution of non-linear hyperbilic equations is presented for instance by LeVeque (2002), Gresho and Sani (2000) and by others. We will continue this issue in Chapter 9, while discussing the simplified forms of the unsteady flow equations.

8.6.3 Solution of the Saint Venant Equations with Shock Wave

When in the considered case of flow discontinuities appear, the computation should be carried out very carefully. First of all, as it was stated previously, the Saint Venant equations must be expressed in the conservative form. Secondly, their solution needs a particular numerical approach. Cunge et al. (1980) present three possible techniques.

The first approach, called the shock fitting method, treats the problem of discontinuity propagation and the unsteady flow between the discontinuities separately. Positions of the traveling discontinuities are calculated using the method of characteristics, whereas the Saint Venant equations which hold in the regions limited by discontinuities, are solved using the standard methods.

The second possible approach, called pseudoviscosity method, requires introduction of a diffusive term into the dynamic equation. This term, having smoothing properties, allows us to control the solution near discontinuity. Its particular form relating intensity of generated diffusion with the wave steepness ensures that this extra term acts strongly only locally, being insignificant far away from the discontinuity. This approach is applicable for non-dissipative schemes. More information on the pseudoviscosity method is given by Potter (1973) and Cunge et al. (1980).

The last possible approach listed by Cunge et al. (1980) bases on the solution using the dissipative methods. In this approach is assumed that the process of smoothing is ensured by the numerical diffusion generated by the applied scheme.

Controlling artificial diffusion one can obtain an acceptable solution. Such technique was applied in Example 8.5 in which the problem of propagation of the shock wave caused by the dam break has been solved using the Preissmann scheme.

The disadvantage of the presented approaches is that they base on the differential equations. This can be avoided by using the finite volume method. The finite volume method has two important features. Firstly, it ensures very good conservation of the transported quantity since it uses the flow equations in the integral form. Secondly, it ensures a relatively simple discretization of complex domain of solution. Of course, in the 1D problem as considered here the last mentioned advantage is not appreciable, whereas it arises distinctly in 2D and 3D problems.

The finite volume method appears effective tool for solving the propagation of shock waves particularly for extreme auxiliary conditions. However it deals with rather idealized situation, when the homogenous equations are solved. Unfortunately the method cannot be simply implemented for more realistic cases. Presence of the source terms in form of the bed and grade line energy slope requires applying of non-trivial approaches to overcome occurring difficulties. Interesting discussion on the finite volume method is presented by Gresho and Sani (2000). Detailed description of the method and its implementation for solution of the 1D shallow water equations is given by LeVeque (2002) among others. This rather complicated algorithm is not presented here, but at the end of this section an example of solution of the dam break problem using the finite volume method will be presented.

The results presented in the last example were obtained using the Preissmann scheme. We know that this classical scheme is a reliable tool for solution of the Saint Venant equations when they are applied for unsteady gradually varied flow. On the other hand, as it was mentioned previously, the differential representation of the flow problem is not valid near a discontinuity. Nevertheless, the Saint Venant equations solved by the classical dissipative schemes are frequently applied (Cunge et al. 1980). As a consequence of this discrepancy the standard schemes appear less reliable than for gradually varied flow. Some computational aspects are illustrated by the presented example.

Example 8.6 Solve the system of equations (8.140) and (8.141) in a wide rectangular channel of length L for the initial and boundary conditions corresponding to surge propagation. They are specified as follows:

- at $t = 0$ the water is at rest filling the channel up to $H_0 = 0.75$ m above the bottom;
- at the upstream end $x = 0$ the flow rate is suddenly increased from 0 to 5 m^3/s/m:

$$q\,(x = 0,t) = \begin{cases} 0 & \text{for} \quad t \leq 0 \\ 5.0\ \mathrm{m^3/s/m} & \text{for} \quad t > 0 \end{cases}$$

- at the downstream end $x = L$ a time constant depth corresponding to the initial one H_0 is imposed: $H(x = L,\, t) = H_0$ for $t \geq 0$ (L is assumed large enough that this end will not be reached by the propagating surge).

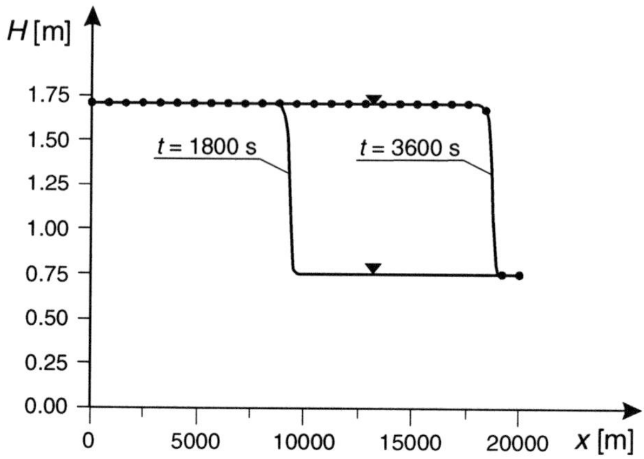

Fig. 8.33 Flow profiles during a surge propagation in horizontal frictionless channel

The system of equations is solved using the difference Preissmann scheme, (see Section 8.2). The computations carried out for $L = 50\,\text{km}$, $\Delta x = 100$ m, $\Delta t = 20\,\text{s}$ and $\theta = 0.67$ furnished the results shown in Fig. 8.33.

As far as the Preissmann scheme is considered, it can be successfully used for computation of the surge propagation in some circumstances only. First of all, the scheme cannot handle initially dry bed. Conversely, the initial depth should be enough large to avoid the failure of computations. Next the scheme does not accept too large and too sudden increase of the boundary condition at the upstream end. Increasing of the flow rate per unit width should not be forced immediately but during approximately $(4 \div 6)\Delta t$. This gives rise to some smearing of the steep front of wave. Elimination of the wiggles connected with scheme's dispersivity requires that some numerical diffusion is introduced. This is performed by choosing of an appropriate value of the weighting parameter θ. If the mentioned conditions are satisfied, the Preissmann scheme is capable to provide quite reasonable results as presented in Fig. 8.33.

Slightly better numerical properties characterize the modified finite element method with upwind effect. To show this approach let us reconsider the system of Saint Venant equations in conservative and homogenous form

$$\frac{\partial H}{\partial t} + \frac{\partial q}{\partial x} = 0 \tag{8.142}$$

$$\frac{\partial q}{\partial t} + \frac{\partial}{\partial x}\left(\frac{q^2}{H}\right) + \frac{1}{2}g\frac{\partial H^2}{\partial x} = 0 \tag{8.143}$$

Approximation of these equations on a uniform grid is carried out as follows:

- At first the advective term in the momentum equation is approximated using difference formulae (5.104):

$$\frac{\partial}{\partial x}\left(\frac{q^2}{H}\right)\bigg|_j = \frac{1}{\Delta x}\left(-\eta\frac{q_{j-1}^2}{H_{j-1}} + (2\eta - 1)\frac{q_j^2}{H_j} + (1-\eta)\frac{q_{j+1}^2}{H_{j+1}}\right) \tag{8.144}$$

where $\eta = 0$ corresponds to the forward difference, $\eta = 0.5$ corresponds to the centred difference, whereas $\eta = 1$ gives the backward difference.

– Next other terms of both equations are approximated with regard to x using the difference scheme basing on the modified finite element method (see Section 8.3.1). This yields the system of ordinary differential equations with regard to time:

- for $j = 1$

$$\omega\frac{dH_j}{dt} + (1-\omega)\frac{dH_{j+1}}{dt} + \frac{-q_j + q_{j+1}}{\Delta x} = 0 \tag{8.145a}$$

$$\omega\frac{dq_j}{dt} + (1-\omega)\frac{dq_{j+1}}{dt} + \frac{1}{\Delta x}\left(-\frac{(q_j)^2}{H_j} + \frac{(q_{j+1})^2}{H_{j+1}}\right) + \frac{g}{2}\frac{-(H_j)^2 - (H_{j+1})^2}{\Delta x} = 0 \tag{8.145b}$$

- for $j = 2, 3, \ldots, M - 1$

$$\frac{1-\omega}{2}\frac{dH_{j-1}}{dt} + \omega\frac{dH_j}{dt} + \frac{1-\omega}{2}\frac{dH_{j+1}}{dt} + \frac{-q_{j-1} + q_{j+1}}{2\Delta x} = 0 \tag{8.145c}$$

$$\frac{1-\omega}{2}\frac{dq_{j-1}}{dt} + \omega\frac{dq_j}{dt} + \frac{1-\omega}{2}\frac{dq_{j+1}}{dt} + \frac{1}{\Delta x}\left(-\eta\frac{(q_{j-1})^2}{H_{j-1}} + (2\eta-1)\frac{(q_j)^2}{H_j} + (1-\eta)\frac{(q_{j+1})^2}{H_{j+1}}\right) + \frac{g}{2}\frac{-(H_{j-1})^2 + (H_{j+1})^2}{2\Delta x} = 0 \tag{8.145d}$$

- for $j = M$

$$(1-\omega)\frac{dH_{j-1}}{dt} + \omega\frac{dH_j}{dt} + \frac{-q_{j-1} + q_j}{\Delta x} = 0 \tag{8.145e}$$

$$(1-\omega)\frac{dq_{j-1}}{dt} + \omega\frac{dq_j}{dt} + \frac{1}{\Delta x}\left(-\frac{(q_{j-1})^2}{H_{j-1}} + \frac{(q_j)^2}{H_j}\right) + \frac{g}{2}\frac{-(H_{j-1})^2 + (H_j)^2}{\Delta x} = 0 \tag{8.145f}$$

where M is total number of nodes.

– The system of Eq. (8.145) is integrated over time using the formulae (8.54). Finally one obtains the following system of algebraic equations:

- for $j = 1$

$$\omega\frac{H_j^{n+1} - H_j^n}{\Delta t} + (1-\omega)\frac{H_{j+1}^{n+1} - H_{j+1}^n}{\Delta t} + (1-\theta)\frac{-q_j^n + q_{j+1}^n}{\Delta x} + \theta\frac{-q_j^{n+1} + q_{j+1}^{n+1}}{\Delta x} = 0 \tag{8.146a}$$

$$\omega\frac{q_j^{n+1}-q_j^n}{\Delta t}+(1-\omega)\frac{q_{j+1}^{n+1}-q_{j+1}^n}{\Delta t}+$$
$$+\frac{1-\theta}{\Delta x}\left(-\frac{\left(q_j^n\right)^2}{H_j^n}+\frac{\left(q_{j+1}^n\right)^2}{H_{j+1}^n}\right)+\frac{\theta}{\Delta x}\left(-\frac{\left(q_j^{n+1}\right)^2}{H_j^{n+1}}+\frac{\left(q_{j+1}^{n+1}\right)^2}{H_{j+1}^n}\right)+$$
$$+\frac{1-\theta}{\Delta x}\frac{g}{2}\left(-\left(H_j^n\right)^2+\left(H_{j+1}^n\right)^2\right)+\frac{\theta}{\Delta x}\frac{g}{2}\left(-\left(H_j^{n+1}\right)^2+\left(H_{j+1}^{n+1}\right)^2\right)=0 \tag{8.146b}$$

- for $j = 2, 3, \ldots, M-1$

$$\frac{1-\omega}{2}\frac{H_{j-1}^{n+1}-H_{j-1}^n}{\Delta t}+\omega\frac{H_j^{n+1}-H_j^n}{\Delta t}+\frac{1-\omega}{2}\frac{H_{j+1}^{n+1}-H_{j+1}^n}{\Delta t}+$$
$$+(1-\theta)\frac{-q_{j-1}^n+q_{j+1}^n}{2\Delta x}+\theta\frac{-q_{j-1}^{n+1}+q_{j+1}^{n+1}}{2\Delta x}=0 \tag{8.146c}$$

$$\frac{1-\omega}{2}\frac{q_{j-1}^{n+1}-q_{j-1}^n}{\Delta t}+\omega\frac{q_j^{n+1}-q_j^n}{\Delta t}+\frac{1-\omega}{2}\frac{q_{j+1}^{n+1}-q_{j+1}^n}{\Delta t}+$$
$$+\frac{1-\theta}{\Delta x}\left(-\eta\frac{\left(q_{j-1}^n\right)^2}{H_{j-1}^n}+(2\eta-1)\frac{\left(q_j^n\right)^2}{H_j^n}+(1-\eta)\frac{\left(q_{j+1}^n\right)^2}{H_{j+1}^n}\right)+$$
$$+\frac{\theta}{\Delta x}\left(-\eta\frac{\left(q_{j-1}^{n+1}\right)^2}{H_{j-1}^n}+(2\eta-1)\frac{\left(q_j^{n+1}\right)^2}{H_j^{n+1}}+(1-\eta)\frac{\left(q_{j+1}^{n+1}\right)^2}{H_{j+1}^n}\right)+$$
$$+\frac{1-\theta}{2\Delta x}\frac{g}{2}\left(-\left(H_{j-1}^n\right)^2+\left(H_{j+1}^n\right)^2\right)+\frac{\theta}{2\Delta x}\frac{g}{2}\left(-\left(H_{j-1}^{n+1}\right)^2+\left(H_{j+1}^{n+1}\right)^2\right)=0 \tag{8.146d}$$

- for $j = M$

$$(1-\omega)\frac{H_{j-1}^{n+1}-H_{j-1}^n}{\Delta t}+\omega\frac{H_j^{n+1}-H_j^n}{\Delta t}+(1-\theta)\frac{-q_{j-1}^n+q_j^n}{\Delta x}+\theta\frac{-q_{j-1}^{n+1}+q_j^{n+1}}{\Delta x}=0 \tag{8.146e}$$

$$(1-\omega)\frac{q_{j-1}^{n+1}-q_{j-1}^n}{\Delta t}+\omega\frac{q_j^{n+1}-q_j^n}{\Delta t}+$$
$$+\frac{1-\theta}{\Delta x}\left(-\frac{\left(q_{j-1}^n\right)^2}{H_{j-1}^n}+\frac{\left(q_j^n\right)^2}{H_j^n}\right)+\frac{\theta}{\Delta x}\left(-\frac{\left(q_{j-1}^{n+1}\right)^2}{H_{j-1}^n}+\frac{\left(q_j^{n+1}\right)^2}{H_j^{n+1}}\right)+$$
$$+\frac{1-\theta}{\Delta x}\frac{g}{2}\left(-\left(H_{j-1}^n\right)^2+\left(H_j^n\right)^2\right)+\frac{\theta}{\Delta x}\frac{g}{2}\left(-\left(H_{j-1}^{n+1}\right)^2+\left(H_j^{n+1}\right)^2\right)=0 \tag{8.146f}$$

This non-linear system of algebraic equations completed by the appropriate boundary conditions is solved using the Newton method.

The properties of the presented method can be examined on the example of the linear wave equations (8.16) and (8.17), using the modified equation approach for accuracy analysis. Following this way finally we obtain the system of modified wave equations corresponding to the solved linear wave equations:

$$\frac{\partial U}{\partial t} + g\frac{\partial H}{\partial x} = D_n \frac{\partial^2 U}{\partial x^2} + E_{n1}\frac{\partial^3 H}{\partial x^3} + \cdots \tag{8.147}$$

$$\frac{\partial H}{\partial t} + \overline{H}\frac{\partial U}{\partial x} = D_n \frac{\partial^2 H}{\partial x^2} + E_{n2}\frac{\partial^3 U}{\partial x^3} + \cdots . \tag{8.148}$$

In Eqs. (8.147) and (8.148) the coefficient of numerical diffusion D_n is given by the expression:

$$D_n = \frac{c \cdot \Delta x}{2}\left(\left(2\theta - 1\right) C_r + \left(2\eta - 1\right)\right), \tag{8.149}$$

whereas the coefficients of numerical dispersion E_{n1} and E_{n2} are defined as follows:

$$E_{n1} = \frac{g \cdot \Delta x^2}{2}\left(\frac{2}{3} - \omega + C_r\left(2\eta - 1\right)\left(1 - \theta\right) + C_r^2\left(\theta - \frac{2}{3}\right)\right) \tag{8.150}$$

$$E_{n2} = \frac{\overline{H} \cdot \Delta x^2}{2}\left(\frac{2}{3} - \omega + C_r\left(2\eta - 1\right)\left(1 - \theta\right) + C_r^2\left(\theta - \frac{2}{3}\right)\right) \tag{8.151}$$

where C_r is the Courant number defined by Eq. (8.18). Note that in this case the numerical diffusivity is controlled by two parameters: θ and η. The first one is related to the time integration, whereas the second one is related to the approximation of the convective term only. The latter parameter allows us to introduce into solution the upwind effect.

Performing similar stability analysis as presented in Section 8.3.2, one can show that the following conditions:

$$\omega \geq 0.5, \tag{8.152a}$$

$$\theta \geq 0.5, \tag{8.152b}$$

$$\eta \geq 0.5 \tag{8.152c}$$

ensure absolute stability of the method.

The numerical experiments dealing with propagation of a surge show that the described method seems more effective than the difference Preissmann scheme. It is possible to increase the flow rate at the boundary immediately from zero to a relatively large amount. Two weighting parameters allow us to control the numerical diffusivity needed for suppressing the wiggles appearing at the steep front. Usually 3–5 iterations should be applied to obtain satisfying results. The mass balance error generated by the method is very low and usually it does not exceed 0.001%. Although the method is absolutely stable for linear problems, in the case of non-linear equations the assumed time step must be limited. Acceptable values of the Courant number varies depending on the circumstances, however it should not exceed 0.6–0.8. The main disadvantage of the method is that it fails for initially shallow water in considered channel, i.e. for dry bed.

To show an example of solution let us apply the described method to solve the dam break problem.

Example 8.7 Equations (8.142) and (8.143) are solved in a wide rectangular channel of length $L = 3{,}000$ m for a discontinuous initial conditions caused by the dam as presented in Fig. 8.31. At the downstream the initial depth is $H_d = 0.50$ m, whereas the upstream depth is $H_u = 10.0$ m. At both channel ends the initially imposed depths are kept in time. Suddenly the dam is removed and the water from the reservoir is released. For the assumed auxiliary conditions the system of Eqs. (8.142) and (8.143) has analytical solution given by Wu et al. (1999). This solution is represented in Fig. 8.34 by the dotted line. The solid line represents the results of numerical solution obtained at $t = 120$ s for $\Delta x = 2.0$ m and $\Delta t = 0.1$ s with the following values of the weighting parameters: $\omega = 0.67$, $\theta = 0.70$ and $\eta = 0.67$.

Both plots are in perfect agreement. Note that $\omega = 0.67$ corresponds to the standard finite element method. Therefore one can suppose that this method gives

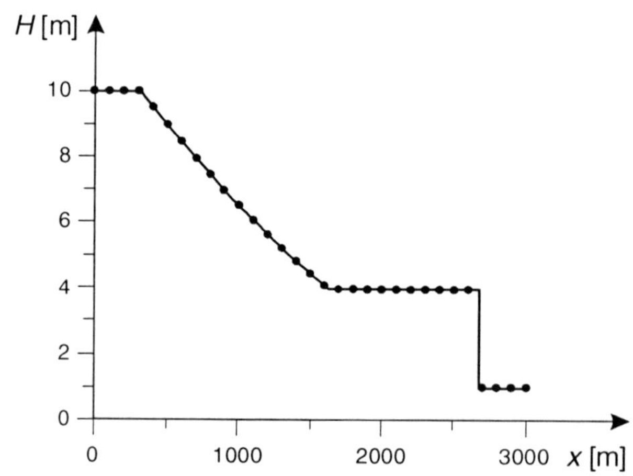

Fig. 8.34 Flow profiles in the channel at $t = 120$ s after removing the dam calculated numerically (*solid line*) and analytically (*dotted line*)

satisfactory results, when some additional dissipation is introduced into solution. In this case it was achieved with $\theta = 0.70$ and $\eta = 0.67$.

Although both discussed methods, i.e. the difference Preissmann scheme and the modified finite element method, appear effective for solving some cases of unsteady flow with discontinuities, they are not reliable for all possible forms of such kind of flow, particularly those corresponding to extreme situations. In such a situation it seems to be better to use the finite volume method, especially suitable for solving the hyperbolic equations with discontinuities.

We will end this section by presentation of the results of dam break problem computed with the finite volume method. Equations (8.142) and (8.143) are solved in a wide horizontal channel described in Example 8.5 for the same data except the initial and boundary conditions down of the dam. A dry state is assumed ($H_d \approx 0.0$ m). For assumed auxiliary conditions, when flow resistance is neglected the system of Eqs. (8.142) and (8.143) has analytical solution given by Ritter (Wu et al. 1999, Chanson 2004). This solution is represented in Fig. 8.35 by the dotted line. In the same figure the solid line represents the results of numerical solution at $t = 70$ s given by the finite volume method.

Comparing both graphs one can find out their perfect agreement. In contrast to the finite difference and element methods this takes place even in the vicinity of channel bed.

Next solution deals with the dry state but taking into account the flow resistance. The results obtained for exactly the same data as assumed previously and for the Manning coefficient assumed to be equal $n_M = 0.025$ are displayed in Fig. 8.36. For comparison they are contrasted with solution from Fig. 8.35 for frictionless flow. Although in this case no exact solution exists, the computed flow profile qualitatively similar to the ones observed in the physical experiments (see Chanson 2004).

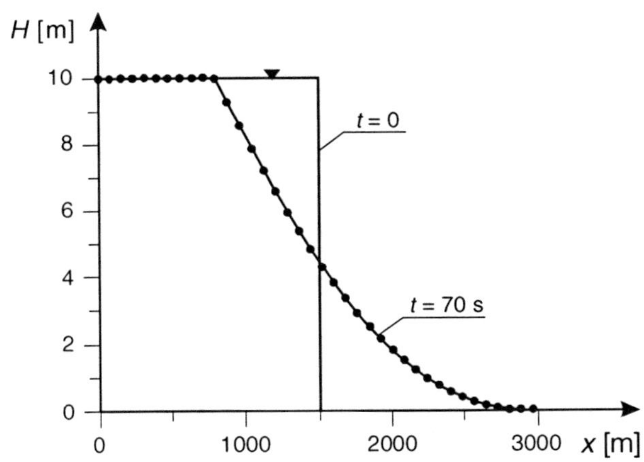

Fig. 8.35 Flow profiles in frictionless channel at $t = 70$ s after removing the dam calculated numerically (*solid line*) and analytically (*dotted line*)

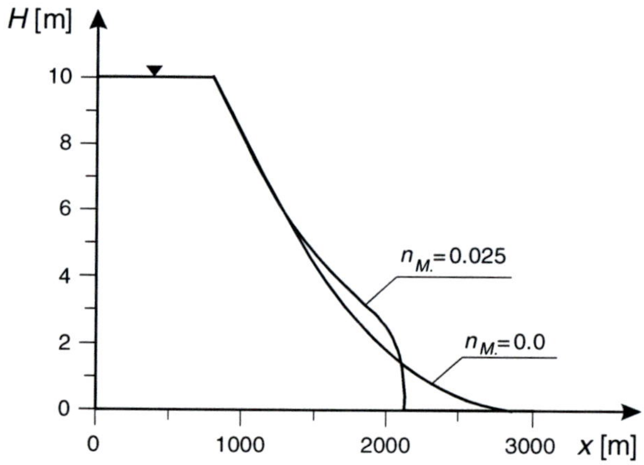

Fig. 8.36 Comparison of the flow profiles in considered channel at $t = 70$ s after removing the dam calculated using the finite volume method without friction and including friction

References

Abbott MB (1979) Computational hydraulics – Elements of the theory of free surface flow. Pitman, London

Abbott MB, Basco DR (1989) Computational fluid dynamics. Longman Scientific and Technical, New York

Abbott MB, Ionescu F (1967) On the numerical computation of nearly – horizontal flows. J. Hydr. Res. 5:97–117

Billingham J, King AC (2000) Wave motion. Cambridge University Press

Chanson H (2004) The hydraulics of open channel flow: An introduction, 2nd edn. Elsevier, Oxford

Cippada S, Ramswamy B, Wheeler MF (1994) Numerical simulation of hydraulic jump. Int. J. Numer. Meth. Eng. 37:1381–1397

Cunge J, Holly FM, Verwey A (1980) Practical aspects of computational river hydraulics. Pitman Publishing, London

Cooley RI, Moin SA (1976) Finie element solution of Saint Venant equations. J. Hydr. Div. ASCE HY6:759–775

Fread DL (1993) Flow routing. In: Maidment DR (ed.) Handbook of hydrology. McGraw-Hill, New York

French RH (1985) Open channel hydraulics. McGraw–Hill, New York

Gresho PM, Sani RL (2000) Incompressible flow and the finite-element method, vol. 1: Advection-diffusion. Wiley, Chichester, England

LeVeque RJ (2002) Finite volume methods for hyperbolic problems. Cambridge University Press

Liggett JA, Cunge JA (1975) Numerical methods of solution of the unsteady flow equations. In: Mahmood K, Yevjewich V (eds.) Unsteady flow in open channels. Water Resources Publishing, Fort Collins, CO, USA

Katopodes N (1984) A dissipative Galerkin scheme for open channel flow. J. Hydr. Engng. ASCE 110 (4):450–466

Korn GA, Korn TM (1968) Mathematical handbook for scientists and engineers, 2nd edn. McGraw–Hill, New York

Mahmood K, Yevjewich V (eds.) (1975) Unsteady flow in open channels. Water Resources Publishing, Fort Collins, CO, USA

Potter D (1973) Computational physics. Wiley, London

Singh AK, Porey PD, Ranga Raju KG (1997) Criterion for location of downstream control for dynamic flood routing. J. Hydrol. 196:66–75

Szymkiewicz R (1991) Finite – element method for the solution of the Saint Venant equations in an open channel network. J. Hydrol. 122:275–287

Szymkiewicz R (1995) Method to solve 1D unsteady transport and flow equations. J. Hydr. Engng. ASCE 121 (5):396–403

Wu C, Huang G, Zheng Y (1999) Theoretical solution of dam-break shock wave. J. Hydr. Engng. ASCE 125 (11):1210–1215

Yalin MS, Ferreira da Silva AM (2001) Fluvial processes. International Association of Hydraulic Engineering and Research, Delft, The Netherlands

Zienkiewicz OC (1972) The finite element method in engineering science. McGraw-Hill, London

Chapter 9
Simplified Equations of the Unsteady Flow in Open Channel

9.1 Simplified Forms of the Saint Venant Equations

The system of Saint Venant equations derived in Chapter 1 in the form of Eqs. (1.77) and (1.78) or Eqs. (1.79) and (1.80) is called the dynamic wave model or the complete dynamic model. This model of unsteady open channel flow gives reliable results if the underlying assumptions are satisfied. On the other hand, the Saint Venant model requires rather complex methods of solution and relatively large number of data characterizing both the channel and the flow conditions. For this reasons hydrologists tried to simplify the system of Saint Venant equations to obtain models, which require less input information.

Consider the dynamic equation (1.77):

$$\frac{\partial U}{\partial t} + U\frac{\partial U}{\partial x} + g\frac{\partial H}{\partial x} = g(s - S), \tag{9.1}$$

where:

t – time,
x – longitudinal coordinate,
H– water depth,
U– flow velocity,
g – gravitational acceleration,
s – channel bottom slope,
S – slope of energy line.

One can notice that its terms have different orders of magnitude. Henderson (1966) showed, that even for a river with considerable longitudinal bed slope, for steep flood waves, i.e. when the inertial force should play relatively significant role, the values of the consecutive terms of Eq. (9.1) differ by about two orders of magnitude. Similar conclusions were presented by Cunge et al. (1980). Evaluation of subsequent terms of Eq. (9.1) carried out for flow with significant variability of its

R. Szymkiewicz, *Numerical Modeling in Open Channel Hydraulics*, Water Science and Technology Library 83, DOI 10.1007/978-90-481-3674-2_9,

parameters both in time and in space, showed that the first and second terms of the left hand side are of the order of $\sim 1.0 \cdot 10^{-4}$ and $\sim 1.5 \cdot 10^{-4}$ respectively. At the same time the bed slope, being in this case of the order of 10^{-3} causes, that the first term of the right side hand of Eq. (9.1) is of the order $\sim 10^{-2}$. If the differences in depth along the channel axis are small then the third term is relatively small comparing with $g \cdot s$. Consequently, the term representing the friction force will be of order of the bed slope, i.e. $s = S$. This means that in the dynamic equation the friction force and the gravitational force dominate. In some cases the differences in orders of magnitude can be even greater than presented above. Thus, it is reasonable to neglect the terms of lesser importance. As a result, one obtains the equations in simpler forms. Simplified equations are still interesting for hydrological practice, since they can be more easily applied to the flood routing problem. There are two simplified forms of the system of Saint Venant equations: the kinematic wave model and the diffusive wave model. Both models base on the original continuity equation and on the appropriately simplified dynamic equation. The theory of the kinematic wave was given by Lighthill and Whitham (1955) whereas the diffusive wave theory was presented by Hayami (1951). The properties and applicability of both mentioned models have been discussed by many authors, e.g. Miller and Cunge (1975), Ponce, Li and Simmons (1978) and Ponce (1990). A comprehensive review is given by Singh (1996). Although not generally appropriate to adequately reproduce the real flood routing process, the linearized forms of the simplified models are applied very often.

Let us recall the system of Saint-Venant equations, which with flow rate and depth as dependent variables, can be written as follows:

$$B\frac{\partial H}{\partial t} + \frac{\partial Q}{\partial x} = 0, \tag{9.2}$$

$$k\left(\frac{\partial Q}{\partial t} + \frac{\partial}{\partial x}\left(\frac{Q^2}{A}\right)\right) + l\left(g \cdot A\frac{\partial H}{\partial x}\right) - g \cdot A \cdot s + g \cdot A \cdot S = 0 \tag{9.3}$$

where:

Q – flow discharge,
A – cross-sectional area of flow,
B – channel width at water surface,
k, l – constants taking the values 1 or 0.

The slope friction S is expressed using the Manning formula:

$$S = \frac{n_M^2 \, |Q| \, Q}{R^{4/3} \cdot A^2} \tag{9.4}$$

where:

n_M – Manning roughness coefficient,
R – hydraulic radius.

The integer constants k and l can take the values of either 0 or 1. For $k = 1$ and $l = 1$ we have full momentum equation which corresponds to the dynamic wave, whereas for $k = 0$ and $l = 1$ Eqs. (9.2) and (9.3) become the diffusive wave model. The kinematic wave model is obtained when $k = 0$ and $l = 0$.

First, let us consider the kinematic wave model. This model is derived when in the momentum equation only the gravitational and friction forces are taken into account. In such a case Eq. (9.3) becomes the equation for steady uniform flow: $s = S$.

Then the kinematic wave model is constituted by the differential continuity equation (9.2) and the Manning equation (9.4), i.e.:

$$\frac{\partial H}{\partial t} + \frac{1}{B}\frac{\partial Q}{\partial x} = 0, \tag{9.5}$$

$$Q = \frac{1}{n_M} R^{1/2} \cdot s^{1/2} \cdot A, \tag{9.6}$$

The momentum equation (9.3) can be simplified in another way leading to the second simpler form of the unsteady flow equations. If apart from the friction and gravitational forces, as previously, we take into account the hydrostatic force as well, then the diffusive wave model will be obtained. It is constituted by the continuity equation (9.2) and simplified momentum equation (9.3):

$$\frac{\partial H}{\partial t} + \frac{1}{B}\frac{\partial Q}{\partial x} = 0, \tag{9.7}$$

$$\frac{\partial H}{\partial x} + S - s = 0, \tag{9.8}$$

The meaning of the introduced assumptions can be illustrated using the so-called rating curve. This curve represents the relation between the water level (or depth) and the discharge rate in the considered channel cross-section. These variables are related via the Manning formula:

$$Q = \frac{1}{n_M} R^{1/2} \cdot S^{1/2} \cdot A. \tag{9.9}$$

In a particular case when the bed slope coincides with the energy line slope, i.e. for the uniform flow, Eq. (9.9) becomes Eq. (9.6). If we calculate the energy line slope S from Eq. (9.1) and substitute in Eq. (9.9), we obtain (Chow et al. 1988):

$$Q = \frac{1}{n_M} R^{2/3} \cdot A \left(s - k \left(\frac{1}{g \cdot A}\frac{\partial Q}{\partial t} + \frac{1}{g \cdot A}\frac{\partial}{\partial x}\left(\frac{Q^2}{A}\right)\right) - l\frac{\partial H}{\partial x}\right)^{1/2} \tag{9.10}$$

In a general case of unsteady flow the relation $Q(H)$ is non-unique, because it depends on $\partial Q/\partial t$, $\partial Q/\partial x$, and $\partial H/\partial x$, Due to this non-uniqueness thewell-known

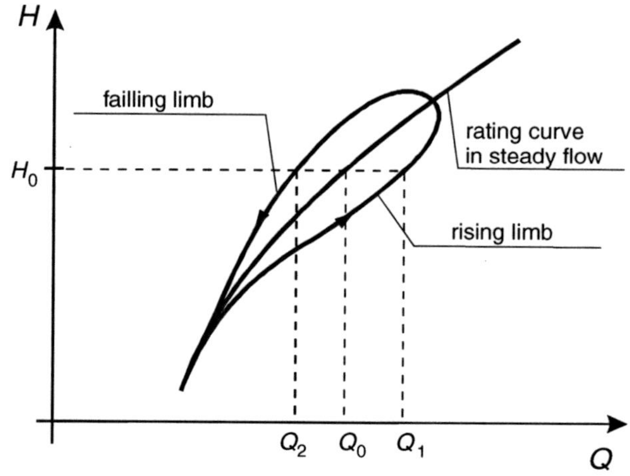

Fig. 9.1 Hysteresis of the function $Q(H)$ during the unsteady flow

phenomenon of the rating curve hysteresis is observed during the passage of a flood wave. At the rising limb of flood wave (Fig. 9.1), where we have $\partial H/\partial x < 0$ and $\partial Q/\partial x < 0$, the flow rate for given depth H_0 is greater comparing with the discharge for steady flow: $Q_1 > Q_0$. Conversely, at the failing limb, when $\partial H/\partial x > 0$ and $\partial Q/\partial x > 0$, the corresponding flow rate will be less than one for the steady flow. A unique relation between Q and H is possible only for steady uniform flow, when the derivatives over x and t disappear and Eq. (9.10) coincides with Eq. (9.6). The diffusive wave model obtained for $k = 0$ and $l = 1$ is able to reproduce to some extent the non-uniqueness of Q-H relation, since it takes into account the non-uniformity of flow, expressed by $\partial H/\partial x$. On the other hand, the kinematic wave model with $k = l = 0$ predicts the same relation $Q(H)$, corresponding to the steady uniform flow, regardless of the actual flow conditions.

In the following example are presented the relations $Q(H)$ obtained for the complete system of Saint Venant equations and its both simplified forms.

Example 9.1 In rectangular open channel of length $L = 80$ km the unsteady flow is considered. Assume the following data:

- bed width of channel is $b = 30$ m,
- bed slope is $s = 0.0005$,
- Manning coefficient is $n_M = 0.035$,
- initial condition:

  at $t = 0$ is determined by the steady uniform flow with $Q_0 = 20$ m^3/s,
- boundary conditions:

  at $x = 0$ the following hydrograph is imposed:

$$Q(x,t) = Q_0 + (Q_{max} - Q_0)\left(\frac{t}{t_{max}}\right)^{\alpha} \exp\left(1 - \left(\frac{t}{t_{max}}\right)^{\alpha}\right) \tag{9.11}$$

where:

Q_0 – baseflow discharge of the inflow,
Q_{max} – peak discharge of the inflow,
t_{max} – time of the peak flow,
α – parameter,

at $x = L$ the depth $H(x, t) = H_n =$ const. (H_n is normal depth corresponding to Q_0) is imposed.

The rating curve is computed at the control cross-section located at mid-length of the channel.

All the systems of equations (i.e. the dynamic, diffusive and kinematic wave models) are solved by the modified finite element method described in Chapter 8. The computations were performed for: $\Delta t = 60\,\text{s}$, $\Delta x = 400$ m, $Q_0 = 20$ m^3/s, $Q_{max} = 200$ m^3/s, $t_{max} = 2\,\text{h}$ and $\alpha = 1.5$.

In Fig. 9.2 the relation $Q(H)$ generated by the system of Saint Venant equation is displayed. The obtained curve has a loop of hysteresis typical for unsteady flow.

The rating curves generated by the simplified forms of the unsteady flow equations are shown in Fig. 9.3. One can see that in the considered case the relation $Q(H)$ for the diffusive wave differs slightly from the one obtained for the complete system of equations. For the kinematic wave the loop of hysteresis disappears and the rating curve becomes a unique relation as it is expected.

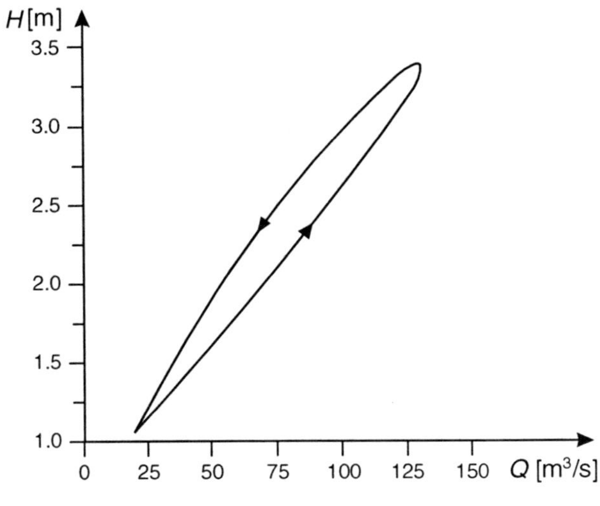

Fig. 9.2 Hysteresis of the function $Q(H)$ for the dynamic wave

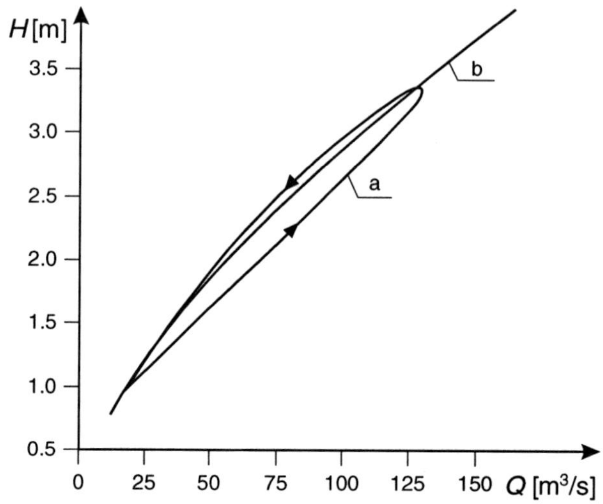

Fig. 9.3 Hysteresis of the function $Q(H)$ for the diffusive wave model (**a**) and for the kinematic wave model (**b**)

9.2 Simplified Flood Routing Models in the Form of Transport Equations

9.2.1 Kinematic Wave Equation

The kinematic wave model in the form of system of equations (9.5) and (9.6) can be reduced to one differential equation with regard to a single unknown function only. The equation for steady flow is rearranged to the following form:

$$A = \alpha \cdot Q^m. \tag{9.12}$$

For the Chezy formula one obtains:

$$\alpha = \frac{1}{\left(\frac{C_C \cdot s^{1/2}}{p^{1/2}}\right)^{2/3}}, \tag{9.13a}$$

$$m = \frac{2}{3} \tag{9.13b}$$

whereas for the Manning formula we have:

$$\alpha = \frac{1}{\left(\frac{s^{1/2}}{n_M \cdot p^{2/3}}\right)^{3/5}}, \tag{9.14a}$$

$$m = \frac{3}{5} \tag{9.14b}$$

If instead of the depth $H(x, t)$ the wetted flow area $A(x, t)$ is introduced as the dependent variable, then the system of equations describing the kinematic wave model takes the following form:

$$\frac{\partial A}{\partial t} + \frac{\partial Q}{\partial x} = 0, \tag{9.15}$$

$$A = a \cdot Q^m \tag{9.16}$$

Let us assume a wide and shallow channel for which the wetted perimeter can be considered constant. Differentiation of Eq. (9.16) with these assumptions yields:

$$\frac{\partial A}{\partial t} = a \cdot m \cdot Q^{m-1} \frac{\partial Q}{\partial t}. \tag{9.17}$$

Substitution of Eq. (9.17) in Eq. (9.15) gives the following equation:

$$\frac{\partial Q}{\partial t} + \frac{1}{a \cdot m \cdot Q^{m-1}} \frac{\partial Q}{\partial x} = 0, \tag{9.18}$$

Equation (9.18) can be rewritten in a more compact form:

$$\frac{\partial Q}{\partial t} + C \frac{\partial Q}{\partial x} = 0, \tag{9.19}$$

where:

$$C = \frac{1}{a \cdot m \cdot Q^{m-1}} \tag{9.20}$$

is the kinematic wave celerity. Substitution of the coefficients a and m given by Eqs. (9.14a) and (9.14b), in Eq. (9.20) yields:

$$C = \left(\frac{s^{1/2}}{n_M \cdot p^{2/3}} \right)^{3/5} \frac{5}{3} Q^{2/5} = \frac{5}{3} U. \tag{9.21}$$

This means, that the kinematic wave moves faster than the water flows in the channel.

9.2.2 Diffusive Wave Equation

Using a similar approach, the diffusive wave model in the form of system of equations (9.7) and (9.8) can be reduced to a single differential equation as well. For different assumptions one obtains various forms of the final equation where the dependent variable can be the depth $H(x, t)$, the water stage $h(x, t)$ or the flow rate $Q(x, t)$. We will derive the classical form of the diffusive wave equation with regard to flow rate. To this end the Manning equation (9.9) is rewritten as

$$Q = K\sqrt{S}, \tag{9.22}$$

where:

$$K = \frac{1}{n_M} R^{2/3} \cdot A \tag{9.23}$$

The coefficient K, called conveyance, is a function of the cross-sectional parameters only. Substitution of Eq. (9.22) in Eq. (9.8) allows us to rewrite the diffusive wave model (9.7) and (9.8) as follows:

$$\frac{\partial H}{\partial t} + \frac{1}{B}\frac{\partial Q}{\partial x} = 0, \tag{9.24}$$

$$\frac{\partial H}{\partial x} + \frac{|Q|\,Q}{K^2} - s = 0, \tag{9.25}$$

In the next step the continuity equation is differentiated with regard to x, whereas the dynamic equation is differentiated with regard to t. Assuming constant width of the channel ($B = \text{const.}$), one obtains:

$$\frac{\partial^2 H}{\partial t \cdot \partial x} + \frac{1}{B}\frac{\partial^2 Q}{\partial x^2} = 0, \tag{9.26}$$

$$\frac{\partial^2 H}{\partial x \cdot \partial t} + \frac{2\,|Q|}{K^2}\frac{\partial Q}{\partial t} - \frac{2\,Q\,|Q|}{K^3}\frac{\partial K}{\partial t} = 0. \tag{9.27}$$

Subtraction of Eq. (9.26) from Eq. (9.27) eliminates the cross derivatives of function $H(x, t)$, yielding:

$$\frac{2\,|Q|}{K^2}\frac{\partial Q}{\partial t} - \frac{1}{B}\frac{\partial^2 Q}{\partial x^2} - \frac{2\,Q\,|Q|}{K^3}\frac{\partial K}{\partial t} = 0. \tag{9.28}$$

Since the conveyance K is a function of the depth H, then one can write:

$$\frac{\partial K}{\partial t} = \frac{\partial K}{\partial H}\frac{\partial H}{\partial t}. \tag{9.29}$$

With the continuity equation (9.24) the above equation becomes:

$$\frac{\partial K}{\partial t} = -\frac{\partial K}{\partial H}\frac{1}{B}\frac{\partial Q}{\partial x}. \tag{9.30}$$

Substitution of Eq. (9.30) in Eq. (9.28) yields:

$$\frac{\partial Q}{\partial t} + \left(\frac{Q}{K \cdot B}\frac{\partial K}{\partial H}\right)\frac{\partial Q}{\partial x} - \frac{K^2}{2B\,|Q|}\frac{\partial^2 Q}{\partial x^2} = 0. \tag{9.31}$$

Introducing new variables:

$$C = \frac{Q}{K \cdot B} \frac{\partial K}{\partial h}, \tag{9.32}$$

$$D = \frac{K^2}{2B\,|Q|} \tag{9.33}$$

Equation (9.31) is rewritten in the following final form:

$$\frac{\partial Q}{\partial t} + C\frac{\partial Q}{\partial x} - D\frac{\partial^2 Q}{\partial x^2} = 0, \tag{9.34}$$

Equation (9.34) is the diffusive wave model in the form of advection-diffusion transport equation in which C is the kinematic wave celerity and D is the coefficient of hydraulic diffusivity.

The most popular form of the diffusive wave model, proposed by Hyami in 1951 (Eagleson 1970), is a particular case of equation (9.30). It is obtained from Eq. (9.30) on condition that:

- the slope of energy line is replaced by the bed slope ($S = s$),
- the considered channel is wide and shallow so one can assume constant wetted perimeter ($p = \text{const.}$).

Taking these assumptions into account one obtains:

$$D = \frac{K^2}{2B \cdot Q} = \frac{1}{2B \cdot Q} \frac{Q^2}{s} = \frac{Q}{2B \cdot s} \tag{9.35}$$

$$C = \frac{Q}{B \cdot K} \frac{\partial K}{\partial H} = \frac{n_M \cdot Q}{B \cdot R^{2/3} \cdot A} \cdot \frac{\partial}{\partial H}\left(\frac{1}{n_M} R^{2/3} \cdot A\right) = \frac{Q}{B \cdot R^{2/3} \cdot A} \frac{\partial}{\partial H}\left(\frac{A^{5/3}}{p^{2/3}}\right) =$$

$$= \frac{Q}{B \cdot R^{2/3} \cdot A} \cdot \frac{1}{p^{2/3}} \frac{5}{3} A^{2/3} \frac{\partial A}{\partial H} = \frac{5}{3}\frac{Q}{A} = \frac{5}{3} U, \tag{9.36}$$

Equation (9.36) represents the kinematic wave celerity, previously derived for the kinematic wave equation (9.19). The diffusive wave model in the form of Eq. (9.34) in which the coefficients are given by Eqs. (9.35) and (9.36) is frequently used.

To show the difference between the kinematic and diffusive wave models, i.e. to notice role of the diffusive term in Eq. (9.34), let us rout the same flood wave using both models.

Example 9.2 In a rectangular straight channel of length $L = 100$ km the propagation of the flood wave is considered. The auxiliary conditions are following:

- initial condition at $t = 0$ is determined by the steady uniform flow with flow discharge Q_0,
- boundary conditions:

  at $x = 0$ a hydrograph in the form of Eq. (9.11) is imposed:

$$Q(x,t) = Q_0 + (Q_{\max} - Q_0)\left(\frac{t}{t_{\max}}\right)^{\alpha} \exp\left(1 - \left(\frac{t}{t_{\max}}\right)^{\alpha}\right)$$

  for the diffusive wave equation at the downstream end $x = L$, additionally is imposed the function $H(x, t) = H_0 = \text{const.}$, where H_0 is normal depth corresponding to Q_0.

Assume the following data:

- bed width of channel is $b = 25$ m,
- bed slope is $s = 0.0005$,
- Manning coefficient is $n_M = 0.030$,
- baseflow $Q_0 = 5$ m^3/s,
- peak flow $Q_{\max} = 100$ m^3/s,
- time to peak $t_{\max} = 4$ h,
- $\alpha = 1.0$.

Solve the linear kinematic and diffusive wave models to compare the results at the control cross-section located at $x = 75$ km of the considered channel. Note that the boundary condition imposed at the downstream end for the diffusive wave equation does not disturb the flow in assumed control.

Since the solved equations are non-linear ones then their linearization is needed. Linearization of Eq. (9.37) is carried out assuming constant values of the parameters C and D. They can be calculated accordingly to Eqs. (9.38) and (9.39) for a constant value of the flow rate taken as arithmetic average from its lowest and highest values i.e. $Q = Q_C = (Q_0+Q_{\max})/2$. For $Q_c = 52.5$ m^3/s one obtains $C = 1.88$ m/s and $D = 2100.0$ m^2/s. The kinematic wave equation is solved using the box scheme applied in Chapter 6 for solving the advection equation, whereas the diffusive wave equation is solved with the modified finite element method applied for solving the advection-diffusion equation in Chapter 7.

The computations performed for $\Delta x = 1{,}000$ m and $\Delta t = 600$ s i.e. with the advective Courant number equal to $C_a = 1.13$ gave the results displayed in Fig. 9.4. They represent typical solutions of the pure advection equation and the advection-diffusion equation with constant parameters.

Note that the kinematic wave model produces practically exact solution ensuring nearly pure translation of the flood wave. On the other hand, as it could be expected, the diffusive term present in the diffusive wave equation attenuates the propagating wave giving rise to decreasing of its peak and to widening of its base.

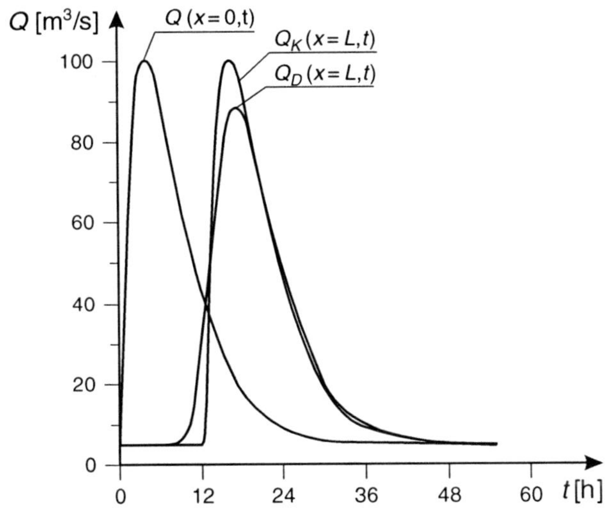

Fig. 9.4 Results of wave routing by the linear kinematic ($Q_K(x = 75\,\text{km},\ t)$) and diffusive ($Q_D(x = 75\,\text{km},\ t)$) wave equations

9.2.3 *Linear and Non-linear Forms of the Kinematic and Diffusive Wave Equations*

In the preceding section it was shown that neglecting the inertial terms the system of Saint Venant equations (9.2) and (9.3) can be transformed into the following advective – diffusive transport equation with the function $Q(x, t)$ as dependent variable:

$$\frac{\partial Q}{\partial t} + C\,(Q)\,\frac{\partial Q}{\partial x} - D\,(Q)\,\frac{\partial^2 Q}{\partial x^2} = 0 \tag{9.37}$$

where:

$C(Q)$ – advective velocity,
$D(Q)$ – hydraulic diffusivity.

For a wide and rectangular channel with the slope of energy line equal to the slope of the channel bottom $C(Q)$) becomes the kinematic wave speed. In such a case the parameters of Eq. (9.37) are expressed as follows:

$$C = \frac{1}{m \cdot \alpha \cdot Q^{m-1}}\ , \tag{9.38}$$

$$D = \frac{Q}{2\,B \cdot s} \tag{9.39}$$

where m is the kinematic wave ratio ($m = 3/5$ for the Manning law friction). Equation (9.37) represents the diffusive wave – the kinematic wave is obtained for $D = 0$. Parameter α is given by Eq. (9.13). Both equations are non-linear since their parameters C and D depend on Q. However, linearized versions of Eq. (9.37) with constant C and D are usually applied, for the sake of simplicity. To show the difference between linear and non-linear forms of this equation, let us consider the following example.

Example 9.3 In the same channel and for the same data as assumed previously in Example 9.2, the flood wave propagation is considered. However, conversely to the preceding example, now the non-linear forms of the kinematic and diffusive waves are solved.

The kinematic wave equation is solved using the box scheme, whereas the diffusive wave equation is solved with the modified finite element method. However, as the approximation of solved equation (9.37) leads to the non-linear systems of algebraic equations the iterative method must be used for its solution. For both models the Newton method generating convergent solution is applied.

In Fig. 9.5 the results of solution of linear and non-linear forms of the kinematic wave equation are shown. The results for linear model are coming from Example 2. One can notice the effect typical for a non-linear equation. The front of the propagating wave becomes steeper, whereas its tail becomes longer. This property results from Eq. (9.38). The advection velocity C increases with the flow rate Q so the wave peak moves faster than the lower parts of the wave. Finally, the propagating wave can break down. More detailed analysis of this phenomenon, based on the theory of characteristics, is given for example by Abbott (1979).

Similar, but less significant effects (see Fig. 9.6) are observed in the solution of the non-linear diffusive wave equation as well. However in this case the propagating wave is simultaneously damped due to the hydraulic diffusion, so its breakdown is rather impossible. Still, the differences between the linear and non-linear versions of the diffusive wave equation cannot be neglected. Therefore, we have to be aware that linearization of Eq. (9.37) while simplifying its numerical solution, changes significantly the results.

While solving the non-linear partial differential equations representing transport of any scalar quantity, the form of solved equation is a problem of great importance. The equation must be written in conservative form. Otherwise one can expect errors in the balance of transported quantity in the numerical solution. As far as Eq. (9.37) is considered, it is a non-linear advection-diffusion transport equation written in the non-conservative form. Therefore one can expect that the presented numerical solutions should be affected by the errors. Indeed, if we thoroughly examine the function $Q(t)$ obtained with non-linear models at the control cross-section, we will notice that the total volume of the water leaving the considered channel reach differs from that entering the reach, the difference exceeding 8 and 6% for the kinematic and diffusive waves, respectively. On the other hand, the numerical solutions of the linear equations satisfy the law of mass conservation perfectly for the non-linear diffusive and kinematic wave equations considered in Example 9.2. Then the numerical solution of Eq. (9.37) seems to be a non-trivial problem.

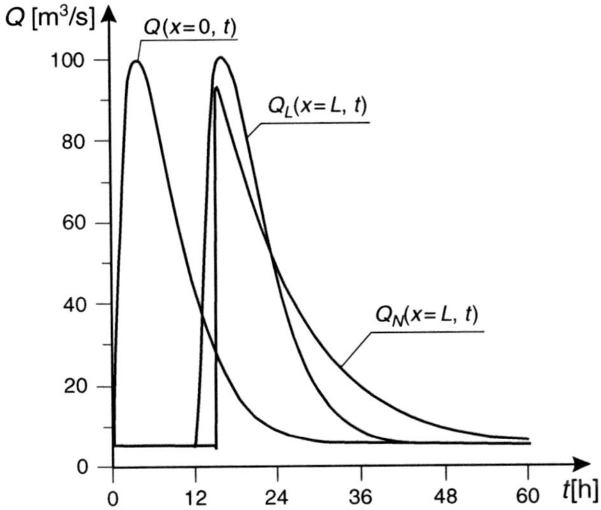

Fig. 9.5 Results of wave routing by the linear ($Q_L(x = 75\,\mathrm{km}, t)$) and non-linear ($Q_N(x = 75\,\mathrm{km}, t)$) kinematic wave equations

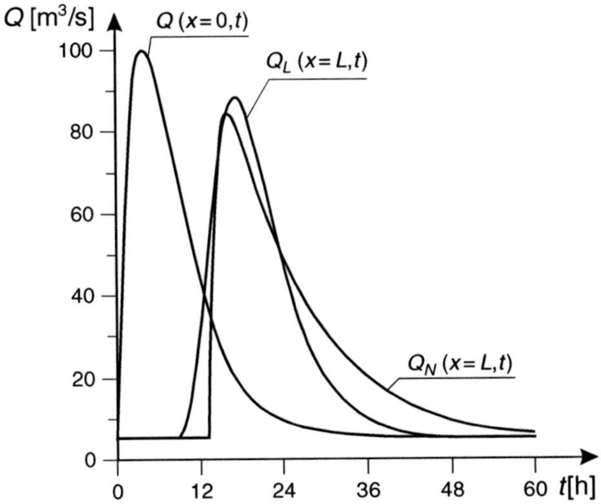

Fig. 9.6 Result of wave routing by the linear ($Q_L(x = 75\,\mathrm{km}, t)$) and non-linear ($Q_N(x = 75\,\mathrm{km}, t)$) diffusive wave equations

9.3 Mass and Momentum Conservation in the Simplified Flood Routing Models in the Form of Transport Equations

9.3.1 The Mass and Momentum Balance Errors

The system of Saint-Venant equations comprises the continuity and dynamic equations which are derived from the mass and momentum conservation principles, respectively. To avoid the mass and momentum balance errors in the numerical solution, both equations have to be written in conservative forms. This problem has been shortly discussed in Chapter 8. As we showed in the preceding section, the linearized forms of simplified models perfectly ensure mass conservation, whereas the non-linear forms of these models may give rise to mass balance error. The numerical calculations presented in Example 9.2 show that this error can be of order of 8%, but even 20% difference in volumes between upstream and downstream hydrograph may be obtained (Cappelaere 1997).

Originally, all simplified models are written in the form of a system of differential equations (i.e. a continuity equation and a simplified momentum equation). Consequently, there are no doubts which quantities are the conservative ones. However the situation is changed at the moment when both equations are combined to obtain a single transport equation. It appears that similarly to other flow problems, the final form of the non-linear simplified models has a fundamental meaning for the mass and momentum conservation in their numerical solutions. On the other hand, one can expect that the non-linear simplified equations, derived from the continuity and dynamic equations, should also fulfil both mass and momentum conservation principles at the same time. For this reason the following question seems to be relevant: how should the simplified models be interpreted from the point of view of the conservation laws? To explain this question let us follow the idea presented by Gasiorowski and Szymkiewicz (2007).

Let us consider a channel reach of length L. Assuming that the flow takes place in a time interval T, which is long enough (after the passing of the flood wave the discharge comes back to its initial value), the mass balance for constant water density takes the following form:

$$\int_0^T (Q(0,t) - Q(L,t))dt = 0 \qquad (9.40)$$

It means that the total volumes passing through the two ends of the channel reach given by the formulas:

$$m_0 = \int_0^T Q\,(0,t)\,dt, \qquad (9.41a)$$

$$m_L = \int_0^T Q\,(L,t)\,dt \tag{9.41b}$$

should be equal. The relative mass balance error can be defined as follows:

$$\Delta E_m = \frac{m_L - m_0}{m_0} \cdot 100\% \tag{9.42}$$

where:

ΔE_m– he mass error expressed in %,
m_0, m_L– olume of inflow and outflow respectively, within the time period of T.

The momentum balance can be represented similarly. The total variation of momentum over a channel reach of length L in time interval T must be equal to the difference of the momentum transported at upstream end $x = 0$ and at the downstream end $x = L$ respectively. If constant water density is assumed, the losses caused by friction, gravitation and pressure are neglected and the time interval T is long enough, this balance will take the following simplified form:

$$\int_0^T \left(\frac{Q^2(0,t)}{A(0,t)} - \frac{Q^2(L,t)}{A(L,t)} \right) dt = 0 \tag{9.43}$$

The total momentum of inflow and outflow is given by the formulas:

$$M_0 = \int_0^T U\,(0,t)\,Q\,(0,t)\,dt = \int_0^T \frac{Q^2\,(0,t)}{A\,(0,t)} dt, \tag{9.44a}$$

$$M_L = \int_0^T U\,(L,t)\,Q\,(L,t)\,dt = \int_0^T \frac{Q^2\,(L,t)}{A\,(L,t)} dt \tag{9.44b}$$

where:

M_0, M_L – total momentum of inflow and outflow, respectively, within the time period of T, divided by water density,
U – flow velocity,
A – cross-sectional area of flow.

The relative momentum balance error can be calculated as follows:

$$\Delta E_M = \frac{M_L - M_0}{M_0} \cdot 100\% \tag{9.45}$$

where: ΔE_M – the momentum error expressed in %.

Equations (9.43), (9.44a) and (9.44b) hold for the system of Saint Venant equations when both functions $Q(x, t)$ and $A(x, t)$ are known. In the case of kinematic and diffusive wave equation the momentum balance can be rearranged using the Manning formula written in the form of Eq. (9.16). Consequently Eq. (9.43) can be rewritten as follows:

$$\int_0^T \left(Q^{2-m}(0,t) - Q^{2-m}(L,t)\right) dt = 0 \tag{9.46}$$

Note that for the linear kinematic wave model with $U(x, t) = \text{const.}$, the momentum balance expressed by Eq. (9.43) takes the form of Eq. (9.40), which represents the mass balance.

Numerical tests confirm that the mass balance error ΔE_m calculated for all considered models written in the form of the systems of conservative equations is very low (of the order of $\sim 10^{-2}\%$), whereas the total momentum of the outflow appears to be less than the momentum of the inflow. It is reasonable, because the effects of friction, gravitation and pressure have been neglected in the momentum balance.

As far as the linear kinematic and diffusive wave models in form of the advection and advection – diffusion equation are considered, one can find out that the mass and momentum principles are satisfied perfectly regardless of the flow conditions. Numerical test confirm that both errors ΔE_m and ΔE_M do not exceed the value of 0.01%.

Comparison of the momentum balance errors in simplified models with those calculated for the full system of Saint-Venant equations reveals an interesting fact. The error ΔE_M ,which for the Saint-Venant system is of the order of a couple per cents, practically disappears for both linear advection and advection-diffusion transport equations. The way of derivation of the simplified equations changes the character of the momentum conservation law. This is due to an additional important hypothesis on the flow process, which is implicitly introduced when the simplified system of Saint Venant equations is transformed into the advection and advection-diffusion equations. The differentiation of the simplified momentum equation with respect to time means that the sum of all forces taken into account in the simplified model is assumed to be constant during the flow process. The value of this sum is determined by the boundary condition imposed at the upstream end of the channel reach. In other words the differentiation of Eq. (9.25) gives rise to an essential modification of the final equation. No losses of the momentum are represented in Eq. (9.37), since during the differentiation the friction force represented by the source term was replaced by a diffusive one.

For the non-linear equations the total mass at the downstream end is not equal to those entering the channel reach. The balance errors ΔE_m and ΔE_M are usually of the order of several per cents. As it was afore-mentioned, Cappelaere (1997) reported this error can reach even about 20%.

9.3.2 Conservative and Non-conservative Forms of the Non-linear Advection-Diffusion Equation

To explain the problem of mass and momentum balance errors which appear for the non-linear flow routing models, let us begin with recalling the basic information on the conservative or non-conservative form in which the solved equations can be written.

The 1-D advective or advective-diffusive transport equation without the source/sink term has the following form:

$$\frac{\partial f}{\partial t} + \frac{\partial}{\partial x}(\phi_A + \phi_D) = 0 \tag{9.47}$$

in which

$$\phi_A = U \cdot f, \tag{9.48a}$$

$$\phi_D = -D\frac{\partial f}{\partial x} \tag{9.48b}$$

where:

ϕ_A – advective flux,
ϕ_D – diffusive flux,
f – scalar function,
U – advective velocity,
D – coefficient of diffusion.

Equation (9.47) has so-called divergence or conservative form because its integration over the domain of solution ($0 \le x \le L$) gives directly the following global conservation law:

$$\frac{\partial}{\partial t}\int_0^L f \cdot dx = \left[U \cdot f - D\frac{\partial f}{\partial x}\right]_0 - \left[U \cdot f - D\frac{\partial f}{\partial x}\right]_L \tag{9.49}$$

The total energy (if f represents temperature) or mass (if f represents concentration of dissolved substance) changes only by the net flux of f out the domain through the boundary $x = 0$ and $x = L$. It should be mentioned that only the divergence form of transport equation assures global conservation (Gresho and Sani 2000).

Comparing Eqs. (9.37) and (9.47) one can notice that Eq. (9.37) is written in a non-conservative form since this equation cannot be transformed into the global conservation law in form of Eq. (9.49). To show the consequences of this difference let us consider the non-linear advection equation of type of Eq. (9.37) with $D = 0$:

$$\frac{\partial f}{\partial t} + a \cdot f^b \cdot \frac{\partial f}{\partial x} = 0 \tag{9.50}$$

where: a, b – constant parameters.

Integration by parts of Eq. (9.50) with regard to x over a channel reach ($0 \le x \le L$) leads to the following expression:

$$\frac{\partial}{\partial t} \int_0^L f \cdot dx = [a \cdot f^b \cdot f]_0 - [a \cdot f^b \cdot f]_L + R_A \tag{9.51}$$

where

$$R_A = \int_0^L f \cdot \frac{\partial (a \cdot f^b)}{\partial x} \cdot dx. \tag{9.52}$$

The extra term R_A results from the non-linearity of the advective term. Equation (9.52) does not represent a global conservation law, because the total quantity represented by function f is changed not only by the net advective flux through the boundaries, as in Eq. (9.49), but by the additional term R_A as well. This term can be considered as the conservation error in the numerical solution. It can be avoided on condition that proper conservative form of the non-linear equation is applied. Obviously, if a constant kinematic speed is assumed ($a \cdot f^b$ = const.), Eq. (9.50) will become conservative and consequently R_A will disappear.

While discussing the conservative properties of the diffusive wave model it is worth to recall the way of its derivation. Let us assume that f represents some conservative quantity as the mass of dissolved matter or energy. The governing equation (Eq. 9.47) for the transport of f by the flowing stream, obtained via balance of conservative quantity for control volume, has the conservative form. Conversely to this equation, Eq. (9.37) was not derived directly from the mass or momentum conservation principle. It was obtained by manipulation on the continuity and momentum equations derived for unsteady open channel flow. Consequently Eq. (9.37) represents a non-conservative form. Equation (9.47) is consistent with Eq. (9.37) only when the advective velocity and the coefficient of diffusion are constant. Unfortunately we can never obtain Eq. (9.47) from Eq. (9.37) and consequently one can expect that Eq. (9.37) will never satisfy the global conservation law.

9.3.3 Possible Forms of the Non-linear Kinematic Wave Equation

In order to emphasize the essential difference between conservative and non-conservative forms of the transport equation in the case of flood routing models, let us consider Eq. (9.37) with $D = 0$:

$$\frac{\partial Q}{\partial t}+\frac{1}{\alpha\cdot m}\cdot Q^{1-m}\frac{\partial Q}{\partial x}=0 \tag{9.53}$$

It represents a non-conservative form of the non-linear kinematic wave equation. Equation (9.53) can be integrated over a channel reach of length L:

$$\int_0^L\frac{\partial Q}{\partial t}dx+\frac{1}{\alpha\cdot m}\int_0^L Q^{1-m}\frac{\partial Q}{\partial x}dx=0 \tag{9.54}$$

In the first term of the above expression the order of integration and differentiation is inversed, whereas the second term is integrated by parts. Consequently Eq. (9.54) is rewritten as follows:

$$\frac{\partial}{\partial t}\int_0^L Q\cdot dx=\frac{1}{\alpha\cdot m}\left(Q_0^{1-m}\cdot Q_0-Q_L^{1-m}\cdot Q_L\right)+R_A \tag{9.55}$$

where:

Q_0 – discharge at the upstream end,
Q_L – discharge at the downstream end.

The term R_A is defined by the following formula:

$$R_A=\frac{1}{\alpha\cdot m}\int_0^L Q\frac{\partial Q^{1-m}}{\partial x}dx \tag{9.56}$$

Now let us consider again non-linear Eq. (9.53), but rewritten in another form:

$$\frac{\partial Q}{\partial t}+\frac{1}{\alpha\cdot m\cdot(2-m)}\cdot\frac{\partial Q^{2-m}}{\partial x}=0 \tag{9.57}$$

One can see that the coefficient in Eq. (9.57) is constant, because α, m = const., whereas the function $Q(x, t)$ is inserted into derivative. In this way Eq. (9.53), which represents a non-conservative form of non-linear kinematic wave, was transformed into a conservative form. As previously, Eq. (9.57) can be integrated over a channel reach:

$$\int_0^L\frac{\partial Q}{\partial t}dx+\int_0^L\frac{1}{\alpha\cdot m\cdot(2-m)}\frac{\partial Q^{2-m}}{\partial x}dx=0 \tag{9.58}$$

which leads to:

$$\frac{\partial}{\partial t}\int_0^L Q \cdot dx = \frac{1}{\alpha \cdot m \cdot (2-m)}(Q_0^{2-m} - Q_L^{2-m}) \tag{9.59}$$

From Eq. (9.59) results that the time-variation of the total quantity Q inside the channel reach of length L is caused only by net flux of Q^{2-m} through upstream and downstream ends of channel, whereas in Eq. (9.55) it depends not only on the net flux of Q^{2-m} through channel boundaries $x = 0$ and $x = L$, but also on the extra term R_A. It means that the global conservation of the quantity represented by function $Q(x, t)$ is not fulfilled.

Equation (9.59) is not the only possible conservative form of the kinematic wave equation. Another one can be derived directly from Eqs. (9.2) and (9.3) with $k = 0$ and $l = 0$. Combining these equations with earlier accepted assumption that $\alpha =$ const., one obtains:

$$\frac{\partial Q^m}{\partial t} + \frac{1}{\alpha}\frac{\partial Q}{\partial x} = 0 \tag{9.60}$$

An integration of Eq. (9.60) over a channel reach yields the following final result:

$$\alpha \cdot \frac{\partial}{\partial t}\int_0^L Q^m \cdot dx = Q_0(t) - Q_L(t) \tag{9.61}$$

Having three different forms of the non-linear kinematic wave equation one can compare their properties. First of all, it should be explained which conservative quantity they preserve. Since all equations were derived using the continuity and momentum equations for open channel flow, it is reasonable to expect that both conservation laws should be satisfied. To answer this question Eqs. (9.55), (9.59) and (9.61) must be integrated over time.

The integration of Eq. (9.55) over the time interval $\langle 0, T \rangle$:

$$\alpha \cdot m \cdot \int_0^T \frac{\partial}{\partial t} \cdot \int_0^L Q \cdot dx \cdot dt = \int_0^T (Q_0^{2-m}(t) - Q_L^{2-m}(t))dt + \int_0^T R_A \cdot dt \tag{9.62}$$

yields:

$$\alpha \cdot m \cdot \int_0^L (Q(x,T) - Q(x,0)) \cdot dx = \int_0^T (Q_0^{2-m}(t) - Q_L^{2-m}(t))dt + \int_0^T R_A \cdot dt \tag{9.63}$$

Making use of the assumption about the time interval T, the integral over channel reach disappears and Eq. (9.63) becomes as follows:

$$\int_0^T (Q_0^{2-m}(t) - Q_L^{2-m}(t))dt = -\int_0^T R_A \cdot dt \tag{9.64}$$

Following a similar way for the conservative forms of the kinematic wave equation one obtains:

- for Eq. (9.59)

$$\int_0^T (Q_0^{2-m}(t) - Q_L^{2-m}(t))dt = 0 \tag{9.65}$$

- for Eq. (9.61)

$$\int_0^T (Q_0(t) - Q_L(t))dt = 0 \tag{9.66}$$

The presented balance formulas allow us to suppose that:

- Equation (9.60) represents the mass conservation principle since Eq. (9.66) coincides with Eq. (9.40);
- Equation (9.57) represents the momentum conservation principle since Eq. (9.65) coincides with Eq. (9.46);
- Equation (9.53) satisfies neither the mass conservation law nor the momentum conservation law since Eq. (9.64) contains an extra term.

All possible forms of the kinematic wave equation are listed in Table 9.1.

Table 9.1 Forms of the kinematic wave equation

Possible form of equation	Conservation of mass	Conservation of momentum
$\frac{\partial Q}{\partial t} + C \cdot \frac{\partial Q}{\partial x} = 0\ (C = \text{const.})$	+	+
$\frac{\partial Q}{\partial t} + \frac{Q^{1-m}}{\alpha \cdot m} \cdot \frac{\partial Q}{\partial x} = 0$	–	–
$\frac{\partial Q}{\partial t} + \frac{1}{\alpha \cdot m \cdot (2-m)} \cdot \frac{\partial Q^{2-m}}{\partial x} = 0$	–	+
$\alpha \frac{\partial Q^m}{\partial t} + \frac{\partial Q}{\partial x} = 0$	+	–

One can conclude that among the possible forms of the kinematic wave model only the linear one respects both conservation laws, whereas the conservative forms of the non-linear equation represent either the mass or momentum conservation principle.

The last considered form of the kinematic wave model (Eq. 9.60) seems suitable with regard to the mass conservation. The global conservation law (Eq. 9.66), corresponding to this equation, is identical with the one for the linear kinematic wave and consequently the mass balance error is always equal to zero. However this equation does not satisfy the momentum conservation principle.

9.3.4 Possible Forms of the Non-linear Diffusive Wave Equation

The diffusive wave model in form of Eq. (9.37):

$$\frac{\partial Q}{\partial t} + \frac{Q^{1-m}}{\alpha \cdot m}\frac{\partial Q}{\partial x} - \frac{Q}{2 \cdot B \cdot s} \cdot \frac{\partial^2 Q}{\partial x^2} = 0 \tag{9.67}$$

is a non-linear advection-diffusion equation written in a non-conservative form. As it was shown earlier, such form of equation generates a significant mass error. To avoid this error it seems reasonable to transform Eq. (9.67) into a conservative form. For this purpose the following relations are used:

$$\frac{Q^{1-m}}{\alpha \cdot m}\frac{\partial Q}{\partial x} = \frac{1}{\alpha \cdot m(2-m)}\frac{\partial Q^{2-m}}{\partial x} \tag{9.68a}$$

$$Q \cdot \frac{\partial^2 Q}{\partial x^2} = \frac{1}{2}\frac{\partial^2 Q^2}{\partial x^2} - \left(\frac{\partial Q}{\partial x}\right)^2 \tag{9.68b}$$

Equation (9.67) becomes as follows:

$$\frac{\partial Q}{\partial t} + \frac{1}{\alpha \cdot m(2-m)}\frac{\partial Q^{2-m}}{\partial x} - \frac{1}{2B \cdot s}\left[\frac{1}{2}\frac{\partial^2 Q^2}{\partial x^2} - \left(\frac{\partial Q}{\partial x}\right)^2\right] = 0 \tag{9.69}$$

Its integration over channel reach yields:

$$\frac{\partial}{\partial t}\int_0^L Q \cdot dx = \left[\frac{C}{2-m} \cdot Q - D\frac{\partial Q}{\partial x}\right]_0 - \left[\frac{C}{2-m} \cdot Q - D \cdot \frac{\partial Q}{\partial x}\right]_L + R_D \tag{9.70}$$

where R_D is an additional term defined as follows:

$$R_D = -\frac{1}{2 \cdot B \cdot s}\int_0^L \left(\frac{\partial Q}{\partial x}\right)^2 \cdot dx \tag{9.71}$$

Since $(\partial Q/\partial x)^2 \geq 0$ the term R_D will decrease the transported quantity monotonically. The only reason of appearance of this term is diffusion. The term R_D acts as a sink and consequently one can expect that the function Q will not be preserved during the advection-diffusion transport process. Consequently, an error in the balance of the transported quantity can be expected.

The second form of the non-linear diffusive wave equation derived from Eq. (9.67) using Eq. (9.68b) is as follows:

$$\alpha \frac{\partial Q^m}{\partial t} + \frac{\partial Q}{\partial x} - \frac{\alpha \cdot m}{2 \cdot B \cdot s}\left[\frac{\partial}{\partial x}\left(Q^m \frac{\partial Q^{m+1}}{\partial x}\right) - \frac{\partial Q^m}{\partial x} \cdot \frac{\partial Q}{\partial x}\right] = 0 \tag{9.72}$$

Integration of this equation with regard to x yields:

$$\alpha \cdot \frac{\partial}{\partial t} \int_0^L Q^m \cdot dx = \left[Q - \frac{D}{C} \cdot \frac{\partial Q}{\partial x}\right]_L - \left[Q - \frac{D}{C} \cdot \frac{\partial Q}{\partial x}\right]_0 + R_D \tag{9.73}$$

In this case the additional term R_D is as follows:

$$R_D = -\frac{\alpha \cdot m}{2 \cdot B \cdot s} \int_0^L \frac{\partial Q^m}{\partial x} \frac{\partial Q}{\partial x} \cdot dx \tag{9.74}$$

Comparing both forms of the non-linear diffusive wave equation one can notice that in the corresponding balance formula the additional terms are always present. The integration of the non-conservative forms (Eqs. 9.69 and 9.72) with regard to x, apart from the net flux through boundary, yields additional term R_D described by Eq. (9.71) or (9.74), respectively. Transformation of Eq. (9.67) into a conservative form eliminates only the effect resulting from the integration of the advective term. The additional term R_D resulting from the integration of the diffusive part of equation is always present. This explains the mass balance error observed in the numerical solutions. Taking into account these facts we can conclude that it is impossible to find a suitable form of the non-linear diffusive wave equation, which could satisfy the mass conservation principle. The reason for this is the way of derivation of the governing equation. Since the non-linear diffusive wave equation is obtained as a combination of the continuity and momentum equation, it cannot be written in a divergent form. The diffusive term in the form:

$$D(Q)\frac{\partial^2 Q}{\partial x^2}$$

existing in Eq. (9.67) cannot be converted into a term of the type:

$$\frac{\partial}{\partial x}\left(D(Q)\frac{\partial Q}{\partial x}\right)$$

i.e. into a divergent form of the diffusive flux. Consequently the global mass conservation will be always affected.

Equations (9.71) and (9.74) show that the balance errors depend on the derivatives of the flow discharge with regard to x ($\partial Q/\partial x$ and $\partial Q^m/\partial x$). Therefore one can expect that the generated error will be greater for a steep wave than for a smooth one.

9.4 Lumped Flood Routing Models

9.4.1 Standard Derivation of the Muskingum Equation

Apart from the previously presented distributed models (i.e. the dynamic, diffusive and kinematic wave equations), a flood routing can be also carried out using lumped models. This family of models is obtained by elimination of the spatial variable x, so they have the form of ordinary differential equations with regard to time. The standard way of derivation is based on spatial integration of the differential continuity equation (9.15):

$$\frac{\partial A}{\partial t} + \frac{\partial Q}{\partial x} = 0. \tag{9.75}$$

Let us consider a channel reach of length Δx as presented in Fig. 9.7. This segment is bounded by two cross-sections located at x_{j-1} and x_j. Equation (9.75) integrated over the reach:

$$\int_{x_{j-1}}^{x_j} \frac{\partial A}{\partial t} dx + \int_{x_{j-1}}^{x_j} \frac{\partial Q}{\partial x} dx = 0 \tag{9.76}$$

gives:

$$\frac{\partial}{\partial t} \int_{x_{j-1}}^{x_j} A \cdot dx + Q|_{x_{j-1}}^{x_j} = 0. \tag{9.77}$$

The first term of Eq. (9.77) represents the variation in time of the water volume S stored by the channel reach of length Δx since

$$\int_{x_{j-1}}^{x_j} A \cdot dx = S. \tag{9.78}$$

The second term of Eq. (9.77) represents the difference between the flow rates at the ends of the reach:

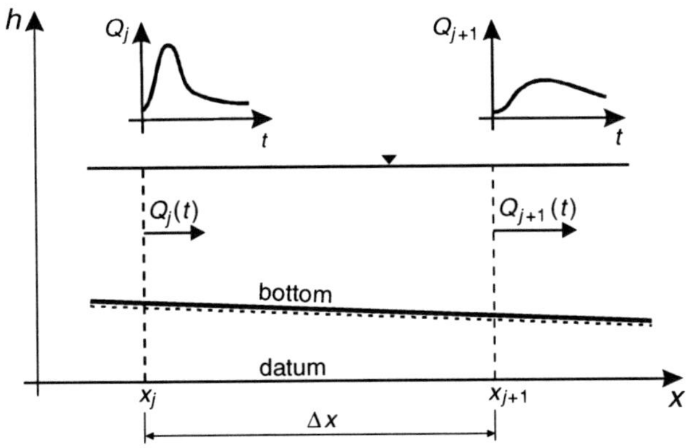

Fig. 9.7 Channel reach of length Δx considered as reservoir

$$Q|_{x_{j-1}}^{x_j} = Q_j(t) - Q_{j-1}(t). \tag{9.79}$$

Substitution of Eqs. (9.78) and (9.79) in Eq. (9.77) yields the well known storage equation:

$$\frac{dS}{dt} = Q_{j-1}(t) - Q_j(t) \tag{9.80}$$

where:

S – volume of water stored by channel reach of length Δx,
$Q_{j-1}(t)$ – discharge entering a channel reach through end x_{j-1},
$Q_j(t)$ – discharge leaving a channel reach through end x_j.

In the next step one function from storage equation is eliminated. For this purpose an additional equation relating storage, inflow and outflow is introduced. The most commonly used formula is (Chow 1959, 1964, Chow et al. 1988):

$$S = K\left(\psi \cdot Q_{j-1}(t) + (1 - \psi)Q_j(t)\right) \tag{9.81}$$

where:

K – constant parameter expressed in time units,
ψ – weighting parameter ranging from 0 to 1.

The parameter K is interpreted as the time of traveling of the flood wave over the channel reach between the cross-sections x_{j-1} and x_j. As the length of the reach is $\Delta x = x_j - x_{j-1}$, and the wave propagates with celerity C, then the constant K may be expressed as follows:

$$K = \frac{\Delta x}{C} \tag{9.82}$$

Conversely to the parameter K, the parameter ψ has no physical interpretation.

Equation (9.81) is differentiated with regard to time and next the obtained result is substituted in the storage equation. Finally one obtains the well known Muskingum model:

$$\psi \frac{d\,Q_{j-1}}{d\,t} + (1-\psi)\frac{d\,Q_j}{d\,t} = \frac{1}{K}\left(Q_{j-1}(t) - Q_j(t)\right) \tag{9.83}$$

Setting $\psi = 0$ one obtains another very popular hydrological lumped model – the linear reservoir model:

$$\frac{dQ_j}{dt} = \frac{1}{K}(Q_{j-1}(t) - Q_j(\mathrm{t})). \tag{9.84}$$

Rather than being represented by a single reservoir, the channel reach is usually conceptualized as a cascade of N reservoirs as shown in Fig. 9.8. The water leaving the preceding reservoir enters the next one. Each reservoir operates according to Eq. (9.83), so in this equation the index of cross-section takes the values $j = 1, 2, 3, \ldots, N$.

9.4.2 Numerical Solution of the Muskingum Equation

Equation (9.83) is an ordinary differential equation, which must be integrated using a numerical method. Since the considered channel reach is divided into N reservoirs we obtain a system of N such equations describing each reservoir. The input to the first reservoir $Q_0(t)$ for $0 \leq t \leq T$ represents the flood wave to be routed. Equation

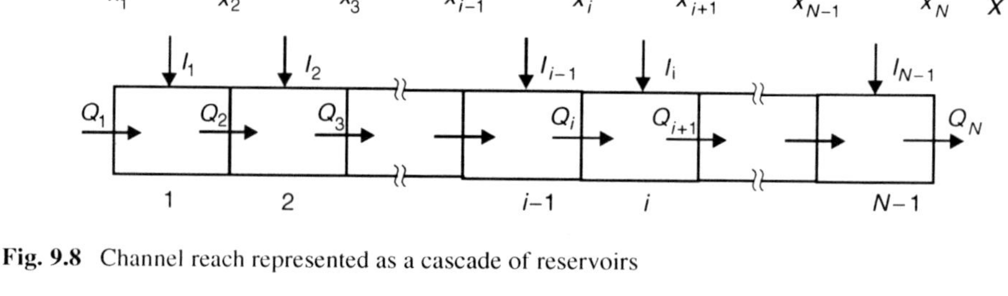

Fig. 9.8 Channel reach represented as a cascade of reservoirs

(9.83) is approximated using the implicit trapezoidal rule (3.21). This yields the following system of algebraic equations:

$$\psi \frac{Q_{j-1}^{n+1} - Q_{j-1}^{n}}{\Delta t} + (1-\psi)\frac{Q_{j}^{n+1} - Q_{j}^{n}}{\Delta t} = \frac{1}{K}\left(\frac{Q_{j-1}^{n+1} + Q_{j-1}^{n}}{2} - \frac{Q_{j}^{n+1} + Q_{j}^{n}}{2}\right) = 0 \quad \text{for } j = 1,2, \cdots N \tag{9.85}$$

where:

Δt – time step,
n – index of time level,
j – index of cross-section.

Since in Eq. (9.85) only one unknown exists, then it can be split into separate equations, from which corresponding unknowns are calculated directly:

$$Q_{j}^{n+1} = \alpha \cdot Q_{j-1}^{n} + \beta \cdot Q_{j}^{n} + \gamma \cdot Q_{j-1}^{n+1} \text{ for } j = 1,2,3, \ldots, N. \tag{9.86}$$

The coefficients α, β and γ are given by the following formulas:

$$\alpha = \frac{K \cdot \psi + 0{,}5\Delta t}{K(1-\psi) + 0{,}5\Delta t}, \tag{9.87a}$$

$$\beta = \frac{K(1-\psi) - 0{,}5\Delta t}{K(1-\psi) + 0{,}5\Delta t}, \tag{9.87b}$$

$$\gamma = \frac{-K \cdot \psi + 0{,}5\Delta t}{K(1-\psi) + 0{,}5\Delta t} \tag{9.87c}$$

The flood wave $Q_0(t)$ specified as the input to the system is transformed using Eq. (9.86) with assumed values of the parameter K and ψ over the time interval $\langle 0, T\rangle$. The output from the last reservoir is the solution of the considered flood routing problem.

Example 9.4 The flood wave given by the following formula:

$$Q(x=0,t) = Q_0 + (Q_{\max} - Q_0)\left(\frac{t}{t_{\max}}\right)^{\alpha} \exp\left(1 - \left(\frac{t}{t_{\max}}\right)^{\alpha}\right)$$

where $Q_0 = 10$ m3/s, $Q_{\max} = 75$ m^3/s, $t_{\max} = 4$ h and $\alpha = 0.85$, propagates in an open channel. Using the Muskingum model with the following data: $N = 4$, $K = 4$ h, $\Delta t = 0.5$ h, let us examine the influence of the value of weighting parameter ψ on the shape of hydrograph leaving the considered channel reach.

The calculations were performed for a couple of values of the weighting parameter Ψ. The results displayed in Fig. 9.9 show that they strongly depend on this parameter. Using the Muskingum model it is possible to obtain oscillating solution, even with negative discharge during the initial phase of wave.

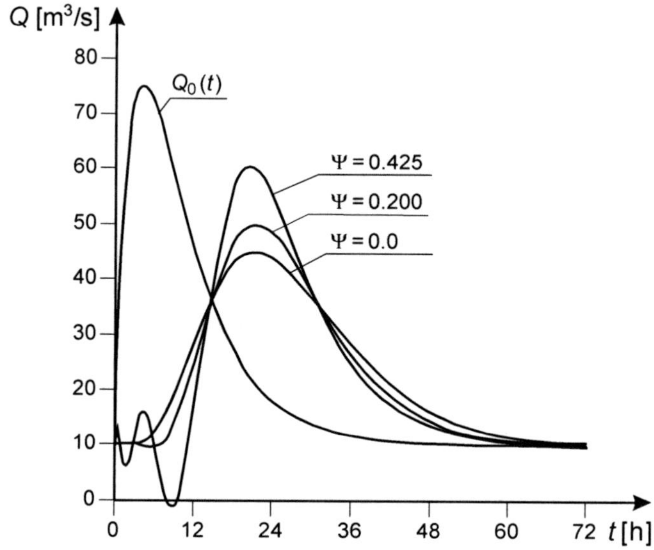

Fig. 9.9 Results of routing of the wave $Q_0(t)$ by the Muskingum model for various values of the parameter ψ

Note that for $\Delta t = K$ and for $\psi = 0.5$ the coefficients of Eq. (9.86) take the following values: $\alpha = 1$, $\beta = 0$, $\gamma = 0$. Therefore this equation is reduced to a very simple relation:

$$Q_j^{n+1} = Q_{j-1}^n \quad \text{for } j = 1, 2, 3, ..., \mathrm{N} \tag{9.88}$$

Let us remember that this formula coincides with the formula (6.117) given by the numerical solution of the linear advection equation using the method of characteristics or using the finite difference up-wind scheme (Eq. 5.108) with the advective Courant number equal to unity. Equation (9.88) ensures exact solution to the problem. Numerical tests confirm this conclusion. In Fig. 9.10 one can see pure translation of flood wave without any deformation of its shape. This property of the Muskingum model gives some insights into its nature.

9.4.3 The Muskingum–Cunge Model

Comparing the Muskingum equation solved by the implicit trapezoidal rule (Eq. 9.85) and the pure advection equation solved by the difference box scheme one can notice their similarity. Equation (6.4) with $\theta = 1/2$ becomes:

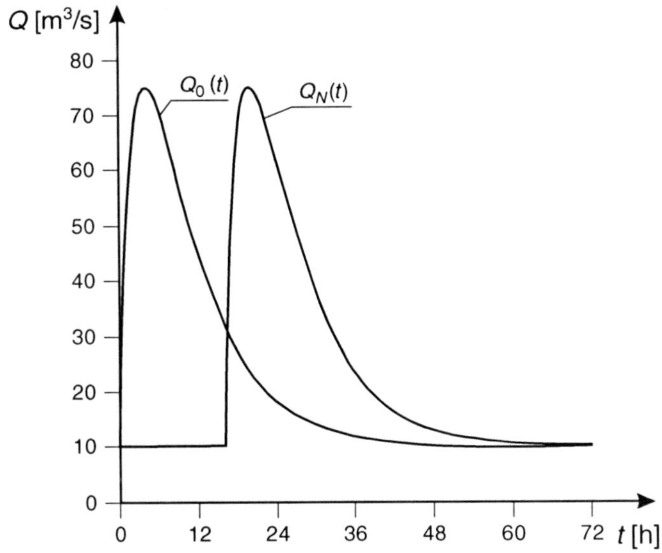

Fig. 9.10 Pure translation of flood wave $Q_0(t)$ produced by the Muskingum model for $\psi = 0.5$ and $\Delta t = K$

$$\psi \frac{Q_{j-1}^{n+1} - Q_{j-1}^{n}}{\Delta t} + (1-\psi)\frac{Q_{j}^{n+1} - Q_{j}^{n}}{\Delta t} = \frac{C}{\Delta x}\left(\frac{Q_{j}^{n} - Q_{j-1}^{n}}{2} + \frac{Q_{j}^{n+1} - Q_{j-1}^{n+1}}{2}\right) = 0 \tag{9.89}$$

Taking into account the relation (9.82) we can see that Eq. (9.89) coincides with Eq. (9.85). For the first time this similarity was noticed by Cunge (1969), who concluded on the kinematic character of the Muskingum model and the numerical nature of the wave attenuation process. The accuracy analysis carried out for Eq. (9.89) using the modified equation approach shows that the applied approximation modifies the kinematic wave model to the form of advection-diffusion equation, known from Chapter 7:

$$\frac{\partial Q}{\partial t} + C\frac{\partial Q}{\partial x} - D_n \frac{\partial^2 Q}{\partial x^2} = 0 \tag{9.90}$$

where D_n is the coefficient of numerical diffusion defined as follows:

$$D_n = \left(\frac{1}{2} - \psi\right) C \cdot \Delta x \tag{9.91}$$

Equation (9.90) indicates that the wave attenuation given by the Muskingum model has the same numerical roots as the kinematic wave equation. Simply, this effect is caused by the numerical diffusion. Cunge (1969) suggested to use such a value of the parameter Ψ which ensures the numerical diffusivity (Eq. 9.91) equal to the hydraulic one present in the diffusive wave equation and given by Eq. (9.39), i.e. $D = D_n$. This condition implies that:

$$\psi = \frac{1}{2} - \frac{Q}{2 \cdot B \cdot s \cdot C \cdot \Delta x} \tag{9.92}$$

or with the relation (9.82):

$$\psi = \frac{1}{2} - \frac{Q}{2 \cdot B \cdot s \cdot K \cdot C^2} \tag{9.93}$$

where:

Q – flow rate,
B – channel width,
s – bottom slope,
K – constant parameter,
C – kinematic celerity.

In such a way Cunge related the value of weighting parameter ψ not only to the flow rate and the kinematic celerity but to the channel parameters as well. Consequently the Muskingum model becomes capable to reproduce the solution of the linear diffusive wave model. This version of the Muskingum equation is commonly called Muskingum–Cunge model (see for instance Chow et al. 1988). However it should be remembered that Cunge's approach does not change the character of the Muskingum model. The wave attenuation is not caused by the physical process but by the numerical error, although its intensity is determined by the parameter ψ dependent on the channel parameters.

Equivalence of the numerical solutions of the diffusive wave model and the Muskingum model is illustrated by the example presented below.

Example 9.5 Consider a rectangular channel of length $L = 70$ km, width $B = 50$ m, bottom slope $s = 0.0004$ and the Manning coefficient $n_M = 0.040$. The flood wave entering the upstream end is given by the following formula:

$$Q(x = 0,t) = Q_0 + (Q_{\max} - Q_0)\left(\frac{t}{t_{\max}}\right)^{\alpha} \exp\left(1 - \left(\frac{t}{t_{\max}}\right)^{\alpha}\right)$$

where $Q_0 = 5$ m^3/s, $Q_{\max} = 50$ m^3/s, $t_{\max} = 6$ h and $\alpha = 1.0$. Let us compare the form of hydrographs leaving the considered channel reach, calculated using the diffusive wave equation and the Muskingum–Cunge model.

Assuming that the propagation is described by Eq. (9.34) with $C =$ const. and $D =$ const., the values of both parameters are calculated for the algebraic average values of flow discharge: $\overline{Q} = (Q_0 + Q_{\max})/2$. For the assumed data these parameters are as follows:

$$C = \frac{5}{3} \cdot \frac{27.5^{1-3/5}}{\frac{3}{5} \cdot \left(\frac{0.04 \cdot 50^{2/3}}{0.0004^{1/2}}\right)} = 0.87\ \mathrm{m/s},$$

$$D = \frac{27.5}{2 \cdot 50 \cdot 0.0004} = 687.5\ \mathrm{m^2/s}$$

The auxiliary conditions are:

- initial condition: $Q(x, t = 0) = Q_0$ for $0 \leq x \leq L$,
- boundary conditions: $Q(x = 0, t) = Q_0$ for $t \geq 0$ and $\left.\frac{\partial Q}{\partial x}\right|_{x=L} = 0$ for $t \geq 0$

Equation (9.34) is solved using the modified finite element method. Computations are performed for $\Delta x = 2{,}000$ m and $\Delta t = 1{,}200$ s. The obtained results $Q(L = 75\ \mathrm{km}, t)$ are shown in Fig. 9.11.

The same wave propagation problem is next solved using the Muskingum–Cunge model (9.83). Equation (9.83) is solved using the implicit trapezoidal rule for the constant parameters K and Ψ given by Eqs. (9.82) and (9.93). As previously their values are calculated for the average flow discharge $\overline{Q}$:

$$K = \frac{\Delta x}{C} = \frac{2000}{0.87} = 2299\ \mathrm{s}$$

$$\Psi = \frac{1}{2} - \frac{\overline{Q}}{2 \cdot B \cdot s \cdot C \cdot \Delta x} = \frac{1}{2} - \frac{27.5}{2 \cdot 50 \cdot 0.0004 \cdot 0.87 \cdot 2000} = 0.105$$

The hydrographs calculated at the downstream end for $\Delta t = 1200$ s are displayed in Fig. 9.11.

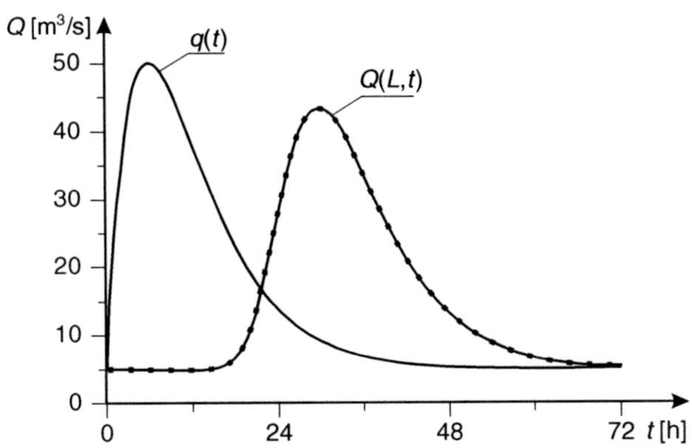

Fig. 9.11 Comparison of the results given by the diffusive wave equation (*solid line*) and the Muskingum–Cunge model (*dotted line*)

Note that both models indeed produce equivalent numerical solutions. Thus instead of the diffusive wave equation one can use the Muskingum–Cunge equation which is simpler to solve numerically

9.4.4 Relation Between the Lumped and Simplified Distributed Models

The similarity of the approximated forms of the kinematic wave equation and the Muskingum equation presented in the preceding section allows us to deduce on the nature of the lumped models. In fact the Muskingum model should be regarded as a semi-discrete form of the kinematic wave equation. Moreover we can estimate directly from Eq. (9.83) the numerical diffusion generated by this model. To this end let us consider the linear kinematic wave model (Eq. 9.37 with $C = \text{const.}$ and $D = 0$). This equation can be discretized in space only. The approximation carried out at point P located between the nodes $j - 1$ and j (Fig. 9.12 and 9.13) gives:

$$\frac{dQ_p}{dt} + C\frac{Q_j - Q_{j-1}}{\Delta x} = 0 \tag{9.94}$$

where Q_p represents the discharge at the point P.

The value of Q_p can be calculated by the linear interpolation between the nodes $j - 1$ and j:

$$Q_P = Q_{j-1} + \frac{Q_j - Q_{j-1}}{x_j - x_{j-1}}\left(x - x_{j-1}\right) \quad \text{for } x_{j-1} \le x \le xj \tag{9.95}$$

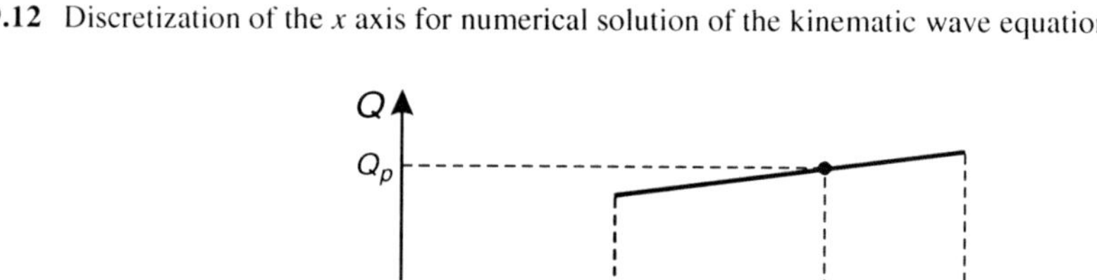

Fig. 9.12 Discretization of the x axis for numerical solution of the kinematic wave equation

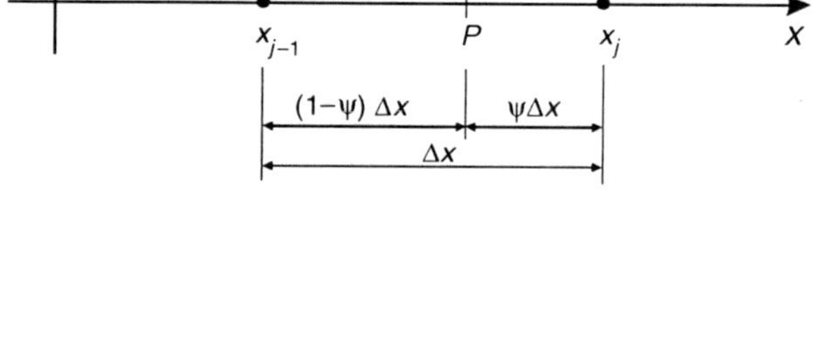

Fig. 9.13 Linear interpolation between neighboring nodes applied for the kinematic wave equation

Equation (9.95) can be rewritten in the following form:

$$Q_p = \psi \cdot Q_{j-1} + (1 - \psi)\, Q_j \tag{9.96}$$

where ψ is defined as:

$$\psi = \frac{x_j - x}{x_j - x_{j-1}} \quad \text{For } x_j - 1 \le x \le x_j \tag{9.97}$$

Note that ψ can be considered as a weighting parameter ranging from 0 to 1.

Substituting Eq. (9.96) in Eq. (9.94) and including Eq. (9.82) yields the Muskingum model in the form of Eq. (9.83):

$$\psi \frac{d\,Q_{j-1}}{d\,t} + (1 - \psi)\frac{d\,Q_j}{d\,t} = \frac{1}{K}\,\left(Q_{j-1}(t) - Q_j(t)\right) \tag{9.98}$$

The performed spatial discretisation of the kinematic wave equation introduces a numerical error caused by the truncation of the Taylor series. In order to estimate this error one can carry out the consistency analysis. The nodal values of Q in Eq. (9.98) are replaced by the Taylor series expansion around point P:

$$Q_{j-1} = Q_p - (1 - \psi)\Delta x \frac{\partial Q_p}{\partial x} + \frac{(1 - \psi)^2 (\Delta x)^2}{2} \frac{\partial^2 Q_p}{\partial x^2} + \cdots \tag{9.99}$$

$$Q_j = Q_p + \psi \cdot \Delta x \frac{\partial Q_p}{\partial x} + \frac{\psi^2 (\Delta x)^2}{2} \frac{\partial^2 Q_p}{\partial x^2} + \cdots \tag{9.100}$$

Substitution of Eqs. (9.99) and (9.100) in Eq. (9.98), including the terms of 2nd order, yields the modified equation in the form:

$$\frac{\partial Q}{\partial t} + \frac{\Delta x}{K} \frac{\partial Q}{\partial x} - \left(\frac{1}{2} - \psi\right) \frac{\Delta x^2}{K} \frac{\partial^2 Q}{\partial x^2} = 0 \tag{9.101}$$

According to the condition of consistency, which must be satisfied by any approximation of the partial differential equation, the modified equation should tend to the governing equation for $\Delta x \to 0$ (Fletcher 1991). For $\Delta x \to 0$ the time of wave translation along the channel reach of the length Δx simultaneously tends to zero ($K \to 0$). Therefore we have

$$\lim_{\substack{\Delta x \to 0 \\ K \to 0}} \frac{\Delta\, x}{K} = C \tag{9.102}$$

and consequently for $\Delta x \to 0$ Eq. (9.101) tends to the kinematic wave equation. This fact proves that the Muskingum model is an approximation of the kinematic wave. This approximation introduces a numerical diffusion. Its coefficient is:

$$D_n = \left(\frac{1}{2} - \psi\right)\frac{\Delta x^2}{K} \tag{9.103}$$

This expression coincides with Eq. (9.91) proposed by Cunge (1969). One can add that the numerical diffusion is caused by the spatial approximation only. An additional diffusion can be generated while integrating Eq. (9.98) over time by a method of the order lower than two. Usually the implicit trapezoidal rule is applied. It ensures an accuracy of 2nd order with regard to t and consequently this approximation is dissipation free.

Summarizing, one can say that the numerical solution of the Muskingum model is in fact equivalent to the numerical solution of the kinematic wave model by the method of lines. In this approach a solution of the partial differential equation is made in two stages. At first it is discretised in space leading to a system of ordinary differential equations over time. Next, this system is integrated using any well known method of the numerical solution of an initial problem for the ordinary differential equations.

The classical derivation of the Muskingum model bases on the storage equation (9.80) completed by the relationship (9.81). Therefore it is interesting to show how numerical diffusion is introduced into this model. To explain this problem let us consider the Muskingum model (Eq. 9.98) rewritten in more general form:

$$K\frac{dQ_p}{dt} = Q_{j-1} - Q_j \tag{9.104}$$

with Q_p defined by Eq. (9.96).

We can show that this equation can be derived directly from the storage equation (9.80) without any additional formula relating storage, inflow and outflow (Szymkiewicz 2002). To do this we have to assume the following:

- the storage S is calculated numerically,
- the equation for uniform steady flow is applied.

Using both above assumptions one can transform Eq. (9.80) into (9.104).

The storage S can be calculated as follows:

$$S = \int_0^{\Delta x} A \cdot dx \approx A_p \cdot \Delta x \tag{9.105}$$

where A_p, being a cross-sectional area at the point P (Fig. 9.12), can be expressed as a function of Q_p using the Manning (or Chézy) formula:

$$A_p = \alpha \cdot (Q_p)^m \tag{9.106}$$

with α and m given by Eqs. (9.14a) and (9.14b), respectively. Therefore the left hand side of Eq. (9.80) can be rewritten as follows:

$$\frac{dS}{d\,t} = \frac{d}{dt}\left(\Delta\, x \cdot A_p\right) = \Delta\, x\frac{dA_p}{dt} = \Delta\, x\frac{d}{dt}(\alpha \cdot (Q_p)^m) \tag{9.107}$$

After differentiating with $\alpha = \text{const.}$ one obtains:

$$\Delta\, x\frac{d}{d\,t}(\alpha \cdot (Q_p)^m) = \Delta\, x\frac{dA_p}{dQ_p}\frac{dQ_p}{dt} = \Delta\, x \cdot \alpha \cdot m \cdot (Q_p)^{m-1}\frac{dQ_p}{dt} \tag{9.108}$$

Because the kinematic wave celerity at the point P is defined as

$$\frac{dQ_p}{dA_p} = C_p = \frac{1}{\alpha \cdot m \cdot (Q_p)^{m-1}} \tag{9.109}$$

the right side of Eq. (9.107) takes the form:

$$\Delta\, x\alpha m(Q_p)^{m-1}\frac{dQ_p}{dt} = \frac{\Delta\, x}{C_p}\frac{dQ_p}{dt} = K\frac{dQ_p}{dt} \tag{9.110}$$

Finally the left hand side of Eq. (9.104) was obtained from the left hand side of Eq. (9.80). Therefore it seems reasonable to accept that an additional formula (9.81) used to derive the lumped models from the storage equation has double meaning. It can be interpreted as a result of the numerical integration of storage and of the application of the steady uniform flow equation. A numerical diffusion is introduced by the numerical calculation of the storage S. Note that the kinematic wave model is based on the same equations i.e. the equation of continuity and the steady flow equation.

As far as the conservative properties of the Muskingum model are considered, its linear version satisfies both mass and momentum conservation principles. However, sometimes the variable parameters K and ψ are introduced into the model. After Tang et al. (1999) this approach generates the mass balance error. Such error can be expected if the governing equation is not written in the appropriate conservative form. As it was shown earlier, the Muskingum model can be considered as the semi-discrete form of the kinematic wave equation. Consequently both models should have similar properties.

9.5 Convolution Integral in Open Channel Hydraulics

9.5.1 Open Channel Reach as a Dynamic System

An open channel reach of length L can be considered as a dynamic system, which is able to transform a flood wave appearing at the upstream end to the one observed at the downstream end (Fig. 9.14).

In other words, an input to the system is turned into an output leaving this system, as it is shown in Fig. 9.15. Our goal is to identify the physical mechanism, which

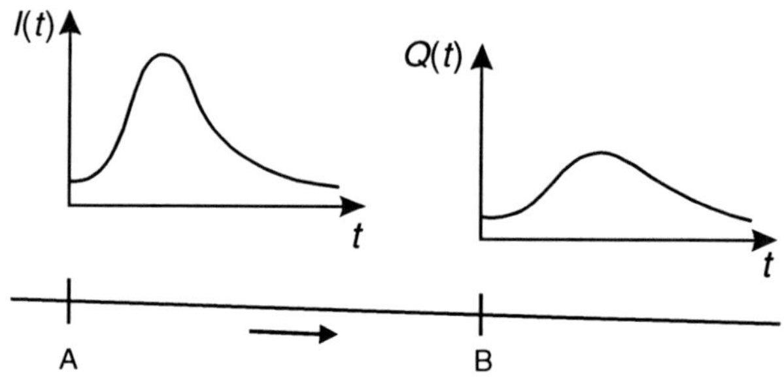

Fig. 9.14 Open channel reach as a dynamic system

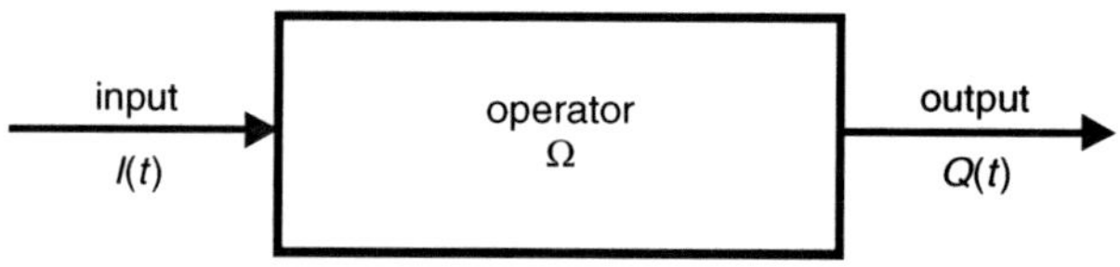

Fig. 9.15 Symbolic representation of the dynamic system

determines the way of transformation, what means that we are looking for the operator Ω representing this process. This operator allows us to compute the output of the system for any input $I(t)$ entering this system, i.e.:

$$Q\,(t) = \Omega\,(I\,(t)) \tag{9.111}$$

The mechanism of transformation can be deduced via the analysis of flow process using the appropriate principles of conservation. Such an approach has been applied in the preceding chapters, in which various mathematical models of unsteady flow were derived. Then the operator Ω took the forms of differential equations as the system of Saint Venant equations, the diffusive wave equation, the kinematic wave equation etc.

In the case of lumped linear systems it is possible to describe their dynamic behavior using either the differential flow equations or an impulse response function, which is the reaction of the system to the input in form of unit quantity entering instantaneously. Both approaches are equivalent. If we consider the open channel flow as a linear process, then it can be described by the previously presented differential equations in linear forms or by an impulse response function. Application of the impulse response function has two advantages. Firstly, such an approach does not require physical analysis of the flow. Secondly, knowing an impulse response function one can compute an output for any input in a simple way, without the difficulties related to the numerical integration of differential equations.

The system's linearity means that the principles of proportionality and superposition are valid. Therefore, if input $I_1(t)$ causes output $Q_1(t)$, whereas $I_2(t)$ gives rise to reaction $Q_2(t)$, then input in the form $I(t) = C_1 \cdot I_1(t) + C_2 \cdot I_2(t)$ will cause the following output $Q(t) = C_1 \cdot Q_1(t) + C_2 \cdot Q_2(t)$, where C_1 and C_2 are arbitrary constants. These principles allow us to decompose a complex input into simpler components, compute the outputs corresponding to the subsequent components and add the results of computations to obtain the final output.

Any lumped linear system can be described by the following ordinary differential equation of mth order (Chow et al. 1988):

$$\begin{aligned} A_m(t)\frac{d^m Q}{dt^m} + \ldots + A_1(t)\frac{dQ}{dt} + A_o(t)\cdot Q = \\ = B_m\frac{d^m I}{dt} + \ldots + B_1(t)\frac{dI}{dt} + B_o(t)\cdot I \end{aligned} \tag{9.112}$$

where:

$I(t)$ – function entering the system,
$Q(t)$ – function leaving the system,
$A_i(t)$, $B_i(t)$ ($i = 0,1, \ldots, m$) – time-variable coefficients describing dynamic properties of the system.

Usually the scale of the time variation of the dynamic properties is incomparable with the one related to the flow process. In such a case one can assume that the coefficients A_i and B_i are constant in time. Then Eq. (9.112) takes the form:

$$A_m\frac{d^m Q}{dt^m} + \ldots + A_1\frac{dQ}{dt} + A_0\cdot Q = B_m\frac{d^m I}{dt^m} + \ldots + B_1\frac{dI}{dt} + B_0\cdot I. \tag{9.113}$$

Equation (9.113) describes the time-invariant system, in which the way of transformation of $I(t)$ into $Q(t)$ is still the same. This means that the same flood wave at the upstream channel end will produce the same wave at the downstream end at any time. In other words, the dynamic properties of channel reach are constant in time.

For the following initial conditions: $I(t = 0) = 0$ and $Q(t = 0) = 0$, Eq. (9.113) has the following solution (Korn and Korn 1968):

$$Q(t) = \int_o^t w(t-\tau)\cdot I(\tau)d\tau, \tag{9.114}$$

where:

$w(t)$ – impulse response function,
τ – dummy parameter.

The integral in Eq. (9.114) is called convolution integral. The function $w(t–\tau)$ represents globally the dynamic properties of the considered system, since it determines the way of transformation an input into an output. This function is a reaction of the system to an input in the form of unit impulse. The unit impulse is a signal of short duration having unit magnitude: $X\cdot\Delta t = 1$. In the limit, when Δt tends to zero, X tends to infinity and the impulse becomes the Dirac delta function given by the following expression (McQuarrie 2003):

$$\delta(t-\tau)=\begin{cases}\infty \text{ for } t=\tau\\ 0 \text{ for } t\neq\tau\end{cases}, \tag{9.115}$$

The Dirac delta function has the following properties:

$$\int_{-\infty}^{+\infty}\delta(t)\cdot dt=1. \tag{9.116}$$

$$\int_{-\infty}^{+\infty}f(t)\,\delta(t-\tau)\,dt=f(\tau), \tag{9.117}$$

where $f(t)$ is an arbitrary function. An essential feature of this function is that instead of any value of $f(t)$ it is capable to introduce the value of $f(\tau)$: $f(t)\rightarrow f(\tau)$. The function $w(t-\tau)$, being the response of system to a unit impulse, is defined as follows (Fig. 9.16):

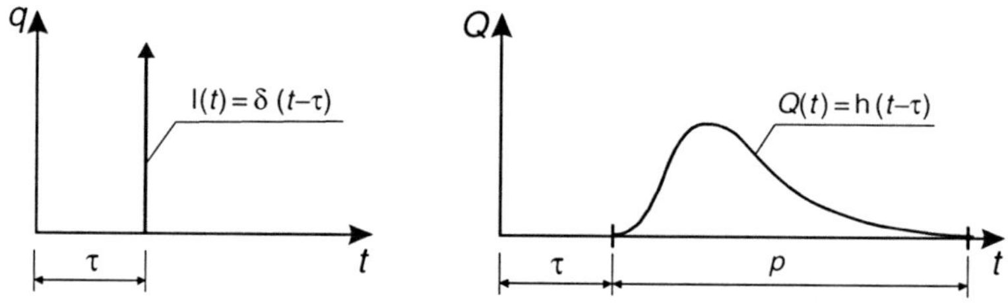

Fig. 9.16 Interpretation of the response function

If

$$I\,(t)\;=\;\delta\,(t) \tag{9.118}$$

then

$$Q\,(t)\;=\;w\,(t) \tag{9.119}$$

Therefore, the response function is an output of the system when an input has the form of Dirac delta function. If the system conserves the mass:

$$\int_{0}^{+\infty}I\,(t)\cdot d\,t=\int_{0}^{+\infty}Q(t)\cdot d\,t, \tag{9.120}$$

then the function $w(t)$ satisfies the following condition:

$$\int_{0}^{+\infty}w(t)d\,t=1. \tag{9.121}$$

Moreover this function:

- has non-negative values only: $w(t) \geq 0$ for $t \geq 0$,
- has one extreme point (maximum),
- is equal to zero for negative argument: $w(t) = 0$ for $t \leq 0$,
- tends to zero as time tends to infinity: $w(t) \rightarrow 0$ for $t \rightarrow \infty$.

If the functions $I(t)$ and $Q(t)$ represent the same quantity and they have the same dimension, the function $w(t)$ is expressed in the inverted time units as s^{-1}, h^{-1} etc.

In hydrology the function $w(t)$ is called Instantaneous Unit Hydrograph (IUH). This term was introduced in the 1960 s as a generalization of the unit hydrograph theory proposed by Sherman in 1932 for direct runoff from catchments (Chow et al. 1988).

For the convolution integral the principle of symmetry is valid. Accordingly to this rule, Eq. (9.114) can be rewritten as:

$$Q(t) = \int_0^t w(\tau) \cdot I(t - \tau) \cdot d\tau. \tag{9.122}$$

Since an input from distant past does not influence the current output, then we can introduce the parameter p representing "the system memory". This is a time interval in which the ordinates of IUH differ significantly from zero (Fig. 9.17).

As $w(t)$ tends to zero for $t \rightarrow \infty$, then the following condition will be satisfied:

$$w(t) \leq \varepsilon \quad \text{for} \quad t \geq p, \tag{9.123}$$

where ε is a small positive number. One can notice, that for $t > p$ the product under the integral (9.122) is close to zero. Then in Eq. (9.122) the upper integral limit can be changed as follows:

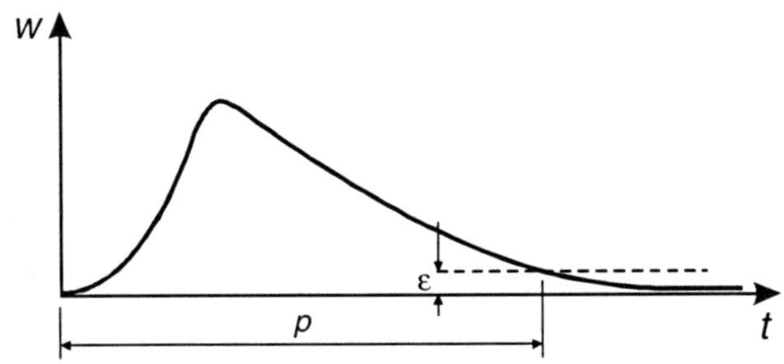

Fig. 9.17 Illustration of the system memory

$$Q(t) = \int_{o}^{p} w(\tau) \cdot I(t - \tau) \cdot d\tau. \tag{9.124}$$

where p is the system memory. Therefore, if we want to compute the function Q at any moment t, we must known the function $I(t)$ not only at this moment, but also in the time interval $\langle t–p, t\rangle$.

As the value of p is determined by the parameter ε, then it is obvious that $w(t)$ does not satisfy the condition (9.121). We have:

$$\int_{0}^{p} w(t)dt \neq 1 \tag{9.125}$$

and consequently the mass balance error will be generated because:

$$\int_{0}^{T} Q(t)dt \neq \int_{0}^{T} I(t)dt \tag{9.126}$$

where T is time of flood wave duration. To eliminate this incompatibility, the ordinates of function $w(t)$ should be rescaled. This can be done in the following way:

– integrate numerically the function $w(t)$ in interval $\langle 0, p\rangle$

$$a = \sum_{j=1}^{P-1} \Delta t \left(\frac{w_j + w_{j+1}}{2} \right) \tag{9.127}$$

where Δt is time step, whereas $P = p/\Delta t{+}1$ is discrete memory;

– calculate the value of correction corresponding to a unit ordinate of function $w(t)$:

$$\Delta = \frac{1 - a}{\sum\limits_{j=1}^{P} w_j} \tag{9.128}$$

– correct the ordinates of the function $w(t)$ proportionally to their values:

$$w_j^* = w_j + \Delta \cdot w_j \quad \text{for} \quad j = 1,2, \cdots P \tag{9.129}$$

where w_j^* is corrected value of w_j.

9.5.2 IUH for Hydrological Models

Determination of the function $h(t)$ for the considered system can be performed using two approaches. The first approach bases on the observed functions $I(t)$ and $Q(t)$ only. No additional information on the system is used. When the function $I(t)$ and $Q(t)$ are known, Eq. (9.124) is the Volterra integral equation of the first type with $w(t)$ as a kernel function (Korn and Korn 1968). Solution of this equation gives the function $w(t)$. There are many techniques to solve this problem. Some of them, oriented to model the direct runoff processes, were proposed by O'Donnell (1960), Chow (1964), Nash (1957), Rodriguez-Iturbe and Valdes (1979) among others. Since these methods are rather less useful for open channel flow modeling they will not be considered here. For the flood routing more suitable is the second possible way of determination of the function $w(t)$. This approach can be applied on condition that the linear differential equation describing the considered system is known. In such a case one can solve this equation for such auxiliary conditions, which ensure solution in the form of impulse response function $w(t)$. This approach is very popular and frequently used.

The first example concerns the diffusive wave equation (9.34) derived in Section 9.2.2:

$$\frac{\partial Q}{\partial t} + C\frac{\partial Q}{\partial x} - D\frac{\partial^2 C}{\partial x^2} = 0. \tag{9.130}$$

With constant wave icelerity C, constant hydraulic diffusivity D and for some specified initial-boundary conditions Eq. (9.130) may be solved analytically. For instance assuming that:

- $Q(x, t = 0) = 0$ for $x \geq 0$,
- $Q(x = 0, t) = \delta(t)$,
- $Q(x = L, t) = 0$ for $L \rightarrow \infty$,

one obtains the following exact solution:

$$Q(x,t) = \frac{1}{2\sqrt{\pi \cdot D}} \cdot \frac{x}{t^{3/2}} \cdot \exp\left(-\frac{(C \cdot t - x)^2}{4 \cdot D \cdot t}\right) \tag{9.131}$$

Equation (9.131) holds for $t > 0$ and for $x \geq 0$. This equation has been obtained by Hayami (Eagleson 1970). Since the assumed conditions correspond to the definition of the IUH, then Eq. (9.131) can be considered as the instantaneous unit hydrograph of a channel reach of length L. Consequently one can write:

$$w(t) = \frac{1}{2\sqrt{\pi \cdot D}} \cdot \frac{L}{t^{3/2}} \cdot \exp\left(-\frac{(C \cdot t - L)^2}{4 \cdot D \cdot t}\right) \text{ for } t > 0 \text{ and } L > 0 \tag{9.132}$$

Let us consider the second simplified model, i.e. the kinematic wave equation (9.19). It is well known that the linear advection equation

$$\frac{\partial Q}{\partial t} + C\frac{\partial Q}{\partial x} = 0 \tag{9.133}$$

implies that the flood wave occurring at the upstream end propagates along channel axis without any shape deformation. Taking into account Eq. (9.117) one can find out, that for the linear kinematic wave equation (9.133) implemented for a channel reach of length L, IUH has the form of Dirac delta function:

$$w(t) = \delta(t - L/C) \tag{9.134}$$

where L is length of channel reach and C is kinematic wave speed. The ratio L/C represents the lag time between the upstream end and considered cross-section. Note that the same result is obtained from Eq. (9.132). For $D = 0$, IUH of the diffusive wave equation becomes the Dirac delta function, which is IUH of the kinematic wave equation.

The flood routing can be carried out with hydrological lumped models as well. They were discussed in the preceding section. One of the most popular of them is the Muskingum equation:

$$\psi\frac{dI}{dt} + (1-\psi)\frac{dQ}{dt} = \frac{1}{K}\,(I - Q) \tag{9.135}$$

where:

$I(t)$ – inflow,
$Q(t)$ – outflow,
ψ – weighting parameter,
K – time of wave translation between the considered cross-sections.

The IUH for the Muskingum model was proposed by Venetis (1969). This equation, obtained by the solution of Eq. (9.135) using the Laplace transformation approach, is following:

$$w(t) = \frac{1}{K(1-\psi)^2}\exp\left(-\frac{t}{K(1-\psi)}\right) - \frac{\psi}{1-\psi}\delta(t) \tag{9.136}$$

The next example of determination of appropriate IUH concerns the linear reservoir described by Eq. (9.135) with $\psi = 0$, i.e. by:

$$\frac{dQ}{dt} = \frac{1}{K}(I(t) - Q(t)). \tag{9.137}$$

Comparing this equation with Eq. (9.113), one can notice that they coincide, since $A_1 = 1, A_0 = 1/K, B_0 = 1/K$, and other coefficients are equal to zero. Then the convolution integral is the solution of Eq. (9.137) as well. The IUH for this equation can be determined by its time integration performed for the following conditions: $Q(t) = 0$ for $t \leq 0$ and $I(t) = \delta(t)$. Following the developments presented for instance

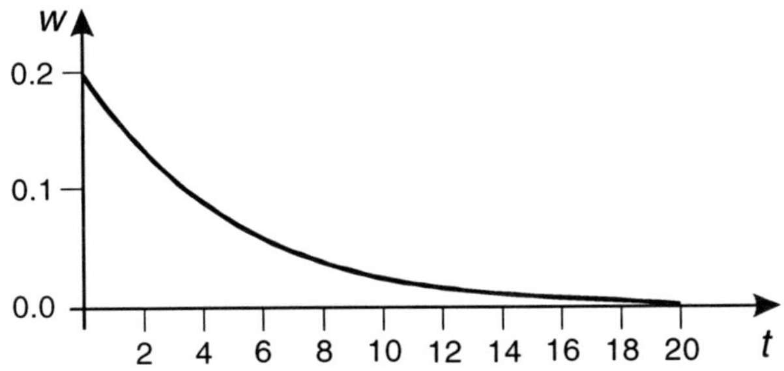

Fig. 9.18 Graph of the function (9.138) for $K = 5\,\text{h}$

by Chow et al. (1988), one obtains:

$$w(t) = \frac{1}{K} \exp\left(-\frac{t}{K}\right) \tag{9.138}$$

An example of the function $w(t)$ given by Eq. (9.138) is show in Fig. 9.18.
Note that Eq. (9.136) with $\psi = 0$ coincides with Eq. (9.138).

Usually to obtain satisfying results it is necessary to apply multiple linear reservoirs in a cascade. The first reservoir is analysed as a single one, so its output is following:

$$Q_1(t) = \frac{1}{K} \exp\left(-\frac{t}{K}\right) \tag{9.139}$$

The second reservoir, having IUH in the form of Eq. (9.138), transforms $Q_1(t)$ into $Q_2(t)$ according to the convolution integral (9.122), i.e.:

$$Q_2(t) = \int_0^t \frac{1}{K} \cdot e^{-\tau/K} \cdot \frac{1}{K} \cdot e^{-(t-\tau)/K} \cdot d\tau = \frac{t}{K^2} \cdot e^{-t/K} \tag{9.140}$$

This function is transformed by 3rd reservoir etc. Application of the same procedure for consecutive reservoirs in series gives (Nash 1957):

$$w(t) = \frac{1}{K \cdot \Gamma(N)} \left(\frac{t}{K}\right)^{N-1} \exp\left(-\frac{t}{K}\right) \tag{9.141}$$

where N is a number of the reservoirs in series, whereas $\Gamma(N)$ is the gamma Euler function. This IUH was proposed by Nash (1957) and is called the Nash model (Chow et al. 1988). In Figs. 9.19 and 9.20 are displayed the graphs of function (9.141) calculated for various values of its parameters K and N.

Unfortunately the IUHs in form of Eqs. (9.136) and (9.141) have some disadvantages which limit their application. Namely it is impossible to achieve satisfying agreement between the results of calculation and experimental data when the time

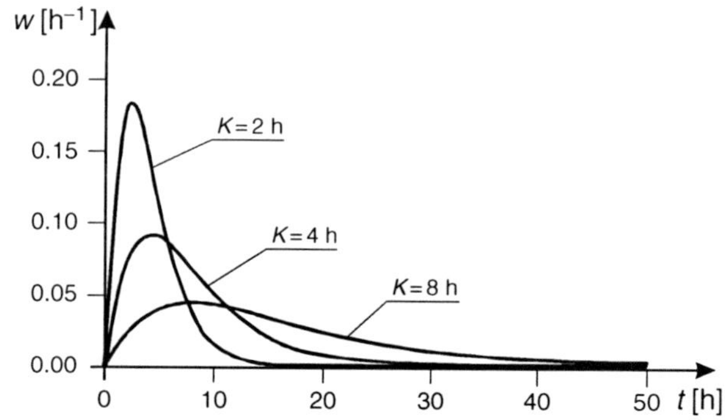

Fig. 9.19 Function $w(t)$ with $N = 2$ for various values of K

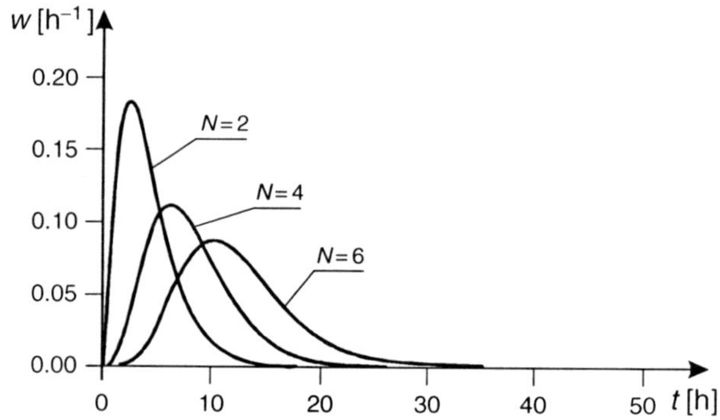

Fig. 9.20 Function $w(t)$ with $K = 2\,\text{h}$ for various values of N

lag of the output is considerable. Both IUHs are unable to reproduce the effect of pure translation. For this reason very often the linear reservoir model is used jointly with the model of linear channel (Dooge 1959, Chow 1964). Moreover the IUH for Muskingum model can produce the negative ordinates. This is caused by the second term of Eq. (9.136) (Strupczewski et al. 1989). This feature disagrees with the definition of the instantaneous unit hydrograph which has to be always non–negative: $w(t) \geq 0$ for $t \geq 0$. Consequently it can produce some unrealistic effects in form of the initial oscillations in the hydrograph calculated at the downstream end. Similar problems can be observed in the numerical solution of the Muskingum model in form of Eq. (9.135).

It is well known that the Muskingum equation integrated in time by the method, which does not produce any numerical diffusion, is capable to ensure a pure translation of flood wave for $\psi = 0.5$. For this reason it seems reasonable to expect the similar effect while using the IUH in form of Eq. (9.136). Therefore this equation should become the Dirac delta function for $\psi = 0.5$. Unfortunately Eq. (9.136) does not fulfil this condition. The discrepancy is caused by the nature of Eq. (9.135). In fact this equation is a semi-discrete form of the kinematic wave equation obtained by its spatial discretization. For this reason it contains a numerical error introduced during the approximation. Because the IUH in form of Eq. (9.136) was derived by an analytical integration of Eq. (9.135) it contains implicitly this error as well. It has the form of numerical diffusion which disappears for $\psi = 0.5$ i.e. when the approximation of spatial derivative is made by centred difference.

9.5.3 An Alternative IUH for Hydrological Lumped Models

It is possible to derive a general instantaneous unit hydrograph for all lumped models which is free from the mentioned disadvantages (Szymkiewicz 2002). To this end the IUH for the diffusive wave (Eq. 9.132) can be applied. As it was obtained by an analytical solution of Eq. (9.130) then it does not contain any numerical error. On the other hand it is known, that the results given by the lumped models are similar to the ones given by the diffusive wave model due to the numerical diffusion.

As it was showed in Section 9.4.4 the Muskingum model is an approximation of the kinematic wave. This approximation introduces a numerical error. The accuracy analysis carried out for the Muskingum equation showed that it modifies the kinematic wave model to the form similar to the diffusive wave model (9.130):

$$\frac{\partial Q}{\partial t} + C\frac{\partial Q}{\partial x} - D_n\frac{\partial^2 Q}{\partial x^2} = 0 \tag{9.142}$$

where D_n is the coefficient of numerical diffusion defined as follows:

$$D_n = \left(\frac{1}{2} - \psi\right) C \cdot \Delta x \tag{9.143}$$

Let us remember that the kinematic character of the Muskingum model as well as the numerical nature of the wave attenuation process was noticed by Cunge (1969), who suggested such value of the parameter Ψ which ensures the numerical diffusivity (Eq. 9.143) equal to the hydraulic one given by Eq. (9.35) i.e. $D = D_n$. Owing to this assumption the Muskingum model can reproduce the solution of the linear diffusive wave model. This version of the Muskingum model is known as Muskingum–Cunge one (Chow et al. 1988).

If the lumped models, being semi-discrete forms of the kinematic wave model, are able to reproduce the solution of the diffusive wave, then one can expect that the reproduction of this solution using the instantaneous unit hydrograph will be possible as well. Therefore one can apply the IUH of the diffusive wave model for the lumped systems. For this purpose let us introduce the parameters typical for the lumped models into Eq. (9.132). The hydraulic diffusivity, the kinematic speed and the length of a channel can be replaced by the following expressions:

$$C = \frac{\Delta x}{K}, \tag{9.144a}$$

$$D_n = \left(\frac{1}{2} - \psi\right) \frac{\Delta x^2}{K}, \tag{9.144b}$$

$$L = N \cdot \Delta x \tag{9.144c}$$

where N corresponds to the number of reservoirs. Substitution of Eq. (9.144) in Eq. (9.132) yields :

$$w(t) = \frac{1}{(2\pi(1-2\psi))^{1/2}} \frac{N}{K} \left(\frac{K}{t}\right)^{3/2} \cdot \exp\left(-\frac{(t-N\cdot K)^2}{2(1-2\psi)K\cdot t}\right) \tag{9.145}$$

This equation can be considered as an instantaneous unit hydrograph of the Muskingum model for a channel reach of length L. With the parameter ψ defined by Eq. (9.92) the formula (9.145) can be considered as an instantaneous unit hydrograph of the Muskingum – Cunge model. Of course, this will take place on condition that ψ is assumed to be constant.

The function (9.145) has the following properties:

- it holds for $\psi \leq 1/2$ only, including the negative values as well;
- for $\psi \to 1/2$ $w(t) \to \delta\,(t - N{\cdot}K)$;
- $w(t) \geq 0$ for $t > 0$;
- the parameter N can be any positive number, not necessarily an integer.

Some of the properties listed above are illustrated in the figures. In Fig. 9.21 the IUHs for $N = 2$, $\psi = 0.4$ and for various values of K are plotted. In Fig. 9.22 the graphs of the IUHs for $K = 3$ h, $\psi = 0$ and for various values of parameter N are plotted.

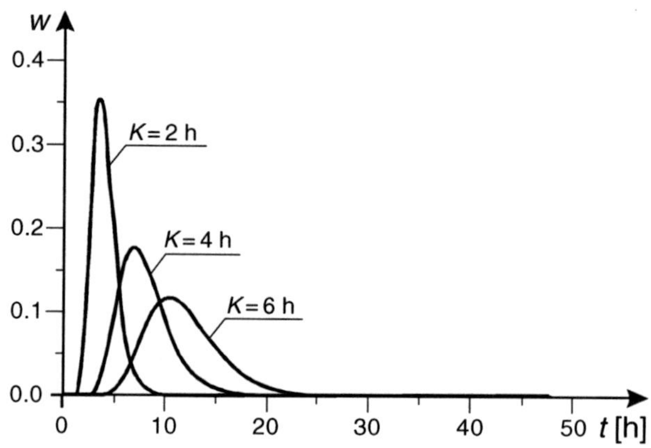

Fig. 9.21 Instantaneous unit hydrographs of the Muskingum model for $N = 2$, $\psi = 0.40$ and for various values of K

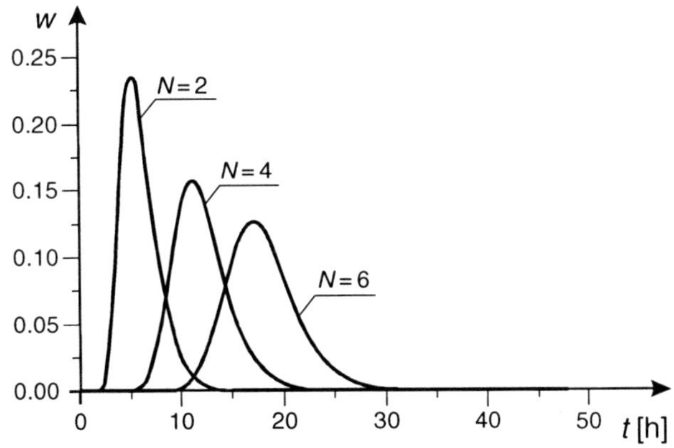

Fig. 9.22 Instantaneous unit hydrographs of the Muskingum model for $K = 3\,\text{h}$, $\psi = 0$ and for various values of N

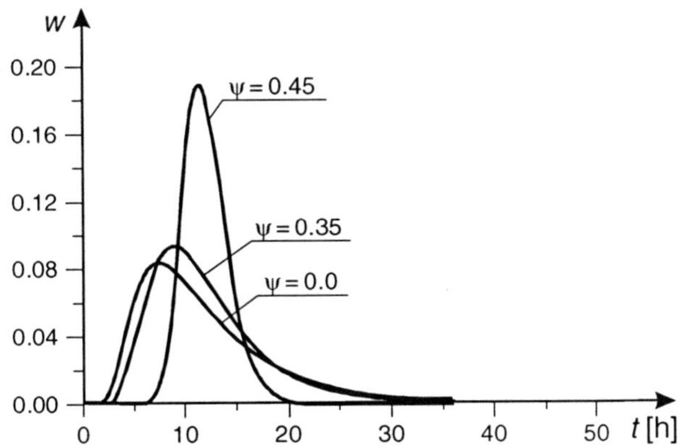

Fig. 9.23 Instantaneous unit hydrographs of the Muskingum model for $N = 3$, $K = 4\,\text{h}$ and for various values of ψ

The shape of $w(t)$ is mainly determined by parameter ψ, whereas its position along the axis t depends on the parameters N and K. With increasing ψ the IUH becomes more and more steep (see Fig. 9.23). For the extreme case when $\psi = 0.5$ its value tends to infinity at the point $t = N \cdot K = 18$. It becomes the Dirac delta function.

Implementation of the IUH requires determining the values of the parameters K, N and ψ. They can be computed using the method of moments or the method of optimisation. Both approaches need the hydrographs observed at the upstream and downstream ends of a channel. The subsequent moments are given by Szymkiewicz (2002):

For $\psi = 0$ Eq. (9.145) becomes an alternative IUH for a cascade of the linear reservoirs:

$$w(t) = \frac{\sqrt{K}}{\sqrt{2\pi}} \frac{N}{t^{3/2}} \exp\left(-\frac{(t - N \cdot K)^2}{2K \cdot t}\right), \tag{9.146}$$

whereas for $N = 1$ it becomes an alternative IUH for a single reservoir:

$$w(t) = \frac{\sqrt{K}}{\sqrt{2\pi}} \frac{1}{t^{3/2}} \exp\left(-\frac{(t - K)^2}{2K \cdot t}\right) \tag{9.147}$$

Both hydrographs presented above differ from the classical ones given by Eqs. (9.138) and (9.141). This difference results from the manner of derivation of both types of IUHs. Classical IUH for the linear reservoir was obtained by integration of Eq. (9.137) resulting from a spatial discretisation of the kinematic wave model. Consequently a mechanism of numerical diffusion was included implicitly. Conversely, Eq. (9.132) represents an analytical solution of the diffusive wave model with regard to both variables x and t. It is interesting that although the classical IUH for a cascade of linear reservoirs in series and the IUH in form of Eq. (9.146) are different, they have the same first and second moments (Szymkiewicz 2002).

Example 9.6 Using the Muskingum model solved by both the finite difference approximation (Eq. 9.86) and the convolution approach with the IUH given by Eq. (9.145), let us carry out flood routing of the following wave imposed at the upstream:

$$q(t) = q_b + (q_m - q_b)\left(\frac{t}{t_m}\right) \exp\left(1 - \frac{t}{t_m}\right) \tag{9.148}$$

with $q_b = 5\ \mathrm{m^3/s}$, $q_m = 75\ \mathrm{m^3/s}$, $t_m = 6\,\mathrm{h}$.

In Fig. 9.24 the results of flood routing using the convolution integral are presented. They are obtained for $K = 6\,\mathrm{h}$ and $N = 3$ and for various values of the parameter ψ. Note that the wave attenuation depends on the value of ψ. The wave damping becomes more pronounced for decreasing ψ. On the other hand, for increasing value of ψ the attenuation of the calculated output is reduced. For $\psi = 0.49$ damping is so small that practically pure translation of the wave is obtained. In this case the lag time of wave centres of gravity is equal to $N \cdot K = 18\,\mathrm{h}$. Regardless of the assumed set of data smooth solution is obtained always.

Conversely, while solving numerically the Muskingum model physically unrealistic results can be very often obtained. Such a situation is presented in Fig. 9.25. For the same flood wave impose at the upstream end the Muskingum model with $N = 3$, $K = 6\,\mathrm{h}$ and $\psi = 0.40$ solved numerically using the classical scheme of the finite difference method (Eq. 9.86) produces the discharge lower than one imposed at upstream end. It takes even negative values.

The examples presented above show considerable flexibility of the IUH in the form of Eq. (9.145). An appropriate set of the parameters K, N and ψ ensure both required effects, i.e. the wave translation and its attenuation. Note that the parameter ψ can take a negative value as well. On the other hand, the parameter N does not have to be an integer. Using IUH and the convolution approach one can avoid

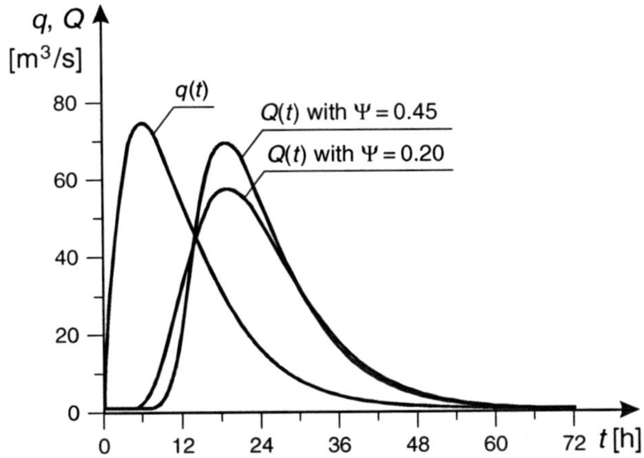

Fig. 9.24 An example of food routing by the Muskingum model using the convolution integral for $N = 3$, $K = 6\,\mathrm{h}$ and for various values of the parameter ψ

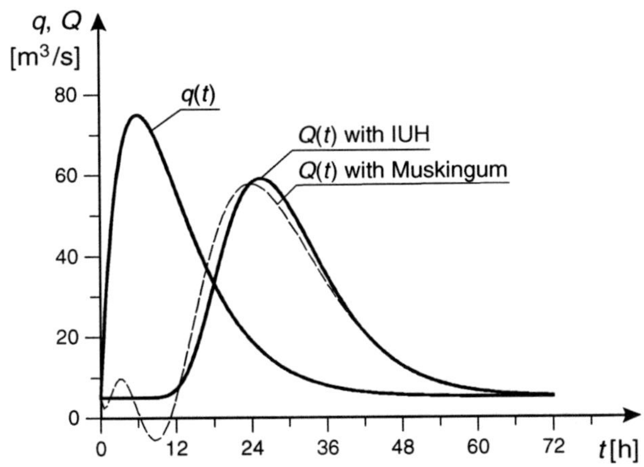

Fig. 9.25 A comparison of the solutions of the Muskingum model by the finite difference method and by the convolution integral with the IUH for $N = 3$, $K = 6\,\mathrm{h}$ and $\psi = 0.40$

all negative numerical effects arising while solution of the ordinary differential equations.

References

Abbott MB (1979) Computational hydraulics-Elements of the theory of free surface flow. Pitman, London

Cappelaere B (1997) Accurate diffusive wave routing. J. Hydr. Engng. ASCE 123 (3):174–181

Chow VT (1964) Handbook of applied hydrology. McGraw-Hill, New York

Chow VT (1959) Open channel hydraulics. Mc Graw-Hill, New York

Chow VT, Maidment DR, Mays LW (1988) Applied hydrology. McGraw-Hill, New York

Cunge JA (1969) On the subject of a flood propagation computation method (Muskingum method). J. Hydr. Res. 7 (2):205—229

Cunge J, Holly FM, Verwey A (1980) Practical aspects of computational river hydraulics. Pitman Publishing, London

Dooge JCI (1959) A general theory of the unit hydrograph. J. Geophys. Res. 64 (1):205—230

Eagleson PS (1970) Dynamic hydrology. McGraw-Hill, New York

Fletcher CAJ (1991) Computational techniques for fluid dynamics, vol. I. Springer-Verlag, Berlin

Gasiorowski D, Szymkiewicz R (2007) Mass and momentum conservation in the simplified flood routing models. J. Hydrol. 346:51–58

Gresho PM, Sani RL (2000) Incompressible flow and the finite–element method, vol. 1: advection-diffusion. Wiley, Chichester, England

Hayami S (1951). On the propagation of flood waves. Kyoto Univ. Disaster Prevent. Res. Inst. Bull. 1:1–16

Henderson FM (1966) Open channel flow. Macmillan Company, New York

Korn GA, Korn TM (1968) Mathematical handbook for scientists and engineers, 2nd edn. McGraw-Hill, New York

Lighthill MJ, Whitham GB (1955) On kinematic waves, I: flood movement in long rivers. Proc. R. Soc. London, Ser. A 229:281–316

McQuarrie DA (2003) Mathematical methods for scientists and engineers. University Science Books, Sausalito, CA

Miller WA, Cunge JA (1975) Simplified equations of unsteady flow. In: Mahmood K, Yevjewich V (eds) Unsteady flow in open channels. Water Resources Publishing, Fort Collins, CO

Nash JE (1957) The form of Instantaneous Unit Hydrograph. IASH Publication no. 45, 3–4: 114–121

O'Donnell T (1960) Instantaneous unit hydrograph derivation by harmonic analysis. IASH Publication no. 51:546–557

Ponce VM (1990) Generalized diffusion wave with inertial effects. Water Resour. Res. 26 (5):1099–1101

Ponce VM, Li RM, Simmons DB (1978) Applicability of kinematic and diffusion models. J. Hydr. Div. ASCE 104 (3):353–360

Rodriguez-Iturbe I, Valdes JB (1979) The geomorpholigic structure of hydrologic response. Water Resour. Res. 15 (6):1409–1420

Singh VP (1996) Kinematic wave modeling in water resources: surface water hydrology. Wiley, New York

Strupczewski WG, Napiórkowski JJ, Dooge JCI (1989) The distributed Muskingum model. J. Hydrol. 111:235–257

Szymkiewicz R (2002) An alternative IUH for the hydrologic lumped models. J. Hydrol. 259:246—253

Tang X, Knight DW, Samuels PG (1999) Variable parameters Muskingum–Cunge method for flood routing in a compound channel. J. Hydr. Res. 37 (5):591—614

Index

R. Szymkiewicz, *Numerical Modeling in Open Channel Hydraulics*, Water Science and Technology Library 83, DOI 10.1007/978-90-481-3674-2,

Breinigsville, PA USA
16 March 2010

234129BV00005B/21/P

9 789048 136735